# FOOD PROCESS ENGINEERING AND QUALITY ASSURANCE

# FOOD PROCESS ENGINEERING AND QUALITY ASSURANCE

*Edited by*
C. O. Mohan, PhD
Elizabeth Carvajal-Millan, PhD
C. N. Ravishankar, PhD
A. K. Haghi, PhD

Apple Academic Press Inc.
3333 Mistwell Crescent
Oakville, ON L6L 0A2 Canada

Apple Academic Press Inc.
9 Spinnaker Way
Waretown, NJ 08758 USA

© 2018 by Apple Academic Press, Inc.

First issued in paperback 2021

*Exclusive worldwide distribution by CRC Press, a member of Taylor & Francis Group*
No claim to original U.S. Government works

ISBN-13: 978-1-77463-653-4 (pbk)
ISBN-13: 978-1-77188-576-8 (hbk)

**Library and Archives Canada Cataloguing in Publication**

Food process engineering and quality assurance / edited by C.O. Mohan, PhD, Elizabeth Carvajal-Millan, PhD, C.N. Ravishankar, PhD, A.K. Haghi, PhD.

Includes bibliographical references and index.
Issued in print and electronic formats.

ISBN 978-1-77188-576-8 (hardcover).--ISBN 978-1-315-23296-6 (PDF)

1. Food industry and trade--Quality control. 2. Processed foods. I. Mohan, C. O., editor II. Carvajal-Millan, Elizabeth, editor III. Ravishankar, C. N., editor IV. Haghi, A. K., editor

TP372.5.F66 2017          338.4'764795          C2017-905789-8          C2017-905790-1

CIP data on file with US Library of Congress

Apple Academic Press also publishes its books in a variety of electronic formats. Some content that appears in print may not be available in electronic format. For information about Apple Academic Press products, visit our website at **www.appleacademicpress.com** and the CRC Press website at **www.crcpress.com**

# CONTENTS

# LIST OF CONTRIBUTORS

**Cristobal N. Aguilar**
Department of Food Research, School of Chemistry, Universidad Autónoma de Coahuila, Blvd. 25280 Saltillo, Coahuila, México, E-mail: cristobal.aguilar@uadec.edu.mx

**A. F. Aguilera-Carbó**
Department of Food Science and Nutrition, Universidad AutónomaAgraria "Antonio Narro" Calzada Antonio Narro, 1923 Buenavista, Saltillo, Coahuila, México

**Miguel A. Aguilar-González**
Center for Research and Advanced Studies of Polythechnique National Institute. CINVESTAV Unidad Saltillo, Avenida Industria Metalúrgica # 1062, 25900, Ramos Arizpe, Coahuila, México, E-mail: mgzlz@hotmail.com, miguel.aguilar@cinvestav.edu.mx

**J. A. Aguirre-Joya**
Universidad Autónoma de Coahuila, School of Chemistry, Food Research Department. Blvd. Venustiano Carranza, 25280, Saltillo, Coahuila, México

**Olga Berenice Alvarez-Pérez**
Universidad Autónoma de Coahuila, School of Chemistry, Food Research Department. Blvd. Venustiano Carranza, 25280, Saltillo, Coahuila, México, Tel.:+528444161238, Fax: +528444159534

**J. A. Ascacio-Valdes**
Department of Food Science and Technology, Universidad Autónoma de Coahuila, Blvd. Venustiano Carranza S/N Col. RepúblicaOriente, 25280 Saltillo, Coahuila, México

**B. L. Barajas**
Departamento de Química Orgánica, Facultad de Ciencias Químicas. Universidad Autónoma de Coahuila. C.P. 25280, Saltillo Coahuila, México

**Ruth E. Belmares**
Department of Food Science and Technology, Universidad Autónoma de Coahuila, Blvd. Venustiano Carranza S/N Col. República Oriente, 25280 Saltillo, Coahuila, México

**J. Bindu**
ICAR-Central Institute of Fisheries Technology (Indian Council of Agricultural Research), Matsyapuri, Cochin – 682029, India

**Spyridoula Bratakou**
Laboratory of Inorganic and Analytical Chemistry, School of Chemical Engineering, Dept 1, Chemical Sciences, National Technical University of Athens, 9 Iroon Polytechniou St., Athens 157 80, Greece

**S. L. Cantú**
Departamento de Química Orgánica, Facultad de Ciencias Químicas. Universidad Autónoma de Coahuila. C.P. 25280, Saltillo Coahuila, México

**Elizabeth Carvajal-Millan**
Research Center for Food and Development, CIAD, A.C., Carretera a La Victoria Km. 0.6, 83000 Hermosillo, Mexico, Tel.: +52-662-289 2400, Fax: +52-662-280 0421. E-mail: ecarvajal@ciad.mx

**Mira Debnath Das**
School of Biochemical Engineering, Indian Institute of Technology (Banaras Hindu University), Varanasi–221005, India

**Efimia K. Dermesonlouoglou**
Laboratory of Food Chemistry and Technology, School of Chemical Engineering, National Technical University of Athens, Iroon Polytechniou 9, 15780 Athens, Greece

**Debashis Dutta**
School of Biochemical Engineering, Indian Institute of Technology (Banaras Hindu University), Varanasi–221005, India

**Omar Garcia-Galindo**
Department of Food Research, School of Chemistry, Universidad Autónoma de Coahuila, Blvd. 25280 Saltillo, Coahuila, México

**Heliodoro de la Garza**
Basic Science Department, Universidad Autónoma Agraria Antonio Narro, Saltillo, 25315, Coahuila, México

**Konstantinos N. Georgopoulos**
Laboratory of Simulation of Industrial Processes, Department of Industrial Management and Technology, School of Maritime and Industry, University of Piraeus, Greece

**J. E. Gerardo-Rodríguez**
Programa de Posgrado en Ciencia y Tecnología de Alimentos, Universidad de Sonora, Rosales y Blvd. Luis Encinas S/N. Centro, C.P. 83000, Hermosillo, Sonora, Mexico

**Maria C. Giannakourou**
Assistant Professor, Technological Educational Institute of Athens, Faculty of Food Technology and Nutrition, Department of Food Technology, Agiou Spiridonos, 12210, Egaleo, Athens, Greece, Tel: 30-210-5385511, E-mail: mgian@teiath.gr

**M. P. González**
Departamento de Materiales Avanzados, Centro de Investigación en Química Aplicada. C.P. 25294, Saltillo Coahuila, México

**Adriana Gutiérrez-Díez**
Universidad Autónoma de Nuevo León–Facultad de Agronomía, Francisco Villa s/n Col. Ex Hacienda el Canadá, C.P. 66050, General Escobedo, Nuevo León, Mexico

**Francisco Hernandez-Centeno**
Universidad Autónoma Agraria Antonio Narro, Calz. Antonio Narro 1923, Buenavista, Saltillo, Coahuila, 25315, México, Tel. +(844) 411 0200, Ext 2009, E-mail: francisco.hdezc@gmail.com

**Stephanos Karapetis**
Laboratory of Inorganic and Analytical Chemistry, School of Chemical Engineering, Dept 1, Chemical Sciences, National Technical University of Athens, 9 Iroon Polytechniou St., Athens 157 80, Greece

**Yogesh Khetra**
Scientist, Dairy Technology Division, ICAR-National Research Institute, Karnal, Haryana, India.

**Nora Ponce de León-Renova**
Centro de Bachillerato Tecnológico Agropecuario 90. 31590 Cd. Cuauhtémoc, Mexico

**Miguel A. De León-Zapata**
Department of Food Science and Technology, Universidad Autónoma de Coahuila, Blvd. Venustiano Carranza, S/N Col. República Oriente, 25280 Saltillo, Coahuila, México

**A. I. Ledesma-Osuna**
Universidad de Sonora, Departamento de Investigación y Posgrado en Alimentos, Universidad de Sonora, Rosales y Blvd. Luis Encinas S/N. Centro, C.P. 83000, Hermosillo, Sonora, Mexico, E-mail:bramirez@guaymas.uson.mx

**L. L. López**
Departamento de Química Orgánica, Facultad de Ciencias Químicas. Universidad Autónoma de Coahuila. C.P. 25280, Saltillo Coahuila, México

**J. López-Cervantes**
Centro de Investigación e Innovación en Biotecnología Agropecuaria, Instituto, Tecnológico de Sonora, Ciudad Obregón, Sonora, México

**Haydee Yajaira López-De la Peña**
Universidad Autónoma Agraria Antonio Narro, Calz. Antonio Narro 1923, Buenavista, Saltillo, Coahuila, 25315, México

**E. Magaña-Barajas**
Universidad Estatal de Sonora, Ley Federal del Trabajo Final, Col. Apolo. C.P. Hermosillo, Sonora, México, Tel: (62) 2857636; E-mail: ely_magbarajas@hotmail.com

**Sivakumar Manickam**
Department of Chemical and Environmental Engineering, Faculty of Engineering, University of Nottingham, Jalan Broga, 43500 Semenyih, Selangor Darul Ehsan, Malaysia, Tel: +6(03) 8924 8156; Fax: +6(03) 8924 8017; E-mail: Sivakumar.Manickam@nottingham.edu.my

**Ana Luisa Martínez-López**
Research Center for Food and Development, CIAD, A.C. Carretera a La Victoria Km. 0.6, Hermosillo, Sonora, 83304, Mexico

**Guillermo Cristian G. Martínez-Ávila**
Universidad Autónoma de Nuevo León, Facultad de Agronomía. Laboratorio de Biotecnología. Campus Ciencias Agropecuarias, 66050 General Escobedo, Nuevo Leon, México

**Jorge Márquez-Escalante**
Research Center for Food and Development, CIAD, A.C. Carretera a La Victoria Km. 0.6, Hermosillo, Sonora, 83304, Mexico

**B. Meenakumari**
Indian Council of Agricultural Research, Krishi Anusandhan Bhawan II, New Delhi–110012, India

**C. O. Mohan**
Scientist, Fish Processing Division, Central Institute of Fisheries Technology, Matsyapuri, Willingdon Island, Cochin – 682029, India, E-mail: comohan@gmail.com

**Arturo Mora-Olivo**
Universidad Autónoma de Tamaulipas-Instituto de Ecología Aplicada. Av. División del Golfo No. 356, Col. Libertad C.P. 87019

**Kasturi Muthoosamy**
Department of Chemical and Environmental Engineering, Faculty of Engineering, University of Nottingham, Jalan Broga, 43500 Semenyih, Selangor Darul Ehsan, Malaysia

**Georgia-Paraskevi Nikoleli**
Laboratory of Inorganic and Analytical Chemistry, School of Chemical Engineering, Dept 1, Chemical Sciences, National Technical University of Athens, 9 Iroon Polytechniou St., Athens 157 80, Greece

**Dimitrios P. Nikolelis**
Laboratory of Environmental Chemistry, Department of Chemistry, University of Athens, Panepistimiopolis-Kouponia, GR-15771 Athens, Greece

**G. Ninan**
ICAR-Central Institute of Fisheries Technology (Indian Council of Agricultural Research), Matsyapuri, Cochin–682029, India

**Satyen Kumar Panda**
Quality Assurance and Management Division, ICAR-Central Institute of Fisheries Technology, Matsyapuri P.O., Willingdon Island, Cochin – 682029, Kerala, India

**Alán Pavlovich-Abril**
Laboratorio de Biopolímeros, Centro de Investigación en Alimentación y Desarrollo, Apartado Postal 1735, Hermosillo, Sonora 83000, México

**R. L. Peralta**
Departamento de Química Orgánica, Facultad de Ciencias Químicas. Universidad Autónoma de Coahuila. C.P. 25280, Saltillo Coahuila, México

**Vasilios N. Psychoyios**
Laboratory of Inorganic and Analytical Chemistry, School of Chemical Engineering, Dept 1, Chemical Sciences, National Technical University of Athens, 9 Iroon Polytechniou St., Athens 157 80, Greece

**P. Narender Raju**
Scientist, Dairy Technology Division, ICAR-National Research Institute, Karnal, Haryana, India

**Nancy Ramírez-Chávez**
Faculty of Agro-Technological Sciences, Autonomous University of Chihuahua, 31125 Chihuahua, Mexico

**M. L. Ramírez**
Departamento de Química Orgánica, Facultad de Ciencias Químicas. Universidad Autónoma de Coahuila. C.P. 25280, Saltillo Coahuila, México

**B. Ramírez-Wong**
Departamento de Investigación y Posgrado en Alimentos, Ave. Rosales y Blvd. Luis Encinas s/n o al Apartado postal 1658, C.P. 83000, Hermosillo, Sonora, México, Tel: (62) 59-22-07 y 59-22-09; Fax: 59-22-08; E-mail: bramirez@guaymas.uson.mx

**Agustín Rascón-Chu**
Centro de Investigación en Alimentación y Desarrollo, A.C. Carretera a la Victoria Km 0.6, Hermosillo, Sonora, A.P. 1735, C.P. 83000, México

**Revathi Raviadaran**
Department of Chemical and Environmental Engineering, Faculty of Engineering, University of Nottingham, Jalan Broga, 43500 Semenyih, Selangor Darul Ehsan, Malaysia

**C. N. Ravishankar**
ICAR-Central Institute of Fisheries Technology (Indian Council of Agricultural Research), Matsyapuri, Cochin–682029, India

**Raul Rodríguez-Herrera**
Department of Food Science and Technology, Universidad Autónoma de Coahuila, Blvd. Venustiano Carranza S/N Col. República Oriente, 25280 Saltillo, Coahuila, México

**Romeo Rojas**
Universidad Autónoma de Nuevo León, School of Agronomy, Francisco Villa s/n, Ex-Hacienda el Canadá, Escobedo, Nuevo León, 66054, México, E-mail: Romeo.Rojasmln@uanl.edu.mx

**Alejandro Romo-Chacón**
Centro de Investigación en Alimentación y Desarrollo. Unidad Cuauhtémoc. Av. Río Conchos S/N. Parque Industrial, Apdo. Postal 781. Cuauhtémoc Chihuahua, México

**Edilia De La Rosa-Manzano**
Universidad Autónoma de Tamaulipas-Instituto de Ecología Aplicada. Av. División del Golfo No. 356, Col. Libertad C.P. 87019

**Latha Sabikhi**
Principal Scientist and Head, Dairy Technology Division, ICAR-National Research Institute, Karnal, Haryana, India

**G. A. Saenz**
Departamento de Química Orgánica, Facultad de Ciencias Químicas. Universidad Autónoma de Coahuila. C.P. 25280, Saltillo Coahuila, México, Tel.: +52 844 416 92 13; Fax: +52 844 415 9534; E-mail: aidesaenz@uadec.edu.mx

**Juan Salmeron-Zamora**
Faculty of Agro-Technological Sciences, Autonomous University of Chihuahua, 31125 Chihuahua, Mexico

**Ernesto J. Sanchez-Alejo**
Universidad Autónoma de Nuevo León, School of Agronomy, Francisco Villa s/n, Ex-Hacienda el Canadá, Escobedo, Nuevo León, 66054, México, E-mail:romeo.rojasmln@uanl.edu.mx

**Saul Saucedo-Pompa**
Universidad Politécnica de Ramos Arizpe, División de Metrología Industrial, Avenida Sigma, Parque Industrial Santa María CP 25900. Ramos Arizpe, Coahuila, México, E-mail: saul.saucedo.p@gmail.com

**M. I. Silvas-García**
Universidad de Sonora, Departamento de Investigación y Posgrado en Alimentos, Universidad de Sonora, Rosales y Blvd. Luis Encinas S/N. Centro, C.P. 83000, Hermosillo, Sonora, Mexico, E-mail:bramirez@guaymas.uson.mx

**Christina G. Siontorou**
Laboratory of Simulation of Industrial Processes, Department of Industrial Management and Technology, School of Maritime and Industry, University of Piraeus, Greece

**Nikolaos G. Stoforos**
Professor, Department of Food Science and Human Nutrition, Agricultural University of Athens, Greece, Tel: 30-210-5294706, E-mail: stoforos@aua.gr

**Alma Rosa Toledo-Guillén**
Research Center for Food and Development, CIAD, A.C., Carretera a La Victoria Km. 0.6, 83000 Hermosillo, Mexico

**Jorge Ariel Torres-Castillo**
Universidad Autónoma de Tamaulipas-Instituto de Ecología Aplicada. Av. División del Golfo No. 356, Col. Libertad C.P. 87019, Tel. +52 (834)3162721, 3181800, Ext. 1606, Ciudad Victoria, Tamaulipas, Mexico, E-mail: jorgearieltorres@hotmail.com

**P. I. Torres-Chávez**
Universidad de Sonora, Departamento de Investigación y Posgrado en Alimentos, Universidad de Sonora, Rosales y Blvd. Luis Encinas S/N. Centro, C.P. 83000, Hermosillo, Sonora, Mexico, E-mail:bramirez@guaymas.uson.mx

**G. J. A. Valdez**
Departamento de Química Orgánica, Facultad de Ciencias Químicas. Universidad Autónoma de Coahuila. C.P. 25280, Saltillo Coahuila, México

**Janeth M. Ventura-Sobrevilla**
Department of Food Science and Technology, Universidad Autónoma de Coahuila, Blvd. Venustiano Carranza S/N Col. República Oriente, 25280 Saltillo, Coahuila, México

**A. A. Zynudheen**
ICAR-Central Institute of Fisheries Technology (Indian Council of Agricultural Research), Matsyapuri, Cochin–682029, India

# LIST OF ABBREVIATIONS

| | |
|---|---|
| ARS | adsorption refrigeration system |
| ALP | alkaline phosphatise |
| AMA | American Medical Association |
| AB | amlodipine besilate |
| APHIS | Animal and Plant Health Inspection Service |
| AFP | antifreeze proteins |
| AO | arabino-oligosaccharides |
| AXOS | arabinoxylan oligosaccharides |
| AX | arabinoxylans |
| ANN | artificial neural networks |
| AMV-RT | avian myeloblastosis virus reverse transcriptase |
| BSE | back scattered electron |
| BOPP | biaxially oriented polypropylene |
| BBA | biochimicaet biophysica acta |
| BSA | bovine serum albumin |
| GB | brilliant green |
| BIS | Bureau of Indian Standards |
| CIC | Canadian Irradiation Center |
| CC | Candesartan cilexetil |
| $CO_2$ | carbon dioxide |
| CIFT | Central Institute of Fisheries Technology |
| CIAD A.C | Centro de Investigación en Alimentación y Desarrollo A.C. |
| COP | coefficient of performance |
| CFU | colony forming units |
| CCAMLR | Commission for the Conservation of Antarctic Marine Living Resources |
| CFD | computational fluid dynamics |
| CLSM | confocal laser scanning microscopy |
| CR | congo red |
| DAS | degree of arabinose substitution |
| DP | degree of polymerisation |
| DIA | Departamento de Investigación en Alimentos |
| DF | dietary fiber |

| DSC | differential scanning calorimetry |
| EA | ellagic acid |
| ESEM | environmental scanning electron microscopy |
| EVOH | ethylene vinyl alcohol |
| EU | European Union |
| ESL | extended shelf life |
| FCQ | Facultad de ciencias químicas |
| FCC | Federal Communications Commission |
| FA | ferulic acid |
| FDM | finite difference |
| FEM | finite element method |
| FVM | finite volume method |
| FCBT | float controlled balance tank |
| FBD | fluidized bed drying |
| FRET | fluorescence resonance energy transfer |
| FAO | Food and Agriculture Organization |
| FDA | Food and Drug Administration |
| FSSP | food spoilage and safety predictor |
| FSANZ | Food Standards Australia New Zealand |
| FTSI | Food Technology Service, Inc. |
| FFS | form fill seal |
| FTIR | fourier transform infrared |
| FOS | fructo-oligo saccharides |
| GO | galacto-oligosaccharides |
| GA | gallic acid |
| GCE | gamma centre of excellence |
| GC | gas chromatography |
| GPPS | general purpose polystyrene |
| GRAS | generally recognized as safe |
| GAX | generator-absorber heat exchanger |
| GLI | gliadins |
| GWP | global warming potential impact |
| GWP | global warming potential |
| GHG | green house gas |
| HA | high amylase |
| HAG | high amylose starch-glycerol |
| HHP | high hydrostatic pressure |
| HIPS | high impact polystyrene |

| | |
|---|---|
| HMW | high molecular weight |
| HPAEC | high performance anion exchange chromatography |
| HPLC | high performance liquid chromatography |
| HPIC | high pressure induced crystallization |
| HPP | high pressure processing |
| HPSF | high pressure shift-assisted freezing |
| HP | high pressure |
| HPF | high pressure-assisted freezing |
| HDPE | high-density polyethylene |
| HIPS | high-impact PS |
| HTST | high temperature short time |
| HC | hydrocarbon |
| HF | hydrofluidisation freezing |
| HF | hydrofluoroolefin |
| HFC | hydrofluoro carbon |
| HLB | hydrophilic-lipophilic balance |
| HT | hydrothermal treatment |
| HPMC | hydroxypropylmethylcellulose |
| HXT | hydroxytyrosol |
| INA | ice nucleation activators |
| ISP | ice structuring proteins |
| IMS | immunomagnetic separation |
| IF | impingement freezing |
| IQF | individual quick frozen |
| ISM | industrial, scientific, and medical |
| IMF | intermediate moisture foods |
| IAEA | International Atomic Energy Agency |
| ICGFI | International Consultative Group on Food Irradiation |
| IFFO | International Fishmeal and Fish Oil Organization |
| INCI | International Nomenclature of Cosmetic Ingredients |
| IPPC | International Plant Protection Convention |
| IFIP | International Project on Food Irradiation |
| IBA | ion beam applications |
| ISFET | ion-sensitive field-effect transistor |
| IPP | isopropyl palmitate |
| JECFI | Joint Expert Committee on Food Irradiation |
| LMS | laccase mediator systems |
| LDGFS | Ley General de Desarrollo Fore Stalsustentable |

| LAPS | light-addressable potentiometric sensor |
| LCM | lignocellulosic material |
| LTLT | low temperature long time |
| MF | magnetic fields |
| MCE | magnetocaloric effect |
| MG | malachite green |
| MPEDA | Marine Products Export Development Authority |
| MIT | Massachusetts Institute of Technology |
| MS | meat systems |
| MVR | mechanical vapour recompression |
| MB | methylene blue |
| MF | micro filtration |
| MT | million tonnes |
| MAP | modified atmospheric packaging |
| MSI | moisture sorption isotherms |
| MVTR | moisture vapor transmission rate |
| NF | nanofiltration |
| NASA | National Aeronautics and Space Administration |
| NMA | National Monitoring Agency |
| $N_2$ | nitrogen |
| NDO | non-digestible oligosaccharides |
| NPA | non-pasteurized angelica |
| NTFP | non-timberforest products |
| NOM | Normas Oficiales Mexicanas |
| NMR | nuclear magnetic resonance |
| NASBA | nucleic acid sequence-based amplification |
| OC | oleoresin capsicum |
| OPP | oriented polypropylene |
| PD | peeled and de veined |
| PUD | peeled and un de veined |
| PIT | phase inversion temperature |
| PHI | phosphohexoseisomerase |
| PDA | photodiode array |
| PL | piplartine |
| PET | polyethylene terephthalate |
| PE | polyethylene |
| PLA | polylactic acid |
| PCR | polymerase chain reaction |

| | |
|---|---|
| PPI | polymeric protein insoluble |
| PPS | polymeric protein soluble |
| PP | polypropylene |
| PS | polystyrene |
| PUFA | polyunsaturated fatty acids |
| PVC | polyvinyl chloride |
| PVDC | polyvinylidene chloride |
| PDA | potato dextrose agar |
| PFA | Prevention of Food Adulteration Act |
| PER | protein efficiency ratio |
| PPP | public private partnership |
| PEF | pulsed electric field |
| RFID | radio frequency identification |
| RF | radiofrequency |
| DT | reduction time |
| RLGDFS | Reglamento de la Ley General de Desarrollo Forestal Sustentable |
| RH | relative humidity |
| RBBR | remazol brilliant blue R |
| RO | reverse osmosis |
| RP-HPLC | reverse phase chromatography |
| SPS | sanitary and phytosanitary |
| SQV | saquinavir |
| SEM | scanning electron microscope |
| SCFA | short chain fatty acid |
| SWCNT | single-walled carbon nanotubes |
| SM | skim milk |
| SMP | skim milk powder |
| SLN | solid lipid nanoparticles |
| SSF | solid state fermentation |
| SNF | solid-not-fat |
| SEA | staphylococcal enterotoxin A |
| SCWH | sub/supercritical water hydrolysis |
| SAW | surface acoustic wave |
| SPR | surface plasmon reference |
| SCM | sweetened condensed milk |
| TBT | technical barriers to trade |
| TDR | thermal destruction rate |

| TVR | thermal vapour recompression |
| TGA | thermogravimetric analysis |
| TTI | time–temperature integrators |
| TFS | tin-free-steel |
| TPC | total polyphenols of candelilla |
| TB | tuberculosis |
| UHT | ultra high temperature |
| UF | ultrafiltration |
| UHP | ultra-high pressure |
| UHT | ultra-high temperature treatment |
| UAF | ultrasound-assisted freezing |
| USAEC | United States Atomic Energy Commission |
| VI | vacuum infusion |
| VCC | vapour compression cycle |
| WVP | water vapor permeability |
| WVTR | water vapor transmission rate |
| WEAX | water-extractable arabinoxylans |
| WUAX | water-unextractable arabinoxylans |
| WPC | whey protein concentrates |
| WPI | whey protein isolates |
| WPM | whole milk powder |
| WM | whole milk |
| WLF | Williams Landel Ferry |
| WHO | World Health Organization |
| WRI | World Resources Institute |
| WTO | World Trade Organization |
| XOS | xylo-oligosaccharides |

# PREFACE

Of all the basic needs, food is the foremost need for everyone; that is the reason it is regarded as the prime endowment among all. Food inspires creativity, and pleasure and provides much-needed peace and fullness. Although the world produces around 50% more food for the entire population, hunger still prevails as nearly one-third of food produced is wasted. According to the UN, one in every nine person goes hungry each night. The Father of the Indian Nation, Mahatma Gandhi, once said, "There are people in the world so hungry that God cannot appear to them except in the form of bread," which is appropriate even today. The processing of food is seen as the best solution to overcome the wastage of food.

This book, *Food Processing Engineering and Quality Assurance*, is an edited collection of chapters by eminent researchers across the globe. It will be of interest to students, professionals, teachers, researchers, and those in the food industry. Our goal was to assimilate the knowledge on food processing technologies on different commodities. It will be of interest to academia and industry and will help to motivate nonprofessionals to learn techniques of food processing and quality assurance.

The book presents topics on food processing and quality assurance representing the recent advances and diversities in the processing of food. The topics published in this book are the contributions from prominent scientists of international repute. Chapters in this book are arranged in different sections for the benefit of readers. Throughout each chapter, we emphasize the basics of food processing technologies and their inter-relationship with the various foods. Topics were selected to give comprehensive coverage of various aspects of advanced food processing technologies along with the traditionally followed food processing technologies. Special emphasis is given to the processing of fish, candelilla, dairy and bakery products. The rapid detection of pathogens and toxins and application of nanotechnology in ensuring the safety are also emphasized. In this book chapters are so arranged that reading linearly from front to back have some merit; however, each chapter can be read separately as they can stand alone as they address specific problems.

In Part I, recent research and developments with applications are presented. This section contains six chapters. In Chapter 1, covalent cross-linking content, rheological and structural characteristics of wheat water-extractable and water-unextractable ferulated arabinoxylan gels are discussed. A comparison between part-baking bread and frozen dough processes are presented in Chapter 2.

The production of xylo-oligosaccharides (AXOS) from biomass waste by hydrothermal treatment has been reviewed in Chapter 3.

In Chapter 4, a study on ferulated arabinoxylans and $\beta$-Glucans as fat replacers in yoghurt and their effects on sensorial properties are presented in detail.

Laccase production by *Trametesversicolor* and *Armillariamellea* using maize bran as support-substrate and its dye decolorization potential as affected by pH is discussed in Chapter 5.

In the last chapter of this section, the freezing process of dough and its effects on structure, viscoelasticity, and bread making are investigated.

In Part II, new technology and processes are presented. In Chapter 7, advances in refrigeration and freezing technologies are reported. In Chapter 8, a simple and multiple emulsions that emphasizes industrial applications and stability assessment are presented. In Chapter 9, the rapid detection of pathogens and toxins are discussed in detail. New developments in food irradiation technology are reviewed in Chapter 10. In Chapter 11 some new prospects of antactic krill utilization in India is reported. The processing and packaging of dairy-based products are introduced in Chapter 12.

The principles of kinetic modeling of safety and quality of foods are reviewed in Chapter 13. In the last chapter, the value addition and preservation of fishery products are reported in detail.

Part III of this book introduces candelilla (*Euphorbia antisiphylitica-*Zucc), where candelilla is a renewable biotic resource in which the natural wax is obtained. Candelilla wax is a highly demanded bioproduct used in a huge range of industries around the world, including foods, cosmetics, paints, energy, and drugs. It represents one of the ten non-timber products with major market value of Mexico, to say nothing of it is endemic of the deserted part of northern Mexico.

Recently, candelilla has the attention of researchers due to its capacity to produce a high quantity of hydrocarbons, antioxidant compounds, and fiber, etc., and also the wax can form biodegradable, natural edible films and/or coatings.

The present section is the outcome of the effort by the Candelilla Reaserch Group of the Universidad Autonóma de Coahuila and a wide quantity of investigators with the intention to generate an international document on the most relevant aspects of candelilla. Its social aspects, the economy of the workers, their living conditions, and the risks that they have to take just to obtain the candelilla wax its unique economic source.

To our knowledge, this section represents the first international document that covers the major aspects of candelilla, such as exploitation, distribution, biological characteristics, nanostructure, market, uses, potential applications, economy, and social impact.

Determinations of microelements present in the wax are reported for the first time, and these microelements are linked to quality characteristics of the final product, demonstrating why candelilla wax is highly appreciated by industries.

Our principal objective in Part III was to generate a scientific/technological section to contribute to the generation and diffusion of the global knowledge about the uses, commercialization, application, and exploitation conditions of candelilla wax; nevertheless this document will be of interest for potential investors from cosmetics, drugs, agroindustries, postharvest technology and fine chemicals, beside others as well as for also for students and general public.

We believe this publication will be a useful guide for students, researchers, academicians, technologists and entrepreneurs engaged in the area of food processing. The editors would like to express their deep appreciation to all the authors for their outstanding contributions to this book. The editors also would like to acknowledge the effort of their peers for their time to read the drafts and provide us with technical corrections and constructive suggestions which are invaluable.

Special thanks to Apple Academic Press for the untiring support and advice throughout the formulation of this book. Also thanks to the CRC Press personnel's for their time and valuable efforts in publishing this book. We thank our entire family for their encouragements, sacrifices and support throughout our journey, which encourages us to contribute more in the future.

# ABOUT THE EDITORS

**C. O. Mohan, PhD**
*Scientist, Fish Processing Division, Central Institute of Fisheries Technology (Indian Council of Agricultural Research), Willingdon Island, Kochi, Kerala, India*

C. O. Mohan, PhD, is currently a Senior Scientist at the Fish Processing Division of the ICAR-Central Institute of Fisheries Technology (Indian Council of Agricultural Research), Willingdon Island, Kochi, Kerala, India. He graduated in fisheries sciences from the College of Fisheries, Mangalore, Karnataka. During his master's and PhD studies, he specialized in fish processing technology at the ICAR-Central Institute of Fisheries Education, Deemed University, Mumbai, India. His areas of interest are thermal processing and active and intelligent packaging. He has guided many students for their masters and PhD degrees. He has published in many national and international journals of repute. He has published 28 peer-reviewed articles, 16 book chapters, 27 conference presentations, and filed two patents. He has been awarded with the Jawaharlal Nehru Award and the Dr. Karunasagar Best Post-Graduate Thesis Award from the Indian Council of Agricultural Research, New Delhi, and the Professional Fisheries Graduates Forum, Mumbai, India.

**Elizabeth Carvajal-Millan, PhD**
*Research Scientist, Research Center for Food and Development (CIAD), Hermosillo, Mexico*

Elizabeth Carvajal-Millan, PhD, is a Research Scientist at the Research Center for Food and Development (CIAD) in Hermosillo, Mexico, since 2005. She obtained her PhD in France at Ecole Nationale Supérieure Agronomique à Montpellier (ENSAM), her MSc degree at CIAD, and undergraduate degree at the University of Sonora, in Mexico. Her research interests are focused on biopolymers, mainly in the extraction and characterization of high value-added polysaccharides from coproducts recovered from the food industry and agriculture, especially ferulated arabinoxylans. In particular, Dr. Carvajal-Millan studies covalent arabinoxylans gels as functional systems for the food and pharmaceutical industries. Globally, Dr. Carvajal-Millan is a pioneer in in-vitro and in-vivo studies on covalent arabinoxylans gels as carriers for oral insulin for the treatment of diabetes type 1. She has published 57 refereed papers, 23 chapters in books, over 80 conference presentations, one patent registered, and two more to be submitted.

### C. N. Ravishankar, PhD
*Director, ICAR-Central Institute of Fisheries Technology (CIFT), Cochin, India*

C. N. Ravishankar, PhD, is at present the Director of ICAR-Central Institute of Fisheries Technology (CIFT) in Cochin, India. He completed his graduate studies in fisheries sciences and specialized in fish processing technology during his master's and PhD degrees from the College of Fisheries, Mangalore, Karnataka, India. He is an expert in the field of fish processing and packaging, and he developed, popularized, and transferred many technologies to the seafood industry. He participated in the First Indian Antarctic Expedition and traveled widely abroad for training and consultancy programs. He has more than 200 international and national publications to his credit, and he has an h-index of 15.0 and has filed 17 patents. He received the Outstanding Team Research award in the field of fish products technology from the Indian Council of Agricultural Research, New Delhi, and the K. Chidambaram Memorial Award from the Fisheries Technocrats Forum as well as a Gold Medal for his PhD work and a Merit Certificate from the Royal Institute of Public Health and Hygiene, London. He was instrumental in establishing the Business Incubation Centre with an office and pilot plant facility for entrepreneurship development in fish and other food products. In addition to his many other activities, he has delivered numerous invited talks on fish preservation techniques, food packaging, business incubation, and other related areas.

## A. K. Haghi, PhD

*Associate Member of University of Ottawa, Canada; Editor-in-Chief,
International Journal of Chemoinformatics and Chemical Engineering;
Editor-in-Chief, Polymers Research Journal*

A. K. Haghi, PhD, holds a BSc in urban and environmental engineering
from the University of North Carolina (USA); a MSc in mechanical engineering from North Carolina A&T State University (USA); a DEA in applied
mechanics, acoustics and materials from the Université de Technologie de
Compiègne (France); and a PhD in engineering sciences from the Université
de Franche-Comté (France). He is the author and editor of 65 books as well
as 1000 published papers in various journals and conference proceedings.
Dr. Haghi has received several grants, consulted for a number of major
corporations, and is a frequent speaker to national and international audi-
ences. Since 1983, he served as a professor at several universities. He is
currently Editor-in-Chief of the *International Journal of Chemoinformatics
and Chemical Engineering* and *Polymers Research Journal* and on the editorial boards of many international journals. He is a member of the Canadian
Research and Development Center of Sciences and Cultures (CRDCSC),
Montreal, Quebec, Canada.

# PART I

# PRINCIPLES AND RECENT RESEARCH DEVELOPMENTS

**CHAPTER 1**

# COVALENT CROSS-LINKING CONTENT, RHEOLOGICAL, AND STRUCTURAL CHARACTERISTICS OF WHEAT WATER-EXTRACTABLE AND WATER-UNEXTRACTABLE FERULATED ARABINOXYLAN GELS

ELIZABETH CARVAJAL-MILLAN, JORGE MÁRQUEZ-ESCALANTE, ANA LUISA MARTHNEZ-LÓPEZ, and AGUSTÍN RASCÓN-CHU

*Research Center for Food and Development, CIAD, A.C. Carretera a La Victoria Km. 0.6, Hermosillo, Sonora, 83304, Mexico, Tel: +52-662-2892400; Fax: +52-662-2800421; E-mail: ecarvajal@.ciad.mx*

## CONTENTS

## ABSTRACT

The aim of this research was to investigate the cross-linking content, rheological, and structural characteristics of water-extractable arabinoxylans (WEAX) and water-unextractable arabinoxylans (WUAX) gels. The intrinsic viscosity, viscosimetric mass and ferulic acid (FA) content were 3.6 and 2.1 dL/g, 440 and 74 kDa and 0.40 and 0.08 µg/mg polysaccharide, for WEAX and WUAX, respectively. The Fourier transform infrared spectrum of WEAX and WUAX presented characteristics bands (1035, 1158 and 897 cm$^{-1}$) related to $\beta(1-4)$ linkages. WEAX and WUAX laccase induced gels at 3% (w/v) registered a di-FA content of 0.076 and 0.008 µg/mg polysaccharide, respectively. Storage (G') and loss (G") moduli were 101 and 20 Pa for WEAX gel and 174 and 19 Pa for WUAX gel. The structural parameters of WEAX and WUAX gels were calculated from swelling experiments. WEAX and WUAX gels mesh size were 123 nm and 48 nm, respectively. As di-FA content was lower in WUAX gels, a more important contribution of physical interactions between polysaccharide chains and/or non-identified ferulate cross-linking structures could be responsible of these rheological and structural differences between WEAX and WUAX gels.

## 1.1 INTRODUCTION

Arabinoxylans are important cereal non-starch polysaccharide constituted of a linear backbone of $\beta$-(1–4)-linked D-xylopyranosyl units to which $\alpha$-L-arabinofuranosyl substituents are attached through O-2 and/or O-3. Some of the arabinose residues are ester linked on (O)-5 by ferulic acid (FA) (3-methoxy, 4 hydroxy cinnamic acid) (Izydorczyk and Biliaderis, 1995). These polysaccharides have been classified as water extractable (WEAX) or water-unextractable (WUAX). WEAX and WUAX form highly viscous solutions and gel through ferulic acid covalent cross-linking upon oxidation by some chemical or enzymatic free-radicals generating agents resulting in the formation of five different di-FA structures (5–5,' 8–5' benzo, 8-O-4,' 8–5' and 8–8' di-FA). This covalent cross-linking has been commonly considered as responsible of the polysaccharide network development even if weak interactions also contribute to the final gel properties (Vansteenkiste et al., 2004). The presence of a trimer structure of FA (tri-FA) has been reported in WEAX and WUAX (Carvajal-Millan et al., 2005a, 2007).

The utilization of water-based gels in industrial applications as texture and stability improvers or as delivery systems generates great interest (Wisniewski et al., 1976; Pothakamury and Canovas, 1995). Most of polysaccharide gels currently used are stabilized by physical interactions (hydrogen bonding and/or ionic); polysaccharide covalent networks such as WEAX or WUAX gels are not common. Covalently cross-linked gels are generally strong, form quickly, they are not temperature dependent on heating and they exhibit no syneresis after long time storage. Furthermore, WEAX and WUAX gels have interesting functional properties, which have not been exploited even though their neutral taste and odor are desirable properties for industrial applications. WEAX and WUAX networks have a high water absorption capacity (up to 100 g of water per gram of polymer) and they are not sensible to electrolytes or pH (Izydorczyk and Biliaderis, 1995).

WEAX and WUAX gels have been already studied (Izydorczyk et al., 1990; Figueroa-Espinoza and Rouau, 1998; Figueroa-Espinoza et al., 1998; Schoonevelds-Bergmans et al., 1999; Carvajal-Millan et al., 2005a, 2007) but their chemical, rheological and structural characteristics of WEAX and WUAX have not been compared elsewhere. In the present study, WEAX and WUAX from a wheat grain cultivar have been extracted, characterized and gelled and then the gel covalent cross-linking content, rheological and structural characteristics have been investigated.

## 1.2  EXPERIMENTAL PART

### 1.2.1  MATERIALS

Wheat flour and wheat bran were kindly supplied by a commercial milling industry in Northern Mexico. Laccase (benzenediol: oxygen oxidoreductase, E.C.1.10.3.2) from *Trametes versicolor* and all other chemical products were purchased from Sigma Chemical, Co. (St. Louis, MO, USA).

### 1.2.2  WEAX AND WUAX EXTRACTION AND CHARACTERIZATION

WEAX were isolated from wheat flour as previously described (Carvajal-Millan et al., 2005a). WUAX were extracted from wheat bran as reported before (Berlanga-Reyes et al., 2011). Sugars were quantified by ion-exclusion chromatography after hydrolysis process (Carvajal-Millan et al., 2005a). Sugars were eluted in a RPM-Monosaccharide $Pb^{+2}$ (8%) column (300 x 7.8 mm) (Phenomenex, Torrance, US) with water at 0.6 mL/min. The column and detector temperatures were 80°C and 40°C, respectively. Mannitol was used as an internal standard. An Alliance Waters e2695 separation module with Waters 2414 RI detector and an Empower Pro Software (Waters, Milford, US) were used. Protein content was determined by the Dumas method (AOAC, 1995) using a Leco-FP 528 nitrogen analyzer. Ferulic acid was quantified by reverse phase chromatography (RP-HPLC) after a de-esterification process. Ferulic acid was eluted in a Supelcosil LC18BD (250 x 46 mm) (Supelco Inc., Bellefont, US) with water:acetic acid:methanol (59:1:40) at 0.6 mL/min and 35°C. A Waters 2998 photodiode array detector (Waters, Milford, US) was used. Detection was by UV absorbance at 320 nm.

Viscosity measurements were made by determination of the flow times of WEAX and WUAX solutions in water (from 0.06 to 0.1% w/v). An Ubbelohde capillary viscometer at 25 ± 0.1°C, immersed in a temperature controlled water bath was used. The intrinsic viscosity ([η]) was estimated from relative viscosity measurements (ηrel) of WEAX solutions by extrapolation of Kraemer and Mead and Fouss curves to "zero" concentration. The viscosimetric molecular weight (Mv) was calculated from the Mark-Houwink relationship, $Mv = ([\eta]/k)1/\alpha$.

FT-IR spectra of dry WEAX and WEAX gel (lyophilized) powder were recorded on a Nicolet FT-IR spectrophotometer (Nicolet Instrument Corp. Madison, WI). The samples were pressed into KBr pellets (2 mg sample/200 mg KBr). A blank KBr disk was used as background. Spectra were recorded between 400 and 4000 $cm^{-1}$.

### 1.2.3   WEAX AND WUAX GELATION

Laccase mediated cross-linking of WEAX and WUAX was performed as reported previously (Carvajal-Millan et al., 2005a). WEAX and WUAX solutions at 3.5% (w/v) were prepared in 0.1 M citrate phosphate buffer pH 5.5. WEAX and WUAX solutions were mixed with 50 μL of laccase (1.675 nkat/mg polysaccharide). Gels were allowed to form for 4 h at 25°C.

## 1.2.4  GELS COVALENT CROSS-LINKING CONTENT

WEAX and WUAX gels di-FA and tri-FA contents were quantified by high performance liquid chromatography (HPLC) after desertification step (Rouau et al., 2003; Vansteenkiste et al., 2004). An Alltima C18 column (250 x 4.6 mm) (Alltech Associates, Inc. Deerfield, IL) and a photodiode array detector Waters 996 (Millipore Co., Milford, MA) were used. Detection was by UV absorbance at 320 nm.

## 1.2.5  GELS RHEOLOGY

Small amplitude oscillatory shear was used to follow the gelation process of WEAX and WUAX solutions (3.0% w/v) by using a strain controlled rheometer (Discovery HR-2, hybrid Rheometer, TA instruments) as previously reported (Carvajal-Millan et al., 2005b). Gelation processes were studied during 90 min at 25°C, 1.0 Hz of frequency and 10% strain (linearity range of viscoelastic behavior). Frequency sweep (0.1 to 10 Hz) was carried out at the end of the network formation at 10% strain at 25°C.

## 1.2.6  GELS STRUCTURE

After laccase addition, WEAX and WUAX solutions were quickly transferred to a 5 ml tip-cut-off syringe (diameter 1.5 cm) and allowed to gel for 90 min at 25°C. After gelation, the gels were removed from the syringes, placed in glass vials and weighted. The gels were allowed to swell in 20 ml of 0.02% (w/v) sodium azide solution to prevent microbial contamination. During 36 h the samples were blotted and weighed. After weighing, a new aliquot of sodium azide solution was added to the gels. Gels were maintained at 25°C during the test. The equilibrium swelling was reached when the weight of the samples changed by no more than 3% (0.06 g). The swelling ratio (q) was calculated as:

$$q = (W_s - W_d)/W_d \tag{1}$$

where $W_s$ is the weight of swollen gels and $W_d$ is the weight of WEAX or WUAX in the gel.

From swelling measurements, the molecular weight between two cross-links (Mc), the cross-linking density $(\rho_c)$ and the mesh size ($\xi$) values of the different WEAX and WUAX gels were calculated. Mc was calculated using the model of Flory and Rehner modified by Peppas and Merrill for gels where the cross-links are introduced in solution (Flory and Rehner, 1943; Peppas and Merrill, 1976). From the Mc values, the $\rho_c$ and $\xi$ in the WEAX gels were calculated as reported before (Peppas et al., 1985, 2000).

### 1.2.7  STATISTICAL ANALYSIS

Chemical and physico-chemical determinations were made in duplicates and the coefficients of variation were lower than 6%. Small deformation measurements were made in triplicates and the coefficients of variation were lower than 9%. Swelling and controlled release tests were made in duplicates, coefficients of variation were lower than 10%. All results are expressed as mean values.

### 1.3   RESULTS AND DISCUSSION

### 1.3.1  WEAX AND WUAX EXTRACTION AND CHARACTERIZATION

Yield of WEAX and WUAX were 0.5 and 4.0% (w/w), respectively, on a dry matter basis. Similar WEAX and WUAX yield values have been reported for wheat grain (Izydorczyk and Biliaderis, 1995; Schooneveld-Bergmans et al., 1999; Berlanga-Reyes et al., 2011). WEAX and WUAX composition is presented in Table 1. The arabinoxylan content of the extracts was estimated from the sum of xylose + arabinose. The arabinoxylan content for WEAX and WUAX content was 68 and 65% (w/w), respectively, on a dry matter basis, which is close to the value reported for other wheat WEAX and WUAX (Schooneveld-Bergmans et al., 1999; Izydorczyk et al., 1990). A residual amount of glucose was also quantified in both samples. The FA content (0.40 and 0.08 µg/mg for WEAX and WUAX, respectively) was in the range reported for this polysaccharide in wheat (Izydorczyk and Biliaderis, 1995; Berlanga-Reyes et al., 2011). Small levels of di-FA were also detected in both samples (0.06 and 0.01 µg/mg µg/mg for WEAX and WUAX, respectively) suggesting that some arabinoxylan chains might be cross-linked as previously reported (Saulnier et al., 1999). The percentages of each one of the different diFA presents in WEAX and WUAX are presented in Figure 1.1. The 8–8' dehydrodimer was not detected in this study. The predominance of 8–5' and 8-O-4' di-FA structures has also been reported in arabinoxylans from wheat and barley flour (Saulnier et al., 1999). The tri-FA 4-O-8,' 5'-5" was detected only in traces. The degree of substitution (arabinose to xylose ratio) was 0.60 and 0.8 for WEAX and WUAX, respectively, which is in the range reported for wheat arabinoxylans (Izydorczyk and Biliaderis, 1995; Berlanga-Reyes et al., 2011). The intrinsic viscosity ([η]) and viscosimetric molecular weight (Mv) values for WEAX and WUAX were 3.6 dL/g and 440 kDa and 2.1 dL/g and 74 kDa, respectively, which are in the range previously reported for other arabinoxylan wheat (Izydorczyk and Biliaderis, 1995; Berlanga-Reyes et al., 2011).

The Fourier transform infrared (FTIR) spectrum of WEAX and WUAX is presented in Figure 1.1. This figure shows mainly a broad absorbance band for polysaccharides at 1200–800 cm$^{-1}$. The main band centered at 1035 cm$^{-1}$ could be assigned to C-OH bending, with shoulders at 1158, and 897 cm$^{-1}$ that were related to the antisymmetric C-O-C stretching mode of the glycosidic link and β(1–4) linkages (Morales-Ortega et al., 2013). The region from 3500 to 1800 cm$^{-1}$ is the fingerprint region of polysaccharides related to arabinoxylans, with two bands (3413 cm$^{-1}$ corresponding to stretching of the OH groups and 2854 cm$^{-1}$ corresponding to the CH$_2$ groups). An absorbance band was observed at 1720 cm$^{-1}$ implying a low degree of esterification with aromatic esters such as ferulic acids.

### 1.3.2  WEAX AND WUAX GELS FORMATION AND CHARACTERIZATION

#### 1.3.2.1   Gels Covalent Cross-Linking Content

WEAX and WUAX gelation was induced by a laccase leading to the formation of 0.076 and 0.008 μg of di-FA per milligram of WEAX or WUAX, respectively. A small amount of tri-FA was registered in WEAX gels while only traces of this cross-linking structure were found in WUAX gels (Table 1.1). In general, the amounts of di-FA and tri-FA produced did not counterbalance the lost in FA. Therefore, at the end of gelation, 30 and 73% of the initial FA in WEAX or WUAX solution disappeared, respectively, with only 63 and 55%, respectively was recovered as di-FA and tri-FA. Low ferulate

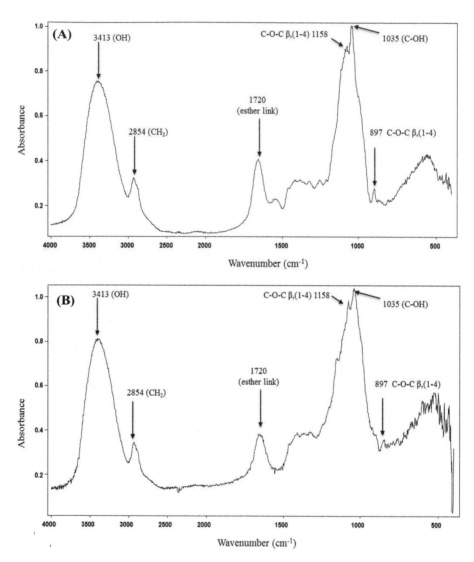

**FIGURE 1.1**   FT-IR spectrum of WEAX (A) and WUAX (B).

recovery after oxidative treatment of arabinoxylans and feruloylated sugar beet pectin has been previously reported and related to the possible formation of higher oligomers of ferulate other than di-FA and tri-FA (Vansteenkiste et al., 2004, Carvajal-Millan et al., 2005a; Berlanga-Reyes et al., 2011; Morales-Ortega et al., 2013).

During WEAX or WUAX gelation, all di-FA did not vary in the same proportion. For WEAX gels, the 8–5' benzofuran form, 8–5,' 8-O-4' and

**TABLE 1.1**  Composition of WEAX and WUAX

|  | WEAX | WUAX |
|---|---|---|
| Arabinose[a] | 25.00 ± 1.5 | 25.00 ± 1.7 |
| Xylose[a] | 43.00 ± 1.2 | 40.00 ± 1.0 |
| Glucose[a] | 3.90 ± 0.3 | 2.70 ± 0.5 |
| Protein[a] | 3.70 ± 0.05 | 4.10 ± 0.10 |
| Ferulic acid[b] | 0.400 ± 0.001 | 0.020 ± 0.001 |
| Diferulic acids[b] | 0.060 ± 0.001 | 0.010 ± 0.001 |
| Triferulic acid[b] | Traces | Traces |

[a] Results are expressed in g/100 g WEAX dry matter.

[b] Phenolics are expressed in µg/mg WEAX dry matter.

All results are obtained from triplicates.

5–5' dimers represented 66, 15, 14 and 5% of the total di-FA amounts respectively. For WUAX gels, the 8–5' benzofuran form, 8–5,' 8-O-4' and 5–5' dimers represented 70, 7, 21 and 2% of the total di-FA amounts respectively (Figure 1.2). The main increase in di-FA concerned the 8–5' benzofuran form. The predominance of 8–5' benzofuran form and 8–5' dimers and absence of the 8–8' structure was also observed in other WEAX and WUAX gels (Vansteenkiste et al., 2004; Carvajal-Millan et al., 2005; Berlanga-Reyes et al., 2011; Morales-Ortega et al., 2013; Martínez-López et al., 2013).

## 1.3.2.2 Gels Rheology

Figures 1.3A and 1.4A show the development of G' and G" moduli versus time of 3% (w/v) WEAX and WUAX solutions undergoing cross-linking process. Storage (G') and loss (G") moduli rise to reach a pseudo plateau region where the G' and G" values were 101 and 20 Pa for WEAX gels and 174 and 19 Pa for WUAX gels, respectively. The gelation time ($t_g$), calculated from the crossover of the G' and G" curves (G' > G") were 12 and 4 min for WEAX and WUAX, respectively. The $t_g$ value indicates the sol/gel transition point and at this point G' = G" or tan δ = G"/G' = 1. The mechanical spectrum of WEAX and WUAX gels (Figures 1.3B and 1.4B), were typical of solid-like materials with a linear G' independent of frequency and G" much smaller than G' and dependent of frequency (Doublier and Cuvelier, 1996). This behavior is similar to that previously reported for arabinoxylan gels cross-linked by laccase or peroxidase/

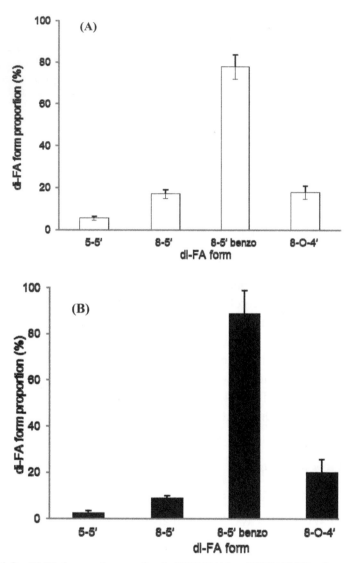

**FIGURE 1.2**   Di-FA form and proportion in WEAX (A) and WUAX (B) gels.

$H_2O_2$ system (Vansteenkiste et al., 2004; Carvajal-Millan et al., 2005a; Morales-Ortega et al., 2013; Martínez-López et al., 2013).

### 1.3.2.3   Gels Structure

The equilibrium swelling of WEAX and WUAX gels was reached between 15–20 h. The swelling ratio (q, g water/g polysaccharide) was

**TABLE 1.2** Covalent cross-linking content in WEAX and WUAX gels

|  | WEAX | WUAX |
|---|---|---|
| di-FA | $0.076 \pm 0.0050$ | $0.0080 \pm 0.0006$ |
| tri-FA | $0.008 \pm 0.0001$ | Traces |

Results are expressed in μg/mg WEAX or WUAX dry matter.

All results are obtained from triplicates.

$188 \pm 7$ and $82 \pm 4$ in WEAX and WUAX gels, respectively. The higher water uptake of WEAX gels could be explained in terms of the existence of longer un-cross-linked polysaccharide chains sections in the network. Uncross-linked polymer chains sections in the gel can expand

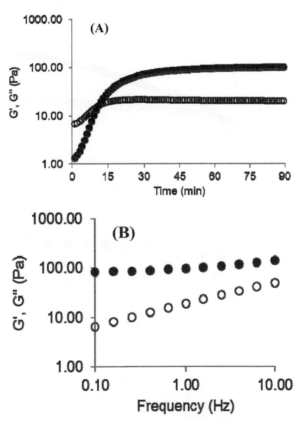

**FIGURE 1.3** (A) Rheological kinetics of 3% (w/v) WEAX solution during laccase induced cross-linking process at 1 Hz and 10% strain. (B) Mechanical spectrum of 3% (w/v) WEAX gel at 10% strain. Measurements at 25°C (G'●, G"○).

**TABLE 1.3**   Structural Characteristics of WEAX and WUAX gels

| Structural parameter | WEAX gel | WUAX gel |
|---|---|---|
| $M_c^a \times 10^3$ (g/mol) | $73 \pm 4.0$ | $20 \pm 3.2$ |
| $\rho_c^b \times 10^{-6}$ (mol/cm³) | $23 \pm 1.3$ | $75 \pm 1.5$ |
| $\xi^c$ (nm) | $123 \pm 4.0$ | $48 \pm 1.0$ |

[a] Molecular weight between two cross-links; [b] Cross-linking density; [c] Mesh size.

All values are means ± standard deviation of tree repetitions.

easily conducting to higher amounts of water uptake (Meyvis et al., 2000). As the covalent cross-links (di-FA, tri-FA) content is lower in WUAX gels, an important involvement of physical interactions between WUAX chains and/or possible higher oligomers of ferulate in the final

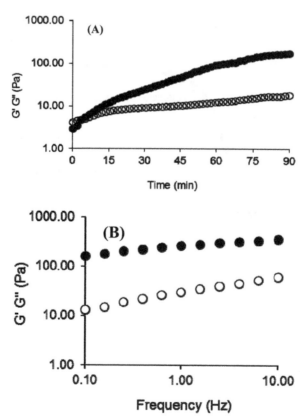

**FIGURE 1.4**   (A) Rheological kinetics of 3% (w/v) WUAX solution during laccase induced cross-linking process at 1 Hz and 10% strain. (B) Mechanical spectrum of 3% (w/v) WEAX gel at 10% strain. Measurements at 25°C (G'●, G"○).

gel structure could be responsible of this evolution as suggested by several authors (Carvajal-Millan et al., 2007; Berlanga-Reyes et al., 2011; Martínez-López et al., 2013).

The molecular weight between two cross-links (Mc), the cross-linking density ($\rho_c$) and the mesh size ($\xi$) values of WEAX and WUAX gels are presented in Table 1.3. Higher $M_c$ and $\xi$ values were found for WEAX gels, which is in agreement with the water uptake at equilibrium swelling discussed above as well as with the rheological characteristics of these gels.

## 1.4 CONCLUSION

By using WEAX or WUAX, gels with different rheological and structural characteristics can be prepared. For WEAX gels di-FA and tri-FA covalent bonds contents appear to play a crucial role in the network structure and rheological behavior. However, for WUAX gels presenting lower di-FA and tri-FA contents, the contribution of physical interactions or possible higher oligomers of ferulate to the gel structure and rheological characteristics appear to be determinant. These differences in the WEAX and WUAX gel could induce changes in the functional properties of the network.

### ACKNOWLEDGMENTS

This research was supported by "Fondo de Infraestructura de CONACYT, Mexico (Grant 226082 to E. Carvajal-Millan)."

### KEYWORDS

- arabinoxylan gels
- dimmers of ferulic acid
- elasticity
- mesh size

## REFERENCES

AOAC. (1995). Official Method of Analysis of AOAC International Method 991.43. 16th ed.; Association of Official Analytical Communities: Arlington, USA.

Berlanga-Reyes, C., Carvajal-Millan, E., Lizardi-Mendoza, J., Islas-Rubio, A. R., & Rascón-Chu, A. (2011). Enzymatic cross-linking of alkali extracted arabinoxylans: Gel rheological and structural characteristics. *International Journal of Molecular Sciences, 12,* 5853–5861.

Carvajal-Millan E., Guigliarelli B., Belle, V., Rouau X., & Micard, V. V. (2005a). Storage stability of arabinoxylan gels. *Carbohydrate Polymers, 59,* 181–188.

Carvajal-Millan, E., Landillon, V., Morel, M. H., Rouau, X., Doublier, J. L., & Micard, V. (2005b). Arabinoxylan gels: Impact of the feruloylation degree on their structure and properties. *Biomacromolecules, 6,* 309–317.

Carvajal-Millan, E., Rascón-Chu, A., Márquez-Escalante, J., Ponce de León, N., Micard, V., & Gardea,A. (2007). Maize bran gum: extraction, characterization and functional properties. *Carbohydrate Polymers, 69,* 280–285.

Doublier, J. L., & Cuvelier, G. (1996). Gums and hydrocolloids: functional aspects, in: A. C. Eliasson (Ed.), Carbohydrates in Food, Marcel Dekker, Inc., New York.

Figueroa-Espinoza, M. C., & Rouau, X. (1998). Oxidative cross-linking of pentosans by a fungal laccase and a horseradish peroxidase: Mechanism of linkage between feruloylated arabinoxylans. *Cereal Chemistry, 75,* 259–265.

Figueroa-Espinoza, M. C., Morel, M. H., & Rouau, X. (1998). Effect of lysine, tyrosine, cysteine and glutathione on the oxidative cross-linking of feruloylated arabinoxylans by a fungal laccase. *Journal of Agricultural and Food Chemistry, 46,* 2583–2589.

Flory, P. J., & Rehner, J. (1943). Statistical mechanics of cross-linked polymer networks. II. Swelling. *The Journal of Chemical Physics, 11,* 521–526.

Izydorczyk, M. S., & Biliaderis, C. G. (1995). Cereal arabinoxylans: Advances in structure and physicochemical properties. *Carbohydrate Polymers, 28,* 33–48.

Izydorczyk, M. S., Biliaderis, C. G., & Bushuk, W. A. (1990). Oxidative gelation studies of water-soluble pentosans from wheat. *Journal of Cereal Science, 11,* 153–169.

Martínez-López, A. L., Carvajal-Millan, E., Rascón-Chu, A., Márquez-Escalante, J., & Martínez-Robinson, K. (2013). Gels of ferulated arabinoxylans extracted from nixtamalized and non-nixtamalized maize bran: Rheological and structural characteristics. CyTA. *Journal of Food, 11,* 22–28.

Meyvis, T. K. L., De Smedt, S. C., Demeester, J., & Hennink, W. E. (2000). Influence of the degradation mechanism of hydrogels on their elastic and swelling properties during degradation. *Macromolecules, 33,* 4717–4725.

Morales-Ortega, A., Carvajal-Millan, E., López-Franco, Y., Rascón-Chu, A., Lizardi-Mendoza, J., Torres-Chavez, P., et al. (2013). Characterization of water extractable arabinoxylans from a spring wheat flour: Rheological properties and microstructure. *Molecules, 18,* 8417–8428.

Peppas, N. A., & Merrill, E. W. (1976). Poly(vinyl alcohol) hydrogels: reinforcement of radiation-crosslinked networks by crystallization. *Journal of Polymer Science, 14,* 441–457.

Peppas, N. A., Bures, P., Leobandung, W., & Ichikawa, H. (2000). Hydrogels in pharmaceutical formulations. *European Journal of Pharmaceutics and Biopharmaceutics, 50,* 27–46.

Peppas, N. A., Moynihan, H. J., & Lucht, L. M. J. (1985). The structure of highly cross-linked poly(2-hydroxyethyl methacrylate) hydrogels. *Journal of Biomedical Material Research, 19,* 397–411.

Pothakamury, U. R., & Barbosa-Canovas, G. V. (1995). Fundamental aspects of controlled release in foods. *Trends in Food Science & Technology*, *6*, 397–406.

Rouau, X., Cheynier, V., Surget, A., Gloux, D., Barron, C., Meudec, E., et al. (2003). A dehydrotrimer of ferulic acid from maize bran. *Phytochemistry*, *63*, 899–903.

Saulnier, L., Crépeau, M. J., Lahaye, M., Thibault, J. F., Garcia-Conesa, M. T., Kroon, P. A., et al. (1999). Isolation and structural determination of two 5,5'-diferuloyl oligosaccharides indicate that maize heteroxylans are covalently cross-linked by oxidatively coupled ferulates. *Carbohydrate Research, 320*, 82–92.

Schooneveld-Bergmans, M. E. F., Dignum, M. J. W., Grabber, J. H., Beldman, G., & Voragen, A. G. J. (1999). Studies on oxidative cross-linking of feruloylated arabinoxylans from wheat flour and wheat bran. *Carbohydrate Polymers, 38*, 309–317.

Vansteenkiste, E., Babot, C., Rouau, X., & Micard, V. (2004). Oxidative gelation of feruloylated arabinoxylan as affected by protein. Influence on Protein Enzymatic Hydrolysis. *Food Hydrocolloids*, *18*, 557–564.

Wisniewski, S. J., Gregonis, D. E., Kim S. W., & Andrade J. D. (1976). Permeation of Water Through Poly(2-hydroxyethyl methacrylate) and Related Polymers. In: Hydrogels for medical and related applications, ACS Symposium Series, v. 31. Gould R. F. (Ed.). American Chemical Society: Washington, DC, pp. 80–87.

# CHAPTER 2

# A COMPARISON BETWEEN PART-BAKING BREAD AND FROZEN DOUGH PROCESSES

J. E. GERARDO-RODRÍGUEZ,[1] B. RAMÍREZ-WONG,[2]
P. I. TORRES-CHÁVEZ,[2] A. I. LEDESMA-OSUNA,[2]
E. CARVAJAL-MILLÁN,[3] J. LÓPEZ-CERVANTES,[4]
and M. I. SILVAS-GARCÍA[2]

[1] Graduate Program in Food Science and Technology, University of Sonora, Rosales y Blvd. Luis Encinas S/N. Centro, C.P. 83000, Hermosillo, Sonora, Mexico

[2] Department of Research and Graduate Studies in Food, University of Sonora, Rosales y Blvd. Luis Encinas S/N. Centro, C.P. 83000, Hermosillo, Sonora, Mexico, E-mail: bramirez@guaymas.uson.mx

[3] Research Centes in Food and Development, A.C. (CIAD, A.C.), Carretera a La Victoria Km 0.6, Hermosillo, C.P. 83304, Sonora, Mexico

[4] Technological Institute of Sonora, Obregyn City, Sonora, Mexico

## CONTENTS

## ABSTRACT

The bakery industry uses the following two primary methods for bread preservation: (i) part-baked bread, and (ii) dough freezing. Both have advantages
and disadvantages that should be considered by each producer. Frozen dough
can be transported more easily due to the stability of the product, while part-
baked bread must be handled more carefully and is bulkier, but the final product
usually has better quality because the yeast is not damaged during the process.
Freezing is typically used in food to preserve freshness, stop enzymatic processes and prevent microbial contamination. The freezing process also damages the food structure. In baked goods, water redistribution during the freezing
process affects the protein structure, starch, and arabinoxylans, which can be
reflected as a decrease in the volume and increased firmness of the product.
Taking into consideration that increased volume and less firmness are desirable
for bread quality, part-baked bread is commercially more advisable for both the
producer and the consumer. The aim of this review is to compare the processes
of part-baked bread and frozen dough to determine the best alternative for preserving bread with a longer shelf life without losing the baking properties of the
product and considering the damage caused by freezing and storage on proteins,
starch and arabinoxylans.

## 2.1   INTRODUCTION

The baking industry is required to improve commercial processes to obtain
better bread quality that can be consumed fresh for a longer time (Gray and

Bemiller, 2003). This reduces costs, expands the market and extends bread shelf life (Selomulyo and Zhou, 2007). For this purpose, the use of new technologies has been very useful, such as dough freezing methods, controlled atmosphere bread packaging (MAP) and part-baked bread methods.

Part-baked bread consists of a short baking step without developing ready-to-eat bread characteristics, but developing a sufficiently rigid structure for storage (Hamdami et al., 2004), with the advantage of being consumed later. Once the loaf is going to be consumed, it only requires that the baking process be finished. This saves time and effort to the sale point and results in benefits to the consumer; however, some quality characteristics are reduced.

When storing part-baked bread by cooling in controlled atmospheres or by freezing, the product suffers damage that is not acceptable to the consumer. It is necessary to consider the following two important aspects when refrigeration is applied: (i) after baking, the product is cooled and packaged, and it can be contaminated with microorganisms capable of growing at refrigerated temperatures; and (ii) although a great number of microorganisms are inactivated during baking, some can survive at the center of the loaf because the temperature is not as high as in the crust (Lainez et al., 2008). Under strict MAP conditions, bakery products are well protected against aerobic spoilage organisms. However, there are still anaerobic spoilers that can cause the early spoilage of bread products (Deschuyffeleer et al., 2011).

Furthermore, when using the freezing process, the gluten network formed to retain yeast gas is damaged by ice crystals. The damage increases with the frozen storage time, which decreases the bread's ability to retain $CO_2$ and develops a lower bread volume (Ribotta et al., 2003). The bread-making industry has benefited from various studies for achieving slow freezing and enzymatic processes by adding various additives, such as pentosanes to improve dough machinability (Martinez-Anaya and Jimenez, 1997), emulsifiers to enhance volume and texture (Twillman and White, 1988), hydrocolloids (Armero and Collar, 1996) and enzymes (Martinez-Anaya et al., 1999), to increase freshness and bread volumes and cause a less firm crumb. Rosell and Gómez (2007) wrote a review about frozen dough and part-baked bread that focused on steps involved in bread making. This review will focus on part-baked bread as an alternative to dough freezing to preserve bread with a longer shelf life without losing the characteristic properties of the product, while also considering the damage caused by freezing and storage on proteins, starch, and arabinoxylans.

## 2.2   FROZEN DOUGH PROCESS VS. PART-BAKED BREAD PROCESS

Due to the importance of bread worldwide in recent decades, studies have been intensified to improve the baking process. This interest is because of the success of current methods of preserving semi-finished products that are used by the baking industry. Two of the most studied and used processes by producers for storage are dough freezing and frozen part-baked bread. Frozen dough method has been studied in recent decades. It has the benefits of expanding the product range, takes less work, improves logistics and reduces costs (Giannou et al., 2005). However, products of this process have decreased quality, including bread volume loss and firmness increases. Freezing causes the death of yeast, which reduces the $CO_2$ that produces bread volume, and weakens the gluten network. This is due to the mechanical breaking of gluten in disulfide bonds by expanding ice crystals upon freezing (Anon et al., 2004; Giannou et al., 2005).

The part-baked bread process has the same advantages and even obtains a final product that can be prepared faster because the only remaining step is baking. Frozen part-baked bread requires little equipment at the store and is faster to prepare because it is only necessary to take it from the freezer and bake in the oven (Carr and Tadini, 2003). This makes it a more attractive option for the consumer and the disadvantages are lower than frozen dough. Additionally, it does not cause yeast deathbecause the fermentation step has already been accomplished before freezing and storage.

## 2.3   CHANGES IN BREAD MAKING FROM FROZEN DOUGH AND PART-BAKED BREAD

### 2.3.1   INGREDIENTS

#### 2.3.1.1   Flour

Flour is the particulate product obtained from wheat dry milling that mainly uses the endosperm fraction (Dubois and Juhué, 2000). It is common to use wheat flour in bread making; it is the only flour with the capacity to form gluten due to its gliadin and glutenin content. Gliadins are monomers and glutenins form high molecular weight (HMW) polymers (Shewry and Halford, 2001). Other components contained in flour are starch (the main component), water and a low content of lipids. Similar to dough freezing, part-baked bread requires strong flour for both processes as the intermediate and final products are exposed to freezing stress. As a consequence, gluten is damaged by crystal growth, especially in the frozen dough because more water is present.

### 2.3.1.2 Water

Water is one of the most critical components in bread making. Water content and distribution are responsible for texture properties, crumb softness, crust crispiness and bread shelf life (Wagner et al., 2007). Water plays a major role in bubble expansion of the gluten by creating the holding capacity of the $CO_2$ produced by the yeast, which also require water for metabolism activation. Moreover, dough development with adequate hydration contributes to good elasticity and extensibility (Cauvain and Young, 2008). During the freezing process, the water contained in frozen dough and part-baked bread is redistributed and forms crystals that damage proteins, which decreases the quality.

### 2.3.1.3 Salt

The presence of salt (sodium chloride) not only contributes to the taste of bread but also plays an important role in fermentation regulation and in dough quality to give consistency and retain water (Miller and Hoseney, 2008), which prevents moisture losses during bread making. Salt also lowers the free water content and extends the product shelf life because it reduces the growth of microorganisms (Cauvain and Young, 2008).

### 2.3.1.4 Yeast

Yeast (*Saccharomyces cerevisiae*) are responsible for the fermentation process, which transforms the sugars into carbon dioxide (giving volume to the bread) and ethanol to avoid heat gain and therefore moisture loss inside the crumb (Mondal and Datta, 2007). Yeast can be used in different forms as either wet yeast, dried yeast or compressed yeast. During the frozen dough process, the amount of yeast added is higher (4%) than the part-baked bread process. The reason for this is to reduce yeast losses due to freezing stress. Yeast that is resistant to freezing is used in some cases; some studies have even intensified resistant strains to environmental stress (Hino et al., 1990). In part-baked bread, the amount of yeast added in traditional bread making is enough to obtain a high-quality product because the yeast does not suffer exposure to freezing temperatures.

### 2.3.1.5   Shortening

The presence of shortening enhances the dough machinability. In agreement with Mousia et al. (2008), shortening regulates $CO_2$ migration from the gluten matrix, resulting in decreased gas loss. Moreover, it also contributes to the bread texture by increasing softness and decreasing moisture loss, which in turn reduces the increase in aging firmness (Hasenhuettl, 2005). Ferreira and Watanabe (1998) studied the effect of vegetable shortening and sugar on frozen part-baked French bread quality. The authors recommended the addition of 10 mg of ascorbic acid/100 g wheat flour and 2 g of shortening/100 g wheat flour to achieve better results. As for frozen dough, addition of shortening (5%) is recommended for part-baked bread to reduce the water redistribution during freezing and decrease bread firmness.

### 2.3.1.6   Sugar

Some formulations for bread making include sugar (sucrose) as a common ingredient. The sugar is hydrolyzed to glucose by yeast to produce ethanol and $CO_2$ that are necessary in baking (Mondal and Datta, 2007). Sugar is most used in sweet bread and gives color to bread after baking. The glucose molecule is involved in Maillard reactions, which transforms sugar into compounds that contribute to bread flavor, color and odor (Purlis and Salvadori, 2007).

### 2.3.1.7   Other Ingredients (Additives)

The addition of other ingredients, such as hydrocolloids and cryoprotectants, is common for frozen products in the bread-making industry because it is necessary to produce bread that satisfies consumer demand. The bread must have certain quality requirements and prolonged storage. The use of preservatives such as salts (sodium propionate) prevents microbiological contamination. To increase bread volume and make the crumb softer, additives such as emulsifiers, oxidants (ascorbic acid), diacetyl tartaric acid (DATEM) and hydrocolloids (guar gum, xanthan gum) are utilized. These additives control dough water mobility (Ribotta et al., 2001). In frozen dough and part-baked bread, it is necessary to add the above compounds to retain water and avoid crystal formation. The unavailability of water helps to maintain bread quality and prevent moisture loss during the thaw.

## 2.3.2 DOUGH PREPARATION

### 2.3.2.1 Ingredients Mixing

The first step in bread making is mixing the ingredients, which consist of wheat flour (base 100%), water (64%), yeast (2%), salt (1.5%) and shortening (2%) (flour and water mainly affect bread crumb texture characteristics). This formulation depends on the bread type (Mondal and Datta, 2007). It is important to homogenize the ingredients (1-minute mixing without water incorporation). Once homogenized, it is essential to incorporate the correct amount of water to develop the dough with a proper texture. Mixing is carried out for 3 minutes after addition of the water (in mixer). Rozylo, (2014) conducted a study that involved several blending processes to determine their effects on the final product. This step is essential for proper dough formation with good plastic properties, including cohesiveness, elasticity, and extensibility (Cauvain and Young, 2008). During this process, small gas bubbles (oxygen and nitrogen) are formed, which are subsequently replaced with carbon dioxide ($CO_2$) produced by the yeast (Mondal and Datta, 2007). However, the freezing preservation processes requires a different formulation and the addition of hydrocolloids to neutralize the crystallization damage. The addition of 4% yeast and 5% shortening is recommended for frozen dough (Silvas-García, 2010). Part-baked bread does not require major changes to the traditional formulation.

### 2.3.2.2 Dough Molding and Forming

The developed dough is molded for loaf formation. The required portions are weighed on a balance in equal pieces. The molded operation can be

performed manually or using a divider (Serna-Saldivar, 1996). The molded
and formed operations may vary by style and bread type. In this step, the
baker's skills are required to obtain a final product with the desired characteristics; this is necessary for consumer acceptance. If a loaf is deformed,
bread could have a decreased volume and display visually unpleasant mal-
formations (Mousia et al., 2008).

### 2.3.3   DOUGH FERMENTATION

During dough fermentation, yeast produces $CO_2$ using carbohydrates as an
energy source. The total gas produced diffuses into the dough, increases the
cell sizes and increases the dough volume, which is required to make good
quality bread (John et al., 2002). Yeasts need activated oxygen that is present
in the cells; once that oxygen is consumed, anaerobic fermentation, which
produces mainly $CO_2$ and ethanol, begins. The $CO_2$ is first solubilized in
the water present in the dough; however, the saturated $CO_2$ is released into
the cells by increasing their sizes, which leads to dough expansion and a
decrease in density (Poitrenaud, 2004; Romano et al., 2007). Usually, dough
pieces tend to ferment in proofers for 1 hour at a controlled relative humidity (85%) and a temperature of 30°C, which optimizes the yeast activity.
The $CO_2$ retained for the gluten network corresponds to 40% of the total
produced by the yeast (Cauvain and Young, 2008). When freezing dough,
it is necessary to carry out a pre-fermentation for a short time (usually 10
minutes) before freezing; this reduces damage because it overpasses its basal
state. Once thawed, dough is fermented for the remaining time to obtain the
desired volume and then baked. Part-baked bread has the advantage that
fermentation is completed before the bread is part-baked and thenthe bread
is subsequently frozen for preservation.

### 2.3.4   PART-BAKING

After fermentation, dough is exposed to the first baking cycle. In this step,
the temperature and part-baked time can vary according the loaf size.
Characteristic changes include an increase in the bread volume. This occurs
in the initial stage of baking and is mainly due to the expansion of the gas
enclosed in the cells in the dough structure (Hamdami et al., 2004). Part-
baked bread reported by Lainez et al. (2008) used a temperature of 240°C for

10 min for a 150 g loaf, while another study by Almeida et al. (2013) used 160°C for 15 min. The aim of this step is to promote starch gelatinization and gluten coagulation without coloring the crust to obtain a semi-finished product with a formed structure. This product can be stored for later use and conserves its freshness until consumption (Roussel and Chiron, 2002). Frozen dough is thawed and baked for the necessary time to reach the appropriate bread characteristics.

## 2.3.5  FREEZING, STORAGE, AND THAWING

Freezing processes are used regularly in industry to preserve food because it retains product freshness by reducing enzymatic activity and water activity (Aw) as well as significantly reducing the development of deteriorative microorganisms. It is widely used by the baking industry for dough as an alternative to part-baked bread. It is recommended that part-baked bread be cooled before freezing to prevent damage and loss of product quality. Bárcenas and Rosell (2007) froze samples until the core reached –40°C, thus ensuring that the entire sample is at the same temperature. Le-Bail et al. (2010) used a set point between –20 and –30°C. The freezing process for part-baked bread may depend on the freezing rate. Almost all food can be frozen with better results at rapid freezing rates because it causes less deterioration. Studies by Silvas-Garcia (2010) on frozen dough have shown better results at slow freezing rates due to the fact that they cause less damage to yeast. However, in part-baked bread, the fermentation step is complete, so a rapid freezing rate is recommended. Nevertheless, more research is needed to confirm this effect.

The aim of freezing a product is to store it for a short period of time without losing freshness. Part-baked bread should be stored with care to maintain its integrity. Once the part-baked bread is frozen, it should be protected with an insulating material to prevent moisture loss or loaf drying. Barcenas et al. (2003) used polyethylene films to prevent sample deterioration; the bread was stored at –18°C at different intervals. Part-baked bread for both commercial and research purposes are generally stored after the core reaches –20°C, though this may vary depending on the freezing rate used. This is achieved using a thermocouple. The industry has already standardized the time to manufacture the product. Storage is performed at the same temperature (–20°C) for the necessary time until thawing for consumption. The freezing–thawing cycles produce dramatic effects on the bread properties (Barcenas et al., 2003).

Thawing is carried out with care to prevent loaf moisture loss and therefore diminish its quality. Some researchers have taken to slowly thawing at 4°C (cooling 4 h). Bárcenas et al. (2003) thawed bread samples after storage at room temperature. After that, part-baked bread is suitable for the rebaking stage.

### 2.3.6   RE-BAKING

The last step in the part-baked bread process is re-baking. Re-baking is carried out to the sales point. In this stage, the loaf is totally baked to be consumed and partial starch gelatinization is finished. Product characteristics that are formed in this step are responsible for the color, smell, and final volume bread. During the part-baked and re-baked steps, moisture flows from the center of the loaf to the exterior of the crumb, the loaf thickens and crust takes on color. At the end of baking, the crust has low water content and a firm, less porous structure. As a result, its apparent density values are not similar to those of the crumb, which has higher water content and is more porous (Zanoni et al., 1995). Usually, re-baking time varies depending on the loaf type and size and the baking and thermal conditions inside the oven. This should be experimentally determined in each case (Fik and Surowka, 2002). Bárcenas et al. (2003) re-baked bread at 195°C for 14 min, completing in a total of 20 min.

### 2.4   GLUTEN IN FROZEN DOUGH AND PART-BAKED BREAD

Gluten is formed by mixing flour and water. The mechanical work produced by the mixer developed this elastic and extensible material; this is due to gluten's two proteins, gliadin and glutenin (Cauvain and Young, 2001). Gluten is necessary for bread making because it is responsible for forming the network necessary for the crumb structure. In frozen dough and stored part-baked bread, the proteins are damaged by ice crystals that form during freezing. The ice crystals weaken the gluten network, decreasing the $CO_2$ retention capacity (Havet and Mankai, 2000). Furthermore, ice crystal formation also redistributes the water that was previously associated with the proteins. This water redistribution increases the hydrophobic bonds, which makes gluten proteins lose their original functionality and conformation (Wagner et al., 2005). These and other factors are responsible for part-baked bread quality deterioration. This damage has been generally reported in frozen dough, though the loss of quality in part-baked bread is derived from the same reasons. However, more studies are necessary to better understand gluten coagulation in bread making.

## 2.5 STARCH IN FROZEN DOUGH AND PART-BAKED BREAD

Wheat starch is mainly responsible for flour water absorption to form dough for bread making. When dough is made or part-baked to produce bread, the starch is partially gelatinized. The starch swells to increase the water absorption due to the increased temperature. Wheat starch gelatinizes at approximately 60°C. However, as not all of the part-baked bread gelatinizes, there is still plenty of water available to freeze and cause damage to the developing crumb quality. When frozen, the water typically usable by gelatinizing starch granules is frozen to form various sized crystals depending on the freezing rate used in the process. Some profound modifications at a molecular level, for instance on starch chain arrangement, have been observed on scanning electron micrographs (Barcenas and Rosell, 2006b). Ribotta et al. (2003) showed that freezing and storage at −18°C for more than 150 days modifies starch properties and influences the rate of retrogradation and retrograded amylopectin. Starch retrogradation appears to be responsible for crumb firming and the interactions between proteins and starch. Part-baked bread at temperatures from −7°C to 4°C experiences maximum retrogradation of starch molecules (Vulicevic et al., 2004). Once the product is ready for the re-baking step, bread should be thawed to room temperature, where the moisture in the product and the water available is reduced to gelatinize the native starch. Because of this, the final product quality is less, which is reflected by a denser crumb structure, firmer and lower moisture content, and being susceptible to a faster retrogradation. Ghiasi, Hoseney, Zeleznak, and Roger (1984) studied the effect of re-baking bread above 55°C and concluded that starch retrogradationand bread firming can be reversed.

## 2.6 ARABINOXYLANS IN FROZEN PART-BAKED BREAD

The arabinoxylans (AX) are compounds presents in some cereals, especially in pericarp. In wheat milling, AXs are mainly found in the sub-product bran flour. When a whole bakery product is prepares from flour with a high content of bran, arabinoxylans will be present in the bread. AXs play a major role in dough elasticity, which is necessary for bread making. Its undesirable effects are ascribed to the characteristic rigid structure of insoluble fiber fractions, which hinders proper dough formation and bread development. It has been suggested that partial hydrolysis of fibrous structures of high molecular masses is beneficial in this regard (Haseborg and Himmelstein, 1988). When part-baked bread is frozen, AXs suffer from changes due to water redistribution that can cause the arabinoxylan-protein and arabinoxylan-ferulic acid bonds to break. It has been reported that increasing AX solubilization has been associated with increased of loaf volume because soluble AX most likely increases the viscosity of the liquid films surrounding the gas cells and, hence, the gas retention capacity of the cells. AXs also lower the cell viscosity and, hence, gas retention properties, which leads to lower cell gas stability and smaller loaf volumes (Courtin and Delcour, 2002). Hydrolysis to overcome water absence may create a simpler polymer and allow for more extensible dough, even without the presence of xylanase. This effect could increase bread quality by contributing to greater elasticity and, therefore, better bread crumb formation. The same effects could be shown, but to a lesser extent, in part-baked bread made with frozen white flour because the endosperm also contains AXs.

## 2.7   FROZEN PART-BAKED BREAD MOISTURE CONTENT

Part-baked bread has free and bound water that is not freezable. Free water freezes below 0°C and forms ice crystals. Usually, this water is responsible for the structural damage to gluten di-sulfate bonds upon crystal formation. The non-freezing water mass fraction can be determined by differential scanning calorimetry (DSC); it is the difference between the total water content and the amount of water detected by the fusion endotherm (Ross, 1978). However, below −18°C, water bound to proteins and part-baked bread components can be detached from its original location and increase crystal sizes. Water is not going to return to its original location; this leads to clustering and a loss in protein conformational structures, which are unique properties that benefit product quality. Starch is also widely affected by this effect due to moisture loss, which leads to an increase in stored bread firmness due to faster retrogradation. Crispy bread crust originates when the starch and gluten matrix are in a glassy state; it has also been associated with low moisture content or low water activity (Stokes and Donald, 2000). Crispness retention can be increased by enhancing the water vapor permeability of the crust. Therefore, the creation of cracks on the crust surface after baking could increase the water vapor permeability and therefore improve crispness (Hirte et al., 2013). Recently, it has been reported that the crust matrixes with highly gelatinized starch and poor gluten distribution showed higher tendencies to crack at the end of part-baking, freezing, and full baking (Hirte et al., 2013). Cracks favor the water vapor permeability of the crust and subsequently are instrumental to crust crispness.

## 2.8   PART-BAKED BREAD FROZEN STORAGE

When dough is part-baked, the loaf is frozen and stored to be re-baked for later consumption. This offers many advantages. However, the disadvantages can be observed in the poor quality and shelf life reduction. This bread quality loss is associated with phenomena that occur during storage at freezing temperatures. While the bread is frozen, temperature variations known as fluctuations occur, which lead to the formation of larger ice crystals that can further damage the gluten network structure formed by proteins. Moreover, this damage decreases crumb cells in size and number and produces a denser and more matted bread (Barcenas and Rosell, 2006), which results in a firmer crumb. After full baking, the full effect of frozen storage is observed in the crumb hardness, with progressive increases with increased frozen storage. This effect is attributed to the damage to the bread constituents due to ice crystal formation and the subsequent protein network breakage and leaching out of intracellular amylose. All of these effects might increase the interactions between the inter- and intra-granular amylose and the formation of an amylose network that increase crumb hardness (Barcenas and Rosell, 2006). Additionally, for longer frozen storage times, the water redistribution is higher, which means the water will not return to its original locations in the bread and will be lost during the thaw. When bread samples from different frozen storage times are aged, they show a significant decrease (P<0.05) in the onset temperature of endotherm retrogradation compared with the time of frozen storage, while the peak and conclusion temperature are barely affected (Bárcenas et al., 2003). Ribotta et al. (2003) obtained similar results when studying the effect of frozen dough storage on retrogradation temperature transition, though only after prolonged storage for 150 or 230 days. All of these processes lead to a less hydrated starch that retrogrades faster by having less water to eject and, therefore, a faster reorganization of amylose and amylopectin polymers, which causes crumb hardening and the flow of low residual moisture towards the bread rind. This may lead to mold growth in the bread if it is not properly stored.

## 2.9   ADDITIVES FOR FROZEN DOUGH AND PART-BAKED BREAD

The baking industry is searching for alternatives to produce more bread with higher quality and at the lowest cost. One of the problems in bread making is the limited shelf life, which leads to loss of quality, low consumption and, therefore, economic losses. Preservation of bread in packaging under controlled conditions as well as using frozen dough and part-baked bread are some alternatives. However, there are additives that improve the quality parameters of bread and reduce the adverse effects on the conservation process, specifically the storage of frozen part-baked bread. Some additives act as gelling and emulsion formers to stabilize the bread crumb structure. Hydrocolloids can give stability to food products during freezing–thawing cycles (Gurkin, 2002), and they help to minimize the adverse effects of freezing and frozen storage on starch-based products (Liehr and Kulicke, 1996). The presence of bread improvers (alpha-amylase, sourdough, hydroxypropyl methylcellulose, and κ-carrageenan) minimizes the adverse effects (increase in the retrogradation temperature range) of frozen storage (Barcenas et al., 2003). A blend of improvers containing acid, ascorbate, alpha-amylase, proteases and hemicellulose has been proposed for extending the stability of frozen part-baked bread (Ribotta and Le Bail, 2007). Barcenas et al. (2004) found that κ-carrageenan was not a good improver for part-baked frozen bread because it has detrimental effects on part-baked bread maintained under frozen conditions. Leon et al. (2000) described the ability of carrageenan to improve bread volume due to its interactions with gluten proteins, even though κ-carrageenan forms rigid gels that are not stable during freezing–thawing cycles (Gurkin, 2002; Ward and Andon, 2002). Hydroxypropyl methylcellulose (HPMC) is a useful bread improver that increases bread volume, improves crumb texture and retards bread staling when it is used in conventional bread making (Guarda et al., 2004). The ability of HPMC to act as a bread improver has been attributed to its hydrophilic structure that allows its interaction with water (Schiraldi, Piazza, and Riva, 1996). HPMC can also increase the interface activity between water and the non-aqueous phases of the bread dough, favoring the formation of emulsions and strong, uniform films (Bell, 1990).

## 2.10  RECENT RESEARCH IN PART-BAKED BREAD

The study of part-baked bread has intensified in recent years due to its importance and benefits in relation to other processes. Figure 2.1 shows the number of articles published in recent years. For example, Borczak et al. (2015) studied the starch digestibility index and antioxidative properties of a part-baked wheat-flour bakery that adds dietary fibers. They found that freezing applied simultaneously to bake-off technology and the addition of oat fiber and inulin to the dough resulted in bakery products with reduced starch digestibility and containing more antioxidants, which appears to be advantageous from a nutritional point of view. Ronda et al. (2014) studied the fermentation time and fiber effects on the recrystallization of starch components and bread staling in frozen part-baked bread. They found that bread characteristics and staling kinetics were significantly affected by

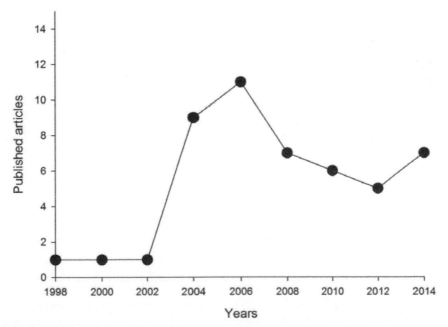

**FIGURE 2.1**    Trend of published research manuscripts related to part-baked bread.

fermentation time and fiber, particularly inulin. Inulin promoted amylopectin recrystallization and delayed water movement from the crumb to the crust during staling. In 2016, Hejrani et al., studied the properties of part-baked frozen bread with guar and xanthan gums. The overall results showed that guar gum had better effects in combination with enzymes (amylase and lipase).

## 2.11   CONCLUSIONS

Frozen part-baked bread is a viable option for preserving the freshness of a product until its consumption. It also allows thawing time after processing for consumption at the sales point. Compared with frozen dough, part-baked bread has the advantage that yeast is no longer a problem because their functions have already been completed at the time of freezing. Different research types are necessary to determine the causes of the decline in quality during the freezing process, which include the types of processes, storage times and use of additives that improve the quality of the obtained products. However, quality problems remain, and therefore, it is necessary to improve the product via further research on the freezing process of bakery products.

## KEYWORDS

- frozen dough
- part-baked bread
- gluten network
- storage time
- freezing rate
- starch
- moisture

## REFERENCES

Anon, M. C., LeBail, A., & Leon, A. E. (2004). Effect of freezing on dough ingredients. En Hui, Cornillon, Legarreta, Lim, Murrell, Nip (Eds.), *Handbook of Frozen Foods*. New York: Marcel Dekker.

Armero, E., & Collar, C. (1996). Antistaling additive effect on fresh wheat bread quality. *Food Sci Technol Int, 2*, 323–333.

Bárcenas, M. E., & Rosell, C. M. (2007). Different approaches for improving the quality and extending the shelf life of the partially baked bread: low temperatures and HPMC addition. *Journal of Food Engineering, 72*, 92–99.

Bárcenas, M. E., Benedito, C., & Rosell, C. M. (2004). Use of hydrocolloids as bread improvers in interrupted breaking process with frozen storage. *Food Hydrocoll, 18*, 769–774.

Bárcenas, M. E., Haros, M., Benedito, C., & Rosell, C. M. (2003). Effect of freezing and frozen storage on the staling of part-baked bread. *Food Research International, 36*, 863–869.

Bell, D. A. (1990). Methylcellulose as a structure enhancer in bread baking. *Cereal Foods World, 35*, 1001–1006.

Borczak, B. Sikora, E., Sikora, M., Kapusta-Duch, J., & Rosell, C. M. (2015). Starch digestibility index and antioxidative properties of partially baked wheat flour bakery with an addition of dietary fiber. *Starch/Stärke, 67*, 913–919.

Carr, L. G., & Tadini, C. C. (2003). Influence of yeast and vegetable shortening on physical and textural parameters of frozen part baked French bread. *Lebensm.-Wiss. u.-Technol, 36*, 609–614.

Cauvain, S. P., & Young, L. S. (2001). Baking Problems Solved. Woodhead Publishing, Cambridge.

Cauvain, S. P., & Young, L. S. (2008). Bakery Food Manufacture and Quality: Water Control and Effects. 2nd Ed., Ames, Iowa.

Courtin, C. M., & Delcour, J. A. (2002). Arabinoxylans and endoxylanases in wheat flour bread-making. *Journal of Cereal Science, 35*(3), 225–243.

Deschuyffeleer, N., Audenaert, K., Samapundo, S., Ameye, S., Eeckhout, M., & Devlieghere, F. (2011). Identification and characterization of yeasts causing chalk mold defects on par-baked bread. *Food Microbiology, 28*, 1019–1027.

Dubois M., & Juhué B. (2000). The important of experimental milling for determining rheological parameters as measured by the alveograph. *Cereal Food World, 45*, 385–388.

Ferreira, P. B. M., & Watanabe, E. (1998). Estudo da formulacao na producao de pao frances pre-assado. XVI Brazilian Congress of Food Science and Technology, CD-Rom, Paper no. 72.

Fik, M., & Surowka, K. (2002). Effect of prebaking and frozen storage on the sensory quality and instrumental texture of bread. *Journal of the Science of Food and Agriculture, 82*, 1268–1275.

Ghiasi, K., Hoseney, R. C., Zeleznak, K., & Roger, D. E. (1984). Effect of waxy barley starch and reheating on firmness of bread crumb. *Cereal Chemistry, 61*, 28–29.

Giannou, V., Tzia, C., & LeBail, A. (2005). Quality and safety of frozen bakery products. In: Sun (Ed.), Handbook of frozen food processing and packaging. New York: Marcel Dekker.

Gray, J. A., & Bemiller, J. N. (2003). Bread staling: molecular basis and control. *Food Science and Food Safety, 2*, 1–21.

Guarda, A., Rosell, C. M., Benedito, C., & Galotto, M. J. (2004). Different hydrocolloids as bread improvers and antistaling agents. *Food Hydrocolloids, 18*, 241–247.

Gurkin, S. (2002). Hydrocolloids Ingredients that add flexibility to tortilla processing. *Cereal Foods World, 47*, 41–43.

Hamdami, N., Monteau, J., & Bail, L. (2004). Heat and mass transfer in par-baked bread during freezing. *Food Research International, 37*, 477–488.

Hartmann, G., Piber, M., & Koehler, P. (2005). Isolation and chemical characterization of water-extractable arabinoxylans from wheat and rye during bread making. *Eur Food Res Technol, 221*, 487–492.

Haseborg, E., & Himmelstein, A. (1988). Quality problems with high-fiber breads solved by use of hemicellulose enzymes. *Cereal Foods World, 33*, 419–422.

Hasenhuettl, G. L, (2005). Fats and fatty oils. Encyclopedia of Chemical Technology, 5th ed. (vol. 12). Wiley, Hoboken, NJ, USA.

Havet, M., Mankai, M., & Le Bail, A. (2000). Influence of the freezing condition on the baking performances of French frozen dough. *Journal of Food Engineering, 45*, 139–145.

Hejrani, T., Sheikholeslami, Z., Mortazavi, A., & Davoodi, M. G. (2016). The properties of part baked frozen bread with guar and xanthan gums. *Food Hydrocolloids, 1–6*.

Hino, A., Mihara, K., Nakashima, K., & Takano, H. (1990). Trehalose levels and survival ratio of freeze-tolerant versus freeze-sensitive yeasts. *Applied and Environmental Microbiology, 56*, 1386–1391.

Hirte, A., Hamer, R. J., Hoffmann, L., & Primo-Martín, C. (2013). Cracks in bread crust cause longer crispness retention. *Journal of Cereal Science, 57*(2), 215–221.

John, P. M. van Duynhoven, Geert, M. P. van Kempen, Robert van Sluis, Bernd Rieger, Peter Weegels, Lucas J. van Vliet, & Klaas Nicolay. (2002). Quantitative assessment of gas cell development during the proofing of dough by magnetic resonance imaging and image analysis. *Cereal Chem, 80*, 390–395.

Lainez, E., Vergara, F., & Bárcenas, M. E. (2008). Quality and microbial stability of partially baked bread during refrigerated storage. *Journal of Food Engineering, 89*, 414–418.

Le-bail, A., Dessev, T., Jury, V., Zuniga, R., Park, T., & Pitroff, M. (2010). Energy demand for selected bread making processes: Conventional versus part baked frozen technologies. *Journal of Food Engineering, 96*, 510–519.

Leon, A., Ribotta, P., Ausar, S., Fernandez, C., Landa, C., & Beltramo, D. (2000). Interactions of different carrageenan isoforms and flour components in bread making. *Journal of Agricultural and Food Chemistry, 48*, 2634–2638.

Liehr, M., & Kulicke, W. M. (1996). Rheological examination of the influence of hydrocolloids on the freeze thaw stability of starch gels. *Starch/Starke, 48*, 52–57.

Lopes Almeida, E., & Kil Chang, Y. (2013). Structural changes in the dough during the pre-baking and re-baking of French bread made with whole wheat flour. *Food Bioprocess Technol, 6*, 2808–2819.

Lopes Almeida, E., Kil Chang, Y., & Steel C. J. (2013). Dietary fiber sources in frozen part-baked bread: Influence on technological quality. *Food Science and Technology, 53*, 262–270.

Marston, P. E. (1983). Moisture content and migration in bread incorporating dried fruit. *Food Technology Australia, 35*, 463–465.

Martinez-Anaya, M. A., & Jimnez, T. (1997). Functionality of enzymes that hydrolyse starch and nonstarch polysaccharide in bread making. *Lebensm Unters Forsch, 205*, 569–583.

Martinez-Anaya, M. A., Devesa, A., Andreu, P., Escriva, C., & Collar, C. (1999). Effects of the combination of starters and enzymes in regulating bread quality and shelf life. *Food Sci Technol Int, 5*, 263–273.

Miller, R. A., & Hoseney, R. C. (2008). Role of salt in baking. *CFW, 53*, 4–6.

Mondal, A., & Datta, A. K. (2007). Bread baking-A review. *J. Food. Eng, 86*, 465–474.

Mousia, G. M., Campbell, S. S., Pandiella, C., & Webb. (2008). Effect of fat level, mixing pressure and temperature on dough expansion capacity during proving. *Journal of Cereal Science, 46*, 139–147.

Poitrenaud, B. (2004). Baker's yeast. In: Yui, Y. H., Meunier-Goddik, L., Hansen, Å. S., Nip, W. K., Stanfield, P. S., Toldrá, F. (Eds.), Handbook of Food and Beverage Fermentation Technology. New York: Marcel Dekker, Inc., pp. 695–719.

Purlis, E., & Salvadori, V. O. (2007). Bread browning kinetics during baking. *Journal of Food Engineering, 80* (4), 1107–1115.

Ribotta, P. D., & Le Bail, A. (2007). Thermo-physical and thermo-mechanical assessment of partially baked bread during chilling and freezing process. Impact of selected enzymes on crumb contraction to prevent crust flaking. *Journal of Food Engineering, 78*, 913–921.

Ribotta, P. D., Leon, A. E., & Añón, M. C. (2001). Effects of yeast freezing in frozen dough. *Cereal Chemistry, 80* (4), 454–458.

Ribotta, P. D., Perez, G. T., Leon, A. E., & Anon, M. C. (2004). Effect of emulsifier and guar gum on micro structural, rheological and baking performance of frozen bread dough. *Food Hydrocolloids, 18*, 305–313.

Romano, A., Toraldo, G., Cavella, S., & Masi, P. (2007). Description of leavening of bread dough with mathematical modeling. *Journal of Food Engineering, 83*, 142–148.

Ronda, F., Quilez, J., Pando, V., & Roos, Y. H. (2014). Fermentation time and fiber effects on recrystallization of starch components and staling of bread from frozen part-baked bread. *Journal of Food Engineering, 131*, 116–123.

Rosell, C. M., & Gomez, M. (2007). Frozen dough and partially baked bread: an update. *Food Rev Int, 23*(3), 303–319.

Ross, K. D. (1978). Differential scanning calorimetry of non-freezable water in solute-macromolecule-water systems. *Journal of Food Science, 43*(6), 1812–1815.

Roussel, P., & Chiron, H. (2002). French Bread: Evolution, Quality, and Production. Mae-Erti Editions, Vesoul.

Schiraldi, A., Piazza, L., & Riva, M. (1996). Bread staling: A calorimetric approach. *Cereal Chemistry, 73*, 32–39.

Selomulyo, V. O., & Zhou, W. (2007). Frozen bread dough: Effects of freezing storage and dough improvers. *J. Cereal Sci., 45*, 1–17.

Serna-Saldívar, S. O. (1996). Chemistry, storage and industrialization of cereals. *México*, D. F: A. G. T. Editor, S. A. p. 247.

Shewry, P. Nigel, G., & Halford. (2001). Cereal seed storage proteins: structures, properties and role in grain utilization. *Journal of Experimental Botany, 53*, 947–958.

Silvas-García, M. I. (2010). Effect of the addition of trehalose on the viscoelastic properties of frozen doughs and quality of French type bread. Department of Research and Graduate Studies in Food.

Stokes, D. J., & Donald, A. M. (2000). In situ mechanical testing of dry and hydrated breadcrumb using environmental SEM. *J. Mat. Sci., 35*, 599–607.

Twillman, T. J., & White, P. J. (1988). Influence of monoglycerides on the textural shelf life and dough rheology of corn tortillas. *Cereal Chem, 65*, 253–257.

Vulicevic, I. R., Abdel-Aal, E. S. M., Mittal, G. S., & Lu, X. (2004). Quality and storage life of par-baked frozen breads. *Lebensm.-Wiss. u.-Technol, 37*, 205–213.

Wagner, M. J., Lucas, T., Le Ray, D., & Trystram G. (2007). Water transport in bread during baking. *Journal of Food Engineering, 78*, 1167–1173.

Ward, F. M., & Andon, S. A. (2002). Hydrocolloids as film formers, adhesives, and gelling agents for bakery and cereal products. *Cereal Foods World, 47*, 52–55.

Zanoni, B., Peri, C., & Gianotti, R. (1995). Determination of the thermal diffusivity of bread as a function of porosity. *Journal of Food Engineering, 26*(4), 497–510.

**CHAPTER 3**

# PRODUCTION OF XYLO-OLIGOSACCHARIDES (XOS) FROM BIOMASS WASTE BY HYDROTHERMAL TREATMENT: AN OVERVIEW OF PREBIOTIC ADVANCES

ALÁN PAVLOVICH-ABRIL

*Biopolymers Laboratory, Research Center for Food and Development, PO Box 1735, Hermosillo, Sonora 83000, Mexico*

## CONTENTS

## 3.1   INTRODUCTION

Plant biomass is a promising raw material for the production of chemicals and food because it is an abundant, renewable and worldwide distributed source of carbon (Ragauskas et al., 2006). The lignocellulosic biomass is generally composed of 40–45% cellulose, 25–35% hemicellulose and 15–30% lignin. Even though plant biomass is one of the most abundant resources of carbon on the planet, one of the main challenges in biomass usage is the efficient depolymerization of cellulose and hemicellulose into its composing mono-mers (Bilgiçli et al., 2007).

Bran, which represents about 15% of the wheat grain weight after mill-ing, is a composite multi-layer material made up of several adhesive tissues: outer pericarp, inner pericarp, testa, nucellar epidermis, and aleurone layer. Wheat bran, as a by-product of wheat flour production, is an underrated constituent that could act as a food ingredient rather than being discarded or used as animal feed (Hemery et al., 2007). It is a rich source of dietary fiber (36.5–52.4 w/w%) (Bilgiçli et al., 2007; Vitaglione et al., 2008) among which arabinoxylans (AX) have been drawing attention due to their dem-onstrated healthy benefits, for example, lowering cholesterol uptake, atten-uating type II diabetes, reducing the risk for cardiovascular diseases and cancers, improving the absorption of certain minerals and increasing fecal bulk (Cao et al., 2011; Vitaglione et al., 2008). Thus, more value would be added to wheat bran by extracting an AX-enriched fraction from it.

More than 60% dry weight of cell walls from wheat bran is composed of AX (Stone and Morell, 2009) and these are present in all layers: 35–47% dm in the outer pericarp, 36–40% dm in the intermediate layer (inner pericarp, seed coat and nucellar epidermis), and 18–28% dm in the aleurone layer, in general, the AX in the pericarp (including outer and inner pericarp) are highly substituted with arabinose (Ara/Xyl > 1) (Barron et al., 2007; Stone and Morell, 2009), whereas those in the aleurone layer are sparsely substi-tuted (Ara/Xyl < 0.5).The majority of wheat bran AX are water-unextract-able (Maes and Delcour, 2002).

In the last two decades, there has been an increasing attention in the pro-cesses taking place during biomass hydrolysis, so, new techniques are being explorer. A physicochemical approach is use water as reaction medium, and there are several processes that use water for breakdown the molecu-lar structure of polysaccharides. Hydrothermal treatment (HT), liquid hot water, and compressed hot water hydrolysis, among others. Although are similarities, each term implies using different operational conditions (Prado et al., 2016). These processes are thinking in generate less sugar degradation compounds than acid and alkali hydrolysis, it does not generate solid residues, it is extremely fast when compared to both acid and enzymatic routes, and it may avoid the pretreatment step required for the enzymatic process. The sub/supercritical water hydrolysis (SCWH) concept was developed in the early 1990s, and is still not fully optimized (Prado et al., 2016).

Hydrolysis of highly polymerized AX leads to formation of arabinoxylan oligosaccharides (AXOS), which are characterized by their average degree of polymerization (avDP) and average degree of arabinose substitution (Snelders et al., 2013). Moreover, AXOS had a higher potential to shift a part of the beneficial sugar fermentation toward the distal colon parts, compared with inulin (Grootaert et al., 2009). Many prebiotics belong to the group of non-digestible oligosaccharides that originating from hydrolysis of polysaccharides, which resist digestion and absorption in the human small intestine and are fermented in the large intestine. A new class of candidate as prebiotics are the AXOS (Carvalho et al., 2012).

## 3.2   BIOMASS WASTE

The main structural carbohydrates of agricultural and food processing waste are cellulose, hemicellulose and starch and other components such lignin, which are converted by supercritical water to value added products and eliminate the problems related to their disposal (Knez et al., 2015)

### 3.2.1   CELLULOSE

Chemical transformation of cellulose to various chemicals has been considered and is strongly dependent on process conditions such as temperature, pressure, residence time and catalyst (Cocero et al., 2013). Cellulose in sub- or super critical water depolymerizes into oligosaccharides, which further hydrolyze to glucose. Glucose can be further degraded to fructose, glycolaldehide and erythrose, glyceraldehyde and dihydroxyacetone. Fructose can be dehydrated to 5-hydroxymethyl furfural, which can be further decomposed to levulinic acid, formic acid, 1,2,4-benzenetriol, and furfural. Glyceraldehyde can be further converted to pyruvaldehyde, lactic acid, and acylic acid (Knez et al., 2015).

### 3.2.2   HEMICELLULOSES

There are heterogeneous polymers of pentoses (xylose and arabinose) and hexoses (mannose, glucose, and galactose) that are substituted with sugar acids. Hemicellulose is important raw material for transformation to biofuels and other products. It is easily hydrolysable to its constituent sugars in water at elevated temperatures, that can be further hydrolyzed and converted to various simple chemicals such as weak carboxylic acids (succinic, fumaric, levulinic, glucaric acids, etc.), polyols (xylitol, arabitol, and sorbitol), furfural or lactones (Carvalheiro et al., 2008).

### 3.2.3   STARCH

Is hydrolyzed more easily than cellulose at temperatures around 100°C and depolymerizes to glucose and further to 5-HMF (Jin and Enomoto, 2011). Nowadays, starch, with an annual market size of more than 60 million tons is a raw material for many purposes, from production of sweeteners for the food industry to production of ethanol (Pavlovic et al., 2013).

### 3.2.4   LIGNIN

Is a cross-linked aromatic biopolymer available in plants in different compositions, molecular weights, and amounts. Lignin has a complex, branched structure mostly made from phenylpropane subunits of trans-p-coumaryl alcohol, trans-p-coniferyl alcohol, and trans-p-sinapyl alcohol (aromatic monomers) (Chen et al., 2001). Lignin decomposition in subcritical water is more difficult than that of other biomass components due to highly cross-linked phenol alcohol structure, bonded together with strong ether (C O C) and C-C bonds. Lignin could be liquefied by the addition of catalysts such as ethanol, phenol, or alkaline salts also at subcritical water conditions, but significant decomposition starts in supercritical water (Pye, 2008). In lignocellulosic biomass the cellulose, hemicelluloses and lignin are cross-linked and connected by hydrogen bonds and van der Waals forces (Zhao et al., 2014) and the dissolution characteristics of natural lignocellulose are different from those of single components. In addition to hemicellulose and cellulose, biomass is rich in pectin.

### 3.2.5 PECTIN

Is defined as a hetero-polysaccharide predominantly containing galacturonic acid residues. The main backbone of pectin contains homogalacturonan ($\alpha$-(1,4)-linked d-galacturonic acid residues) and a hetero-polymer consisting of d-galacturonic acid and L-rhamnose like rhamnogalacturonan I. Sugar beet pectin contains arabinan ($\alpha$-(1,5)-linked L-arabinofuranosyl residues) and/or galactan ($\beta$-(1,4)-linked d-galactose residues) as side chains of rhamnogalacturonan I. It is known that ferulic acid, is linked to arabinan and galactan side chains of beet pectin (Saulnier, and Thibault, 1999). Martínez et al. (2010) reported that sugar beet pectin with arabinan and galactan is partially hydrolyzed, leading to the effective release of arabino-oligosaccharides (AOs) and galacto-oligosaccharides (GOs) by HT treatment. Also, previous report have shown that almost all of the ferulic acid ester-linked to arabinan and/or galactan is recovered as ferulate ester of their oligosaccharides, under HT conditions at approximately 170°C (Takano et al., 2010). Sato et al. (2013), using HT technology, achieved more than 80% of the arabinan content of beet fiber, which was recovered as AOs and FA-AOs. Almost all the ferulic acid contained in the beet fiber was recovered as ferulate esters of oligosaccharides (Knez et al., 2015)

The chemical reactions in HT conversion of real biomass, agricultural and food waste, are even more complex; the conversion mechanisms depend on process parameters and are still not fully understood. The complex composition of some biomass wastes may represent a drawback because of the large amount of char formed during HT reforming or the formation of viscous oils which may lead to malfunction of reforming equipment. These problems can be overcome by the use of higher water to biomass ratio, by the application of higher temperatures or by the presence of an appropriate catalyst (e.g., metal catalysts, activated carbon, oxides) (Du et al., 2013).

The widespread varieties of biomass waste that can be processed by sub- and critical water underline the versatility of the technology. Additionally, the method points to a more efficient use of raw material and of production wastes, being favorable to the general economy of the process. The state of research of HT conversions of agricultural and food processing waste has been reviewed by Pavlovic et al. (2013). Experiments on real biomass mostly do not meet yields and selectivity as those done with model substances (Pavlovic et al., 2013). Therefore, the theoretical principles and the processes require further investigation. Xylo-oligosaccharides (XOS) production from LCM depends on two treatment steps. The first step is the xylan extraction from LCM, which includes a chemical pre-treatment.

## 3.3  CEREAL BRAN

The term 'bran' is applied to a range of products derived from cereal grains, and usually is related to the outer layers of the grain or caryopsis. Moreover, the use of cereal bran instead of dietary fiber (DF) has the economic advantage of using agricultural by-products without the added costs of DF extraction. Most cereal bran products contain high levels of insoluble fiber, ash, vitamins, lipids and pigments (Jayadeep et al., 2009).

The majority of cereal bran and some legume hulls are good sources of DF (Table 3.1); the insoluble DF contributions of bran are oat, rice, wheat, and corn. The insoluble fraction of fiber has been related to intestinal

**TABLE 3.1**   DF Content (%) of Some Bran Cereal and Hull Legume Sources

| Source of fiber | Total DF content | Analytical method | References |
|---|---|---|---|
| Wheat bran | 44.46 | Enzymatic gravimetric method | Prosky et al. (1988) |
| Corn bran | 87.86 | Enzymatic gravimetric method | Prosky et al. (1988) |
| Rice bran | 27.04 | Enzymatic gravimetric method | Abdul-Hamid and Luan (2000) |
| Hulled barley | 20.40 | Enzymatic gravimetric method | Marconi et al. (2000) |
| Oat bran | 19.50 | Enzymatic gravimetric method | Kahlon and Woodruff (2003) |
| Rye bran | 37.70 | Gas-chromatographic-colorimetric gravimetric method | Rakha et al. (2010) |
| Pea hull | 88.90 | Enzymatic gravimetric method | Dalgetty and Baik (2006) |
| Lentil hull | 86.70 | Enzymatic gravimetric method | Dalgetty and Baik (2006) |
| Chickpea hull | 74.80 | Enzymatic gravimetric method | Dalgetty and Baik (2006) |
| Soya hull | 79.50 | Enzymatic gravimetric method | Dikeman et al. (2006) |

*Source*: Pavlovich-Abril et al. (2012).

regulation, whereas soluble fiber is associated with decreased cholesterol levels and the absorption of intestinal glucose (Rodríguez et al., 2006). Extensive research over the last three decades supports the beneficial role of DF in health and nutrition via a reduction in chronic diseases such as cardiovascular disease, certain forms of cancer and constipation (Schaafsma, 2004).

## 3.4 WHEAT BRAN

Wheat bran and aleurone layer are matrices with a very complex structure. In these fractions, at the molecular level, DF and other bioactive compounds are mostly present in bound forms, not as free constituents. DF are primarily polysaccharides present on cell walls in the outermost layers. Other bioactive compounds are also embedded in these complex matrices, either encapsulated in the cells and/or cell-wall structures or chemically bound at the molecular level. For example, ferulic acid (FA), which is the major phenolic acid of wheat bran, is mainly ester-linked with AX and embedded in bran and aleurone cell walls. The physiological and health effects of wheat grain and its fraction have been studied, but the importance of their structure, from macro via micro to molecular levels, is not often mentioned or taken into account in the scientific literature (Sibakova et al., 2015). Food with greater nutritional properties could be obtained through proper assembly of the hierarchical structures (Zuñiga and Troncoso, 2012). These structures could be obtained by targeted disintegration or reconstruction of one or more components from the microscopic to macroscopic levels.

Arabinoxylans and β-glucans are the nutritionally relevant cell wall polysaccharides of wheat grain and fractions. Wheat AX are formed by a linear backbone of (1→4)-linked b-D-xylopyranosyl units and generally divided into two groups: water-extractable AX (WE-AX) and water-unextractable AX (WU-AX). In wheat grain, the WE-AX and WU-AX contents are around 0.51 and 1.84 g/100 g flour, respectively (Saulnier et al., 2007), but the proportion of WU-AX is even higher in wheat bran and aleurone (around 95% w/w) (Rosa et al., 2013). β-glucans account to 0.4–0.8 g/100 g wheat flour and most of them (80–90%) are water-unextractable (Nemeth et al., 2010). FA is the most abundant polyphenol in wheat grain, representing over 95% of all the phenolic acids. FA is mostly present in an insoluble form, esterified to AX, whereas sinapic and vanillic acid are mostly found in the conjugated form that is, linked with oligosaccharides (Li et al., 2008). As summarized in Table 3.2, wheat bran contains relatively high AXs in various fractions. So, wheat bran is a cost effective source of AXs, which is a by-product of cereal processing.

## 3.5 HYDROTHERMAL TREATMENT (HT)

According to the new criteria proposed for chemical processes and synthesis, which are the basis for so called sustainable chemistry or green chemistry, the new designed processes should include principles such as lower energy consumption, process intensification and simplification, use of renewable feedstocks, use of safe and environmentally friendly solvents.

Chemical and biochemical processes are generally performed in solutions. Conventional organic solvents are potential environmental pollutants and therefore research is oriented towards either solvent free processes or solvents with lower environmental impact – like supercritical water. "Green solvents for the future" as supercritical fluids offer the possibility to obtain new products with special characteristics or to design new processes, which are environmentally friendly and sustainable, these being a great challenge for chemical engineers (Knez et al., 2015). Lately, the process has been applied to biomass waste, following various industrial processes.

Water as a reaction medium presents advantages over other solvents because it is a non-expensive and environmentally friendly solvent. In addition, the medium identity can be tuned by changing the temperature and pressure in order to favor the desired reactions without using any catalyst. The use of sub and supercritical water has been proposed as a promising

**TABLE 3.2** Arabinoxylan Contents of Various Cereals and Cereal By-Products (Dry Basis)

| Tissues | Total AXs (%) | WEAXs (%) | References |
|---|---|---|---|
| Whole grain | 5.77 | 0.59 | Hashimoto et al. (1987) |
| Bran | 21.4 | – | Courtin and Delcour (2001) |
| Bran | 25 | 1 | Hollmann and Lindhauer (2005) |
| Bran | 19 | – | Bataillon et al. (1998) |
| Bran | 19.38 | 0.88 | Hashimoto et al. (1987)[a] |
| Durum wheat | 4.07–6.02 | 0.37–0.56 | Lempereur et al. (1997) |

*Source*: Adapted from Zhang et al. (2014).

solvent to process biomass due to its special properties that are very promising to perform the hydrolysis reactions (Arai et al., 2009). Supercritical water refers to the state of water under pressure and temperature conditions above its critical point. The critical point of water is 374°C and 22.1 MPa. Near its critical point, a solvent drastically changes its physical properties by simply modifying the pressure and being favorable to the general economy of the process temperature. This behavior is a promising alternative to manage the selectivity in chemical reactions (Cantero et al., 2015). Due to unique properties of water at elevated temperatures versatile chemicals and fuels in gaseous, liquid, or solid state can be produced. In these processes water may act as solvent, reactant or a catalyst. During the treatment of the biomass in sub- or supercritical water numerous reactions can occur parallel in the system, for example, hydrolysis, dehydration, decarboxylation, aromatization, condensation, depolymerization and/ or polymerization, hydrogenation/dehydrogenation, rearrangement, isomerization reaction (Cantero et al., 2015)

HT liquefaction is performed at medium temperatures and high pressures (250–370°C, 10–25 MPa) at which biomass hydrolyses and decomposes to unstable small components, which further re-polymerize and produce highly viscous water-insoluble bio-oil, water soluble substances, char, and light gasses. The reaction mechanism is influenced by temperature, pressure, reaction time, type of feedstock and catalysts. In general, both acids and bases catalyze liquefaction of biomass (Cantero et al., 2015).

In supercritical water hexoses produced from cellulose decompose rapidly into non-fermentable fragmentation products. In subcritical water (at temperatures from 200 to 374°C), the decomposition of hexoses is much slower than in supercritical water (Ehara and Saka, 2005). Compared with the conventional technologies including acid treatment and enzymatic hydrolysis, the combined supercritical/subcritical technology demonstrates obvious advantages, such as much higher reaction rate, not requiring additional catalyst, and no inhibitory reaction of intermediates (Zhao et al., 2009).

A combined supercritical/subcritical process for pre-treatment and hydrolysis of lignocellulosic waste has been projected; lignocelluloses are pretreated and hydrolyzed in supercritical water to strip the lignin and to produce oligosaccharides from cellulose, followed by a secondary hydrolysis in subcritical water to convert the oligosaccharides into fermentable hexoses (Zhao et al., 2009). However, these process lead to production of the several residual materials.

Bio-oil (biocrude) is a hydrophobic mixture of over several hundred oxygenated compounds of various molecular weights, originating from the decomposition of three main constituents in biomass: cellulose, hemicelluloses, and lignin (Akhtar and Amin, 2011). Although the product distribution in bio-oil varies with the composition of the raw material and process conditions, the same groups of compounds are detected in almost all bio-oils (Zhang et al., 2008). HT gasification is a process in which biomass reacts with water at high temperatures and pressures to form gaseous products, mainly $CH_4$, $H_2$, CO, $CO_2$, and C1–C4 carbon gases. As side products, also some bio-oil, char, and tar are formed, which decrease the yield of gases. HT gasification can be performed at temperatures (300–500°C) and is named catalytic wet gasification (in sub- and near-critical water) or at higher temperatures (500–800°C) and is named supercritical water gasification (Knez et al., 2015).

### 3.5.1   AUTOHYDROLYSIS

Implicates the deacetylation of xylan by the thermal hydrolysis of hemicellulose to produce acetic acid (Garrote et al., 2002). When lignocellulosic material (LCM) containing xylan is submitted to autohydrolysis under mild operating conditions, xylo-oligosaccharides are the main products of the hydrolysis of hemicellulose. The basic concept of the autohydrolisis is simple, it's the breakdown of the polysaccharides into fermentable oligo or monosaccharides. Hydrolysis leads to cleavage of the ether and ester bonds by adding of one molecule of water for each broken linkage, which end in simpler sugar (Carvalho et al., 2012).

AX may also be released from its insoluble matrix by autohydrolysis, then the structure of the carbohydrates released are different from those released by enzyme (Rose and Inglett, 2010). For instance, autohydrolysis breaks down the AX into many size pieces ranging from monosaccharides on up to polysaccharides with a DP of 750 or more (Rose and Inglett, 2010) while enzymatic treatment releases mostly shorter chain oligosaccharides (Swennen et al., 2006). Yang et al. (2014) identified a high-pressure HT of cereal bran that resulted in fragmentation of the cell wall. Although, the concept is simple, the biomass it is formed by a lignocellulose complex, composed by cellulose, hemicellulose and lignin, which is particularly to break because is chemically bonded to hemicellulose and cellulose, increasing the resistance to hydrolysis (Prado et al., 2016). Additionally, autohydrolysis requires specialized equipment that needs to be functioned at high temperatures (Zhu et al., 2006).

The products obtained from LCM by this process comprise a range of unwanted components such as lignin, monosaccharides, furfural, and others, thus requiring further purification (Zhu et al., 2006). XOS produced by autohydrolysis contains a significant proportion of compounds with a high degree of polymerization (Nabarlatz et al., 2007). While individual AXOS molecules are structurally characterized by a degree of polymerization (DP) and degree of arabinose substitution (DAS), AXOS products, comprising a mixture of different molecules, are typically characterized by an average DP (avDP) and an average DAS (avDAS) (Snelders et al., 2013).

## 3.6 PHYSICOCHEMICAL CHARACTERISTIC OF XYLO-OLIGOSACCHARIDES

AXOS consist of xylose backbone ($\beta$-1, 4/1, 3/1, 2) with arabinose in the side chain substituted at either or 2nd, 3rd, and 5th positions of xylose and most of these oligosaccharides do consist of ester linked phenolic acids such as ferulic, coumaric and caffeic acids.

AXOS stabilities can differ greatly depending on the types of oligosaccharide and sugar residues, linkages, ring forms and anomeric configurations. Generally, $\beta$-linkages are stronger than $\alpha$-linkages, and hexoses are more strongly linked than pentoses. Most oligosaccharides can be hydrolyzed, resulting in the loss of nutritional and physicochemical properties at pH < 4.0, when treated at high temperatures for short time periods, or when subjected to prolonged storage under room conditions. For example, fructo-oligosaccharides (FOS) it is related that in a 10% solution of pH 3.5 less than 10% is hydrolyzed after heat treatments of 5 min at 45°C or 60 min at 70°C. After two days at 30°C less than 50% is hydrolyzed (Voragen, 1998). Nevertheless, the AXOS are stable over a wide range of pH (2.5–8.0), even the relatively low pH value of gastric juice, and temperatures up to 100°C. This is an advantage compared with other non-digestible oligosaccharides such as FOS and inulin (Bhat, 1998).The prebiotic properties of AXOS products depend on their structural characteristics. Especially, the (av)DP has a substantial influence on the prebiotic properties of AXOS (Craeveld et al., 2008; Pollet et al., 2012). AXOS with the same avDP but different avDAS have either been reported to be equally fermentable (Craeveld et al., 2008) or to be more fermentable when less arabinose substituents are present (Pollet et al., 2012). So, next to chain length, the level of the arabinose substituents presents a second structural characteristic that can be used to fine-tune AXOS functionality (Snelders et al., 2013).

As functional properties are related to structural characteristics, detailed structural characterization of AXOS is important. As AXOS are so varied, it is impossible to quantify every single molecule of complex AXOS mixtures. Gas chromatography (GC) analysis only allows determining structural characteristics on an average basis for the population of different molecular entities. Yet, other methods can assess the molecular structure, a high performance anion exchange chromatography (HPAEC) method has been tested to characterize the DP distribution of xylan backbone of AXOS, based on quantification of XOS after the removal of arabinose substituents. Enzymatic release of arabinose proved specific, removing 80% to 90% of the arabinoses, while leaving the xylan backbones uncleaved. HPAEC analysis of AXOS pretreated by enzymatic hydrolysis provides insight in the DP distribution of the xylan backbones of the different AXOS entities. HPAEC analysis therefore is a useful and complementary addition to the standard GC and RP-HPLC methods that allow accurate determination of the AXOS level, avDP, avDAS and FA content of XOS and AXOS (Snelders et al., 2013).

The stability and sensory properties of oligosaccharides are strongly dependent on their molecular structure. Heat stability and shelf-life measurements showed that the short chain non digestible oligosaccharides (NDO) preparations, XOS and FOS (average DP of 3 and 5, respectively), are more sensitive to alkaline decomposition than were longer chain AXOS (average DP of 15). The higher sensitivity of the former to alkaline peeling is not surprising, as short chain NDO, per weight basis, have much more reducing ends than have their longer chain analogs. AXOS shows the most interesting stability properties, especially under extreme pH and temperature conditions. However, applications involving low or high pH, high temperature processing or prolonged storage under ambient conditions will (to some extent) decompose NDO structures, resulting in a loss of nutritional and physicochemical properties (Courtin et al., 2009).

Sarbini et al. (2013) indicated that low-molecular-mass oligosaccharides obtained by pectin hydrolysis are often more selectively fermented by *bifidobacteria* and *lactobacilli* than are their parent high-molecular-weight carbohydrates. According to Sanchez et al. (2009) who used a human colon model system, the chain length of the AXOS is critical in determining the site of fermentation because shorterchain-length molecules are fermented in the proximal colon, whereas longer molecules (average DP = 29) reach the distal colon and increase short chain fatty acid (SCFA) concentrations. Gómez et al. (2015) show that both substituted and unsubstituted oligosaccharides of xylose and arabinose were confirmed in purified extract by MALDI-TOF MS. And the AXOS proved their prebiotic activity by in vitro fermentation assays using fecal inocula from elderly people. Experimental results showed that AXOS fermentation caused an increased *bifidobacteria* and *lactobacilli* populations with respect to the negative control cultures as well as the accumulation of SCFA, confirming a stimulatory effect. Moreover, taking into account the variations in the bacterial populations and the production of organic acids, it was concluded that the results obtained with AXOS are slightly better than the ones achieved with FOS.

### 3.6.1   PREBIOTIC

The prebiotic defined as "a selectively fermented ingredient that allows specific changes, both in the composition and/or activity in the gastrointestinal microbiota that confers benefits upon host wellbeing and health" (Roberfroid, 2007). According to this definition, candidate prebiotics must accomplish the cited criteria which are to be proven by in *vitro* and finally in *vivo* tests. In recent years, there is a sizable interest in the use of prebiotics as functional foods in order to modulate/modify the composition of the colonic microbiota to provide health and benefits to the host (Silk et al., 2009). The majority of the health benefit claimed by the prebiotics are associated with optimized colonic function and metabolism, such as an increase in the expression or change in the composition of short-chain fatty acids, increased fecal weight, a reduction in luminal colon pH, a decrease in nitrogenous end products and reductive enzymes, an increased expression of the binding proteins or on certain biomarkers in the field of lipid and mineral metabolism and immune system modulation (Qiang et al., 2009). In the last two decades, lots of studies have investigated the production of oligosaccharides as food ingredients that promote a state of wellbeing, and health of the consumers. Furthermore, the current development of commercial prebiotic oligosaccharides and probiotic bacteria has led naturally to a new concept, that of symbiotic one, combining probiotics and prebiotics. They can be obtained by direct extraction from natural sources, or produced by chemical processes hydrolyzing polysaccharides, or by enzymatic and chemical synthesis from disaccharides (Mussamatto and Mancilha, 2007). Most of them are synthesized or isolated from plant and algae polysaccharides depolymerization such as FOS, galactooligosaccharides, isomalto-oligosaccharides and AXOS. So, it is well accepted that each prebiotics and probiotic health benefit is strain-specific or oligosaccharide-specific, and these cannot be extrapolated to others. Recently, several clinical trails and in vitro studies have provided significant results supporting the health benefits of prebiotics. Particularly for enhancing the intestinal health, and for the irritable bowel syndrome. In this latter, the effect of probiotic and prebiotic remains ambiguous and requires more investigations in order to be confirmed or to be corroborated (Saad et al., 2013).

## 3.7  PREBIOTIC EFFECTS OF AXOS

The intestine is an organ that consists of a vast superficial area and permits vital interactions with micro-organisms living within the intestine, referred to as gut microbiota. The gut microbiota exerts significant effects on the host physiology, such as digestion and vitamin synthesis, control of energy homeostasis, and the immune system (Cani et al., 2013) and inhibition of pathogen colonization (Wardwell et al., 2011).

The human gastrointestinal tract is a sophisticated environment and carbohydrates are found to be the first limiting nutrient for many bacterial species in the intestinal tract and thus the type of carbohydrates available govern the growth of the gut microflora (Manisseri and Gudipati, 2010). A large group of these carbohydrates are glycans, resistant to digestion by human enzymes and relies on microbial enzymes for their digestion. The fermentation of these glycans by microbes yield energy for microbial growth, and the end products such SCFA, mainly acetate, propionate and butyrate have profound effects on the health of the host (Tremaroli and Backhed, 2012). Butyrate acts mostly as the energy substrate for the colonic epithelium because it is the main energy source of colonocyte (Koropatkin et al., 2012). Acetate and propionate are absorbed into the blood stream and travel to the liver where they get incorporated into lipid and glucose metabolism, respectively (Rombeau and Kripke, 1990). The functional association among the intestinal microbiota, intestinal epithelial cells and the host immune system helps maintain the balance between tolerance and immunity to pathogenic or non-pathogenic microbes, or food ingredient. Wang et al. (2011) produce AXOS by wheat bran hydrolysis and, was found to be effective in improving blood glucose and lipid levels in liver and heart of rats with supplementation of a high-fat diet. They proposed that dietary supplementation of AXOS could mitigate the damage of oxidative stress induced by high-fat diet through modulating lipid metabolism and antioxidant defense system and providing a beneficial result for human health. Depending on their botanical source, and the hydrolysis process, AXOS can have different structural complexities to begin with. Many gut microbes have evolved to contain enzymes, receptors and transporters that achieve efficient degradation and utilization of these complex AX molecules. Due to their variation in DP, DS, and ferulic acid substitution AX hydrolysates are considerably diverse and complex molecules compared to many other dietary fibers. Thus, AXOS plays a role in modifying the microbial composition in the gut and exerting immunological responses (Mendis and Simsek, 2014). From Table 3.3 it can be noticed that in the last years the numbers of studies on agricultural waste in the production of AXOS, using the HT has increased.

## 3.8 CONCLUSION

The main conclusion to be drawn is that the AXOS production for prebiotic purposes by biomass waste transformation using hydrothermal treatment is an efficient way to achieve it. The use of by-product as cereal bran in particular the wheat bran has a great potential to do so, since there is a major source of LCM, because it is available in most flour millers, it is cheap and easy to use, and requires little or no pre-treatment prior to the transformation process. In this assay, it has been tried to set out the main hydrothermal treatments, and the results of these studies, providing us with a better understanding of the use of these techniques in cereal bran as biomass waste. Subsequent research should focus on strengthen practical and economic methods from LCM reducing production cost and increasing the accessibility of AXOS to progress to greater accessibility to them, as prebiotics.

**TABLE 3.3** Hydrothermal Processes and Operating Condition for LCM Conversion to AXOS

| Hydrothermal processes | LCM source | Operation condition, P/T | Time of residence | AXOS yield (%) | References |
|---|---|---|---|---|---|
| Batch reactor | Beet fiber | 180–160°C 0.3 MPa | 7–12 min | 84.0 | Sato et al., 2013 |
| Pressure reactor | Corn straw | 215°C~ 1.9 MPa | 27 seconds | 72.2 | Moniz et al., 2016 |
| Pressure reactor | Corn and wheat straw | 280–390°C | ~ 15 seconds | 12.5 | Zhao et al., 2009 |
| Pressure reactor | Brewer's spent grain | 180–195°C~ 1.0 MPa | 12.5 min | 77.0 | Gómez et al., 2015 |
| Pressure reactor* | Wheat bran | 400°C 25 MPa | ~0.5 seconds | 22.0 | Cantero et al., 2015 |
| Pressure reactor | Corn and Wheat bran | 180–200°C~ 1.2 MPa | – | 44–49 | Yang et al., 2014 |
| Autoclave+ | Rice bran Cassava pulp | 135°C 0.22 MPa | 180 min | 45.5–47.5 XOS mixture | Kurdi and Hansawasdi, 2015 |
| Pressure reactor | A cherimola seed | 150–230°C ~ 2.0 MPs | – | 74.5 | Branco et al., 2015 |
| Pressure reactor | Wheat bran | 170–220°C ~1.5 MPa | – | 70.0 | Rose and Inglett, 2010 |

* Super critical water; + Low pressure and temperature.

## KEYWORDS

- arabinoxylan
- AXOS
- biomass waste
- cereal bran
- hydrothermal treatment
- prebiotic

# REFERENCES

Akhtar, J., & Amin, N. A. S. (2011). A review on process conditions for optimum bio-oil yield in hydrothermal liquefaction of biomass, *Renewable and Sustainable Energy Reviews, 15*, 1615–1624.

Amen-Chen, C., Pakdel, H., & Roy, C. (2001). Production of monomeric phenols by thermochemical conversion of biomass: a review, *Bioresource Technology, 79*, 277–299.

Arai, K., Smith, R. L., & Aida, T. M. (2009). Decentralized chemical processes with supercritical fluid technology for sustainable society. *The Journal of Supercritical Fluids 47*, 628–635.

Azebedo-Carvalho, A. F., Olivia-Neto, P., Fernandes da Silva, D., & Pastore, G. M. (2012). Xylo-oligosaccharides from lignocellulosic material: Chemical structure health benefits and production by chemical and enzymatic hydrolysis. *Food Research International, 51*, 75–85.

Barron, C., Surget, A., & Rouau, X. (2007). Relative amounts of tissues in mature wheat (*Triticum aestivum* L.) grain and their carbohydrate and phenolic acid composition. *J. Cereal Sci., 45* (1), 88–96.

Bechtel, D. B., Abecassis, J., Shewry, P. R., & Evers, A. D. (2009). Chapter 3: Development, structure, and mechanical properties of the wheat grain. *Wheat: Chemistry and Technology*. AACC International, Inc., pp. 51–95.

Bhat, M. K. (1998). Oligosaccharides as functional food ingredients and their role in improving the nutritional quality of human food and health. *Recent Research Developments in Agricultural and Food Chemistry, 2*, 787–802.

Bilgiçli, N., Ibanoglu, S., & Herken, E. N. (2007). Effect of dietary fiber addition on the selected nutritional properties of cookies. *Journal of Food Engineering, 78*, 86–89

Bobleter, O. (1994). Hydrothermal degradation of polymers derived from plants. *Progress in Polymers Science, 19*, 797–841.

Branco, P. C., Dionísio, A. M., Torrado, I., Carvalheiro, F., Castilho, P. C., & Duarte, L. C. (2015). Auto hydrolysis of *Annona cherimola* Mill. Seed: Optimization, modeling and products characterization. *Biochemical Engineering Journal, 104*, 2–9.

Brouns, F., Hemery, Y., Price, R., & Anson, N. M. (2012). Wheat aleurone: Separation, composition, health aspects, and potential food Use. *Critical Reviews in Food Science and Nutrition, 52*, 553–568.

Cani, P. D., Everard, A., & Duparc, T. (2013). Gut microbiota, enteroendocrine functions and metabolism. *Current Opinion in Pharmacology Gastrointestinal Endocrine and Metabolic Diseases, 13*, 935–940.

Cantero, D. A., Martínez, C., Bermejo, M. D., & Cocero, M. J. (2015). Simultaneous and selective recovery of cellulose and hemicellulose fractions from wheat bran by supercritical water hydrolysis. *Green Chemistry, 17*, 610–618.

Cao, L., Liu, X., Qian, T., Sun, G., Guo, Y., Chang, F., et al. (2011). Antitumor and immunomodulatory activity of arabinoxylans: a major constituent of wheat bran. *Int. J. Biol. Macromol, 48*(1), 160–164.

Carvalheiro, F., Duarte, L. C., & Gírio, F. M. (2008). Hemicellulose biorefineries: A review on biomass pretreatments, *Journal of Scientific and Industrial Research, 67*, 849–864.

Cocero, M. J., Cantero, M. D., Bermejo, J. S., Queiroz, J. S., Mato, F., Yedro, F., et al. (2013). Biomass depolymerization by supercritical water, in: D. Skala, A. Dekanski (Eds.), 6th international symposium on high pressure processes technology, *Association of Chemical Engineers of Serbia*, Belgrade, Serbia, pp. 28–32.

Courtin, C. M., Swennen, K., Verjans, P., & Delcour, J. A. (2009). Heat and pH stability of prebiotic arabinoxylo oligo saccharides, xylo oligo saccharides and fructo oligo saccharides. *Food Chemistry, 112*, 831–837

Du, L., Wang, Z., Li, S., Song, W., & Lin, W. (2013). A comparison of monomeric phenols produced from lignin by fast pyrolysis and hydrothermal conversions, *International Journal of Chemical Reactor Engineering*, 11, 135.

Ehara, K., & Saka, S. (2002). A comparative study on chemical conversion of cellulose between the batch-type and flow-type systems in supercritical water. *Cellulose, 9*, 301–311.

Ehara, K., & Saka, S. (2005). Decomposition behavior of cellulose in supercritical water, subcritical water, and their combined treatments. *Journal of Wood Science, 51*, 148–153.

Fincher, G. B., & Stone B. A. (1986). Cell walls and their components in cereal grain technology. *Advances in Cereal Science and Technology, 8*, 207–295

Garrote, G., Dominguez, H., & Parajó, J. C. (2002). Autohydrolysis of corncob: Study of non-isothermal operation for xylo-oligosaccharide production. *Journal of Food Engineering, 52*, 211–218.

Gómez, B., Míguez, B., Veiga, A., Parajó, J. C., & Alonso, J. L. (2015). Production, Purification, and *in vitro* evaluation of the prebiotic potential of arabinoxylooligosaccharides from brewer's spent grain. *Journal of Agricultural and Food Chemistry, 63*, 8429–8438.

Grootaert, C., Van den Abbeele, P., Marzorati, M., Broekaert, W. F., Courtin, C. M., Delcour, J. A., et al. (2009). Comparison of prebiotic effect of arabinoxylan oligosaccharides and inulin in a simulator of the human intestinal microbial ecosystem. *FEMS Microbiol Ecol, 69*, 231–242.

Hemery, Y., Rouau, X., Lullien-Pellerin, V., Barron, C., & Abecassis, J. (2007). Dry processes to develop wheat fractions and products with enhanced nutritional quality. *Journal of Cereal Science, 46*(3), 327–347.

Jayadeep, A., Singh, V., Sathyendra-Rao, B. V., Srinivas, A., & Ali, S. Z. (2009). Effect of physical processing of commercial de-oiled rice bran on particle size distribution, and content of chemical and biofunctional components. *Food Bioprocess Technology, 2*, 57–67.

Jin, F., & Enomoto, H. (2011). Rapid and highly selective conversion of biomass into value added products in hydrothermal conditions: Chemistry of acid/base-catalyzed and oxidation reactions, *Energy and Environmental Science, 4*, 382–397.

Knez, Z., Markocic, E. M., Hrncic, K., Ravber, M., & Skerget, M. (2015). High pressure water reforming of biomass for energy and chemical: a short review. *The Journal of Supercritical Fluids, 96*, 46–52.

Koropatkin, N. M., Cameron, E. A., & Martens, E. C. (2012). How glycan metabolism shapes the human gut microbiota. *Nature Reviews Microbiology, 10*, 323–335.

Kurdi, P., & Hansawasdi, C. (2015). Assessment of the prebiotic potential of oligosaccharides mixtures from rice and cassava pulp. *LWT Food Science and Technology, 63*, 1288–1293.

Li, L., Shewry, P. R., & Ward, J. L. (2008). Phenolic acids in wheat varieties in the healthgrain diversity screen. *Journal of Agricultural and Food Chemistry, 56*, 9732–9739

Maes, C., & Delcour, J. A. (2002). Structural characterisation of waterextractable and waterunextractable arabinoxylans in wheat bran. *J Cereal Sci, 35*, 315–326.

Manisseri, C., & Gudipati, M. (2010). Bioactive xylo-oligosaccharides from wheat bran soluble polysaccharides. *LWT Food Science and Technology, 43*, 421–430.

Martínez, M., Gullón, B., Yañez R., Alonso, J. L., & Parajó, J C. (2010). Kinetic assessment on the autohydrolysis of pectin-rich by-products, *Chemical Engineering Journal, 162,* 480–486.

Mendis, M., & Simsek, S. (2014). Arabinoxylans and human health. *Food Hydrocolloids, 42*(2), 239–243.

Mendis, M., Leclerc, E., & Simsek, S. (2016). Arabinoxylans, gut microbiota and immunity. *Carbohydrates Polymers, 139,* 159–166.

Moniz, P., Ling Ho, A., Duarte L. C., Kolida, S., Rastall, R. A., Pereira, H., et al. (2016). Assessment of the bifidogenic effect of substituted xylo-oligosaccharides obtained from corn straw. *Carbohydrates Polymers, 136,* 466–473.

Mussamatto, S. I., & Mancilha, I. M. (2007). Non-digestible oligosaccharides: A review. *Carbohydrate Polymers, 68,* 587–597.

Nabarlatz, D., Ebringerová, A., & Montané, D. (2007). Autohydrolysis of agricultural by-products for the production of xylo-oligosaccharides. *Carbohydrate Polymers, 69,* 20–28.

Nemeth, C., Freeman, J., Jones, H. D., Sparks, C., Pellny, T. K., Wilkinson, M. D., et al. (2010). Down-regulation of the CSLF6 gene results in decreased (1,3;1,4)-b-D-glucan in endosperm of wheat. *Plant Physiology, 152,* 1209–1218.

Pavlovic, I., Knez, Z., & Skerget, M. (2013). Hydrothermal reactions of agricultural and food processing wastes in sub and supercritical water: a review of fundamentals, mechanisms, and state of research, *Journal of Agricultural and Food Chemistry, 61,* 8003–8025.

Pavlovich-Abril, A., Rouzaud-Sández, O., Torres, P., & Robles-Sánchez, R. M. (2012). Cereal bran and wholegrain as a source of dietary fiber: Technological and health aspects. *International Journal of Food Sciences and Nutrition, 63,* 882–892.

Pollet, A., Van Craeyveld, V., Van de Wiele, T., Verstraete, W., Delcour, J. A., & Courtin, C. M. (2012). *In vitro* fermentation of arabinoxylan oligosaccharides and low molecular mass arabinoxylan with different structural properties from Wheat (*Triticum aestivum*) bran and psyllium (Plantago ovata Forsk) seed husk. *Journal of Agricultural and Food Chemistry, 60,* 946–954.

Prado, J. M., Lachos-Perez, D., Forster-Carneiro, T., & Rostango, M. (2016). Sub-and super-critical water hydrolysis of agricultural and food residues for the production fermentable sugars: A review. *Food and Bioproducts Processing, 98,* 95–12.

Pye, E. K. (2008). Industrial lignin production and applications, in: B. Kamm, P. R. Gruber, M. Kamm (Eds.), Biorefineries-Industrial Processes and Products, Wiley-VCH Verlag GmbH, Weinheim, Germany, 165–200.

Qiang, X., YongLie, C., & QianBing, W. (2009). Health benefit application of functional oligosaccharides. *Carbohydrate Polymers, 77,* 435–441.

Ragauskas, A. J., Williams, C. K., Davison, B. H., Britovsek, G., Cairney, J., Eckert, C. A., et al. (2006). The Path Forward for Biofuels and Biomaterials. *Science, 311,* 484–489.

Roberfroid, M. (2007). Prebiotics: the concept revisited. *Journal of Nutrition, 137,* 830–837

Rodríguez, R., Jiménez, A., Fernández-Bolaños, J., Guillén, R., & Heredia, A. (2006). Dietary fiber from vegetable products as source of functional ingredients. *Food Science and Technology, 17,* 3–15.

Rombeau, J. L., & Kripke, S. A. (1990). Metabolic and intestinal effects of short-chain fatty acids. *Journal of Parenteral and Enteral Nutrition, 14*(5 suppl), 181S–185S.

Rosa, N. N., Aura, A. M., Saulnier, L., Holopainen, U., Poutanen, K., & Micard, V. (2013). Effects of disintegration on in vitro fermentation and conversion patterns of wheat aleurone in a metabolical colon model. *Journal of Agricultural and Food Chemistry,* *61,* 5805–5816.

Rose, D. J., & Inglett, G. E. (2010). Two-stage hydrothermal processing of wheat (*Triticum aestivum*) bran for the production of feruloylated arabinoxylooligosaccharides. *J. Agric. Food Chem, 58,* 6427–6432.

Saad, N., Delattre, C., Urdaci, M., Schmitter, J. M., & Bressolier, P. (2013). An overview of the last advances in probiotic and prebiotic field. *LWT Food Science Technology, 50,* 1–16.

Sanchez, J. I., Marzorati, M., Grootaert, C., Baran, M., Van Craeyveld, V., Courtin, C. M., et al. (2009). Arabinoxylan-oligosaccharides (AXOS) affect the protein/carbohydrate fermentation balance and microbial population dynamics of the Simulator of Human Intestinal Microbial Ecosystem. *Microb. Biotechnol, 2,* 101–113.

Sarbini, S. R., Kolida, S., Naeye, T., Einerhand, A. W., Gibson, G. R., & Rastall, R. A. (2013). The prebiotic effect of $\alpha$-1,2 branched, low molecular weight dextran in the batch and continuous fecal fermentation system. *J. Funct. Foods, 5,* 1938–1946.

Sato, N., Takano, Y., Mizuno, M., Nozakia, K., Umemura, S., Matsuzawa, T., et al. (2013). Production of feruloylated arabino-oligosaccharides (FA-AOs) from beet fiber by hydrothermal treatment. *The Journal of Supercritical Fluids, 79,* 84–91.

Saulnier, L., & Thibault, J. F. (1999). Ferulic acid and diferulic acid as components of sugar beet pectins and maize bran heteroxylan, *Journal of the Science of Food and Agriculture, 79,* 396–402.

Saulnier, L., Sado, P., Branlard, P., Charmet, G., & Guillon, F. (2007). Wheat arabinoxylans: Exploiting variation in amount and composition to develop enhanced varieties. *Journal of Cereal Science, 46,* 261–281.

Schaafsma, G. (2004). Health claims, options for dietary fiber. In: Van der Kamp, J. W., Asp, N. G., Miller Jones, J., & Schaafsma, G., (Eds.). Dietary fiber: Bioactive carbohydrates for food and feed. Wageningen, The Netherlands: Wageningen Academic Publishers, pp. 27–38.

Sibakova, N. R., Poutanenb, K., & Micarda, V. (2015). How does wheat grain, bran and aleurone structure impact their nutritional and technological properties? *Trends in Food science and Technology, 41,* 118–134.

Silk, D. B., Davis, A., Vulevic, J., Tzortzis, G., & Gibson, G. R. (2009). Clinical trial: The effects of a trans-galactooligosaccharide prebiotic on fecal microbiota and symptoms in irritable bowel syndrome. *Alimentary Pharmacology and Therapeutics, 29,* 508–518.

Snelders, J., Dornez, E., Broekaert, W. F., Delcour, J. A., & Courting, C. M. (2013). Determination of the xylan backbone distribution of arabinoxylan-oligosacchrides. *Bioactive carbohydrates and Dietary Fibre, 2,* 84–91.

Stone, B., & Morell, M. K. (2009) Chapter 9: Carbohydrates. Wheat: Chemistry and Technology. *AACC International, Inc.,* 299–362.

Swennen, K., Courtin, C. M., Lindemans, G. C. J. E., & Delcour, J. A. (2006). Large scale production and characterization of wheat bran arabinoxylooligosaccharides. *Journal of Science Food Agricultural, 86,* 1722–1731.

Takano, Y., Sato, N., Mizuno, M., Nozaki, K., Matsuzawa, T., Kanda, T., et al. (2010). Development of the recovery technology of soluble sugars from beet fiber by the hydrothermal treatment, in: Proceedings of 59th Annual meeting of the Japanese Society Applied Glycoscience, *Shizuoka, Japan,* Bp2–19 (in Japanese).

Tremaroli, V., & Backhed, F. (2012). Functional interactions between the gut microbiota and host metabolism. *Nature, 489*(7415), 242–249.

Van Craeveld, V., Swennen, K., Dornez, E., Van de Wiele, T., Marzorati, M., & Verstraete, W. (2008). Structurally different wheat-derived arabinoxyloologosaccharides have different prebiotic and fermentation properties in rats. *Journal of Nutrition, 138,* 2348–2355.

Van Hung, P., Maeda, T., Miyatake, K., & Morita, N. (2009). Total phenolic compounds and antioxidant capacity of wheat graded flours by polishing method. *Food Research International, 42,* 185–190.

Vardakou, M., Nueno Palop, C., Christakopoulos, P., Faulds, C. B., Gasson, M. A., & Narbad, A. (2008). Evaluation of the prebiotic properties of wheat arabinoxylan fractions and induction of hydrolase activity in gut microflora.International. *Journal of Food Microbiology, 123,* 166–170.

Vitaglione, P., Napolitano A., & Fogliano, V. (2008). Cereal dietary fiber: A natural functional ingredient to deliver phenolic compounds into the gut. *Trends Food Sci. Technol, 19* (9), 451–463.

Voragen, A. G. J. (1998). Technological aspects of functional food-related carbohydrates. *Trends in Food Science & Technology, 9,* 328–335.

Wang, J., Cao, Y., & Wang, C., Sun, B. (2011). Wheat bran xylooligosaccharides improve blood lipid metabolism and antioxidant status in rats fed a high-fat diet. *Carbohydrate Polymers, 86,* 1192–1197.

Wardwell, L. H., Huttenhower, C., & Garrett, W. S. (2011). Current concepts of the intestinal microbiota and the pathogenesis of infection. *Current Infectious Disease Reports, 13,* 28–34.

Yang, J., Maldonado-Gómez, M. X., Hutkins, R. W., & Rose D. J. (2014). Production and *in vitro* fermentation of soluble, non-digestible, feruloylated oligo-and polysaccharides from maize and wheat brans. *J. Agric. Food Chem, 62,* 159–166.

Zhang, B., von Keitz, M., & Valentas, K. (2008). Maximizing the liquid fuel yield in a biorefining process, *Biotechnology and Bioengineering, 101,* 903–912.

Zhang, Z., & Smith, C., Li, W. (2014). Extraction and modification technology of arabinoxylans from cereal by-products. *A critical review, Food Research International, 65,* 423–436.

Zhao, Y., Lu, W. J., Wang, H. T., & Yang, J. L. (2009). Fermentable hexose production from corn stalks and wheat straw with combined supercritical and subcritical hydrothermal technology. *Bioresource Technology, 100,* 5884–5889.

Zhao, Y., Lu, W., Chen, J., Zhang, X., & Wang, H. (2014). Research progress on hydrothermal dissolution and hydrolysis of lignocellulose and lignocellulosic waste, *Frontiers of Environmental Science and Engineering, 8,* 151–161.

Zhu, S., Wu, Y., Zhang, X., Li, H., & Gao, M. (2006). The effect of microwave irradiation on enzymatic hydrolysis of rice straw. *Bioresource Technology, 97,* 1964–1968.

Zuñiga, R. N., & Troncoso, E. (2012). Improving nutrition through the design of food matrices. In: B. Valdez (Ed.), Scientific, Health and Social Aspects of the Food Industry. Rijeka: InTech Europe. pp. 295–320.

# CHAPTER 4

# FERULATED ARABINOXYLANS AND β-GLUCANS AS FAT REPLACERS IN YOGHURT AND THEIR EFFECTS ON SENSORIAL PROPERTIES

NANCY RAMÍREZ-CHÁVEZ,[1] ELIZABETH CARVAJAL-MILLAN,[2]
JUAN SALMERON-ZAMORA,[1] AGUSTÍN RASCÓN-CHU,[2]
ALMA ROSA TOLEDO-GUILLÉN,[2] and
NORA PONCE DE LEÓN-RENOVA[3]

[1]Faculty of Agro-Technological Sciences, Autonomous University of Chihuahua, 31125 Chihuahua, Mexico

[2]Research Center for Food and Development, CIAD, A.C., Carretera a La Victoria Km. 0.6, 83304 Hermosillo, Mexico, Tel.: +52-662-289 2400, Fax: +52-662-280 0421. E-mail: ecarvajal@ciad.mx

[3]Agricultural and Livestock High School 90. 31590 Cd. Cuauhtémoc, Mexico

## CONTENTS

## ABSTRACT

The effect of ferulated arabinoxylans and β-glucans on the physicochemical, microbial and sensory properties of non-fat set yogurt during storage was investigated. The pH values and mean lactic acid bacteria counts of control and food hydrocolloids-added yogurt ranged from 4.6 to 3.8 and from $2.7 \times 10^{10}$ to $9.30 \times 10^{10}$ CFU/mL, respectively, when stored at 4°C for 21 days. The a* and b* values were not significantly influenced by the addition of food hydrocolloids; however, the L* values significantly decreased with the addition of ferulated arabinoxylans at 0.2% (w/v). Yoghurts containing 0.2% (w/v) of ferulated arabinoxylans or β-glucans were scored as the most accepted by consumers. It is concluded that a concentration of 0.2% (w/v) of ferulated arabinoxylans or β-glucans could be used to produce fat-reduced yogurts without significantly adverse effects on the physicochemical, microbial, and sensory properties of the product.

## 4.1 INTRODUCTION

Reduced-fat yogurts are receiving increasing attention because of the increasing demand for low calorie products (Mistry and Hassan, 1992). These non-fat set yogurts can be produced by replacing partially the fat content of the milk with low calorie products known as fat replacers. Non-dairy ingredients, especially polysaccharides can be used as fat replacers (Oh et al, 2007). On the other hand, the growing incidence of chronic diseases and obesity and the demonstrated link between the intake of dietary fiber and various health benefits have increased consumer interest in foods enriched in dietary fiber (Önning, 2007). The nutritional value of arabinoxylans as fiber component has not been investigated to the same extent as other polysaccharides. However, some studies revealed positive effects of water-soluble maize arabinoxylans on cecal fermentation, production of short-chain fatty acids, reduction of serum cholesterol, and improved adsorption of calcium and magnesium (Hopkins et al., 2003; Lopez et al., 1999). Arabinoxylans are not digested in the small intestine but provide fermentable carbon sources for bacteria that inhabit the large bowel (Hopkins et al., 2003). In clinical studies oat β-glucan was shown to reduce serum cholesterol levels and attenuate postprandial blood glucose and insulin responses in a viscosity related fashion (Cui, 2001). Arabinoxylan is a polysaccharide constituted of a linear backbone of β-(1→4)-linked D-xylopyranosyl units to which α-L-arabinofuranosyl substituents are attached through O-3 and/or O-2,3 positions of the xylose residues (Izydorczyk and

Biliaderis, 1995). Some of the arabinose residues are ester linked on (O)-5 to ferulic acid (Smith and Hartley, 1983). β-glucan is a linear polysaccharide made up entirely of sequences of (1→4)-linked D-glucopyranosyl units separated by single (1→3)-β-linked units (Izydorczyk and Biliaderis, 2007).

This study has been focused on the effect of two food hydrocolloids, ferulated arabinoxylans and β-glucans as fat replacers, on the physico-chemical, microbial, and sensory properties of non-fat set yogurt.

## 4.2 MATERIALS AND METHODS

### 4.2.1 MATERIALS

Ferulated arabinoxylans and β-glucans were extracted and character-ized as reported before (Carvajal-Millan et al., 2007; Ramos-Chavira et al., 2009). All chemicals were analytical grade purchased from Sigma Chemical, Co. (St. Louis, MO, USA).

### 4.2.2 YOGHURT PREPARATION

Fat-reduced yogurt was made by blending milk and ferulated arabinox-ylans or β-glucans as fat-replacers at different concentrations (0.05, 0.10 and 0.2% w/v), to produce mixes with 3.2% fat content in the control and 1.6% in the fat-replacers-supplemented mix. After blending, the mixes were homogenized with a hand-operated homogenizer (Fisher Scientific Co., Eden prairie, MN, USA). Each mix was pasteurized at 90°C for 10 min, cooled to 40°C and inoculated with 2% yogurt starter (commercial plain yogurt). Mixes were incubated at 40°C for approximately 3 h and then cooled to 4°C and hold for 21 days.

### 4.2.3 pH

The pH values of yogurt samples were measured using a pH meter (Orion 900A, Boston, MA).

### 4.2.4 LACTIC ACID BACTERIA

One milliliter of yogurt samples was diluted with 9 mL of sterile peptone and water. Subsequent dilutions of each sample were plated in triplicate on Man Rogosa and Sharpe agar combined with 0.004% bromophenol blue and incubated at 37°C for 48 h.

### 4.2.5  WATER HOLDING CAPACITY

A 10 g sample was centrifuged (3,000 rpm, 60 min, 10°C). The supernatant was removed within 10 min and the wet weight of the pellet was recorded. The water holding capacity was expressed as percentage of pellet weight relative to the original weight of yogurt (Parnell-Clunies et al., 1986).

### 4.2.6  COLOR

The yogurt samples stored for 0, 6, 12, and 21 days were measured for their color by a colorimeter (Minolta Data Processor DP-301, Chroma Meter CR-300 Series, Japan), using the CIE L*C*h scale values.

### 4.2.7  TEXTURE

Firmness of fresh and stored yogurts was measured with a TA.XT2i Texture Analyzer (RHEO Stable Micro Systems, Haslemere, England). The yogurts were deformed by compression at a constant speed of 1.0 mm/s to a distance of 4 mm from the sample surface using a cylindrical plunger (diameter 15 mm). The peak height at 4 mm compression was called yogurt firmness.

### 4.2.8  SENSORY EVALUATION

Yoghurt samples were examined for overall acceptability by 30 panelists. The panelists were asked to evaluate fresh and 21 days old samples. A hedonic scale of 1 (dislike extremely) to 10 (like extremely) was used.

### 4.2.9  STATISTICAL ANALYSIS

The experiments were performed in triplicates. Analysis of Variance (ANOVA) was performed using the Duncan's multiple-range test to compare treatment means. Significance was defined at $p \leq 0.05$.

## 4.3 RESULTS AND DISCUSSION

The pH values of control and food hydrocolloids-added yogurt ranged from 4.6 to 3.8 during storage (Figure 4.1). There were no significant

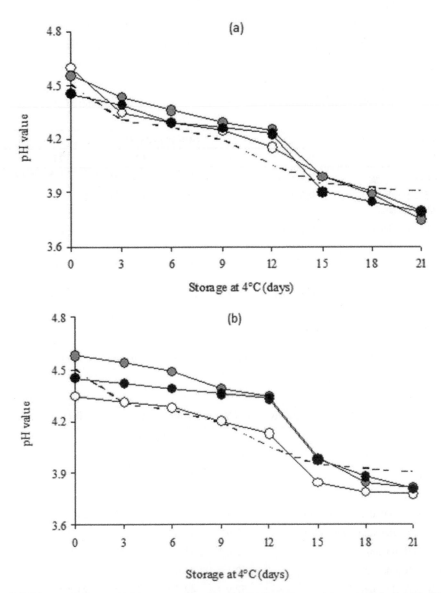

**FIGURE 4.1** Changes in pH in yogurts samples during storage. (a): control (--), 0.05 (O), 0.1 (●) and 0.2 (●) % in ferulated arabinoxylans. (b): control (--), 0.05 (O), 0.1 (●) and 0.2 (●) % in β-glucans.

differences in pH between control and treated yogurts throughout storage. The pH value of all yogurts diminished during storage, but this decrease was not significant for all samples ($p \leq 0.05$). The mean lactic acid bacteria counts in yogurt samples ranged from $2.7 \times 10^8$ to $9.30 \times 10^{10}$ CFU/mL (Table 4.1).

Water holding capacity is a method for indirect evaluation of network homogeneity. There were any differences in water holding capacity of the samples (Table 4.2). It was probably because the yogurt was homogenized

**TABLE 4.1**  Lactic Acid Bacteria Counts (CFU/mL) in Yogurt Samples Stored at 4°C for 21 Days

| Sample | Storage (days) | | | |
|---|---|---|---|---|
| | 0 | 6 | 12 | 21 |
| Control | $9.2 \times 10^{10}$ * | $7.6 \times 10^{10}$ * | $4.9 \times 10^9$ * | $3.2 \times 10^8$ * |
| 0.05% ferulated arabinoxylans | $9.3 \times 10^{10}$ * | $6.9 \times 10^{10}$ * | $4.5 \times 10^9$ * | $2.7 \times 10^8$ * |
| 0.10% ferulated arabinoxylans | $9.2 \times 10^{10}$ * | $7.0 \times 10^{10}$ * | $4.3 \times 10^9$ * | $2.9 \times 10^8$ * |
| 0.20% ferulated arabinoxylans | $9.0 \times 10^{10}$ * | $6.9 \times 10^{10}$ * | $3.9 \times 10^9$ * | $2.8 \times 10^8$ * |
| 0.05% β-glucans | $9.1 \times 10^{10}$ * | $6.9 \times 10^{10}$ * | $4.6 \times 10^9$ * | $3.1 \times 10^8$ * |
| 0.10% β-glucans | $9.3 \times 10^{10}$ * | $7.8 \times 10^{10}$ * | $4.2 \times 10^9$ * | $2.8 \times 10^8$ * |
| 0.20% β-glucans | $9.0 \times 10^{10}$ * | $7.2 \times 10^{10}$ * | $4.1 \times 10^9$ * | $2.7 \times 10^8$ * |

* Means in a column with different letters are significantly different ($p \leq 0.05$).

All results are obtained from triplicates.

**TABLE 4.2**  Water Holding Capacity (%) of Yogurt Samples Stored at 4°C for 21 Days

| Sample | Storage (days) | | | |
|---|---|---|---|---|
| | 0 | 6 | 12 | 21 |
| Control | 53.3 * | 43.0 * | 32.5 * | 22.1 * |
| 0.05% ferulated arabinoxylans | 53.4 * | 43.0 * | 32.6 * | 21.8 * |
| 0.10% ferulated arabinoxylans | 52.8 * | 42.0 * | 31.9 * | 21.2 * |
| 0.20% ferulated arabinoxylans | 52.9 * | 42.1 * | 31.6 * | 21.0 * |
| 0.05% β-glucans | 52.0 * | 41.8 * | 31.2 * | 21.0 * |
| 0.10% β-glucans | 52.1 * | 41.6 * | 31.3 * | 21.0 * |
| 0.20% β-glucans | 52.9 * | 42.3 * | 32.0 * | 21.8 * |

* Means in a column with different letters are significantly different ($p \leq 0.05$). All results are obtained from triplicates.

before fermentation. Homogenization produces small-sized fat globules and more protein is absorbed on the surface of the fat globules, leading to increased ability to immobilize water (Sodini et al., 2004).

In color, the a* and b* values were not significantly influenced by the addition of food hydrocolloids (data not shown); however, the L* values significantly decreased with the addition of the greatest concentration (0.2%) of ferulated arabinoxylans at 0 days of storage (Table 4.3). The storage time did not significantly ($p \leq 0.05$) influence the lightness of yogurt samples.

Textural characteristics of the yogurts are presented in Figure 4.2. The gel structure of food hydrocolloids-added yogurt was harder than that of the control yogurt. The firmness of yogurts was reduced with storage time. The textural changes would be mainly caused by degradation of the gel structure.

In the term of sensorial attributes, yogurt samples containing 0.2% of ferulated arabinoxylans or β-glucans were scored as the most accepted by consumers (Table 4.4). Based on the data obtained from the current study, it is concluded that a concentration of 0.2% of ferulated arabinoxylans or β-glucans could be used to produce a fat-reduced yogurt without significantly adverse effects on the sensory properties.

Good quality low fat yogurts can be made with ferulated arabinoxylans or β-glucans as fat replacers. Based on the data obtained from the current study, it is concluded that a concentration of 0.2% of ferulated arabinoxylans or

**TABLE 4.3**   Changes of L* Values of Yogurt Samples Stored At 4°C for 21 Days

| Sample | Storage (days) | | | |
|---|---|---|---|---|
| | 0 | 6 | 12 | 21 |
| Control | 88.5 * | 88.4 * | 88.3 * | 88.0 * |
| 0.05% ferulated arabinoxylans | 87.6 * | 87.3 * | 87.2 * | 87.1 * |
| 0.10% ferulated arabinoxylans | 87.4 * | 87.1 * | 87.0 * | 87.0 * |
| 0.20% ferulated arabinoxylans | 85.6 b | 85.4 b | 85.3 b | 85.2 b |
| 0.05% β-glucans | 88.4 * | 88.3 * | 88.3 * | 88.1 * |
| 0.10% β-glucans | 88.6 * | 88.4 * | 88.3 * | 88.0 * |
| 0.20% β-glucans | 87.9 * | 87.9 * | 87.6 * | 87.0 * |

*Means in a column with different letters are significantly different ($p \leq 0.05$).

All results are obtained from triplicates.

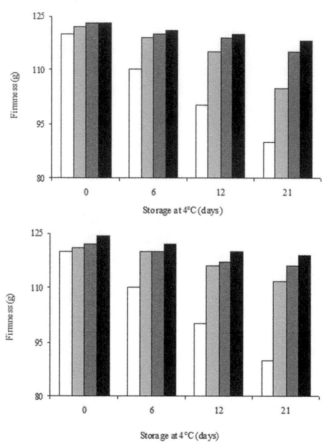

**FIGURE 4.2**   Firmness of yogurts samples during storage. (a): control (□), 0.05 (■), 0.1 (■) and 0.2 (■) % in ferulated arabinoxylans. (b): control (□), 0.05 (■), 0.1 (■) and 0.2 (■) % in β-glucans.

β-glucans could be used to produce a fat-reduced yogurt without significantly adverse effects on the physicochemical, microbial, and sensory properties.

Ferulated arabinoxylans and β-glucans continue to be investigated as fat-replacers and new information about their potential application in the food industry is being generated. Nevertheless, more research is needed to elucidate several questions, especially those concerning the effect of temperature on the texture properties.

## ACKNOWLEDGMENTS

This research was supported by "Fondo de Infraestructura de CONACYT, Mexico (Grant 226082 to E. Carvajal-Millan)."

**TABLE 4.4** Overall Acceptance of for Yogurt Samples Stored at 4°C for 21 Days

| Sample | Storage (days) | | | |
| --- | --- | --- | --- | --- |
| | 0 | 6 | 12 | 21 |
| Control | 4.2 ** | 4.0 ** | 4.1 ** | 4.2 ** |
| 0.05% ferulated arabinoxylans | 4.2 ** | 4.0 ** | 4.4 ** | 4.1 ** |
| 0.10% ferulated arabinoxylans | 4.3 ** | 4.4 ** | 4.0 ** | 4.1 ** |
| 0.20% ferulated arabinoxylans | 5.2 * | 5.1 * | 5.0 * | 5.0 * |
| 0.05% β-glucans | 4.1 ** | 4.2 ** | 4.3 ** | 4.1 ** |
| 0.10% β-glucans | 4.3 ** | 4.3 ** | 4.2 ** | 4.3 ** |
| 0.20% β-glucans | 5.1 * | 5.0 * | 5.0 * | 4.9 * |

* Means in a column with different letters are significantly different ($p \leq 0.05$).

** Scale of overall acceptance: 1 = dislike extremely, 7 = like extremely.

All results are obtained from triplicates.

## KEYWORDS

- **fat-reduced yogurt**
- **hydrocolloids**
- **polysaccharides**
- **sensory properties**

## REFERENCES

Carvajal-Millan, E., Rascón-Chu, A., Marquez-Escalante, J., Ponce de León, N., Micard, V., & Gardea, A. (2007). Maize bran gum: Characterization and functional properties. *Carbohydr. Polym, 69*, 280–285.

Cui, S. W. (2001). Polysaccharide gums from agricultural products. Processing, structures and functionality. Technomic Publishing Co. Inc., Pennsylvania, USA.

Hopkins, M. J., Englyst, H. N., Macfarlane, S., Furrie, E., Macfarlane, G. T., & McBain, A. J. (2003). Degradation of cross-linked arabinoxylans by the intestinal microbiota in children. *Appl. Environ. Microbiol, 69*, 6354–6360.

Izydorczyk, M. S., & Biliaderis, C. G. (1995). Cereal arabinoxylans: Advances in structure and physicochemical properties. *Carbohydr. Polym, 28*, 33–48.

Izydorczyk, M. S., & Biliaderis, C. G. (2007). Arabinoxylans: Technological and nutritional functional plant polysaccharides, 249–290. In: Functional Food Carbohydrates. Izydorczyk, M. S., Biliaderis, C. G. (ed.). CRC Press, Taylor and Francis Group, Boca Raton, FL, USA.

Lopez, H. W., Levrat, M. A., & Guy, C. (1999). Effects of soluble corn bran arabinoxylans on cecal digestion, lipid metabolism, and mineral balance (Ca, Mg) in rats. *J. Nutr. Biochem, 10*, 500–509.

Mistry V. V., & Hassan H. N. (1992). Manufacture of nonfat yogurt from a high milk protein powder. *J. Dairy Sci, 75*, 947–957.

Oh H. E., Anema, S. G., Wong, M., Pinder, D. N., & Hemar, Y. (2007). Effect of potato starch addition on the acid gelation of milk. *Int. Dairy J., 17*, 808–815.

Önning G. (2007). Carbohydrates and the risk of cardiovascular disease. 291–319. In: Functional Food Carbohydrates. Biliaderis, C. G., Izydorczyk, M. S. (ed.). CRC Press, Inc., Boca Raton, FL, USA.

Parnell-Clunies, E. M., Kakuda, Y., Mullen, K., Arnott, D. R., & deMan, J. M. (1986). Physical properties of yogurt: a comparison of vat versus continuous heating systems of milk. *J. Dairy Sci., 69*, 2593–2603.

Ramos-Chavira, N., Carvajal-Millan, E., Marquez-Escalante, J., Santana-Rodriguez, V., Rascon-Chu, A., & Salmerón-Zamora, J. (2009). Characterization and functional properties of an oat gum extracted from a drought harvested *A. sativa. Food Sci. Biotechnol, 18*, 900–903.

Smith, M. M., & Hartley, R. D. (1983). Occurrence and nature of ferulic acid substitution of cell-wall polysaccharides in graminaceous plants. *Carbohydr, Res, 118*, 65–80.

Sodini, I., Remeuf, F., Haddad, S., & Corrieu, G. (2004). The relative effect of milk base, starter, and process on yogurt texture: a review. *Crit. Rev. Food Sci. Nutr, 44*, 113–137.

# CHAPTER 5

# LACCASE PRODUCTION BY *TRAMETES VERSICOLOR* AND *ARMILLARIA MELLEA* USING MAIZE BRAN AS SUPPORT-SUBSTRATE AND ITS DYE DECOLORIZATION POTENTIAL AS AFFECTED BY pH

AGUSTÍN RASCÓN-CHU,[1] ANA L. MARTÍNEZ-LÓPEZ,[1] and ALEJANDRO ROMO-CHACÓN[2]

[1]*Research Center for Food and Development (CIAD, AC). Carretera a La Victoria Km 0.6, Hermosillo, Sonora, 83304, Mexico*

[2]*Research Center for Food and Development (CIAD, AC). CIAD-Cuauhtémoc. Av. Río Conchos S/N. Parque Industrial, Cuauhtémoc, Chihuahua, 31570, Mexico*

## CONTENTS

## ABSTRACT

In this study, growth and laccase production of *Armillaria mellea* and *Trametes versicolor* in solid-state fermentation were monitored using maize bran as a support, and surface secreted extract was used for synthetic dyes discoloration. The maximum laccase activity was obtained *ca.* 25 d of culture for both, but a higher specific activity of 751 U/mg for *T. versicolor,* vs. 487 U/mg for *A. mellea* was observed. The growth was monitored for 984 h by weight measurements. Laccase activity was determined using syringaldazyne as substrate. In addition, the discoloration of Remazol Brilliant Blue R (RBBR), Methylene Blue (MB), Brilliant Green (GB) and Congo Red (RC) was carried out at pH 5.4 and 4.5, (0.016 U/ml laccase activity) with raw extracts from both fungi, and commercial laccase from *T. versicolor* as control. In this work, *T. versicolor* and *A. mellea* were able to thrive, produce and excrete the enzyme laccase, while grown on maize bran. Synthetic dyes are differentially decolorated within small changes of pH values from 4.5 to 5.4.

## 5.1   INTRODUCTION

Maize bran is a by-product of the commercial maize dry milling process. This sub-product is an important source of soluble and insoluble carbohydrates also contain acid and derivatives from splint lignocellulose (Hespell, 1998). Therefore, it can be used as a potential substrate for production of secondary metabolites of industrial importance (Rosales et al., 2007). Currently, various substrates have been studied as a source of production of metabolites from wastes: the brewing industry (Rodriguez et al., 2004), the industry of palm (Vikineswary et al., 2006), as well as orange peelings (Rosales et al., 2007) and banana skin (Osma et al., 2007). These substrates can be used by white rot fungi, with the ability to secrete enzymes during development, such as laccases (benzenediol: oxygen oxidoreductases, EC 1.10.3.2), which are related in delignification processes (Rodriguez & Toca, 2006), phytopathogenic (Iyer and Chattoo, 2003) and formation of fruiting bodies (Xing et al., 2006), are also widely used at industrial and environmental characteristics and catalytic specificity toward the substrate (Zhang et al., 2006). Among the industrial applications of laccases are different water treatment residues from industry, and detoxification of industrial effluents (Abadulla et al., 2000). The ability of laccases to degrade various chemical structures of synthetic dyes has been widely studied (Zille et al., 2005; Rodriguez et al., 2004; Osma et al., 2007; Lu et al., 2007) this ability depends on the source of microorganisms and the substrate is used for production (Han et al., 2005). Synthetic dyes from textile industries are of special interest due some are released in wastewaters and many azo dye-derived amines produced under anaerobic conditions might be carcinogenic and/ or mutagenic (Banat et al., 1996; O'Neill et al., 2000; Brüschweiler et al., 2014). The purpose of this study was to evaluate the potential of maize bran as a substrate for the development of *Trametes versicolor* and *Armillaria mellea* and its effect on the production of laccases. As well, it was studied the ability of decolorization of different dyes by the extracellular enzymes secreted during solid state fermentation (SSF) by both fungi.

## 5.2  MATERIALS AND METHODS

### 5.2.1  MICROORGANISMS

The wild strains of the basidiomycete T. versicolor, and A. mellea were isolated from fruiting bodies collected from dead wood of trees in the region of Northern Mexico and were grown on potato dextrose agar (PDA) plates at $29 \pm 2°C$ for seven days. Stocks cultures were maintained through periodic transfer at 4°C PDA slants.

### 5.2.2  SUPPORT-SUBSTRATE

Maize bran was used as support-substrate for laccase production by both fungal strains under SSF conditions. The chemical composition of maize bran was determined by AOAC, proteins 10%, lipids 4%, starch 9–23%, and ash 2% (Hespell, 1998). The support-substrate was autoclaved at 121°C for 15 min until used.

### 5.2.3  CULTURE CONDITIONS

The cultures were grown in sterile polypropylene boxes (size 136 x 119 x 76 cm) containing 28 g of maize bran and 50 ml of basal medium. The medium used had the following composition (w/v): 0.1% $KH_2PO_4$, 0.05% $MgSO_4$, 0.001% $CaCl_2$, 0.001% $FeSO_4$, 0.25% asparagine, 0.0015% phenylalanine, 0.00275% adenine (Lu et al., 2007) and 1 mM of $CuSO_4$ to stimulate the secretion of extracellular enzymes. Inoculation was carried out directly in the boxes. Two agar plugs (diameter, 6 mm) from the outer of fungal colony growing on PDA plates (6 to 7 d) were used as the inoculum. The cultures were incubated statically under passive aeration atmosphere at $29 \pm 2°C$ and 85% humidity in complete darkness for 41 days. All experiments were performed twice and the samples of each one were analyzed by triplicate.

### 5.2.4   ANALYTICAL DETERMINATIONS

Laccase activity was determined spectrophotometrically as reported by Carvajal-Millan et al. (2005), using siryngaldazine 0.25 mM as substrate, on citrate-phosphate 100 mM buffer solution at pH 5.5 and extracellular liquid in a final reaction volume of 1 ml. The enzymatic reaction was followed for 2 minutes at 530 nm. One activity unit was defined as the amount of enzyme that oxidized 1 μmol syringaldazine per minute. The activity was expressed in U/ml.

### 5.2.5   1DETERMINATION OF PROTEIN

The protein concentration was determined by the method of Bradford (1976), using bovine serum albumin (BSA) as standard.

### 5.2.6   DYES DECOLORIZATION

The decolorization of Congo Red (CR; diazo dye; CI 22120; MW 696.66; $\lambda_{max}$ 498 nm), Remazol Brilliant Blue R (RBBR, anthraquinonic dye; CI 61 200; MW; $\lambda_{max}$ 595 nm), Methylene Blue (MB; Thiazin dye, CI 52015; MW 373.9; $\lambda_{max}$ 665 nm) and Brilliant Green (BG; triarylmethane dye; CI 42040; MW 482.6; $\lambda_{max}$ 625 nm), by the extracellular liquid secreted by *A. mellea* and *T. versicolor*, and commercial laccase (E.C. 1.10.3.2) was studied. All chemical products were purchased from Sigma Chemical, Co. (St Louis, MO, USA).

The reaction mixture (final volume 5 ml) consisted of 100 mM sodium citrate buffer (pH 4.5 and 5.4), enzyme solution (final concentration 0.016 U/

ml) and aqueous solution dye to give the final concentration of 0.1 g/l for CR, 0.007 g/l for BG, 0.15 g/l for RBBR and 0.01 g/l for MB. The experiments were incubated at 29°C for indicated time periods, in static conditions and in complete darkness. Dye decolorization was measured by monitoring the decrease in absorbance on a spectrophotometer at peak of maximum wavelength of each dye and expressed in terms of percentages. A control test containing the same amount of heat-denatured laccase was performed in parallel. The assays were made by triplicate.

## 5.3 RESULTS AND DISCUSSION

The major aim of this study was to demonstrate the capability of a waste to support a productive fermentation process. A waste from the maize industrialization process, mainly bran, was used for laccase production knowing certain ferulic acid content may act as an inducer. Two wild strains of basidia were cultured on solid state fermentation (SSF) conditions and monitored.

### 5.3.1 SUBSTRATE DEGRADATION AND BIOMASS DEVELOPMENT

Figure 5.1, shows weigh loss of substrate as percentage *vs.* culture age. The profile registered, clearly follows a pattern similar to biomass development for both strains. The substrate degradation by both strains show an almost superimposable pattern. Furthermore, the follow of growth and enzyme activity for the two wild strains is shown in Figure 5.2. As mentioned above the weight loss and biomass development follow a similar and complementary pattern due, we assume, conversion to biomass of the waste used.

### 5.3.2 LACCASE PRODUCTION

Biomass development and laccase production shown as enzyme activity vs. age of culture expressed in days are presented, and show how *T. versicolor* reached its maximum value (13 U/ml; Figure 2a,) on the 25th day of culture. On the other hand, laccase production began on the 14th day, and reached its maximum on the 25th day of culture for *A. mellea* (8.34 U/ml; Figure 5.2b). These results are similar to previous reports, as Osma et

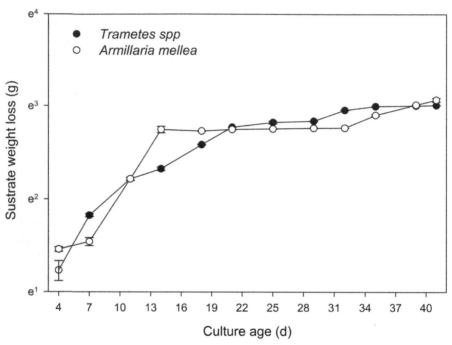

**FIGURE 5.1** Weigh loss of the substrate-support by growth of *T. versicolor* (●) and *A. mellea* (○) as a function of culture age.

al. (2007) with 1.570 U/ml using banana skin; and Rosales et al. (2007) with 31.78 U/ml with orange peelings, our results on maize bran would be on the average enzyme activity on the overall.

Comparison of substrate consumption shows no difference between species and biomass production does not appear with any difference. Nonetheless, the enzymatic activity is slightly higher for wile *T. versicolor*, as shown in Table 5.1 for the whole fermentation process. The latter could be explained, based on the profile shown in Figures 5.2a and 5.2b; where *A. mellea* shows a lower maximum peak on enzymatic activity in a similar extent of time than wile *T. versiclor*. Table 5.1 resumes the evaluation of both and by the end of the SSF process; the accumulated specific activity is almost one and a half times higher for wile *T. versicolor*, with 751 U/mg of protein than *A. mellea*.

In addition, the pattern followed by enzyme activity records, suggest an inductive production of laccase after 16 d of culture for both strains; though more extended for the *T. versicolor*. This production was mainly from surface goutlets recovered and tested for enzymatic activity with syringaldazine.

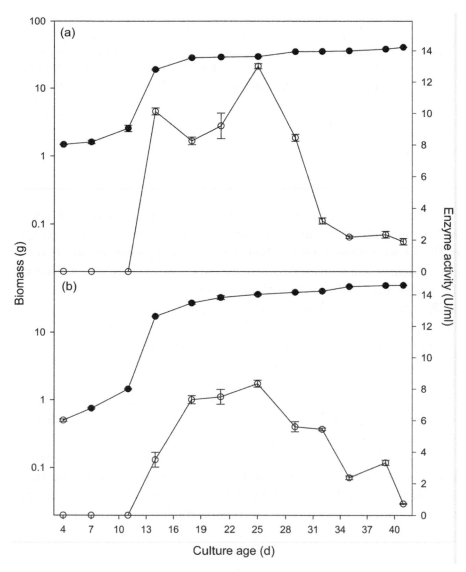

**FIGURE 5.2** Laccase activity for *T. versicolor* and *A. mellea* in a SSF batch. Biomass production (●) and Laccase activity (○) for *T. versicolor* and *A. mellea* according to age of culture. The values are triplicates.

Apparently, there is a period of growth where laccase is secreted in a growth associated manner; nevertheless, under SSF culture conditions certain changes in biomass development triggered surface goutlets with a high enzimatic content which we monitored for 41 d. Their specific activity was registered throughout the process and our evidence suggests that this response

**TABLE 5.1**  Comparative Results for *A. mellea* and *T. versicolor* on Maize Bran Based Medium for Laccase Activity Production After 984 h Culture Age

| Strain | Culture Age (h) | X (g) | μ (1/h) | Y (x/s) (mg/g) | Y (Lac/s) (U/g) | Y (Lac/x) (U/g) | Enzyme Activity (U/ml) | Specific Activity (U/ mg of protein) |
|---|---|---|---|---|---|---|---|---|
| *T. versicolor* | 984 | 41 | 133 | 574 | 0.854 | 1.5 | 59 | 751 |
| *A. mellea* | 984 | 49 | 110 | 694 | 0.618 | 0.89 | 44 | 487 |

is due to fruiting body formation and lignification, as a normal evolution of each fungus. This behavior is concordant to the phenotype observed specially for the fruiting bodies, where -apparently- lignification for *T. versicolor*, seems more important than in the *A. mellea* wild isolates. Observing the values attained for yield of biomass by amount of substrate ($Y_{x/s}$), *A. mellea* shows a higher value with 694 mg/g and *T. versicolor*, follows with 574 mg/g. Conversely, the enzymatic yield for wile *T. versicolor*, is higher (0.854 U/g) than that of *A. mellea* isolates (0.62 U/g). The yield of enzyme per gram of biomass ($Y_{lac/x}$) was 1.5 U/g for *T. versicolor*, while 0.89 U/g for *A. mellea* (Table 5.1). Apparently, the metabolism of these organisms differs in that *A. mellea* trends to form more biomass, and *T. versicolor*, to protect itself by lignification of the fruiting bodies. Further experiments shall be conducted in this regard to obtain more conclusive results.

The results discussed above demonstrated the capability of maize industrialization waste to support growth, and enzyme production. The latter, measured mostly as laccase activity, though several other potential enzymes might be present. Anyway, previous reports mentioned that laccase activity and dye decolorization are related to glucose and ammonium starvation or to induction by ferulic acid (Zilly et al., 2002).

### 5.3.3   DECOLORIZATION OF DIFFERENT DYES BY LACCASE FROM T. VERSICOLOR AND A. MELLEA

The raw extracts were used to test the decolorization of different dyes with a structural common feature of phenolic rings. Four different dyes varying in structure complexity were used. For comparison, a commercial laccase from *T. versicolor* was introduced as positive control. Assesing the percentage of decolorization, significant differences were found, and interaction dye- pH and type of laccase used for the overall experiment. As shown, Remazol Brilliant Blue R (RBBR), Methylene Blue (MB), Brilliant Green (GB) and Congo Red (RC) were subjected to the action of extracellular liquid from maize bran cultures of *T. versicolor*, *A. mellea* and commercial laccase from *T. versicolor*. Raw extracts and commercial laccase were adjusted to 0.016 U/ml expressed as laccase using syringaldazyne as substrate. The major effect was observed on dye RC with average values of 35 and 75% for pH 4.5 and 5.4, respectively. Interestingly, for the rest of structure model dyes at pH 4.5 *Trametes* decolorization percentage was superior in the average (Figure 5.3).

Congo Red resistance to decolorization has been previously reported by Yang et al. (2009) and Mendoza et al. (2014). When assaying Congo Red (RC, a diazo structure), the results showed a dramatic influence of pH within relatively close values of 4.5 and 5.4. Neither treatment reaches a 35% decolorization value at pH 4.5; while the values of decolorization reached above 70% for all different enzymatic treatments with no mediator compounds present at pH 5.4. Conversely, Rodríguez et al. (2004) who reported

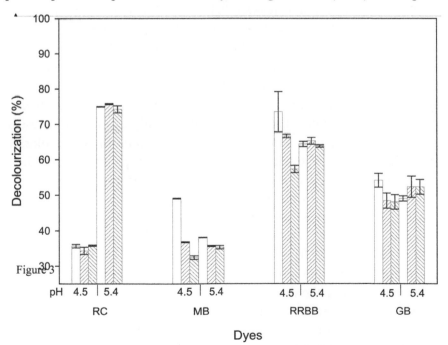

**FIGURE 5.3** Decolorization of different dyes by laccase raw extract on biomass surface from *T. versicolor* and *A. mellea* as affected by pH.

decolorization of several synthetic dyes was more effective at pH 4 from *T. versicolor*. Though some authors suggest that the source of the enzymes is the most critical fact, we agree that dye structure as influenced by pH, as well as the difference in laccase isoenzymes produced by each strain, might as well play a critical role to decolorization susceptibility. Previous work on *Trametes modesta* suggested that variation of decolorization extent is possibly due to the structural variation of these dyes (Nyanhongo et al., 2002). Moreover, Murugesan et al. (2006) also showed the potential of laccase produced by *Pleurotus sajor-caju* to be used for decolorization of industrial effluents testing for different groups of azo dyes. Conversely to Zille et al. (2005), who reported the unacceptable limitation of the polymerizing action of laccases on colored effluents with azo dyes. Further research on reaction conditions should be adressed in order to get more conclusive results beyond laboratory *in vitro* evidence, enhalting the potential use of laccases for bioremediation of azo dye structure like pollutants.

Among the dyes tested, Methylene blue (MB, a thiazin structure) showed low values of decolorization percentage even though different, they were all within 32 and 38%, except for *T. versicolor,* raw extract at pH 4.5, but still with a low average value ca. 49%, regarding other dyes tested in this study. Hypothesizing, these results could be further enhanced if mediator agents were used as evidence reported elsewhere clearly show the effectiveness of laccase mediator systems (LMS) on recalcitrant dyes. Hu et al. (2009) demonstrated that LMS were related to the types of mediator, the dye structure and decolorization condition. However costs and toxicity rest a major drawback for mediators. Zilly et al. (2002) tested the ability of a Brazilian strain of *Pleurotus pulmonarius* to decolorize structurally different synthetic dyes (including azo, triphenylmethane, heterocyclic and polymeric dyes) in solid and submerged cultures. No decolorization of Methylene Blue and Poly R 478 was observed. Of the four phenol-oxidizing enzymes tested in culture filtrates (lignin peroxidase, manganese peroxidase, aryl alcohol oxidase, laccase), *P. pulmonarius* produced only laccase. Though decolorization values are relatively low, our results still far more promising for MB. Younes et al. (2007) tested the decolorization of the dyes basic blue, methylene blue, maxilon blue, neolane yellow, malachite green and gentian purple with records of less than 10% decolorization after treatment with the enzyme. Furthermore, the decolorization of gentian purple, methylene blue, malachite green, neutral red, pink Bengal was not improved by the addition of HBT as a mediator.

The latter evidence enhance therefore the importance of our results on MB without mediators.

RBBR (a major anthraquinone structure) was also subjected to enzymatic action and showed a particular profile. At pH 5.4 no differences were found between the commercial laccase and the raw extracts from the two wild strains approximately 64%. Interestingly, at pH 4.5 the decolorization percentage separates for all three, and commercial laccase drops significantly below 57%, while *A. mellea* extract rest steadily around 66% and *T. versicolor,* extract goes up significantly above 74% though a far more extended variation. Evidently, these differences are no further explainable but for the presence of isoenzymes in the raw extracts. In fact, RBBR decolorization by all three enzymatic solutions at both pH 4.5 and 5.4, with no mediator present is remarkable as this dye is used as basic material for polymeric dye production, and represents toxic environmental pollutants. Finally, Brilliant green (GB, a triphenylmethane structure), showed values within 48 and 54% decolorization activity. Though there were some differences among them, these were of no consequence for pragmatic purposes. Triphenylmethane dyes are aromatic xenobiotic compounds, and are used extensively in the textile industry to dye nylon, wool, silk and cotton, and in paper and leather industries. Some of these dyes are used as biological stains. Decolorization of triphenylmethane dyes was found to be carried out by laccase in extracellular fluid from *Cyathus bulleri* (Vasdev et al., 1995). In studies carried out by Li et al. (2009) the dye decolorization potential of crude laccase from the white rot fungus *Rigidoporus lignosus* W1 was demonstrated on an anthraquinone dye, RBBR, and a triphenylmethane dye, MG. In this regard, our raw extracts from wild strains cultured on maize industrialization wastes, are potential candidates for a biotechnological application for the textile industry wastes as well.

There is a diverse response registered for the raw extracts and the commercial laccase treatment. Up to this point of our research, it was evident that the fact that raw extracts contained more than one enzyme, would explain this performance as comparison was referred to a commercial purified laccase and further purification steps of the extracts could reveal new enzymes acting within this pH values for this biotechnological application. Interestingly, when assaying the diazo structure (RC) the results showed a totally different behavior. Indistinctively of the treatments, the discoloration percentages were higher at pH 5.4 including the commercial laccase. This evidence confirms the relationship between structure complexity, and pH in the substrate susceptibility to enzymatic action. RC has the less complex structure within those assayed and a strong affinity to cellulose. However, the cellulose industry has lesser use to congo red, due to its tendency to change color, to run with humidity, and its toxicity. The overall evidence shows the high potential of the raw extracts for synthetic dye decolorization, especially for anthraquinonic and diazo dyes.

Eventhough our enzyme production is within the average regarding other authors, their capacity to decolorize dyes as methylene blue far more efficiently as reported previously, and without mediators is a fact to consider. Controversy on azo dye decolorization is still undergoing, but in vitro evidence on decolorization with or without mediators is undeniable for most synthetic dyes. Further research should be carried out to optimize conditions and conceive an acceptable process for industrial application. Most authors agree that efforts should be made for cheap overproduction of laccase, modification for more robust laccases and alternatives developed for actual expensive, toxic, synthetic mediators. Concerning low cost, our results propose a waste of parsimonious value for substrate, and raw enzymatic extracts to avoid purification costs, as well as wild strains free of royalties.

Further research is on their way to fully purify the enzymes present in this raw extracts to explain conclusively this results; independently that the aim of this work was limited to the application of raw materials as maize wastes extracts to diminish as possible an expensive treatment process.

## 5.4   CONCLUSIONS

Biotechnological applications of enzymatic procedures often have the disadvantage of being expensive due to enzyme preparations. One of many strategies to dodge this inconvenience is the selection of new sources of enzymes along with more efficient and cheaper substrates for enzyme production. In this work, *T. versicolor*, and *A. mellea* were able to thrive produce and excrete the enzyme laccase while grown on maize industrialization wastes. The maximum activity of the enzyme was determined within 25 days of culture. The laccase extracted was capable of decoloring different dyes representative for different phenolic structures. The anthraquinonic and diazo dyes were those structures showing the major susceptibility to enzymatic decolorization with a clear dependency on pH. The latter opens the way for further research on biotechnological application of the enzyme extracts of these fungi, as well as some other species which still are undergoing similar studies.

## KEYWORDS

- **decolorization**
- **dyes**
- **laccase**
- **maize bran**

## REFERENCES

Abadulla, E., Tzanov, T., Costa, S., Heinz, K. R., Carvaco, P. A., & Güibitz G. M. (2000). Decolorization and detoxification of textile dyes with a laccase from *Trametes hirsuta*

Banat, I. M., Nigam, P., Singh, D., & Marchant, R. (1996). Microbial decolorization of textiledye-containing effluents: *a review. Bioresour Technol, 58*, 217–227.

Bradford, M. M. (1976). A rapid and sensitive method for the quantitation of microgram quantities of protein utilizing the principle of protein-dye binding. *Analytical Biochemistry, 72*, 248–254.

Brüschweiler, B. J., Küng, S., Bürgi, D., Muralt, L., & Nyfeler, E. (2014). Identification of non-regulated aromatic amines of toxicological concern which can be cleaved from azo dyes used in clothing textiles. *Regulatory Toxicology and Pharmacology, 69*, 263–272.

Carvajal-Millan, E., Guigliarelli, B., Belle, V., Rouau, X., & Micard, V. (2005). Storage stability of laccase induced arabinoxylan gels. *Carbohydrate Polymers, 59*, 181–188.

Han, M. N., Choi, H. T., & Song, H. G. (2005). Purification and characterization of laccase from white rot fungus Trametes versicolor. *The Journal of Microbiology, 43*, 555–560.

Hu MR., Chao, Y-P., Zhang, G. Q., Xue, Z. Q., & Qian, S. (2009). Laccase-mediator system in the decolorization of different types of recalcitrant dyes. *J. Ind Microbiol Biotechnol, 36*, 45–51.

Hespell, R. B. (1998). Extraction and characterization of hemicellulose from the corn fiber produced by corn wet-milling processes. *Journal of Agriculture Food Chemistry, 46*, 2615–2619.

Iyer, G., & Chattoo, B. B. (2003). Purification and characterization of laccase from the rice blast fungus, Magnaporthe grisea. *FEMS Microbiology Letters, 227*, 121–126.

Li, L., Dai, W., Yu, P., Zhao, J. and Qu, Y. (2009). Decolorisation of synthetic dyes by crude laccase from Rigidoporus lignosus W1. *Journal of Chemical Technology and Biotechnology, 84*, 399–404.

Lu, L., Zhao, M., Zhang, B. B., Yu, S. Y., Bian, X. J., Wang, W., et al. (2007). Purification and characterization of laccase from Pycnoporus sanguineus and decolorization of an anthraquinone dye by the enzyme. *Applied Microbiology and Biotechnology, 74*, 1232–1239.

Mendoza, L., Ibrahim, V., Álvarez, M. T., Hatti-Kaul, R., & Mamo, G. (2014). Laccase production by *Galerina* sp. and its application in dye decolorization. *Journal of Yeast and Fungal Research, 5*(2), 13–22.

Murugesan K., Arulmani, M., Nam, I. H., Kim, Y. M., Chang Y. S., & Kalaichelvan, P. T. (2006). Purification and characterization of laccase produced by a white rot fungus Pleurotus sajor-caju under submerged culture condition and its potential in decolorization of azo dyes. *Appl Microbiol Biotechnol, 72,* 939–946.

Nyanhongo G. S., Gomez, J., Gubitz, G., Zvauya, R., Read, J. S., & Steiner, W. (2002). Production of laccase by a newly isolated strain of Trametes modesta. *Bioresource Technology, 84,* 259–263.

O'Neill, C., Lopez, A., Esteves, S., Hawkes, F., Hawkes, D. L., & Wilcox, S. (2000). Azo-dye degradation in an anaerobic-aerobic treatment system operating on simulated textile effluent. *Appl. Biochem. Biotechnol, 53,* 249–254.

Osma, J. F., Toca Herrera, J. L., & Rodríguez, C. S. (2007). Banana skin: A novel waste for laccase production by Trametes pubescens under solid-state conditions. *Application to Synthetic Dye Discoloration: Dyes and Pigments, 75,* 32–37.

Rodriguez, C. S., & Toca J. (2006). Industrial and biotechnological applications of laccases: A review. *Biotechnology Advances, 24,* 500–513.

Rodríguez, C. S., Rosales, E., Gundín, M., & Sanromám, M. A. (2004). Exploitation of a waste from the brewing industry for laccase production by two Trametes species. *Journal of Food Engineering, 64,* 423–428.

Rosales E., Rodríguez-Couto, S., & Sanromán, M. A. (2007). Increased laccase production by Trametes hirsuta grown on ground orange peelings. *Enzyme and Microbial Technology, 40,* 1286–1290.

Vasdev, K., Kuhad, R. Ch., & Saxena, R. K. (1995). Decolorization of triphenylmethane dyes by the bird's nest fungus Cyathus bulleri. *Current Microbiology, 30*(5), 269–272.

Vikineswary S., Abdullah, N., Renuvathani, M., Sekaran, M., Pandey, A., & Jones, E. B. G. (2006). Productivity of laccase in solid substrate fermentation of selected agro-residues by Pycnoporus sanguineus. *Bioresource Technology, 97,* 171–177.

Xing, Z. T., Cheng, J. H., Tan, Q., & Pan, Y. J. (2006). Effect of nutritional parameters on laccase production by the culinary and medical mushroom, Grifola frondosa. *World Journal of Microbiology and Biotechnology, 22,* 799–806.

Yang, X., Zhao, X., Liu, C., Zheng, Y., & Qian, S. (2009). Decolorization of azo, triphenyl-methane and anthraquinone dyes by a newly isolated Trametes sp. SQ01 and its laccase. Process. Biochem. 44, 1185–1189.Younes S. B., Mechichi, T., Sayadi, S. (2007). Purification and characterization of the laccase secreted by the white rot fungus Perenniporia tephropora and its role in the decolorization of synthetic dyes. *Journal of Applied Microbiology, 102,* 1033–1042.

Zhang, M., Wu, F., Wei, Z., Xiao, Y., & Gong, W. (2006). Characterization and decolorization ability of a laccase from Panus rudis. *Enzyme and Microbial Technology, 39,* 92–97

Zille A., Górnacka, B., Rehorek A., & Cavaco-Paulo A. (2005). Degradation of azodyes by Trametes villosa laccase over long periods of oxidative conditions. *Applied and Environmental Microbiology, 71,* 6711–6718.

Zilly A, Souza, C. G., Barbosa-Tessmann, I. P., & Peralta, R. M. (2002). Decolorization of industrial dyes by a Brazilian strain of Pleurotus pulmonarius producing laccase as the sole phenol-oxidizing enzyme. *Folia Microbiol (Praha), 47,* 273–277.

CHAPTER 6

# THE FREEZING PROCESS OF DOUGH AND ITS EFFECTS ON STRUCTURE, VISCOELASTICITY, AND BREADMAKING

E. MAGAÑA-BARAJAS[1] and B. RAMÍREZ-WONG[2]

[1]*Universidad Estatal de Sonora, Ley Federal del Trabajo Final, Col. Apolo. C.P. Hermosillo, Sonora, Mexico, Tel.: (62) 2857636; E-mail: ely_magbarajas@hotmail.com*

[2]*Departamento de Investigaciyn y Posgrado en Alimentos, Ave. Rosales y Blvd. Luis Encinas s/n o al Apartado postal 1658, C.P. 83000, Hermosillo, Sonora, Múxico, Tel: (62) 59-22-07 y 59-22-09; Fax: (62) 59-22-07 y 59-22-08; E-mail: bramirez@guaymas.uson.mx*

## CONTENTS

## ABSTRACT

The freezing process has been one of the main methods in food preservation, which is also used in wheat dough. It is important to know the mechanism of freezing of dough at structural level and during process. That is relevant to detect the negative effects of this technology and to propose improvements from the formulation to its production steps. Obviously the sensory evaluation characteristics of bread are reduced due to dough freezing. This is evidenced by getting a lower bread volume and higher firmness. One of the main factors affecting bread quality is the hydrolysis of gluten proteins, reflecting dough weakening and poor gas holding capacity. The objective of this review was to understand the effect of freezing process on the structure, and viscoelasticity of dough and bread quality in order to propose improvements of the final product.

## 6.1 INTRODUCTION

Freezing of dough is an alternative process which offers always fresh baked products and organoleptic characteristics similar to freshly. This preservation medium limits a number of deteriorative reactions such as the growth of microorganisms, enzymatic and oxidative reactions due to a reduction in water activity. The implementation of the freezing step in process of making bread still presents complications. The bread volume loss is one of the major problems associated with this technology, attributed to structural changes in dough and reducing the viability of yeast (Le-Bail et al., 1999; Yi et al., 2009). On the other hand, wheat dough is a viscoelastic material. This feature is provided by the gluten network formed by hydrophobic interactions among protein polymers of flour (mainly glutenins and gliadins) (Godón and Herard, 1984). Lu and Grant (1999) observed that glutenin fraction mainly affects baking characteristics of frozen dough, and has less influence in gliadin fractions and starch.

During storage of dough at freezing temperature, changes in the number, size and shape of ice crystal occur; phenomenon named "recrystallization," causing dough weakening in its structure (Baier-Schenk et al., 2005a). The ice crystal grows during storage, but the water balance remains constant. Gant et al. (1990) observed breaks in the cell membrane of the gas attributed to mechanical damage to the gluten network caused by ice crystal formation (Esselink et al., 2003). They also argue that dehydration occurs in frozen dough caused by the redistribution of water by increasing ice crystal size in the gas cell. Water transfer occurs first followed by free water of the water bound to proteins, promoting dehydration and re-distribution. During defrosting (re-hydration), water transfer occurs in the opposite direction, coalescence different than originally occupied in proteins, linking at different sites.

Several authors discuss that frozen dough has a more elastic behavior than fresh dough, due to a lower water content in its formulation (El-Hady et al., 1996; Sahlstrøm et al., 1999). The elastic and viscous moduli (G' and G") of frozen dough decreases with time, prevailing viscous behavior compared to fresh dough (higher values of Tan δ = G"/G') (Ribotta, 2004), reducing its gas retention capacity (Selomulyo and Zhou, 2007). The loss of elasticity in the frozen dough is attributed to fractionation of the polymer chains of the proteins present. Using SDS gel electrophoresis (sodium dodecyl sulfate) was observed an increase in the solubility of the protein, probably due to a reduction of the size of the polymer glutenin (Kennedy, 2000). Leray et al. (2010) concluded that during storage at freezing temperature, changes are observed in the frozen dough viscoelasticity which can be related reduction in bread volume (Ribotta et al., 2001).

Prolonged storage (more than one month) produce a reduction in loaf volume and an increase of the fermentation time (Aibara et al., 2005), because the yeast viability declines in the frozen dough while increases damage the gluten structure (Varriano-Marston et al., 1980; Berglund et al., 1991). Giannou and Tzia (2007) stored dough under freezing conditions up to nine months, and observed that in the first months the most dramatic changes in the gluten and yeast occur.

There are some information around frozen dough from the point of view of the relationship of its thermal properties, pre -baked and physicochemical properties and the quality of bread (Lind, 1991; Rosell and Gómez, 2007; Silvas-García et al., 2013).

The aim of this chapter is focuses on understanding the phenomenon of freezing the dough since the engineering point of view and the relationship of changes in dough (protein denaturation and visceolasticity) and bread quality.

## 6.2   FREEZING PROCESS OF DOUGH

The popularity of freezing process as a conservative method of food has increased in the past five decades, being the case of the baking industry. The use of frozen dough is an alternative to the immediate production of freshly prepared bread. The main goal is to get similar products to fresh products at any desired time (Selomulyo and Zhou, 2007). From the 90's, the freezing of bakery products was industrialized, in order to reduce manufacturing times, expand the range of consumer products (Sluimer, 2005; Giannou et al., 2003), extend the shelf life, and facilitate logistics (Baier-Schenk et al., 2005b). The methodology used is varied; the dough can be frozen as such or molded. Pre-fermented molded dough has been popularly used for a decade, because it reduces the time of making bread after thawing (Le-Bail et al., 2010; Sluimer, 2005). This composition has low thermal conductivity by its porous structure, increased resistance of yeast, and higher recovery of gas lost during freezing (Le-Bail et al., 2010; Zounis et al., 2002).

Food freezing is a method of preserving associated with the unavailability of water to chemical, biochemical, and microbiological reactions. During freezing, the speed of aging reactions is reduced by decreasing the water activity (WA). A product with a WA near '1' is highly perishable and has a higher freezing point. In bakery products, this term is related to the ability of interaction between water and the remaining ingredients, and its effect on the final product. Bread and leavened products have a WA of 0.95 to 0.99, taking values of 0.65 to 0.66 in the crumb. Water migration, textural changes, and storage stability of bread are determined by this parameter In this case, it is also reduced (approximately $-5°C$) or suspended yeast activity. The shelf life can be extended to 16–24 weeks.

Thermodynamically, water transformation is observed at the point of convergence of changes in Gibbs energy and chemical potential in the transition temperature. At this point, the water phase changes from liquid to solid state, by removing the latent and sensitive heats to sub-zero temperatures (Cauvain and Young, 2008). It is understood as sensitive heat, the energy required to decrease the water temperature to $0°C$ without a phase change. On the contrary, the latent heat or "latent heat of fusion," is the energy required to change the phase of liquid water to solid or crystalline (Cauvain and Young, 2008; Xu et al., 2009a) (334 kJ/kg) which is known as enthalpy. The crystallization of water is a phase of first order transition (Rao, 1999) for presenting a continuous Gibbs energy from one phase to another (Laidler and Meiser, 2005). Liquid water molecules have a tetrahedral structure, which changes to hexahedral during freezing (Cauvain and Young, 2008).

In foods, the required melting temperature is lower than in pure water $(0°C)$ and dependent on its composition. In most food, soluble ingredients such as salt and sugar with lower melting points at $0°C$ (Rao, 1999) are present. During freezing, water is removed from the dough, increasing polymer concentration, decreasing its melting point (Räsänen et al., 1998). At $<0°C$, a glassy state is formed composed of proteins, moisture, and high concentrations of solutes. The range of crystallization of the bread dough occurs from –3°C to –20°C (Sluimer, 2005). This coincides with other authors (Cauvain and Young, 2008), who report that at –10°C, chemical and physical reactions are reduced to the maximum in the bread dough.

Suitable freezing system depends on the characteristics of the product to be obtained. In some cases, it is appropriate to use a rapid freezing, achieved with large temperature gradients, where the ice crystal is smaller than that obtained by slow freezing. The selection of the type of freezing will depend on the impact on product features. Several authors agree that the slow freezing is the most suitable for the wheat dough because it is less damaging to the wheat gluten structure and viability of yeast (Codón et al., 2003; El-Hady et al., 1996; Silvas-García et al., 2014). The increase in the rate of freezing, increases the damage to the gluten net, reducing as much as 9% the bread volume obtained (Havet et al., 2000).

### 6.2.1 FREEZABLE WATER AND NON-FREEZABLE WATER

There are five types of water according to the quantity and form in foods: free, dehydration, monolayer, capillary, and unfreezable. In the frozen dough is set out, generally, the association of water as free (freezable water) and linked (water non-freezable) (Rizvi and Benado, 1984; Ruan et al., 1999).

Free water fraction of the total product water represents the freezable water (Xu et al., 2009a). In the dough, is the quantity of water that is not bound to gluten. Within this category is the water: free, dehydration, monolayer, and capillary. This fraction of water in dough increases versus the storage time (Lu and Grant, 1999). Some authors found that 62% of the water present in the frozen dough has mobility. The distribution of water in doughs was evaluated with different moisture content, using nuclear magnetic resonance (NMR) (Ruan et al., 1999). They found that in doughs with more than 23% of moisture, all binding sites between the solids and the water have been occupied, and in dough with >35% humidity, free water appears. They also found that from 40% humidity appears the structure of the bi-continuous network of the dough.

This type of water is not crystallized when the dough reaches –18°C, corresponding to ≈ 29% by weight and will depend on their moisture content (Ruan et al., 1999). Bound water is simultaneously united and secured, and presents greater mobility than freezable, being free in the energy sense (Cauvain and

Young, 2008). For long periods of storage at sub-zero temperatures, can freeze. This transition of the water in crystal occurs in several stages, including from nucleation until re-crystallization (Rao and Hartel, 1998).

## 6.2.2 WATER CRYSTALLIZATION PHENOMENON

The transition from liquid/solid water phase causes crystallization, which determines the freezing process efficiency and final product characteristics (Kiani et al., 2011). During the crystallization phenomenon, liquid water molecule subject to sub-zero temperatures fixes its orientation of tetrahedral to hexahedral (Cauvain and Young, 2008). This evolution occurs from nucleation until recrystallization. The freezing rate increases at low temperatures being inversely proportional to the size of the crystal formed. In semisolid systems, the type, size, and orientation of the crystal is variable. In the outer layers, the heat transfer rate is higher than the internal. This causes the formation of larger crystals in the center of the material. The saturation point where water and solutes present in the system change to the solid phase is named eutectic point. The gluten frozen dough is weakened by crystallization, reducing its quality (Xu et al., 2009a).

### 6.2.2.1 Nucleation

The probability of growth and distribution of ice crystal is affected by the phenomenon of nucleation (Kiani et al., 2011). In this first stage, the transfer rate of initial heat is high, decreasing as the heat is extracted, and increasing again to receive the first crystal (Cauvain and Young, 2008). The form, distribution, and size of the crystal have a close relationship with the nucleation stage (Kiani et al., 2011). At this stage, is spontaneously created a pattern in the system affected by the solutes and surface properties, among others (Kiani et al., 2011). There are two types of nucleation: homogeneous and heterogeneous. Secondary nucleation occurs with fragmentation of an existing glass, creating new centers, then spread.

#### 6.2.2.1.1 Homogeneous

This type of nucleation is applied to pure materials, and the group of nuclei originates within the same liquid phase (Rao and Hartel, 1998). The temperature used to form the core in the very heart of phase is less than that used in the heterogeneous nucleation (> –40°C) (Cauvain and Young, 2008). There is an average size required to initiate crystal growth. The theory of crystal growth indicates that most of the group nuclei are thermodynamically similar, depending their growth only on the energy of addition of a molecule added to the group. This depends on two concepts: the free energy of the surface affects the growth of the group, while the free energy of the majority favors.

### 6.2.2.1.2   Heterogeneous

This type of nucleation is related to composite materials known systems and follows the same theory of homogeneous nucleation. The nuclei appear on solid surfaces of the own system or not. Fewer nuclei and a higher temperature is required to initiate propagation, compared to the homogeneous (Kiani et al., 2011). This category includes foods such as dough of wheat. The appearance of cores is limited by decreasing the required temperature. Freezing speeds originate a greater number of nuclei. Gant et al. (Gant et al., 1990) suggest the onset of ice crystal on the interface of the gas cell, promoting heterogeneous nucleation. Baier-Schenck et al. (Baier-Schenk et al., 2005b) agree that nucleation in frozen dough is more prevalent in the wall of the gas cell, but its mechanism is not yet specified. Ultrasound technology has been used to understand and manipulate the nucleation process, being able to extrapolate its application in the frozen food industry (Kiani et al., 2011).

### 6.2.2.2   Ice Crystal Growth

Adding molecules to the ice crystal is known as crystal growth process (Rao and Hartel, 1998). The first principle governing this process is the transfer of heat. The heat removed from the liquid water requires 300 J/g of energy to transform its status to solid. The second principle is the mass transfer. The water molecule itself undergoes a structural rearrangement. The solutes are forced to migrate from the crystallized species. The heat transfer rate is greater for quick freezing that uses temperatures below the slow freezing (–18°C), favoring the formation of small ice crystals. The ice crystal size affects yeast viability and the structure of the gluten network. Crystal growth causes dehydration of gluten reducing the hardness and elasticity of the dough (Cauvain and Young, 2008). Small crystals embedded in the protein network, migrate to the crystal placed in the gas cell by promoting crystal growth.

### 6.2.2.3   Recrystallization

In systems such as dough, besides the crystallization phenomenon, there is a redistribution of the various species that alter its structure. During storage, occurs changes in the number, size, and shape of ice crystal; phenomenon named "recrystallization" (Baier-Schenk et al., 2005a). Fennema (Fennema, 1996) mentions three different mechanisms of recrystallization: isomass, immigration, and additive. This phenomenon increases the damage on the frozen dough. Irregular and high volume ice crystal is rounded without mass transfer, occurring the isomass recrystallization. Pressure changes promote dissolution of small crystals that are adsorbed on the surface of larger crystals, happening the migratory recrystallization. Additive recrystallization is the union of two ice crystals via migration by surface contact, causing a larger new ice crystal. In pre-frozen sourdough, ice crystals are placed on the gas cells, causing heterogeneous distribution of the water, altering the baking quality. During defrosting (re-hydration) water transfer occurs in the opposite direction, joining in different sites to the initially occupied in the proteins, changing again its conformation.

### 6.3   EFFECT OF FREEZING PROCESS ON SOME PROPERTIES

The temperature and the time used for freezing the dough and the temperature used during storage, modify the quality of the final product. Some of these effects are related to the alteration of the rheological properties of frozen dough and the yeast viability (Le-Bail et al., 1999). The main alterations observed in the bread due to freezing are: low volume and high strength of its crumb, coupled with its rapidly aging (Yi et al., 2009).

### 6.3.1   CHANGES IN DOUGH

#### 6.3.1.1   Re-Distribution of Water

The presence of different solutes in the dough involved in the low melting point with respect to water. Water is considered a plasticizer and unifying of the present components. During freezing, dough is exposed to cold temperatures and being no homogeneous, and it is hard to predict its behavior. Generally, the outer layers are the first ones to freeze and distributed radially towards the center of dough. Free water is the first to freeze, changing its location. Approximately 45% of dough weight corresponds to the water (Baier-Schenk et al., 2005b). Water dispersion in components of dough is 46% starch, 31% protein, and 23% pentosans. Gluten proteins are hydrated to form the gluten, being glutenin the most hydrophilic. The interaction of water with flour causes lowering its melting point. In frozen dough, approximately 53% of the water is melted (Baier-Schenk et al., 2005b).

During freezing of dough, the free water followed by the bound water form ice crystals causing a redistribution of the water. During storage, the amount of freezable water increases (El-Hady et al., 1996). During defrosting, water integrates again to the dough but differently. Changes in the dough by the freezing process, are attributed to the re-distribution of water (Sluimer, 2005). Baier-Schenk et al. (2005b) observed a formation of preferential ice crystals on the gas cells, causing a cryo-concentration of the reversible protein during defrosting. These researchers concluded that ice crystal damages to a lesser extent the gluten network, compared with the re-distribution of water in the frozen-thawed dough. Overall, water is considered stationary in the frozen dough. However, a small amount of water is free to flow and promote in deteriorative reactions of bread quality.

Water contributes to the dough expansion and volume of the bread, favoring the organoleptic characteristics of the product. Moisture loss from the frozen dough is one of the major problems associated with spoilage, causing a redistribution of water in the dough system. Normally, water migrates to the surface of the dough, which presents the lowest relative humidity (RH). Second, loss of moisture at the surface given by the change phase of liquid water to steam (sublimation), during cooling, at a rate of 2 g of water/°C, due to the vapor pressure 0 of the water to sub-zero temperatures. Third, the relative humidity (RH) of the environment is much lower than the average RH of the frozen dough (92.5%), causing a gradient in the interface, and consequently, water transfer occurs to the environment (Cauvain and Young, 2008). Another factor is the air speed in the environment, which should not exceed 0.25 m/s avoiding using impermeable packaging (Sluimer, 2005).

Moisture loss in dough is intensified with temperature and storage time due to the high-pressure air saturated vapor. Giannou and Tzia (2007) observed a weight loss of the frozen dough during storage. Frozen dough was stored at different ranging temperatures and monitoring the temperature of the dough in the center and surface (Phimolsiripol et al., 2008). These authors also agreed to the weight loss in the frozen dough at all temperatures used during storage. They related the weight loss occurred, with others authors (Laguerre and Flick, 2007) theory of weight loss by evaporation in packaged foods. These authors believe that the existence of steam pressure gradients and temperature among the dough, air, and packaging, promote the diffusion of water steam of dough to the package or vice versa. The release of gas in the dough contributes to a lesser extent in the weight loss of the frozen dough during storage (Cauvain and Young, 2008).

Thawed dough has a different relocation of water, so its rehydration is incomplete (Cauvain and Young, 2008). To reduce this effect in fresh and frozen dough, the rehydration is aided during fermentation, exposing to air saturated with water steam in combination of temperature named relative humidity (RH). Exposure of dough to high RH prevents further dehydration. The RH is commonly known as the ratio of partial pressure of the steam/ steam saturation pressure, at the same temperature.

During fermentation of the thawed dough exists a temperature gradient between this material and the environment. Heat transfer occurs first between the dough and the environment, which grants part of his latent heat to dough, inducing condensation of water on the surface. This causes a temperature gradient within the dough directing the heat transfer to its center, facilitating the activity of the yeast. At this stage, it recovers 0.34% of the weight before freezing. RH in the dough, higher at the air, becomes permeable surface providing its degasification (Cauvain and Young, 2008). The gradual thawing of the dough reduces water condensation on the surface. Clearly, the properties of dough and bread vary during frozen storage, its different properties of dough and bread (Sluimer, 2005). Selomulyo and Zhou (2005) worked on the effect of storage and some additives in frozen bread quality, associated them to the rheological changes of the frozen dough, which can be caused by the alteration of the structure and associations of proteins, changing its fermentative capacity and quality characteristics of bread.

### 6.3.1.2   Protein Denaturation

Wheat dough is a viscoelastic material, and is the only one obtained from cereals, able to retain gas, resulting after processing, and a spongy product named bread. The gluten network is formed by hydrophobic interactions among protein polymers of flour (mainly glutenins and gliadins), and increase during mixing (Godón and Herard, 1984). Cornec et al. (1993) characterized by high-resolution liquid chromatography by molecular exclusion (SE-HPLC) the sub-fractions of fresh gluten, evaluating individual rheology and its relationship with the dough viscoelasticity. They found that gliadin has a plasticizing effect in the dough, while glutenins of high molecular weight (HMW-GS), govern the behavior of the gluten network. The rheological properties differences between the native gluten of two different commercial flours from different baking quality was evaluated and compared to the gluten reconstituted by the glutenin and gliadin fractions of each flour (Janssen et al., 1996). Gluten reconstituted from both flours presented a more elastic behavior compared to control, probably due to variation in the range of glutenin/gliadin and the source of each fraction.

A number of processes provoke disorder and reorganization of protein polymers, causing their denaturation. The change in the native form of proteins is an endothermic reaction in the presence of water (Rao, 1999). The conformation of the proteins are modified by breaking its primary arrangement. Generally, the hydrophobic groups of the amino acids (nonpolar) are located in the interior of the protein being exposed the polar ones ($-OH$, $-COO$ and $NH^{3+}$) that interact by hydrogen bonding with water (Laidler and Meiser, 2005). Low levels of water increase the temperature stability of proteins (Rao, 1999). In frozen dough, protein rearrangement is caused by mechanical damage during the formation and growth of ice crystal, and not by exposure to heat. Berglund et al. (1991) attributed the structural changes in the dough to the damage occurred in the gluten network during freezing and thawing.

Some authors evaluated structural changes in the hydrated and gluten stored at subzero temperatures were evaluated. They observed that ice formation and recrystallization phenomenon causes disruptions in the gluten structure, altering their morphology (Kontogiorgos et al., 2008). Others researches observed by laser scanning microscopy changes in the isolated gluten during freeze-thaw cycles. These authors showed changes in gluten fibrils given by cryo-concentration of the protein polymers by melting water. The changes observed in the dough not fermented, were reversible during thawing, but changed the distribution of water (Baier-Schenk et al., 2005a, 2005b). Several authors explain that the reversibility of the structural rearrangement of the gluten depends on how it was originated and the type of links between the polymers present (Baier-Schenk et al., 2005b).

During the freezing process, the gluten network is mechanically damaged by the formation of ice crystals that cause breaks in the gas cell membrane (Gant et al., 1990), combined with dehydration caused by the redistribution of water by increasing the crystal size within the gas cell (Esselink et al., 2003). The ice crystal is born in the gas cell where starts water transfer. The recrystallization causes a redistribution of the water, occurring first, transfer free water and then, bound water to protein polymers. The ice crystal grows during storage, but the water balance remains constant. This growth occurs at the expense of dehydration of proteins changing their original conformation. During freezing, the ice crystal itself can break inter or intramolecular bonds giving rise to short chain polymers, or new long chains caused by regrouping them. Furthermore, during storage occur changes in the number, size, and shape of ice crystal; phenomenon named "recrystallization" (Baier-Schenk et al., 2005a). This phenomenon increases the damage on frozen dough. Phimolsiripol et al. (2008) attribute the loss of gas retention in the frozen dough to the damage occurred in the gluten network by re-crystallization phenomenon. During thawing (rehydration), water transfer occurs in reverse, joining at different sites in the protein initially occupied, changing its conformation again and causing the weakening of dough structure (Esselink et al., 2003).

The loss of elasticity in the frozen dough is attributed to fractionation of the polymer chains of the proteins present. Some authors observed the effect of freezing conditions in the baking quality of frozen dough for French bread. Results of these authors showed a loss in the elasticity of the dough associated with breaks in the gluten-forming polymers (Havet et al., 2000). Using SDS gel electrophoresis (for short), an increase was observed in the solubility of the protein, probably by reducing the size of the polymer glutenin (Havet et al., 2000; Ribotta et al., 2001). They associated this change with the loss of elasticity of frozen dough and the capacity of gas retention. In general the SPP fraction was the most affected during the frozen dough process. Magaña-Barajas et al. (2014) found a negative correlation between polymeric protein soluble (PPS) and gliadins (GLI) ($r = -0.96$), which indicate the increase in the amount of GLI by the higher levels of PPS hydrolysis. Others authors observed a less reduction of polymeric protein insoluble (PPI) in dough frozen at fast freezing rate than the slow freezing rate (Silvas-García et al., 2014). The glutenin fraction, mainly affects baking characteristics of the frozen dough, affecting in a less extent the gliadin fractions and starch. These authors suggest to use strong flours (high glutenin) to produce frozen dough (Lu and Grant, 1999). Borneo and Khan (1999) found a direct and inverse relationship of the glutenin and albumins + globulins fractions with bread volume. By SEM images can observed that the fibrils of the gluten in the dough frozen are short, thick and non-homogeneous with many bend.

### 6.3.1.3   Fermentative Capacity

The yeast viability decreases with the freezing and storage time. About 9% of yeast viability is lost during freezing of dough (Phimolsiripol et al., 2008). Yeast cells are damaged releasing protein material inside, contributing to the weakening of dough (Holmes and Hoseney, 1987; Sluimer, 2005). Williams and Paullen (Cauvain and Young, 2008) found the compressed yeast of about 70% of water in free state. During freezing process, occurs a water transfer from within the yeast cell to the system known as osmosis. This causes a pressure gradient between the wall of the yeast cell with a system named osmotic pressure, facilitating the expulsion of water from inside the cell.

Yeast efficiency is determined by the rate of production of $CO_2$. In frozen dough the production rate of $CO_2$ is lower than in fresh dough, decreasing with time. This happens for the low activity and the amount of yeast present, among others factors (Autio and Sinda, 1992; El-Hady et al., 1996). Phimolsiripol et al. (Phimolsiripol et al., 2008) observed a decrease of 7% in the production of $CO_2$ stored in the frozen dough for a day compared to the fresh dough. They also observed that both, the cumulative production and the rate of $CO_2$ production decrease together with respect to time. The total volume of $CO_2$ in the frozen dough is reduced during freezing and storage. This reduction is attributed to the breakdown of the structure of the gluten network caused by the crystal formation and growth (Kontogiorgos et al., 2008; Phimolsiripol et al., 2008).

The diffusion of nutrients and by-products of metabolism of yeast is restricted to damage the gluten structure (Magaña-Barajas et al., 2014). This results in an increase in crumb firmness and volume decreased of the bread (Aibara et al., 2001). Le-Bail et al. (Le-Bail et al., 1999) found that low expansion of dough prior to freezing, favors bread volume. The optimal time for pre-fermentation for frozen dough is $\leq 20$ min (Sluimer, 2005). It

is recommended thawing in two stages: the first at refrigeration temperature (4°C), and the second at room temperature to increase the activity of
the yeast. Holmes and Hoseney (Holmes and Hoseney, 1987) found that
the high temperatures used during rehydration reduces the damage in yeast.
They believe that the optimum temperature for yeast rehydration is 38°C.

### 6.3.1.4   Viscoelasticity

In some cases, water content is reduced in the formulation for frozen
dough, increasing its elasticity (Holmes and Hoseney, 1987; Sahlstrøm et
al., 1999). The dough strength decreases during freezing, reflecting lower
bread volume. Sluimer (2005) agreed that thawed dough is more extensible and weaker than fresh dough. The weakening of the frozen dough
is associated with the presence of the metabolic products of yeast and the
damage caused to the gluten network during crystal formation and growth
(Selomulyo and Zhou, 2007).

During storage, the elastic and viscous moduli (G' y G") of frozen
dough decrease with time, prevailing a viscous behavior compared to
fresh dough (Angioloni et al., 2008; Magaña-Barajas et al., 2014; Ribotta,
2004; Silvas-García et al., 2014), and reducing its gas retention capacity (Selomulyo and Zhou, 2007). Angioloni et al. (Angioloni et al., 2008)
observed an increase in the value of the phase angle ($\delta$=G"/G') in the stored
frozen dough, relating with breakthroughs of polymer bonds, which coincides with others authors (Autio and Sinda, 1992). These authors attribute
the loss of elasticity of dough to mechanical damage caused by ice crystal
growth on the gluten network, resulting in a network with more breaks
and less continuous, disintegrated of the starch granules, resulting in the
weakening of the dough. Varriano-Marston et al. (1980) hypothesized that
the weakening of frozen dough is due to the ice crystallization. The loss
of elasticity is then, assigned to fractionation of the polymer chains of
proteins caused by ice crystal. This fractionation can occur for the formation of ice crystals and the presence of certain substances such as glutathione (Angioloni et al., 2008; Phimolsiripol et al., 2008). Magaña-Barajas et
al. (2011) found a decrease in elastic behavior of frozen dough associate
with the increase in solubility of SPP. This coincided with Sílvas-García
et al. (2014), they also found that slow freezing rate affect more the elastic
behavior el frozen dough than fast freezing rate.

Leray et al. (2010) concluded that during storage, changes are observed in frozen dough viscoelasticity related with the reduction in bread volume (Ribotta et al., 2001). The rheological changes in the frozen dough are associated with the deterioration of its baking quality (Selomulyo and Zhou, 2007).

## 6.3.2 CHANGES IN BREAD

During freezing structural changes occur in proteins and rheology of dough, associated with the unavailability of water. Baking quality is altered by the amount of water that does not return to the system during baking (Selomulyo and Zhou, 2007).

### 6.3.2.1 Volume

The volume of bread is one of the main parameters of quality and consumer acceptance (Aibara et al., 2001; Sharadanant and Khan, 2003). It is considered an indicator of the ability of the dough to retain gas during baking (Giannou and Tzia, 2007). The decrease in bread volume is one of the major problems associated with the freezing of dough. This reduction in bread volume is added to the reduction in yeast viability and changes in the gluten structure of the dough (Cauvain and Young, 2008; Esselink et al., 2003; Gant et al., 1990; Phimolsiripol et al., 2008). Magaña-Barajas et al. (2014) found a negative effect of frozen storage of dough in specific volume of bread. This is associates with the changes in PPS and the weakening of dough. They recommended no more of 21 days of frozen storage of the dough. In this case, the slow freezing rate affect less than fast freezing rate on the specific volume of bread made with frozen dough (Silvas-García et al., 2014).

Crystal formation occurs within the cell gas. During storage, water migrates contributing to crystal growth, resulting in weakening in the structure of the frozen dough and reduction in bread volume (Selomulyo and Zhou, 2007). Phimolsiripol et al. (2008) reported a 9% decrease in volume of frozen dough bread, regarding bread made with fresh dough, a difference that increases with storage time. Bread volume is directly related to the degree of expansion of the dough. Cauvain and Young (mention three sources that cause the expansion of the gases within dough:

- Thermal expansion of gas. Ruled by Law of Gay-Laussac gas: the volume of a gas at constant pressure is directly proportional to absolute temperature (Maron and Prutton, 1980). As the volume increases 1/273 times its original volume per $1°K = 1°C$.
- Release of $CO_2$. In bread, yeast is the main contributor of $CO_2$. This acts from the early stages of baking even before its inactivation ($\approx 45°C$), contributing to a lesser degree in thermal expansion. The potential of the $CO_2$ produced in the fermentation, depends on its concentration and the porosity of the dough.
- Water vapor. The water begins to evaporate when the dough exceeds $100°C$, aiding expansion. Collaboration water vapor depends on the amount of water present, its location and the heating rate.

In frozen dough, gas solubility is altered. The solubility of $CO_2$ (Le-Bail et al., 2010) and helium gas increases when decreasing temperature, as opposed to the $N_2$ (Sluimer, 2005). Le-Bail et al. (2010) considers a reduction of 26.5% of the dough pressure, assuming constant the amount of gas inside the cell. They observed a direct relationship between the reductions of the Henry coefficient ($CO_2$ mass/$CO_2$ in the gas cells) with temperature. The diffusion of gases in the frozen dough is another important factor in their quality. The concentration of gases in the small cells is increased by high hydrostatic pressure to be disseminated to the larger cells. $CO_2$ diffuses faster than the $N_2$ and helium by its high solubility and concentration. This generates the coalescence of the smaller gas cells, resulting in a lower volume bread and more irregular and compact crumb, than that obtained from fresh dough. In the pre-fermented frozen dough, nitrogen acts as a stabilizer (Sluimer, 2005). During thawing and fermentation, part of gases diffuses into the gas cells adding to the generated by yeast (Le-Bail et al., 2010). Magaña-Barajas et al. (2014) associate the loss of specific volume with the weakening of dough by the increase de PPS during the storage.

Long periods of storage (more than one month) reduced loaf volume and increase the fermentation time (Aibara et al., 2005), because yeast viability in the frozen dough decreases while increases damage in the gluten structure (Berglund et al., 1991; Giannou and Tzia, 2007; Varriano-Marston et al., 1980).

## 6.3.2.2  Firmness

The bread fresh feeling is attributed to the water lubricating function. The bread has an average moisture level compared with other baked goods. Crust has a lower moisture content than the crumb (Cauvain and Young, 2008). Frozen bread firmness increases with storage (Berglund and Shelton, 1993; Phimolsiripol et al., 2008). The increased firmness of bread relates to water migration out of the product that occurs by diffusion, vapor phase transfer and syneresis. The moisture diffusion occurs through the cells that act as capillaries, being higher in airier crumb. The water vapor transfer is initiated by the moisture gradient between the surface of the product and the environment, still evident in packaged products. In the frozen and thawed bread, syneresis is identified by the condensation of water on its surface, obtained by the release of the water trapped in the formed starch gel.

The content of fatty ingredients causes a barrier reducing the coefficient of water migration into the dough. However, frozen bread suffers previous severe dehydration that are specific to their processing and storage. This causes the production of harder bread crumb from frozen dough and hardening is greater after processing. This hardening is increased at 4°C during defrosting, temperature where occur at maximum the reactions of aging of bread. The crumb crust is less crunchy, darker (directly related to the time of pre-fermentation), thick, and leathery (Sluimer, 2005).

Phimolsiripol et al. (2008) observed an increase in the firmness of bread from frozen dough after 14 days storage. This increase is probably due to lower bread volume caused by gluten weakening, and its moisture loss (Berglund and Shelton, 1993; Inoue and Bushuk, 1992). Some authors observed a significative increase of firmness of bread since the seven-day of frozen storage. These authors found associate this changes with the high solubilities of protein (Magaña-Barajas et al., 2014). Ribotta et al. (2003b) observed an increase in firmness of bread by changes in amylase and amylopectin during frozen storage. These changes increase the rate of starch retrogradation, collaborating with loss of softness of bread.

### 6.3.2.3 Others

The repetitive freeze-thaw cycles cause "freeze burns" in dough, showing a white ring between the crumb and crust of bread. Another phenomenon observed in rolls and pre-baked bread is the "husked," that occurs when the crust and crumb emerge. This arises for different melting points, because of heterogeneous solute concentrations. Dehydration also causes cracks on the surface of the dough, appearing breaks during thawing and baking, thereby, reducing the volume of bread. In the crust appear undesirable stripes for the water condensation. This phenomenon is attributed to the de-gassing or the excessive production of gas before freezing. Part of the water extracted from dough is frozen on the surface, causing white spots detected after baking. Browning reactions or Maillard are limited by decreasing the concentration of sugars in these small areas (Cauvain and Young, 2008).

The need for fresh products of consistent quality is essential for the consumer. This demand has been rectified using frozen dough, but the bread quality is greatly reduced, during the storage time (Huang et al., 2008a, 2008b; Wolt and Appolonia, 1984). The use of additives in the baking industry aims to maintain or enhance the quality and freshness of the product, and ease its processing and handling. In order to minimize the negative effects of the freezing process of bread dough, they have used various additives such as emulsifiers, vital gluten, cysteine, ascorbic acid, hydrocolloids, gelatin, ethanol, antifreeze proteins, among others.

## 6.4 CONCLUSIONS

The use of freezing process as bread preservation medium adversely affects the properties of dough and bread quality. Freezing is a complicated process to discuss from the standpoint of thermodynamic and physicochemical. However, as observed in this work is important to understand from the thermodynamic point of view the process of water freezing and translate to dough to interpret the changes that occur in the system and facilitate the proposed solutions. Freezing dough summarized a change of state of material, consulted derivative research analysis considered that water is the main factor that alters its rheological and structural properties, negatively affecting its baking quality. During ice crystal formation and recrystallization water loss in dough occurs promoting relocation. Crystal growth and favor the dehydration gluten weakening caused by the breaking of that shape, especially in increasing the solubility of the glutenin fraction is the hydrophilic polymer chains. Decreasing the amount of glutenin polymer and the presence of secondary metabolites of the yeast in frozen dough generates irreversible changes in its rheology. The dough has a more viscouselastic behavior affecting their gas retaining capacity and therefore its volume and bread firmness. Altered the structure of gluten is unable to retain all of the $CO_2$ produced by obtaining a lower volume bread. Firmness increases for two reasons: first by decreasing the amount and size of the gas cells, and the second water loss during frozen storage. Throughout the review it was that the changes caused by the freezing of water in the dough are inter-related. Freezing mainly because structural changes that alter the rheology of the dough, fermentation capacity and baking quality. Besides the use of certain additives, it has been proven to modify some stages of freezing as is the pre-fermentation or pre-cooked. The freezing rate is another important parameter observed that fast and slow least alters the quality of bread and gluten, respectively. There are several investigations into the understanding of the process of the frozen dough and its effects, though much remains to be studied. It has begun to thoroughly evaluate changes in protein polymers frozen dough finding a direct effect on the changes in the dough and bread. The water is at the nanopores formed by the network of protein, lipids and arabinoxylans, is considered important to analyze the changes that occur in these components and associate them with the quality of dough and bread.

## KEYWORDS

- bread volume
- frozen dough
- protein solubility
- viscoelasticity

## REFERENCES

Aibara, S., Mishimura, K., & Esaki, K. (2001). Effects of shortening on the loaf volume of frozen dough bread. *Food Sci. and Biotech, 10*, 521–528.

Aibara, S., Ogawa, N., & Hirose, M. (2005). Microstructures of bread dough and the effects of shortening on frozen dough. *BBB, 69*, 397–402.

Angioloni, A., Balestra, F., Pinnavaia, G. G., & Dalla Rosa, M. (2008). Small and large deformation test for the evaluation of frozen dough viscoelastic behavior. *J. Food Eng, 87*, 527–531.

Autio, K., & Sinda, E. (1992). Frozen doughs: Rheological changes and yeast viability. *Cereal Chem, 69*, 409–413.

Baier-Schenk, A., Handschin, S., & Conde-Petit, B. (2005a). Ice in prefermented frozen bread dough an investigation based on calorimetry and microscopy. *Cereal Chem, 82*, 251–255.

Baier-Schenk, A., Handschin, S., Von Shönau, M., Bitterman, A. G., Bächi, T., & Conde-Petit, B. (2005b). In situ observation of the freezing process in wheat dough by confocal laser scanning microscopy (CLSM): Formation of ice and changes in the gluten network. *J. Cereal Sci., 42*, 255–260.

Berglund, P. T., & Shelton, D. R. (1993). Effect of frozen storage duration on firming properties of breads baked from frozen dough. *Cereal Food World, 38*, 89–93.

Berglund, P. T., Shelton, D. R., & Freeman, T. P. (1991). Frozen bread dough ultrastructure as affected by duration of frozen storage and freeze-thaw cycles. *Cereal Chem. 68*, 105–107.

Borneo, R., & Khan, K. (1999). Protein changes during various stages of bread-making of four spring wheat's: Quantification by size-exclusion HPLC. *Cereal Chem, 76*, 711–717.

Cauvain, S., & Young, L. (2008). Bakery food manufacture and quality. Water Control and Effects. 2nd Ed., Ames, Iowa.

Codón, A. C., Rincón, A. M., Moreno-Mateos, M. A., Delgado-Jarana, J., Rey, M., Limón, C., et al. (2003). New Saccharomyces cerevisiae baker's yeast displaying enhanced resistance to freezing. *J. Agr. Food Chem, 51*, 483–491.

Cornec, M., Popineu, Y., & Lefebvre, J. (1993). Characterization of gluten subfraction by SE-HPLC and dynamic rheological analysis in shear. *J. Cereal Sci, 19*, 131–139.

El-Hady, E. A., El-Samahy, S. K., Seibel, W., & Brümmer, J. M. (1996). Changes in gas production and retention in non-prefermented frozen wheat doughs. *Cereal Chem, 73*, 472–477.

Esselink, F. J., Van Aalst, H., Maliepaard, M., & Van Duynhoven, P. M. (2003). Long-term storage effect in frozen dough by spectroscopy and microscopy. *Cereal Chem, 80*, 396–403.

Fennema, O. R. (1996). Food Chemistry. 3rd Ed., Marcel Dekker Inc.

Gant, Z., Angold, R. E., Wiliams, M. R., Ellis, P. R., Vaughan, J. G., & Galliard, T. (1990). The microstructure and gas retention of bread dough. *J. Cereal Sci., 12*, 15–24.

Giannou, V., & Tzia, C., Frozen dough bread: Quality and textural behavior during prolonged storage-prediction of final product characteristics. *J. Food Eng, 79*, 929–934.

Giannou, V., Kessoglou, T., & Tzia, C. (2003). Quality and safety characteristics of bread made from frozen dough. *Trends in Food Sci. Tech, 14*, 99–108.

Godón, B., & Herard, J. (1984). Extractability of wheat proteins. Effects on the protein associations of the mixing of dough in presence of various active compounds [molecular interactions]. *Sci. Aliment, 4*, 287–303.

Havet, M., Mankai, M., & Le-Bail, A. (2000). Influence of the freezing condition on the baking performances of French frozen dough. *J. Food Eng. 45*, 139–145.

Holmes, J. T., & Hoseney, R. C. (1987). Frozen dough: Freezing and thawing rates and the potential of using a combination of yeast and chemical leavening. *Cereal Chem, 64*, 348–351.

Huang, W. N., Yuan Y. L., Kim, Y. S., & Chung, O. K. (2008b). Effects of transglutaminase on rheology, microstructure, and baking properties of frozen dough. *Cereal Chem, 85*, 301–306.

Huang, W., Kim, Y., Li, X., & Rayas-Duarte, P. (2008a). Rheofermentometer parameters and bread specific volume of frozen sweet dough influenced by ingredients and dough mixing temperature. *J. Cereal Sci., 48*, 639–646.

Inoue, Y., & Bushuk, W. (1992). Studies on frozen doughs. II. Flour quality requirements for bread production from frozen dough. *Cereal Chem, 69*, 423–428.

Janssen, A. M., van Vliet, T., & Vereijken, J. M. (1996). Rheological behavior of wheat glutens at small and large deformations. Effect of gluten composition. *J. Cereal Sci, 23,* 33–42.

Kennedy, C. J. (2000). Freezing process foods: En Kennedy, C. J. (Ed.), managing frozen foods. Woodhead Publishing, Cambridge.

Kiani, H., Zhang, Z., Delgado, A., & Sun, D. (2011). Ultrasound assisted nucleation of some liquid and solid model foods during freezing. *Food Res. Int, 44,* 2915–2921.

Kontogiorgos, V., Goff, H. D., & Kasapis, S. (2008). Effect of aging and ice-structuring proteins on the physical properties of frozen flour-water mixtures. *Food Hydrocolloid, 22,* 1135–1147.

Laguerre, O., & Flick, D. (2007). Frost formation on frozen products preserved in domestic freezers. *J. Food Eng, 79,* 124–136.

Laidler, K. J., & Meiser, J. H. (2005). Fisicoquímica. 1st Ed., Editoral Continental.

Le-Bail, A., Grinand, C., Le Cleach, S., Martínez, S., & (1999). Quilin, E., Influence on storage conditions on frozen French bread dough. *J. Food Eng, 39,* 289–291.

Le-Bail, A., Nicolitch, C., & Vuillod, C. (2010). Fermented frozen dough: Impact of pre-fermentation time and of freezing rate for a pre-fermeted frozen dough on final volume of bread. *Food Bioprocess Tech, 3,* 197–203.

Leray, G., Oliete, B., Mezaise, S., Chevallier, S., & De Lamballerie, M. (2010). Effects of freezing and frozen storage conditions on the rheological properties of different formulations of non-yeasted wheat and gluten free bread dough. *J. Food Eng. 100,* 70–76.

Lind, I. (1991). The measurement and prediction of thermal properties of food during freezing and thawing–A review with particular reference to meat and dough. *J. Food Eng, 13,* 285–319.

Lu W., & Grant, L. A. (1999). Role of flour fractions in breadmaking quality of frozen dough. *Cereal Chem, 76,* 663–667.

Magaña-Barajas, E., Ramírez-Wong, B., Torres-Chavez, P. I., Sánchez-Machado, D. I., & López-Cervantes, J. (2014). Changes in protein solubility, fermentative capacity, viscoelasticity, and breadmaking of frozen dough. *Africal J. of Biotechnol, 13,* 2058–2071.

Magaña-Barajas, E., Ramírez-Wong, B., Torres, P. I., Sánchez-Machado, D. I., & Y López-Cervantes, J. (2011). Efecto del contenido de proteína, grasa y levadura en las propiedades viscoelásticas de la masa y la calidad del pan tipo francés. *Interciencia, 36,* 248–255.

Maron, S. H., & Prutton, C. F. (1980). Fundamentos de Fisicoquímica, 1era (Ed.), Limusa: México, D. F.

Phimolsiripol, Y., Siripatrawan, U., Tulyathan, V., & Cleland, D. (2008). Effects of freezing and temperature fluctuations during frozen storage on frozen dough and bread quality. *J. Food Eng, 84,* 48–56.

Rao, M. A. (1999). Rheology of Fluid and Semisolid Foods: Principles and Applications. Gaithesburg, Maryland. Aspen.

Rao, M. A., & Hartel, R. W. (1998). Phase/state transitions in foods. Chemical, Structural and Rheological Changes. New York, NY.

Räsänen, J., Blanshard, J. M. V., Mitchel, J. R., Derbyshire, W., & Autio, K. (1998). Properties of frozen wheat dough at subzero temperatures. *J. Cereal Sci., 28,* 1–14.

Ribotta, P. D., León, A. E., & Añón, M. C. (2001). Effect of freezing and frozen storage of doughs on bread quality. *J. Agric. Food Chem, 49,* 913–918.

Ribotta, P. D., León, A. E., & Añón, M. C. (2003b). Effects of freezing and frozen storage on the gelatinization and retrogradation of amylopectin in dough baked in a differential scanning calorimeter. *Food Res. Int, 36,* 357–363.

Ribotta, P. D., León, A. E., & Añón, M. C. (2004). Effect of emulsifier and guar gum on micro structural, rheological and baking performance of frozen bread dough. *Food Hydrocolloid. 18*, 305–313.

Rizvi, S. S. H., & Benado A. L. (1984). Thermodynamic properties of dehydrated foods. *Technol, 38*, 83–92.

Rosell, C. M., & Gómez, M. (2007). Frozen Dough and Partially Baked Bread: An update. *Food Reviews International, 23*, 303–319.

Ruan, R. R., Wang, X., Chen, P. L., Fulcher, R. G., Pesheck, P., & Chakrabarti, S. (1999). Study of water in dough using nuclear magnetic resonance. *Cereal Chem, 76*, 213–235.

Sahlstrøm, S., Nielsen, A. O., Færgestad, E. N., Leap, P., Park, W. J., & Ellekjær, M. R. (1999). Effect of dough processing conditions and datem on Norwegian hearth bread prepared from frozen dough. *Cereal Chem, 76*, 38–44.

Selomulyo, V. O., & Zhou, W. (2007). Frozen bread dough: Effects of freezing storage and dough improvers. *J. Cereal Sci, 45*, 1–17.

Sharadanant, R., & Khan, K. (2003). Effect of hydrophilic gums on frozen dough. *Cereal Food World, 40*, 827–831.

Silvas-García M. I., Ramírez-Wong, B., Torres-Chávez, P. I., Carvajal-Millan, E., Bello-Pérez, L. A., & Barrón-Hoyos, J. M. (2013). Cambios fisicoquímicos en masa congelada y su efecto en la calidad del pan: Una Revisión. *Interciencia, 38*, 332–338.

Silvas-García, M. I., Ramírez-Wong, B., Torres-Chávez P. I., Carvajal-Millan E., Bello-Pérez L. A., Barrón-Hoyos J. M., &Quintero-Ramos, A. (2014). Effect of freezing rate and storage time on gluten protein solubility, and Dough and Bread properties. *J. Food Process Eng, 37*, 237–347.

Sluimer, P. (2005). Principles of breadmaking. Functionality of raw materials and process steps. American Association of Cereal Chemists. St. Paul, MN.

Varriano-Marston, E., Hsu, K. H., & Mahdi, J. (1980). Rheological and structural changes in frozen dough. *Baker's Digest, 54*, 32–34, 41.

Wolt, M. J., & D'Appolonia, B. L. (1984). Factors involved in the stability of frozen dough. I. The influence of yeast reducing compounds on frozen-dough stability. *Cereal Chem, 61*, 209–212.

Xu, H., Huang, W., Jia, C., Kim, Y., & Liu H. (2009a). Evaluation of water holding capacity and breadmaking properties for frozen dough containing ice structuring proteins from Winter wheat. *J. Cereal Sci., 49*, 250–253.

Yi, J., Johnson, J. W., & Kerr, W. L. (2009). Properties of bread made from frozen dough containing waxy wheat flour. *J. Cereal Sci, 50*, 364–369.

Zounis, S., Quail, K. J., Wootton, M., & Dickson, M. R. (2002). Studying frozen dough structure using low-temperature scanning electron microscopy. *J. Cereal Sci, 35*, 135–147.

# PART II

# NEW TECHNOLOGY AND PROCESSES

# CHAPTER 7

# ADVANCES IN REFRIGERATION AND FREEZING TECHNOLOGIES

MARIA C. GIANNAKOUROU and
EFIMIA K. DERMESONLOUOGLOU

¹*Associate Professor, Technological Educational Institute of Athens, Faculty of Food Technology and Nutrition, Department of Food Technology, Agiou Spiridonos, 12210,Egaleo, Athens, Greece, Tel.: 30-210-5385511, E-mail: mgian@teiath.gr*

²*Laboratory of Food Chemistry and Technology, School of Chemical Engineering, National Technical University of Athens, Iroon Polytechniou 9, 15780 Athens, Greece*

## CONTENTS

## 7.1   INTRODUCTION

Food preservation technologies aim at hindering processes that cause either microbiological spoilage or/and deterioration of quality attributes of food products. This quality decline may be due to physical phenomena such as loss of water by evaporation or loss of texture, chemical/biochemical phenomena such as enzymatic degradation or microbial processes such as the growth of bacteria and molds (Pham, 2014).

Amongst other conventional methods of food preservation, low-temperature processes are well established the last years, since they cause minimum damage, especially of fresh, high-value foods. Therefore, freezing and chilling are considered as ideal food preservation techniques to produce safe and nutritious foods of superior quality, with extended shelf life. When talking about refrigeration, it is understood that the product continues to deteriorate, even at slower rates, since microorganisms are not eliminated and thus chilling conditions should be carefully maintained throughout its whole life cycle. Thus, refrigeration technology is a more integrated sector, including not only initial chilling and freezing, but also cold storage, distribution, transport, retail display and domestic storage.

Chilling is the unit operation in which the temperature of a food is reduced to between $-1°C$ and $8°C$, depending on the product's nature, structure and special characteristics. It is used to reduce the rate of biochemical and microbiological changes, impedes temperature-dependent activities, such as respiration, transpiration, and ethylene production, and inhibits enzymatic reactions, thus extending shelf life (Fellows, 2009). The main advantage of this technology is that it maintains the sensory, texture characteristics and nutritional properties of foods and, as a result, chilled foods are well accepted by consumers as being convenient, easy to prepare, and close to 'natural.' For all these reasons, chilling is the most popular method for the preservation of fresh foods, especially meat, fish, dairy products, fruit, vegetables, and ready-made meals (Dermesonluoglu et al., 2015).

On the other hand, freezing is the process in which the temperature of a food is reduced below its freezing point and a proportion of the water undergoes a phase change from liquid to solid, through ice crystallization. As a result, water mobility is significantly hindered and the concentration of dissolved solutes in unfrozen water lowers the water activity level ($a_w$). In this case, preservation and shelf life extension is accomplished by the combined effect of low temperatures and reduced water activity.

Freezing main benefits are correlated to minor changes to nutritional or sensory qualities of foods when correct freezing and storage procedures are followed.

When describing the advantages of the freezing process, the effects of both the freezing rate (that are related to the freezing process and equipment used) and the conditions of the subsequent frozen storage must be studied in a combined way. A very well-designed, effective freezing process that gives way to a high-quality frozen product can be easily counterbalanced and obscured by an inadequate frozen storage (Giannakourou, 2015).

Established chilling and freezing technologies have been frequently reviewed the last decade (Dermesonluoglu et al., 2015; Fellows, 2009; Giannakourou, 2015) and it is not in the scope of this chapter to overview in detail the state-of the art of these traditional preservation methods. Although most of these have undergone significant changes/modifications in order to improve their efficiency, their principles and main applications have remained the same.

Current consumer needs for minimal processing and high sensory and nutritional quality, as radical changes in requirements for environmentally friendly procedures, with the minimum energy and sources consumption have led to the need to develop novel, alternative methods of low-temperature preservation. In this context, during last decades, research is intensive in either upgrading existing methodologies or designing and implementing new approaches of chilling and freezing food products, using single step techniques or hybrid systems. In this chapter, the most important novel chilling and freezing technologies are described in separate sections, including their relative status compared to traditional methods, their advantages and drawbacks and their potential implementation in the food sector. Important issues on improvement of the current chill chain are also discussed.

## 7.2 DEVELOPMENTS IN TECHNIQUES AND EQUIPMENT

### 7.2.1 FREEZING

In literature, there are several ways proposed in order to categorize freezing methods, with the most popular being the means used to remove heat from foods. In this context, the most popular methods include air-, plate-, liquid immersion- and cryogenic freezing. An overview of the most common freezing methods, their advantages and disadvantages, are briefly presented in Table 7.1.

For decades, the efforts to optimize freezing processes have been focused on improving the efficiency of heat removal using different strategies and techniques such as lowering the cooling medium temperature, enhancing the surface heat transfer coefficient, or reducing the size of the product to be frozen (Fikiin 2009; Reid 2000). During freezing, due to water gradient, there is osmotic phenomena taking place, leading to water transfer from the interior of the cell into the extracellular medium, and ice crystal formation is the main phenomenon observed. Depending on the speed of heat removal, either slow freezing is obtained, leading only to extracellular ice crystals, or fast freezing leading to both intracellular and extracellular ice. The rate of change of temperature depends partly upon external conditions, such as the equipment used, and upon the size of the object to be frozen.

First attempts have been made so as to increase heat transfer during conventional freezing. Impingement technologies are being used to increase heat transfer 3–5 times that of a conventional tunnel utilizing axial flow fans. This process has been well suited for products that require very rapid surface freezing and chilling. Attempts have also been made to reduce cooling times by increasing the surface heat transfer coefficient, for example, by using radiative plates in conjunction with blast-air. The utilization of a dynamic dispersion medium as cooling medium has also been proposed as a way for intensifying air-blast freezing (Evans, 2012). However, more accelerated chilling systems rely on the maintenance of very low temperatures (−15 to −70°C) during the initial stages of the chilling process. This has been achieved by cryogenic liquids (Estrada-Flores, 2012). The potential benefits of more rapid freezing have led interest to cryogenic techniques. The most common method is by direct spraying of liquid nitrogen onto a food product when it is conveyed through an insulated tunnel. Newer designs have multiple liquid nitrogen spray zones, providing better control and eliminating the need for gas transfer fans. Immersion chilling and freezing that involve direct contact between food pieces and a concentrated solution have been proposed. Immersion chilling and freezing has not been developed on an industrial scale, mainly because of an inadequate control of mass transfer between the product and the refrigeration solution (Table 7.1).

**TABLE 7.1** Categories of Common Freezing Techniques with Different Cooling Media: Advantages and Disadvantages: An Overview

| Main categories | Equipment/Advantages | Disadvantages |
|---|---|---|
| *Air freezing:* | *Batch blast freezers:* | Moisture from the food is transferred to the air building up as ice on the refrigeration coils, high dehydration losses ($<5\%$) due to the large volumes of recycled air; Freezer burn and oxidative changes to unpackaged/individually quick frozen (IQF) foods. |
| It uses cold air of about $-23$ to $-30°C$, that circulates either naturally (natural convection, heat transfer coefficient: 5–10 W/(m²K) or with the aid of fans (forced convection, heat transfer coefficient: 20–30 W/(m²K) | Simple and commonly used compact equipment, low capital cost, high throughput (200–1500 kg/h), flexible in terms of food shape and size. | |
| | *Belt freezers (spiral freezers).* | |
| | Continuous flexible mesh belt which is formed into spiral tiers and carries food up through a refrigerated chamber. | |
| • Batch or continuous. | Require relatively small floor-space and have high capacity, automatic loading and unloading, low maintenance costs and flexibility to freeze a wide range of foods (including pizzas, cakes, pies, ice cream, whole fish and chicken portions). | Restricted to products of small particles of uniform size and shape (such as peas, green beans, strawberry slices, corn kernels, diced carrots). |
| | Fluidized bed freezers: (modified blast freezers): Force cold air up and under the product at high enough velocity to "fluidize" the product | |
| | Low requirements for room space, great overall heat transfer surface of the fluidized foods, high surface heat transfer coefficients of small particles of uniform size and shape, easy and rapid freezing due to the small sizes and thermal resistance of Individual Quick Frozen (IQF) products, good quality of the final product with low possibility of freezer burn. | |

**TABLE 7.1** (Continued)

| Main categories | Equipment/Advantages | Disadvantages |
|---|---|---|
| *Plate freezing:*<br>The refrigerant or the cooling medium is separated from the materials to be frozen by a conducting material, often a steel plate (contact freezing) (heat transfer coefficient: 50–100 W/ (m²K)<br><br>• Batch or continuous. | Relatively low initial cost, good quality retention.<br><br>*Horizontal or double-plated systems:* Further increase of the heat transfer rate to obtain higher quality, suitable for processed food products (such as fish products).<br><br>Vertical systems are suitable for unpackaged material with a flexible shape (such as fish). Products can be filled in between the plates by weight or volume.<br><br>Scraped-surface systems are suitable for ice cream where the liquid or semisolid food is frozen on the surface of the freezer vessel, and the rotor scrapes the frozen portion from the wall. | Slow freezing rate, difficulty in maintaining good contact between the materials being frozen and the heat exchange surface, not suitable for all food products (flat packages or flat food portions such as blocks of fish flesh, beef burgers, dinner meals). |
| *Liquid immersion freezing:*<br>Food is immersed in a low-temperature solution to achieve fast temperature reduction through direct heat exchange.<br><br>Food, usually prepackaged, is sprayed with the cold solution, that remains in a liquid form throughout the process (propylenoglycol, glycerol, salt solutions CaCl₂ or NaCl, sugar or alcohol solutions) (heat transfer coefficient: 50–100 W/(m²K)) | Low investment and operational costs, high heat transfer rate, easy to introduce in an on-line continuous procedure, fine ice crystal system in foods, suitable for products of non-uniform shape, suitable for orange juice condensates, poultry (especially in the initial stages of freezing, fish tissues.<br><br>*Brine freezers:* Use of super-saturated solutions for maximum surface contact by immersing the product into a liquid freezing agent, especially for irregular shapes (such as crabs).<br><br>*Scraped-surface freezers:* Suitable for liquid or semi-solid foods (such as ice cream). In ice cream manufacture, the rotor scrapes food from the wall of the freezer barrel and simultaneously incorporates air or air can be injected into the product.<br><br>Freezing is very fast and up to 50% of the water is frozen within a few seconds, very small ice crystals, which are not detectable in the mouth and thus gives a smooth creamy consistency to the product. | Limited applicability due to inadequate control of mass transfer (uncontrolled solid uptake from the cooling medium), industrial applications remain centered on sodium chloride solutions used with products such as fish, the selection of the liquid (need to protect the food due to direct contact with the refrigerant), high danger of freezer burn, additional costs (need to wash or otherwise remove the media form package surfaces). |

| Main categories | Equipment/Advantages | Disadvantages |
| --- | --- | --- |
| *Cryogenic Freezing:*<br><br>The immediate contact of the food with the refrigerant (or cryogen), while the latter changes phase during freezing (liquid nitrogen changes from a liquid to a gas, while solid $CO_2$ changes from the solid to the gas phase) (heat transfer coefficient: <200 W/(m$^2$ K)) | Low capital costs (approximately 30% of the capital cost of mechanical systems), smaller weight losses from dehydration of the product (0.5% compared with 1.0–8.0% in mechanical air-blast systems), rapid freezing which results in smaller changes to the sensory and nutritional characteristics of the product, the exclusion of oxygen during freezing, rapid startup and no defrost time, low power consumption<br><br>Cryogenic gases can also be used to produce a hard, frozen crust on a soft product to allow for easier handling, packaging or further processing. | High cost of cryogens and their environmental and safety impact.<br><br>The setup cost of a cryogenic freezing system is approximately one-fourth of the cost of its mechanical counterpart; however, the operating costs are almost eight times.<br><br>Limited use to high-value products, such as shrimps, crab legs. |

An alternative classification based on the rate of movement of the ice front are slow freezers and sharp freezers (0.2 cm/h) (including still-air freezers and cold stores), quick freezers (0.5–3 cm/h) (including air-blast and plate freezers), rapid freezers (5–10 cm/h) (including fluidized-bed freezers and ultra rapid freezers) (10–100 cm/h), and cryogenic freezers (Fellows 2000, 2009; Fikiin 2009; Giannakourou & Giannou, 2015; Hui et al., 2004; James & James, 2003; Karel & Lund, 2003; Lucas & Raoult-Wack, 1998; North & Lovatt, 2012; Rahman & Velez-Ruiz ,2007; Ramaswamy & Marcotte, 2006).

Freezing is one of the most widely used preservation method for the production of end products of high sensory and nutritional quality. However, besides being inappropriate for specific biological tissues, in several cases, it can cause significant deterioration of the initial raw material, mainly due to the crystallization process (James et al., 2015; Kiani & Sun, 2011; Kiani et al., 2013b). It is well known that the formation of large ice crystals (Fellows 2009) can severely damage the structure of frozen tissue. The size, shape, and distribution of the ice crystals depend on the rate of ice nucleation and crystal growth, that is, crystallization is highly dependent on freezing kinetics and, therefore, it is important to design an optimized procedure to minimize tissue injuries. In this context, the main focus of novel methods proposed for food freezing is the increased heat removal rates (e.g., impingement, cryogenic technologies, pressure shift freezing, etc.) in order to produce tiny and numerous ice crystals, avoiding food quality degradation. Other alternative ways to achieve the same crystal structure is the lowering of the refrigeration medium temperature or the reduction of food items to be frozen (Fikiin 2009; Otero et al., 2016). As Fikiin (2009) points out, there are several issues that innovations within frozen food sector attempt to address, such as that, despite the numerous technologies presented, only a few have already been successfully implemented in industry (Otero et al., 2016). Moreover, classical methods can be significantly improved by incorporating mathematical modeling and studying systematically substantial freezing mechanisms (e.g., heat transfer, fluid flow, ice crystal formation, etc.). All the above need to be looked from a practical viewpoint, aiming at introducing new freezing equipment, more convenient, energy saving and competitive on the global markets. Additionally, some innovative techniques aim at modifying food physical/chemical structure. In the following subsection, some of the most representative novel freezing technologies are reviewed (Table 7.2) and some examples of their application in recent literature will be briefly presented.

**TABLE 7.2**   Review Studies on the Use of Novel Techniques of Freezing for Foods*

| Novel technique | Key review studies |
| --- | --- |
| 1. Magnetic fields freezing | Otero et al., 2016; Woo and Mujumdar, 2010 |
| 2. Ultrasound assisted freezing | Cheng et al., 2015; Tao and Sun, 2015 |
| 3. High pressure assisted freezing | Fernández et al., 2006; Le Bail et al., 2002a; Norton and Sun, 2008; Otero and Sanz ,2006 |
| 4. Dehydrofreezing | James et al., 2014a |
| 5. Electroacoustic freezing | Le Bail et al., 2012 |
| 6. Impingement freezing | Salvadori and Mascheroni, 2002 |
| 7. Hydrofluidization | Fikiin, 2009 |
| 8. Microwave assisted freezing | WäpplingRaaholt et al., 2015; Xanthakis et al., 2014d |
| 9. Radio frequency freezing | Marra et al., 2015 |
| 10. Pulsed electric fields (PEF) | Barba et al., 2015; Siemer et al., 2015 |
| 11. Antifreeze proteins (AFPs) or ice structuring proteins (ISPs) | Ramløv and Johnsen, 2014; Ustun and Turhan, 2015; Wang and Sun, 2012 |
| 12. Ice nucleation proteins (INPs) | Kawahara, 2002 |
| 13. Cryoprotectants (the case of trehalose) | Le Bail et al., 2012 |

*James et al. (2015).

## 7.2.1.1   Magnetic Fields (MFs)

The effort of food freezing industry is to improve heat removal rates and being able to control the nucleation procedure. One of the techniques proposed to accomplish the above is the application of magnetic fields (Woo & Mujumdar, 2010). In conventional equipment, during freezing, undesirable water migration and mass transfer take place within foods causing dehydration and quality degradation. Magnetic fields allow for retaining water within cells, by delaying ice crystallization and leading to supercooling below the initial freezing point (Fellows, 2009). Actually, MFs have the unique advantage of being implemented and actually applied in food freezing industry (Ryoho Freeze Systems Co., 2011; ABI Co., 2012). The equipment is described established by several patents (e.g., Mihara and Nakagawa, 2012; Norio and Satoru, 2001; Owada, 2007; Owada and Kurita, 2001; Owada and Saito, 2010). However, recently, a serious concern is arisen (Wowk, 2012) concerning the Cells Alive System, already patented by a Japanese Company (ABI Corporation Ltd, Chiba, Japan), given the small-filed strengths used. This system is not a shelf standing equipment, but is applied as adjunct to an existing freezing process, and is claimed to delay formation of ice crystals through oscillating magnetic field action on polarized water molecules (Owada, 2007; Owada and Kurita, 2001), leading to a smaller ice crystal size, and thus limited tissue disintegration. An overview of the existing patents developed for the application of MFs to improve freezing of foods, living cells, chemicals, pharmaceuticals, etc., is illustrated in Otero et al. (2016), claiming that MFs influence water crystallization and allow large supercooling.

In order to understand how magnetic field influence ice crystals growth, let's consider some properties of water molecules. Water molecule, consisting of 2 hydrogen atoms covalently bonded to one atom of oxygen, is a diamagnetic material and hence is susceptible to be magnetized by a magnetic field. Having strong polarity, water molecules interact with each other, and form intermolecular hydrogen bonds, which are much weaker than covalent bonds. These H-bonding intermolecular forces account for most of the anomalous properties of water, such as volume expansion due to freezing, high freezing point, etc. (Otero et al., 2016). Under a magnetic field, the hydrogen bonds between the water molecules are stronger, giving a more ordered and stable configuration. The above observation was confirmed in recent literature by analyzing magnetized and non-magnetized water by spectroscopic techniques (Iwasaka and Ueno, 1998) attempting to explain the modifications depicted in water spectra, as a result of the magnetically induced electric fields on electric dipoles in ice-water boundaries, corresponding to water samples exhibiting more hydrogen bonding (Iwasaka et al., 2011; Pang et al., 2012).

An important finding is that the main influencing parameters on the changes observed include MF strength and frequency, the exposition time and the temperature, as well as that this effect reaches a maximum after a given time (Otero et al., 2016). Another point that needs to be mentioned is the so-called memory effect of magnetized water, that is, MF effects do not disappear immediately after MF removal (Pang and Deng, 2008, 2009; Pang et al., 2012).

Concerning MFs application on food freezing, literature data is rather scarce, especially when static magnetic fields are studied. Two studies focusing on freezing of foodstuffs, when oscillating weak magnetic fields are applied, reported no actual effect (Suzuki et al., 2009; Watanabe et al.,

2011). An overview of the experimental data studying the effect of MFs on food freezing is given in Otero et al. (2016).

### 7.2.1.2 Ultrasound-Assisted Freezing (UAF)

The application of ultrasound, and more specifically, power ultrasound to food freezing is a relatively new subject, although its advantageous use in other cases has attracted considerable interest in food science and technology due to its promising effects in food processing and preservation (Delgado and Sun 2012).Power ultrasound, a form of ultrasound wave with low frequency (18–20 kHz to 100 kHz) and high intensity (generally higher than 1 W/cm$^2$), has proved to be extremely useful in crystallization processes, since, under its influence, a much faster and uniform ice formation occurs, also leading to a much shorter freezing time. Power ultrasound plays an important role in the initiation of nuclei and also in the subsequent crystallization. Thorough reviews on ultrasonics' application in food freezing have been published by many authors (Cheng et al., 2015; Delgado and Sun, 2012; Kiani et al., 2014; Tao and Sun, 2015, Xin et al., 2015; Zheng and Sun, 2006).

It is not intended here to provide detailed information regarding the fundamentals of ultrasound. An outline of the basic principles will be presented to better understand the benefits of its application to food freezing.

The transmitting of the sound waves across the aqueous phase can cause the occurrence of cavitation if its amplitude exceeds a certain level. Therefore, the ultrasound creates cavitation bubbles throughout the product, which promotes a more even ice nucleation and fragments ice crystals already present into smaller crystals (James et al., 2015). Cavitation has a double role: it can lead to the production of gas bubbles but also the occurrence of microstreaming. While the former can promote ice nucleation, the latter is able to accelerate the heat and mass transfer process related to the freezing process. Ice crystals will fracture when subjecting to alternating acoustic stress, consequently leading to products of smaller crystal size distribution, which is one of the most important targets of the freezing industry regarding the retention of high quality of frozen food stuffs (Kiani et al., 2013b; Zheng and Sun, 2006). In addition, ultrasound can inactivate some enzymes (Islam et al., 2014), avoiding the necessity of blanching of some delicate tissues prior to freezing.

Recent experimental studies indicate that power ultrasound accelerate the freezing process of fresh food products, mainly through its ability in enhancing the heat and mass transfer process (Islam et al., 2015; Kiani et al., 2013a, 2015; Xu et al., 2015). Another area of concern is the quality improvement observed in frozen products, pretreated with ultrasound (Tu et al., 2015; Xu et al., 2015). An example of effective use of this method is in the ice cream industry, where ice crystal size is adequately reduced, high heat transfer rates are accomplished thanks to elimination of encrustation phenomena, etc. (Awad et al., 2012; Zheng and Sun 2006).

Summarizing, the main aspects of the application of power ultrasound in food freezing process include the initiation of nucleation, the control of size of ice crystals, the acceleration of freezing rate and the improvement of quality of frozen foods (Cheng et al., 2015).

Apart from freezing, this technique has been proposed for freeze concentration and freeze drying processes in order to control crystal size distribution of frozen food products. Since ultrasound technology is used to transmit mechanical energy and accelerate mass transfer processes, it is an interesting technology in order to complement standard freeze-drying processes (Schössler et al., 2012). Freezing is a crucial step in freeze drying because it determines the size and distribution of ice crystals in frozen materials, influencing the efficiency of drying process and the quality of final freeze dried products (Cheng et al., 2015). Hence, power ultrasound can be a means of improving this freezing step. In this context, numerous publications are recently available elucidating this potential of ultrasound pretreatment (Cheng et al., 2014, 2015; Garcia-Perez et al., 2012; Santacatalina et al., 2015; Wang et al., 2013).

At this point, one should point out that all published studies regarding ultrasound-assisted freezing actually refer to its application during immersion freezing and at a bench top scale (James et al., 2015), mainly investigating the effect of ultrasound process parameters on heat transfer phenomena and quality attributes. Little information is provided regarding this method commercialization. Kiani et al. (2014) suggest some possible ways for introducing ultrasonic devices to commercial freezing equipments, indicating different locations of the process, where the ultrasonic device can be attached to the freezer. However, as no widespread industrial application is known, more research is required to verify its feasibility and practicality (Delgado and Sun 2012).

### 7.2.1.3 High Pressure Shift-Freezing (High Pressure Assisted Freezing)

Applying high pressure (HP) (typically of 200 to 400 MPa) and particular 'pressure-shift' freezing and thawing (Le Bail et al., 2002b) on several food products has been the subject of numerous studies in recent literature (Cheftel et al., 2000, 2002; Picart et al., 2004, 2005). According to Le Bail et al. (2002a), three areas of food processing are of particular interest in HP including food texture, preservation and phase change. HP freezing and thawing processes are related to these three applications. In the same published work, details are presented regarding the six different freezing and thawing processes based on phase transition under pressure. Taken into account the numerous researches on this subject the last years, in this chapter we will be limited to a simple overview of the method, mentioning the most important aspects.

HP treatment includes the cooling of a sample under pressure until it reaches its phase change temperature at the pressure of interest, that is, freezing process is accomplished under a constant temperature (Le Bail et al., 2002a) The main advantage is that, in contrast to what happens at atmospheric pressure, water volume does not increase, and, most importantly, ice exists in a 'vitreous' non-crystalline state, which leads to reduced tissue damage (James et al., 2015). At a pressure of 200 MPa the freezing point drops to about −22°C (Fikiin, 2009), which enables a depth of vitrification of about 200 μm, so that objects with a thickness of up to 0.4–0.6 mm could be well frozen.

According to Fikiin (2009), the main potential advantages of high-pressure freezing are as follows:
- freezing point depression and reduced latent heat of phase change;
- short freezing times and resulting benefits (e.g., microcrystalline or vitreous ice);
- inactivation of micro-organisms and enzymes, and structure modifications with no essential changes of nutritional and sensory quality.

What actually happens is that the use of high pressure leads to high degrees of supercooling possible resulting in even and fast ice nucleation and growth all over the sample on pressure release. As a result, in contrast with the conventional methods in which an ice front moving through the sample is produced, fine ice crystals are formed, improving the quality of the final frozen product (Kiani and Sun, 2011; Norton and Sun, 2008). The application of high hydrostatic pressure to control and enhance the freezing process has been an interesting subject of research in recent decades (Urrutia et al., 2007; Van Buggenhout et al., 2006).

There are three distinct forms of freezing using high pressure, high pressure-assisted freezing (HPF), high pressure shift-assisted freezing (HPSF), and high pressure induced crystallization (HPIC) or pressure-induced solid-solid phase transition (Fernández et al., 2006, James et al., 2015). Pressure-assisted means phase transition under constant pressure, higher than the atmospheric, pressure-shift means phase transition due to a pressure release, and pressure-induced means phase transition initiated by a pressure increase and continued at constant pressure (Knorr et al., 1998). At HPSF, releasing the pressure once the temperature of the food reduces to the modified freezing point results in a high supercooling effect and the ice nucleation rate is greatly increased. The main advantage is that the initial formation of ice is instantaneous and homogeneous throughout the whole volume of the product. Therefore, high pressure shift freezing can be especially useful to freeze foods with large dimensions where the effects of freeze cracking caused by thermal gradients can become harmful (Norton & Sun 2008). It is generally accepted that HPSF is the more advantageous high-pressure freezing process, thanks mainly to the level of supercooling reached throughout the sample after expansion (Fernández et al., 2006). A lot of studies have been published to demonstrate the effectiveness of HPSF, especially when compared to conventional techniques (Alizadeh et al., 2009; Choi et al., 2016; Fernández et al., 2006, 2008, Hong and Choi 2015, Otero et al., 2009; Su et al., 2014; Tironi et al., 2010; Volkert et al., 2012). On the other hand, High Pressure Induced Crystallization (HPCI) aims at changing ice I to ice III and back to ice I. According to Van Buggenhout et al. (2007), this kind of transient phase transitions could provide a valuable tool for reducing microbial load in the frozen state, and be of large benefit for frozen food applications in which a thermal decontamination step (before freezing) is to be avoided due to quality injuries.

Despite the advantageous quality of high-pressure frozen products, more research is necessary, not only on the theoretical base regarding the influence of low temperature/high pressure on food tissue, but also on the optimization of equipment/process design in order to be implemented as a common industrial practice. At a commercial level, Japan is at the forefront of HPSF application for food processing, with the United States and Europe also investigating the commercialization of this technology. In the widespread use of HPSF technology, the biggest obstacle is the high capital costs, due to the application of very low temperatures. Furthermore, constant and precise monitoring is also necessary for improving product quality and stability of the operation (Sun and Zheng, 2006; Otero and Sanz, 2012) actually recommended that for method wider application, it is necessary first to optimize the efficiency in the heat removal under pressure since the long pre-cooling step, before nucleation, is the limiting factor which must be reduced to make the process competitive.

### 7.2.1.4   Dehydrofreezing

Dehydrofreezing is a well-established commercial method to reduce cost of shipping, handling, and storage of fruits and vegetables (Sun and Zheng, 2006) and it consists of freezing a partially dehydrated food. In the particular case where this dehydration step is accomplished through osmosis of the tissue, the whole procedure may also be termed as osmo-dehydrofreezing (James et al., 2014a). Although this process is not new, as far as its theoretical background is concerned, dehydrofreezing has attracted a lot of research interest the last decades and a lot of work has been published since (James et al., 2014a) offer a very thorough review of this interesting technique, addressing issues such as the different pre-treatments (physical or osmotic based) and freezing methods, their comparative evaluation, the potential commercial production and development, and finally, the effects of dehydrofreezing compared to conventional freezing. In this chapter, we will focus on dehydrofreezing, when an osmotic step is performed prior to freezing.

Since fresh fruits and vegetables contain more water and have much more delicate tissue structure than meat, they are likely to undergo significant damage during freezing (Xin et al., 2015). Thus, removing partially water before freezing allows the ice crystals to form and ice to expand in a more uniform and consistent way, minimizing cellular disintegration (James et al., 2015). Furthermore, a reduction in moisture content would reduce the amount of water to be frozen, thus lowering refrigeration load during freezing.

During osmodehydrofreezing, the osmotic step before freezing aims at the lowering of the water activity and the partial dehydration of the food matrix; at the same time, by carefully designing the composition of the osmotic solution in which the food is immersed, a significant selective enrichment can be obtained and novel food products with functional properties may be produced (Giannakourou, 2015). In most works, a wide range of carbohydrates and salts have been used, as the main ingredients of the osmotic solutions and the effect of several controlling parameters (temperature, concentration of the osmotic solution, mixing, etc.) was assessed on mass transfer mechanism and quality attributes of several fruits and vegetables, prior to freezing (Blanda et al., 2009; Zhao et al., 2014, 2016) Another issue of concern of recent literature is the comparison of osmodehydrofreezing with conventional freezing of plant tissue, through systematic storage studies of osmotreated and untreated frozen samples (Dermesonlouoglou et al., 2007a, 2007b, 2007c, 2008, 2015; Giannakourou et al., 2001a), revealing the significantly shelf life extension accomplished through the application of the osmotic step prior to freezing.

Furthermore, a number of pre-treatments have been proposed in order to improve and maximize osmotic dehydration efficiency, including application of vacuum (Wu et al., 2009), ultrasound (Xin et al., 2014; Xu et al., 2014, 2015), pulsed electric fields (Parniakov et al., 2015, 2016), etc.

As far as commercial application is concerned, not much detail is available on respective osmodehydrofreezing systems. According to James et al. (2014a), diced and sliced vegetables and fruits intended for use in soups, potato pies, stews, casseroles, salsas and salads, and as pizza toppings are the most popular dehydrofrozen products for the time being. One innovation in commercial dehydrofreezing involves the agreement between EnWave Corporation and Bonduelle, a Canada-based frozen vegetable producer in order for the latter to develop and launch dehydrofrozen vegetables using EnWave Corporation's Radiant Energy Vacuum ("REV™") technology (Anon, 2013; Durance, 2013). This technology is a form of vacuum-microwave drying, and has the advantage of being rapid and operating at low temperatures.

### 7.2.1.5 Electrostatic Assisted Freezing

Electrostatic field-assisted freezing is a very recent proposal that investigates the impact of electrical disturbance in terms of phase change. Supercooling of several foods with application of an external electric field is studied and the preliminary results seem really promising regarding the area of improvement of the size of ice crystals (Le Bail et al., 2012; Xanthakis et al., 2014a, 2014b).

The principle of this method is that by applying an external electric field, the orientation of polar molecules such as water may be modified, allowing for the control of supercooling and ice crystallization. The main focus of recent literature is to investigate the possible improvement of the size of ice crystals in frozen food and the initiation of nucleation in super-cooled liquids (Le Bail et al., 2012; Woo and Mujumdar 2010). However, most published work is limited to model systems and this technique is only in its first steps to be applied in real food systems in a laboratory scale (Xanthakis et al., 2013).

### 7.2.1.6 Impingement Freezing (IF)

The impingement technique involves a or more than one high velocity (up to 50 m/s) air jet (or jets) directed on the product surface, in order to cause a change, that is, to diminish the thickness, or break up the static boundary layer of air surrounding the food product (Góral and Kluza 2009, 2012). As a result, heat exchange rate between the frozen product and the surrounding air, which becomes more turbulent, is significantly increased, causing a rapid growth in the freezing rate (Sarkar et al., 2004; Sarkar and Singh 2004a, 2004b). Most studies published on this area concern the calculation of heat and mass transfer coefficients, rather than describing IF advantages over the traditional freezing procedures (Erdogdu et al., 2005; Sarkar et al., 2004; Sarkar and Singh 2003, 2004b). Salvadori and Mascheroni (2002) analyzed impingement freezers performance and concluded that 'this operation helps to reduce processing times considerably, giving freezing times similar to those provided by cryogenic equipment.' In addition, operating costs are similar to those of traditional mechanical equipment, leading to the conclusion that using this impingement technology, the high cost of the additional refrigeration of air can be substantially counterbalanced. In recent literature, an effort is presented to design an efficient impingement system, where jets are designed to fluidize vegetables, in a pilot scale equipment. The specific design of the equipment is an issue of interest, frequently addressed in various papers (Góral and Kluza, 2009, 2012).

## 7.2.1.7   Hydrofluidization Freezing (HF)

The hydrofluidization freezing is actually a type of immersion freezing and was suggested and patented to combine the advantages of both air fluidization and immersion food-freezing techniques (Fikiin 1992, 2009). HF freezing uses a circulating system that pumps the refrigerating liquid upwards, through orifices or nozzles, in a refrigerating vessel, thereby creating agitating jets. These form a fluidized bed of highly turbulent liquid and moving products, and thus evoke extremely high surface heat transfer coefficients (Fikiin 2009; James et al., 2015). The main advantages are: the use of equipment of small size and the improved IQF freezing (Individual Quick freezing of small particles). The combination of these high coefficients with the use of small food samples leads to processes in which the transport phenomena within the food is affected by the fluid flow over the food samples (Giannakourou, 2015).

Several works have been published studying mostly heat and mass transfer coefficients, as long as the design of such hydrofluidization systems (Belis et al., 2015; Peralta et al., 2009, 2012), which however are not yet commercially introduced in food market.

As fluidizing agents, unfreezable liquid refrigerating media (brines, sugar-ethanol cooling media) or pumpable ice slurries, described as mixtures of microcrystals of ice in a carrier liquid that do not freeze under the operating conditions used (Torres-De María et al., 2005), can be alternatively used (Fikiin, 2009). Sodium chloride, ethanol, ethylenoglycol and propylene glycol are the four most commonly used freezing point depressants in industry (Bellas and Tassou, 2005; Kauffeld et al., 2010).

As far as heat transfer coefficients obtained by the HF method, values of $h$ up to 900 $W \times m^{-2} \times K^{-1}$ were obtained by Fikiin (2009) for freezing of fish and vegetables by HF, which compares to heat transfer coefficients, previously reported for forced convection immersion chilling (680–2800 $W \times m^{-2} \times K^{-1}$) and rotating drum spray freezing (280–850 $W \times m^{-2} \times K^{-1}$) of vegetables, both in aqueous media (Verboven et al., 2003). The goal of the ice slurry involvement is to provide an enormously high surface heat transfer coefficient (of the order of 1000–2000 $W \times m^{-2} \times K^{-1}$ or more), excessively short freezing time and uniform temperature distribution in the whole volume of the freezing apparatus.

According to Fikiin, 2009, besides the aforementioned advantages, there are more important issues obtained through HF:

- the critical zone of water crystallization (from $-1°C$ to $-8°C$) is quickly passed through, protecting cellular tissues from dramatic injury;
- as product freezes immediately in a solid crust, osmotic transfer is significantly limited and appearance is of high quality;
- by appropriately selecting composition of the HF media, new products can be formulated, with extended shelf life and improved sensory/nutritional characteristics;
- the operation is continuous, easy to maintain, convenient for automation, and further processing or packaging of the HF-frozen products is considerably easier;
- when ice-slurry are used as HF agents, they may easily be integrated into systems for thermal energy storage, accumulating ice slurry during the night at cheap electricity charges;
- the HF freezers use environmentally-friendly secondary coolants (for instance, syrup type aqueous solutions and ice slurries) and the refrigerant is closed in a small isolated system, in contrast to the conventional and harmful air fluidization technology of HCFCs and expensive HFCs where there is a much greater risk for emission to the environment.

### 7.2.1.8   Microwave Assisted Freezing

In Wäppling-Raaholt et al. (2015) an overview is provided on the microwave preservation and processing applications in the food industry. The main principle-governing microwave assisted freezing is the influence of lower frequency microwaves (915 MHz vs. 2450 MHz) (MW) on water dipole rotation, and thus subsequent ice nucleation and formation. Jackson et al. (1997) studied the coupled interaction of microwave radiation and cryoprotectant concentration on ice formation. They developed an experimental setup which was able to cool rapidly by plunging the samples into liquid nitrogen while simultaneously irradiating it with a powerful coherent field, and found an enhanced vitrification in cryopreservation media at a relative slow cooling rate. Although results seemed promising, little relevant work has been published since Xanthakis et al. (2014c) designed a novel experimental setup in order to apply microwave radiation during freezing of a real food, in a similar way that Anese et al. (2012) applied radiofrequency technology to assist pork freezing. Therefore, pork tenderloin was considered for microwave assisted freezing on microstructure. The results revealed that apart from controlling the nucleation temperature, the final quality of the frozen product was improved as the damage of the meat microstructure was significantly improved under the tested conditions. However, since recent literature lacks evidence of advantageous application of microwave-assisted freezing (in common with radiofrequency assisted freezing) further investigation is needed in the direction of application of this innovative freezing process to other food, in order to compare to conventional systems and assess the feasibility of a commercial use.

### 7.2.1.9   Radiofrequency Assisted Freezing (RF Freezing)

Marra et al. (2015) present a brief overview of the radiofrequency (RF) heating mechanism. Similar to microwave-assisted freezing, the basic principle of this novel technique lies on the effect of low voltage RF pulses on ice nucleation and formation during freezing (Anese et al., 2012). In this work, which is the only one available on real food tissue, freezing of pork meat was performed in a RF pilot equipment modified to allow food immersion in a liquid nitrogen spray. During freezing, pulsed RF treatments were applied along with cryogenic fluid flow (cryogenic freezing) and air freezing as controls. Results on the microstructure of meat revealed that the tissue exhibited a better cellular structure when RF was applied, with smaller ice crystals, mainly located at intracellular level. This was attributed to the ability of RF to depress the freezing point thus producing more nucleation sites.

At the moment, this methodology is in a very early, laboratory level stage and further research is needed to assess its benefits over the classical methods and propose its commercialization at RF assisted freezing is possible by using low voltage pulses (2 kV).

### 7.2.1.10   Pulsed Electric Fields (PEF)

In recent literature, there are numerous studies investigating the feasibility of pulsed electric fields (PEF) for different applications in food industry. Barba et al. (2015) have recently reviewed the current applications and the potential developments regarding this innovative technique. Principles of action of this novel technology are detailed in Siemer et al. (2015).

Pulsed electric fields (PEF) technology consists on an electrical treatment of short time (from several nanoseconds to several milliseconds) with pulse electric field strength from 100 to 300 V/cm to 20–80 kV/cm. There are two main areas of concern: at high electric fields (>20 kV/cm), PEF application is studied as an alternative to traditional thermal processing to inactivate microorganisms and quality related enzymes, with the advantage of slightly deteriorating sensorial, nutritional and health-promoting attributes of liquid food products (Sánchez-Vega et al., 2015). On the other hand, at low electric fields, the cell membrane is electrically perforated (electroporation or electropermeabilization) and loses its semi-permeability temporarily or permanently (Barba et al., 2014; Deng et al., 2014), leading to enhanced mass transfer and freezing rates and, also, cause modification of main properties of the food matrix. The character of electroporation depends on the processing parameters and properties of the treated material (Wiktor et al., 2015). Therefore, due to its mechanism and its influence on cell tissue, PEF is studied as an alternative way to enhance the freezing process. In this context, it has been evaluated as a technique to improve the uptake of cryoprotectant and texturizing agents (James et al., 2015). Shayanfar et al. (2013, 2014) studied the interaction of pulsed electric fields (PEF) with different cryoprotectant (e.g., glycerol and trehalose) and texturizing agents ($CaCl_2$) in quality retention plant products (carrot discs and potato strips) and found that 'increasing the permeability properties by PEF may lead to better accessibility of intracellular materials to freezing and thus reducing the freezing time, leading to better maintaining the texture after thawing.' Recently, Parniakov et al. (2016) applied a PEF-assisted osmotic dehydration on apple juice -glycerol solutions using different concentrations of glycerol and time of dehydration, and the main characteristics of freezing-thawing processes and texture of were evaluated. Results showed that PEF-assisted osmotic dehydration results in noticeable acceleration of the freezing/thawing processes as well as strengthening of texture for the osmotically dehydrated PEF-treated tissues after defrosting as compared with those for untreated samples. Several other studies have been recently published, investigating the advantages of applying PEF technology, as a sole or combined with another pretreatment before freezing of fruits and vegetables (Dymek et al., 2015; Mok et al., 2015; Parniakov et al., 2015; Wiktor et al., 2015).

Although PEF technique seems rather effective as a freezing pretreatment, mainly due to the cell disintegration achieved, we agree with James et al. (2015) that further research is needed to test the potential applicability in an industrial level, since all the work published till now refers to laboratory scale. PEF could be better considered as a combined pretreatment, and thus its future development is also strongly related to the interest in cryoprotectants and texturizing agents, a combination that has been demonstrated to offer the maximum of benefits.

### 7.2.1.11  Antifreeze Proteins (AFPs) or Ice Structuring Proteins (ISPs)

The interest in the use of antifreeze proteins started in the 70s, when they were found in the blood of fish living in polar and northern coastal waters, and served to lower the freezing point of their blood to below the freezing point of seawater, without significantly increasing the osmotic pressure of the plasma (Ustun and Turhan 2015). Recently, the term 'ice structuring proteins' has been alternatively proposed to describe in a more accurate way their functionality, which is mostly related to the control of the size, shape and aggregation of ice crystals (Hassas-Roudsari and Goff 2012; James et al., 2015), and not to the inhibition of freezing, implied by their name 'antifreeze proteins.'

The properties, characteristics and the potential applications of ISPs have been thoroughly presented and reviewed since early 80s (Buckley and Lillford, 2009; Crevel et al., 2002; Griffith and Ewart, 1995; Hassas-Roudsari and Goff, 2012; Ramløv and Johnsen, 2014; Ustun and Turhan, 2015; Wang and Sun, 2012; Venketesh and Dayananda, 2008). Apart from fish, proteins with similar activity and some with crystal growth inhibition activity were discovered in other biological systems such as animals, insects and plants. Besides lowering the freezing temperature, AFPs have been found to significantly hindericerecrystallization and prevent large ice crystal growth, maintaining therefore initial quality during frozen storage and thawing. There are two main types of AFPs: glycoproteins and nonglycoproteins, with the latter and being classified essentially according to Davies and Hew (1990) into types I to IV and antifreeze glycoproteins (AFGPs) (Venketesh and Dayananda, 2008). A review on the antifreezing activity of AFGP has been published by Carvajal-Rondanelli et al. (2011). There are also insect and plant-derived AFPs. These macromolecular AFPs present a great diversity in terms of their protein structures and composition of the amino acids (Ustun and Turhan, 2015).

During recent years, numerous studies have been published on the extraction and application of AFPs, with the main goal to demonstrate that the incorporation of these proteins in frozen foods could protect them from freeze damage, and extend their shelf life. The main categories of food studied till now include frozen dairy products, and especially ice cream (Regand and Goff, 2006; Soukoulis and Ian, 2014), meat, fish and frozen dough products (Ding et al., 2014; Zhang et al., 2008, 2015). On the other hand, the source of AF proteins was investigated and among other vegetative tissues, carrot was found to produce and accumulate those molecules, in order to protect its own tissue during winter (Ding et al., 2014; Zhang et al., 2008). In a recent work, Ding et al. (2015) studied the effect of barley antifreeze protein on the thermal properties and water state of dough during freezing and freeze-thaw cycles, and concluded that this AFP was highly effective in doughs by improving their thermal properties and restrict water mobility and distribution. Cruz et al. (2009) studied the use of an AFP-I in a vacuum impregnation process of watercress, in order to test the potential benefits on the quality of this frozen leafy vegetable microstructure and texture. They concluded that 'AFP-I application, as a food additive, can have a major impact in frozen watercress quality since it reduced the size of the ice crystals, avoiding less damage to the microstructure, and thus improving its texture after thawing.'

Despite being a subject that has attracted a lot of theoretical interest in recent years, when seeking for experimental studies on real food matrices, there are actually few works that have studied the advantages of using AFPs for improving frozen food quality during freezing and subsequent storage. Further research on a wider range of foodstuffs is deemed necessary in order to assess the feasibility of extending AFPs use in a commercial level (James et al., 2015).

### 7.2.1.12 Ice Nucleation Proteins (INPs)

During freezing, ice formation initiates by ice nucleation that can be promoted by the presence of foreign particles and act as ice nucleation activators (INAs) (Wang and Sun, 2012). The mechanism in which they act on frozen foods is totally different from that of AFPs. INPs have been found to be produced by certain Gram-negative bacteria that promote the nucleation of ice at temperatures higher than the initial freezing point. They raise the temperature of ice nucleation and reduce the degree of supercooling, decreasing the freezing time (Zhang et al., 2008). Moreover, INAs have been shown to create large and long ice crystals in ordered directions (Kiani and Sun, 2011), enhancing the texture of frozen foods. A review concerning 'The Structures and Functions of Ice Crystal-Controlling Proteins from Bacteria' has been published by Kawahara (2002).

There are several works on the extraction and the expression of INP's of different species of ice nucleation bacteria, showing that the majority of INPs are produced by certain strains of a few plant-associating bacterial species, particularly members of the genera *Pseudomonas, Panteola* (Erwinia), and *Xanthomonas* (Lagzian et al., 2014; Li et al., 2012; Vanderveer et al., 2014).

As far as their application in the food sector, many INA bacterial cells have been studied for their effect on overall quality of frozen foods during cold storage (Hassas-Roudsari and Goff, 2012; Li and Lee, 1995). For example, samples of egg white when freezing at −10°C underwent supercooling lower than −6°C, but when INA bacterial cells (*Erwiniaananas*) were added, the samples showed only a slight degree of supercooling (Arai and Watanabe, 1986; Watanabe and Arai, 1987).

Despite the positive results concerning the effectiveness of INA bacteria on the quality of frozen foods, one major concern to their applications in the food industry refers to their source, i.e., bacteria and thus safety issues must be first solved. Bacterial ice nucleators must be environmentally safe, non-toxic and non-pathogenic and palatable (Li and Lee, 1995). If the whole bacterial cells are added, it is necessary to ensure that inedible microorganism is completely removed from food prior to consumption (Wang and Sun, 2012).

### 7.2.1.13  Cryoprotectant Agents

Cryoprotectants are compounds used to prevent the stresses that occur in the lipid bilayers during the freezing that can lead to physical and chemical damage of the food tissue. Although it may not be considered a novel process, recently new approaches have been proposed, involving either new substances, or combinations with new techniques (James et al., 2015). There are several explanations for the protective mechanism, which are mainly related to the structure of these substances that contain multiple hydroxyl groups or, alternatively, explaining the phenomenon based on the vitrification theory (Degner et al., 2014), as it is presented in detail in a recent review published by Morais et al. (2016). Vitrification and cryostabilization theory is thoroughly presented by Zaritzky (2012). The most common cryoprotectant agents applied in food matrices, include sugars, polyols, some amino acids, other inorganic salts, such as potassium phosphate and ammonium sulfate, as presented by Jaczynski et al. (2012), especially for cryoprotection of aquatic products. As they also state 'however, many of them are not feasible for use because of high cost, food regulations, or adverse sensory properties.'

One of the newly proposed cryoprotectant, is trehalose, a stable, colorless, odor-free, and non-reducing disaccharide consisting of two glucose units linked in an $\alpha,\alpha$-1,1-glycosidic bond. It is easily found in nature in many survival forms of organisms, including bacteria, yeast, fungi, insects, invertebrates, and plants. It can serve as a source of energy and carbon and may participate in the actions of metabolism. As Le Bail et al. (2012) suggest, it can serve as an efficient cryoprotectant for frozen bakery products. Trehalose has been also studied in combination with glycerol and a Pulsed Electric Field system as a pretreatment prior to freezing potatoes and carrots (Shayanfar et al., 2013, 2014) and found to significantly improve final product's texture. Numerous studies have been presented examining the use of trehalose under Vacuum Infusion (VI), in order to improve solute uptake of vegetable (e.g., spinach leaves) or fruit tissues (Panarese et al., 2014, 2016; Velickova et al., 2013), or even a combination of VI with PEF for further improving the freezing tolerance of spinach (Phoon et al., 2008). Being already an approved food ingredient (Decision 2001/721/EC), trehalose is already commercially used in a variety of food matrices (cereals, confectionary, bakery products) as a reduced calorie sweetener, a anti-browning and anti-retrogradation of starch agent, and a cryoprotective substance, that lowers glass transition temperature of food, and, thus, stabilizes the tissue against freezing damage (Le Bail et al., 2012).

## 7.2.2  REFRIGERATION

Chilling is the unit operation in which the temperature of a food is reduced to between −1°C and 8°C, depending on the product's nature, structure and special characteristics (Fellows, 2000):

- 1°C to +1°C (fresh fish, meats, sausages and ground meats, smoked meats, breaded fish);
- 0°C to +5°C (pasteurized canned meat, milk, cream, yogurt, prepared salads, sandwiches, baked goods, fresh pasta, fresh soups and sauces, pizzas, pastries, unbaked dough);
- 0°C to +8°C (fully cooked meats and fish pies, cooked or uncooked cured meats, butter, margarine, hard cheese, cooked rice, fruit juices, soft fruits).

Chilling equipment is classified by the method used to remove heat into mechanical refrigerators and cryogenic systems presented in Table 7.3 (Dermesonlouoglou et al., 2016). Apart from the particular characteristics of each product, the selection of the appropriate equipment depends on several factors, such as the temperature reduction requirements, the refrigeration load, the desirable cooling rate, and equipment and operating costs. Therefore, in cases where multiple products with diverse optimal cooling conditions are received, the use of different means may be required. As far as the capital cost is concerned, liquid ice coolers can be the most expensive, followed by vacuum coolers, forced-air coolers, hydro-coolers and room coolers. Moreover, vacuum coolers exhibit the highest energy cost, followed by hydro-coolers, ice coolers and forced-air coolers. Labor, maintenance and other equipment costs should also be contemplated when comparing different cooling systems, especially if special packaging (waxed boxes or reusable plastic containers) is required. Operating conditions (environmental temperature, humidity, etc.) and product parameters (type, size, shape, composition) should also be considered (Pham, 2014). Chilling can be defined by pre-cooling and—following the pre-cooling—the cold chain throughout transportation, storage and preferably during retailing and domestic use. The main purpose of pre-cooling is to rapidly remove the undesired heat from freshly harvested/slaughtered goods in order to retard the rate of products' degradation. Pre-cooling differs from cold storage, where the temperature is simply maintained at a predetermined level, raising however, the energy cost of the cooling process. In order to minimize the refrigeration cost, there is a common practice not to pre-cool products to the ideal temperature but to reduce it at half or 7/8 the desired value (usually the operating temperature of the coolant) and then transfer products to a cold room for further cooling. As a general rule, the 7/8 cooling time requires three times longer when compared with half-cooling time (Dermesonlouoglou et al., 2016).

**TABLE 7.3**  Main Categories of Common Chilling Systems: Advantages and Disadvantages: An Overview

| Categories | Advantages | Disadvantages |
|---|---|---|
| 1. *Mechanical refrigerators:* | | |
| The mechanical refrigeration system is characterized by the circulation of a fluid (refrigerant) in a closed cycle and exhibits four basic components, a compressor, a condenser, an expansion valve and an evaporator. The refrigerators' parts (different types are available) are manufactured from materials presenting high thermal conductivity, such as copper, in order to achieve high rates of heat transfer and high thermal efficiencies. The chilling medium in mechanically cooled chillers may be air, water or metal surfaces/Batch or continuous | | |
| Room cooling | Simple, commonly applied in products sensitive to surface moisture with a relatively long shelf life, use for the curing of produce, cooling and storage in the same place (requiring less handling) | It necessitates a relatively large empty floor space between stacked containers, enough open spaces between products and good air speed to achieve the optimal cooling effect otherwise cooling is unreasonably long, food should be adequately chilled when loaded onto the vehicle, as the refrigeration plant is only designed to hold food at the required temperature and cannot provide additional cooling of incompletely chilled food, the cost of chill storage is high and to reduce costs, large stores may have a centralized plant to circulate refrigerant to all cabinets. |
| Products are exposed to cold air in a refrigerated room or transportation truck. Cold air is usually discharged into the room horizontally just below the ceiling, sweeps the ceiling and returns to the cooling coils after circulating through the produce on the floor by the refrigeration fans (air velocity: 60–120 m/min around and between cooling containers; cooling can also be accelerated through the use of ceiling jets or cooling bays). | Eutectic plate systems: Salt solutions (e.g., potassium chloride, sodium chloride or ammonium chloride) are frozen to their eutectic temperature and air is circulated across the plates, to absorb heat from the vehicle trailer. The plates are regenerated by re-freezing in an external freezer (refrigerated vehicles, especially for local distribution). | |

**TABLE 7.3** (Continued)

| Categories | Advantages | Disadvantages |
|---|---|---|
| *Forced-air cooling:* | | |
| High velocity cold air is distributed (horizontally or vertical) by air ducts and forced to flow through the containers in direct contact with each piece of produce. In this way, an air pressure difference is developed between the opposite faces of stacks of vented containers carrying heat away and resulting in rapid and even cooling. | *Tunnel-type, cold wall, serpentine cooling, evaporative forced-air cooling:* Most versatile and widely used, suitable for bulk produce and palletized products, 10 times faster than room cooling enabling better control of the cooling process. | There is a high potential for water loss from the fresh produce, due to air movement, unless high airflow rates are applied and humidity is kept near 100% (Condensation problems on the produce can be minimized as well by placing a cover on top of the stack of containers). |
| *Hydro-cooling:* | | |
| The produce, either in bulk or in containers, is carried on a conveyor through a shower of water | *Conveyor hydro-coolers:* The most economic method, simple and efficient, employed in fruits and vegetables (such as celery, asparagus, peas, radishes, carrots, peaches) | Requires particular attention to water quality and sanitation (use of mild disinfectants). Requires containers tolerant to water and chemicals rising therefore packaging cost, immediate re-handling of the products is usually necessary by shipping or transferring them to a cold storage room. |
| Fresh products are immersed in, flooded with, sprinkled or sprayed with cold water at a temperature of around 0°C, batch or continuous modes. | For optimal cooling and energy saving, hydro-coolers should also be adequately insulated. | |
| *Ice cooling:* | | |
| Ice, in crushed or fine granular form, or ice-slush is packed around or placed in direct contact with the produce | The oldest and fastest method used for chilling field-packed vegetables (such as broccoli, cabbage, cantaloupes, peaches, root crops as well as poultry and fish), it may result in an increase in the weight (such as poultry) | Water sanitation (use of chlorine). The containers employed must be resistant to water for a prolonged time as well, and should have drainage vents for the melted water, considerably higher labor cost and freight load as the typical weight of ice for initial cooling is equivalent to 30% of the product weight. |
| The cooling effect is maintained during and after transportation for sufficient time since it is often a marketing requirement for several products that ice remains until they are received by the retailer. | Ice cooling provides a high relative humidity environment for fresh produce. | |

| Categories | Advantages | Disadvantages |
|---|---|---|
| *Vacuum cooling:* | | |
| Fresh produce is loaded in a vacuum chamber, equipped with a vacuum device (mechanical vacuum pump or steam-jet pump) and a condenser (necessary when vacuum is achieved through a mechanical vacuum pump), and air is drawn out creating a high vacuum (4–4.6 mm Hg). This reduction in the atmospheric pressure causes a substantial decrease in the boiling temperature of water generating moisture evaporation which withdraws heat from the products. The extent of cooling is proportional to the water evaporated (evaporation of 1 kg of water removes approximately 2200 kJ), while the temperature is lowered about 5°C for each 1% reduction in water content. | The fastest and most uniform method recommended for chilling leafy and floral vegetables (such as lettuce, celery) with high surface to volume ratio and ability to readily release internal water, portable and readily available for use or built in productions plants with large annual workload and as closely as possible to the harvest area (if not portable). | Expensive method, requires high capital investment and skilled operators, designed for batch operation only, water hygiene, moisture loss (2 to 3%) induced during cooling (it can be overcome by pre-wetting before vacuum or by using a water spray system during cooling (hydro-vacuum cooling). |

2. *Cryogenic chillers:*

Use of solid carbon dioxide (dry-ice pellets: a higher boiling and sublimation point than nitrogen, and therefore a less severe effect on the food), liquid carbon dioxide or, most commonly in chilling operations, liquid nitrogen. These can either be used to chill another refrigerant or directly chill the products/Batch or continuous

Significantly higher cost, cryogenic refrigerators have a limited range of applications (i.e., in mechanically formed meat products, cryogenic grinding and dough products).

An extended presentation on current practices, principles, changes caused to food tissues and applications of food refrigeration are provided by Pham (2014) and Dermesonlouoglou et al. (2016). It is not in the scope of this chapter to present a thorough overview of the traditional methods, models and quality modifications induced by chilling; focus will be given on recent developments and trends concerning this popular preservation method, for a lot of food products.

The refrigeration industry went through a major boost with the arrival of refrigerants like CFC-R12, HCFC-R22 (1935) and CFC-R502 (1961), HFC R134 (1993), etc. The HFCs have no ozone depletion advantages over R22; however, they are still accused of global warming potential impact (GWP). The environmental protection is one more challenging issue. Refrigeration techniques have to meet strict requirements regarding two major domains: Initially the destruction of the ozone layer (CFC, HCFC) and then the intensification of greenhouse gases (CFC, HCFC and HFC). These two preoccupations led to Montreal and Kyoto protocols that caused relevant industries looking for alternative solutions. Therefore, these international regulations showed the necessity for new refrigeration technologies, with more efficient refrigerants and equipment (Sari and Balli, 2014).

In Tassou et al. (2010) a detailed review of emerging technologies for food refrigeration applications is presented and the main aspects will be briefly described in the following section. As authors state, the increasing cost of electricity and the current environmental concern to minimize the impact and carbon footprint of food processes has renewed interest in alternative techniques and the development of new and innovative technologies that could offer both economic and environmental advantages over the conventional vapor compression cycle in the future. In this review paper each technology is described, the state of its development is outlined, and possible applications and barriers to its implementation are also presented. The most important alternative approaches proposed are overviewed in Figure 7.1 and detailed in the following section.

### 7.2.2.1   Refrigeration Cycle Improvements

In Mota-Babiloni et al. (2015b) some basic cycle improvements and new developments are presented. Subcooling has been extensively utilized low-temperature refrigeration systems wherein a simple vapor-compression refrigeration system is altered to save energy. According to Qureshi and Zubair

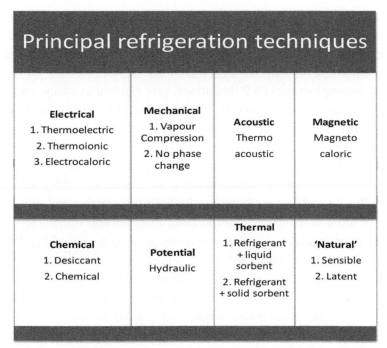

**FIGURE 7.1** Classification of main novel food refrigeration techniques [adapted from Steven Brown and Domanski (2014)].

(2013b), subcooling technologies are as follows: (a) ambient subcooling, (b) suction-line heat exchanger usage as heat sink, (c) systems with an external heat sink, and (d) mechanical subcooling. Recent literature (Qureshi et al., 2013, Qureshi and Zubair, 2012, 2013a) concludes that using mechanical subcooling systems can help to increase coefficient of performance (COP) of simple vapor compression cycles, leading also to an increase of cooling capacity and possibly energy savings, in systems that can be easily commercialized.

### 7.2.2.2 New Refrigerants with Lower Emissions

After the signature of Kyoto Protocol in 1997, HydroFluoroCarbon (HFC) refrigerants were signaled as direct greenhouse gases because of their high Global Warming Potential (GWP) (though they have zero Ozone Depletion Potential, ODP), and the last environmental regulations are focused on banning these fluids (Mota-Babiloni et al., 2015a). With EU Regulation No 517/2014 (Regulation EU No 517/20, 2014) the European Commission aims at limiting the total amount of the most important F-gases that can be sold in the EU from 2015 onwards and phasing them down in steps to one-fifth of 2014 sales in 2030. R134a, R404A and R410A (GWP of 1430, 3922 and 2088 respectively are the refrigerants most affected by the regulation due to their common use in refrigeration applications. Besides looking for refrigerants that accomplish the GWP limitations, safe fluids that imply low energy consumption in vapor compression systems should be used. Therefore, the required characteristics for a refrigerant are: environmental acceptability, chemical stability, materials compatibility, refrigeration-cycle performance, non-flammability and non-toxicity, boiling point, etc. (Calm, 2008).

Since there is a lot of published work on alternative refrigerants, more energy efficient and less environmentally harmful, at this point only a brief outline of the state-of-the art will be described. Several options have been proposed to replace the refrigerants currently used, as described by Mohanraj et al. (2009): hydrocarbons, efficient but highly flammable (Palm 2008); natural refrigerants, focusing on ammonia (Pearson, 2008) (R717), $CO_2$ (R744) (Paul and Zhongjie, 2015) and hydrofluoroolefin (HFO) (Mota-Babiloni et al., 2014, 2016). Research is still intensive in an effort to find promising alternatives to the refrigerants that are ruled out by the recent EU Regulation.

Another encouraging potential is the use of nanorefrigerants, studied by (Alawi et al., 2015; Nair et al., 2016). A nanorefrigerant is a special type of refrigerant in which the nanoparticles are suspended and well-dispersed in the base. As described in the former review paper, 'nanorefrigerant is one kind of nanofluids, in which the host fluid is conventional pure refrigerant. Experimental studies showed that the nanorefrigerant has higher thermal conductivity than the host refrigerant and the refrigeration system using nanorefrigerant has better performance than that of using conventional pure refrigerant. However, the aggregation and sedimentation of nanoparticles in the nanorefrigerant may reduce the stability of nanorefrigerant and limit the application of nanorefrigerant in the refrigeration system.' As stated by Nair et al. (2016), research should focus on using nanorefrigerants within smaller and lighter refrigeration systems, which would consume less compressor power, that is, they will be more energy efficient.

### 7.2.2.3   Sorption Refrigeration–Adsorption Systems

Sorption refrigeration seems really promising since it uses low grade energy, has proven to have zero ozone depletion potential (ODP) and global warming potential (GWP), which is important for the environmental protection (Zhou et al., 2016). Sorption refrigeration technologies are thermally (instead of mechanically driven) governed systems, in which the mechanical compressor of the conventional vapor compression cycle is replaced by a 'thermal compressor' and a sorbent (Rezk et al., 2014). The sorbent can be either solid in the case of adsorption systems or liquid for absorption systems. Typical working fluids include ammonia/water and lithium bromide/water for absorption systems, and activated carbon/ammonia, activated carbon/methanol, silica gel/water, zeolite/water, metal chlorides/ammonia, or composite adsorbents for adsorption systems (Steven Brown and Domanski 2014). The adsorption refrigeration relies on the phenomena of evaporation and condensation of a refrigerant, which involve adsorption or chemical reaction as well as heat transfer. A representative adsorption refrigeration system (ARS) is a solid adsorbent bed to adsorb and desorb a refrigerant to gain the cooling effect (Shmroukh et al., 2015). Alternatively, to reduce electricity consumption while obtaining a low refrigeration temperature, an absorption/compression refrigeration system was developed combining effectively an absorption sub cycle and a compression sub cycle driven by heat and power, respectively (Chen et al., 2015).

Different types of adsorption/absorption chillers are detailed in Deng et al. (2011), where innovative concepts are also proposed. Absorption chillers represent a large part of chillers sold some markets (e.g., Japan and China). The two most important novelties in absorption cooling have been the introduction of triple-effect chillers and the introduction of generator-absorber heat exchanger (GAX) technology (Brown and Domanski, 2014).

As far as application in the food sector is concerned, as Tassou et al. (2010) state applications in the food sector will be primarily in areas where waste heat is available to drive the adsorption system. Such applications can be found in food factories and transport refrigeration. Other possible applications are where solar energy is abundant.

### 7.2.2.4  Ejector Refrigeration Systems

Ejector or jet pump refrigeration is a thermally governed technique that has a much lower coefficient of performance (COP) than vapor compression systems but offer advantages of simplicity and no moving parts. Their most important advantage is their capability to produce refrigeration using waste heat or solar energy as a heat source at temperatures above 80°C (Tassou et al., 2010). The inclusion of an ejector instead of an expansion valve in the refrigeration system can also result in energy efficiency increase (Wang et al., 2016), and also having the benefit of simplicity in construction, installation and maintenance, as stated by Sarkar (2012) and Besagni et al. (2016) in their recent review. Recent research in principles, design and potential applications of compact and miniature mechanical vapor compression refrigeration systems are reviewed in (Barbosa Jr. et al., 2012). To improve their performance, more complex systems have been proposed (Wang et al., 2016), as well as their combined use with vapor compression and absorption systems (Tassou et al., 2010). In the food industry, the use of this technology is advantageous in cases where there is waste heat, for example, in the food processing sector or for transport refrigeration.

### 7.2.2.5   Tri-Generation

Tri-generation refers to generation of three forms of output energy: electricity and use of the waste heat for both heating and cooling, starting from the combustion of conventional or alternative fuels or from solar heat collectors (Iodice et al., 2016). Tri-generation is also known as CCHP (Combined Cooling, Heating and Power) or CHRP (Combined Heating, Refrigeration and Power). In a practical level, tri-generation systems are CHP (Combined Heat and Power) or co-generation systems, integrated with a thermally driven refrigeration system to provide cooling as well as electrical power and heating. Many food industries use cogeneration plants with either gas engines or turbines to cover their steam, hot water and electrical demands. The combination of an absorption refrigeration with a cogeneration plant allows to use all generated heat for the production of cooling. Examples of this technique in the food sector are presented by Bassols et al. (2002). Suamir and Tassou (2013) modeled supermarket conventional and integrated $CO_2$ refrigeration and trigeneration energy systems, with significant energy and emission savings, with 3.2 years being the payback period. They also studied the use of biofuels to drive the trigeneration plant, which seems attractive in terms of overall reduction in GHG (greenhouse gasses) emissions. However, authors state that 'the cost of biofuels can be higher than that of conventional fuels which will have a negative impact on the economic attractiveness of the system.' Cooling produced by trigeneration systems ensure subcritical operation throughout the year condensing $CO_2$ fluid (Suamir et al., 2012). These systems have been widely used in food industries, and thus developments mainly involve improvements of individual subsystems, such as gas and diesel engines, microturbines, fuel cells, system integration and control.

Apart from the successful integration of tri-generation plants in current food industry, its use has been lately extended to supermarkets, mainly used for space cooling. Several works have been published regarding trigeneration technique application in the food retail sector (Ge et al., 2013; Suamir et al., 2012; Sugiartha et al., 2009).

## 7.2.2.6   Stirling Cycle Refrigeration

The Stirling regenerative refrigeration cycle consists of four thermodynamic processes: compression process, hot to cold process, expansion process and cold to hot process (Gao et al., 2016). The Stirling cycle cooler is a closed-cycle regenerative thermal machine. As stated by Tassou et al. (2010), this type of equipment can operate down to cryogenic temperatures and hence is useful in many food refrigeration applications. The main drawbacks concern the low cooling capacities, the lower COP and higher cost compared to vapor compression. Jianfeng et al. (2013) investigated the influencing factors of the Stirling refrigeration on the cooling and churning functions to butter processing. Results showed that both refrigerant volume and mechanical vibration frequency have significant effects on the cooling effect, optimal parameters are given for refrigeration and that, based also on quality assessment application of the Stirling refrigeration seems promising in butter processing industry. Although the method has disadvantages in energy consumption and cost compared with traditional methods, authors state that 'the proposed method presents a new idea and theoretical basis for the application of Stirling refrigeration technology in food processing.'

## 7.2.2.7   Thermoelectric Refrigeration

Thermoelectric refrigeration is based on Peltier effect, which causes the junction of two dissimilar conducting materials; hence a temperature difference will develop across the two junctions, that is, one junction will become colder and the other one hotter. Thermoelectric materials are characterized by a so-called figure of merit Z, which is a function of material properties and the absolute temperature (Brown and Domanski, 2014; Tassou et al., 2010). This method of cooling offers a number of benefits over traditional vapor compression refrigeration, such as: more compact, robust and noiseless equipment, due to minimum moving parts, better temperature control, and the lack of refrigerants. However, the main drawback is the increased energy consumption, that is, process efficiency (Astrain et al., 2016). Several publications have been recently presented aiming mainly at modeling the process and finding ways of improving thermoelectric refrigeration efficiency (Lv et al., 2016; Martinez et al., 2016; Qian et al., 2016; Reddy, 2016).

Regarding application in the food sector, thermoelectric cooling products include compact refrigerators (15–70 L) for hotel rooms (mini bar), transport trucks, recreational vehicles; wine coolers; portable picnic coolers; beverage can coolers and drinking water coolers (Riffat and Ma 2003). Min and Rowe (2006) studied a number of prototype thermoelectric refrigerators for their cooling performances evaluated in terms of the coefficient-of-performance (COP), heat-pumping capacity and cooling-down rate. The results showed a COP around 0.3–0.5 for a typical operating temperature at 5°C with ambient at 25°C and authors concluded that an increase in COP value is possible through improvements in module contact resistances, thermal interfaces and the effectiveness of heat exchangers. Reviewing current literature, regarding this technique's future development, it can be assessed, with the efficiency much below that of vapor compression technology, thermoelectric cooling may only be attractive in specific applications, for example, where compactness or portability are needed such as certain military and space applications, certain medical applications, recreational cooling, etc. (Brown and Domanski, 2014). In this context, there is a major need of finding and commercializing new, improved materials, other than the state-of-the-art materials currently applied, such as $Bi_2Te_3$ (bismuth telluride) and $Sb_2Te_3$ (antimony telluride with focus on nanostructures, for example, quantum dot superlattices (Harman et al., 2002).

### 7.2.2.8  Thermoacoustic Refrigeration

The physical principle of this method is described by Brown and Domanski (2014): 'Thermoacoustic cooling is based on the conversion of acoustic energy to thermal energy. The presence of an acoustic wave (standing or traveling) expands and contracts a working fluid (gas). As the gas expands, its pressure and temperature are reduced; likewise, as the gas contracts, its pressure and temperature are increased. In addition, the pressure differences lead to movement of the working gas. To achieve cooling and heating, the working gas must be coupled to an external heat transfer fluid through heat exchangers.' Thermoacoustic devices are typically characterized as either 'standing-wave' or 'traveling-wave.' A thermoacoustic cooling device may function on a electrical or thermal base. In the former case, a mechanical compressor is used, while the latter simply uses heat as driving source. Luo et al. (2006) studied a thermally-driven thermoacoustic refrigerator system without any moving part and results showed a good potential. The prospect of the refrigeration system in room-temperature cooling such as food refrigeration and air-conditioning, since a cooling capacity of about 270 W at 20°C and 405 W at 0°C was measured.

According to Tassou et al. (2010) and Brown & Domanski (2014), so far this system application is limited to prototypes and design concepts in a research level. Some main issues of concern include low cooling capacities, large size and heat exchanger inefficiencies. Research effort is currently focused on the development of flow-through designs which will reduce the use of heat exchangers and on thermoacoustically-driven pulse tube cryocoolers because of their reliability and physical size.

### 7.2.2.9    Magnetic Refrigeration

Magnetic refrigeration system is considered actually as an attractive alternative for conventional vapor compression refrigeration systems and has been initially applied to maintain very low temperatures. Sari and Balli (2014) presented a very interesting review on the principles, advantages and potential applications of magnetic cooling. Recently, the magnetocaloric effect (MCE) has attracted interest because of numerous advantages of magnetic refrigeration as compared with conventional cooling systems, such as the ecological benefit (no ozone depleting and direct greenhouse CFC or HCFC gases) and the high energy efficiency. Magnetocaloric effect is a basic property of magnetic solids, characterized by a temperature change due to a change in the strength of an external magnetic field. Magnetic refrigeration requires the combination of a magnetic source of rather high strength and a material with sufficiently high refrigerant capacity.

In the late 1990s the development of a prototype proving the feasibility of the magnetic refrigeration in the room temperature has advanced a lot the use of this technique (Zimm et al., 1998). Since then, an important research activity is conducted around the world for the development of magnetic heating and cooling prototypes (Yu et al., 2010), refrigeration systems (Jacobs et al., 2014) and the application of new magnetocaloric materials (Gschneidner Jr. and Pecharsky, 2008; Sari and Balli, 2014).

Taking into account the principles of this technique, it is concluded that it has the potential for use in the food refrigeration, covering the whole temperature range, from near room temperature operation down to cryogenic temperatures. Despite the presence of numerous prototypes and the ongoing research realized in this area, the literature lacks reliable experimental data for comparing magnetic refrigeration and vapor compression technology (Steven Brown & Domanski 2014). We agree with aforementioned authors that 'however, significant breakthroughs in magnetocaloric materials and design concepts are needed to bring magnetic cooling to the marketplace and make it competitive with vapor compression technology.'

### 7.2.2.10   Refrigeration from Solar Energy

Kim and Infante Ferreira (2008) offer a thorough review of the different technologies that are available to deliver refrigeration from solar energy, including not only sorption cooling technologies but also solar electric, thermo-mechanical, sorption and also some newly emerging technologies. At the same time, in this review paper, another aim was to compare the potential of these different technologies in delivering competitive sustainable solutions. Numerous other publications illustrate the prospects of this technique and present an overview of solar cooling technologies (Abdulateef et al., 2009; Nkwetta and Sandercock, 2016), while other studies describe real case studies of its application in the food sector (Allouhi et al., 2014; Best et al., 2013; Suamir 2014) At this point, there will be a brief overview of the methods that use solar energy to create cooling, since the main principles, potential applications and special concerns have been already described in the former sections (Figure 7.2).

(a) **Solar electric refrigeration.** This refrigeration system consists of photovoltaic panels and an electrical refrigeration device. Solar cells are basically semiconductors whose efficiency and cost vary

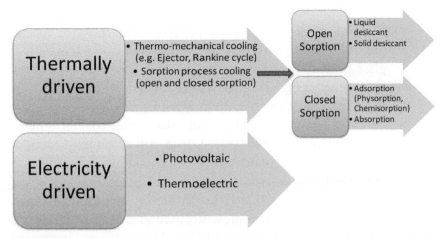

**FIGURE 7.2** Classification of most popular solar cooling systems [adapted from Zeyghami et al., 2015)].

depending on the material and the manufacturing methods they are made from. Most of the solar cells commercially available in the market are made from silicon. The biggest benefit of using solar panels is the simplicity in construction and the high overall efficiency when combined with a conventional vapor compression system (Ferreira and Kim, 2014; Kim and Ferreira, 2008). Solar electric vapor compression refrigeration systems are found scarcely in literature (Fong et al., 2011; Karellas and Braimakis, 2016). According to Kim and Ferreira (2008) there are some issues that need to be addressed in order to promote broader commercialization of this technique. Firstly, the systems should be equipped with some means, for example, electric battery, mixed use of solar- and grid-electricity or a variable-capacity compressor to cope with the varying electricity production rate with time and secondly, the price of a solar photovoltaic panel should be further decreased to be competitive with other alternative solar cooling technologies.

(b) **Solar thermal refrigeration**

    (i) *Thermomechanical refrigeration.* A recent review is available on solar thermo-mechanical refrigeration (Zeyghami et al., 2015). In solar thermal cooling systems, the heat from a solar collector is used to produce mechanical power to compress the refrigerant vapor in a conventional Vapor Compression Cycle (VCC) or as a heat source for the generator of a sorption cooling cycle. Flat-plate solar collectors are the most frequent type, consisting of a metallic absorber and an insulated casing topped with glass plate(s). Evacuated collectors have less heat loss and perform better at high temperatures (Kim and Ferreira 2008). The most common thermo-mechanical solar refrigeration systems are the solar rankine cooling and solar ejector cooling systems. As Zeyghami et al. (2015) conclude in their review paper that 'research in thermo-mechanical solar cooling technologies is growing. But to be able to compete with other cooling systems, they should be able to provide the cooling effect with the same efficiency as those of the presently used solar electric and absorption cooling systems. To accomplish this, an increase in research and development (R&D) will be required to speed up the innovation process.'

(ii) *Sorption refrigeration.* Sorption refrigeration uses physical or chemical attraction between a pair of substances to produce refrigeration effect. A sorption system principally transforms thermal energy directly into cooling power. Among the pair of substances, the substance with lower boiling temperature is called sorbate and the other is called sorbent. The sorbate plays the role of refrigerant. Absorption refers to a sorption process where a liquid or solid sorbent absorbs refrigerant molecules and changes physically and/or chemically. Adsorption (either physical or chemical) refers to a solid sorbent that attracts refrigerant molecules on its surface by physical or chemical force, without changing its form in the process. These processes are employed in closed sorption systems, as described in detail in reviews of Fernandes et al. (2014) and Sarbu & Sebarchievici (2015). Finally, desiccation refers to an open sorption process where a sorbent, that is, a desiccant, absorbs moisture from humid air (Kim and Ferreira, 2008).

### 7.2.3   SUPERCHILLING

In traditional chilling methods, no freezing should take place on the surfaces of products, and thus, the driving force for heat removal is very low.

This makes traditional industrial chilling very time consuming and challenging in modern in-line production. As described in the review published by Magnussen et al. (2008): 'the terms *superchilling* and *partial freezing* describe a process where a minor part of the products water content is frozen and the temperature of the foodstuff is lowered, often 1–2°C below the initial freezing point of the product. After initial surface freezing, the ice distribution reaches a specific value and the product obtains a uniform temperature at which it is maintained during storage and distribution.' Two are the main advantages of this process that makes it attractive to the industrial application: cold is accumulated within the product and the necessary holding time inside the chilling equipment is shortened compared to traditional chilling. Assuming there is still some ice content left after equilibrium, the product will have an amount of cold to absorb heat from the surroundings without changing the product temperature too much. In the same publication, there are details concerning disadvantages of the method, and its application in food processing/distribution. According to authors, the main challenges refer to optimize mathematical models and obtain simulation tools, integrate Computational Fluid Dynamics (CFD) analysis to determine mass and heat transfer phenomena, define the degree of superchilling in order to improve quality attributes and finally, design a flexible procedure that would be industrially appealing.

Other interesting reviews are published by Kaale et al. (2011) and Kaale & Eikevik (2014), detailing ice crystallization during the process and the subsequent storage. This information is of great importance for industry, in order to estimate the refrigeration requirements for a superchilling system, design the required equipment and decide on how to minimize quality losses. A recent approach regarding the prediction of superchilling is proposed in Kaale et al. (2014) using either analytical or numerical methods—finite difference method (FDM), finite elements method (FEM), or finite volume method (FVM)—of the heat transfer equations. Analytical methods may offer an exact solution but have significant limitations. They are mainly for one-dimensional cases with simple initial and boundary conditions and constant thermal properties. On the other hand, numerical methods are versatile and can account for changing thermal properties and food heterogeneity. The same research group has studied deeply the advantages of superchilling on preserving and storing efficiently salmon filets (Kaale et al., 2013a, 2013b; Kaale and Eikevik, 2013, 2015).

Reviewing the existing literature, there is a need for improved methods to control the superchilling process, by being able to precisely the degree of superchilling. A major concern is to proceed to commercialization of the process, by optimizing the existing equipment and in this case, superchilled storage might be also used to add value to commercial foods, in a quality viewpoint.

## 7.3 COLD CHAIN: LATEST DEVELOPMENTS AND INTEGRATED MONITORING

Continuously monitoring and maintaining the cold chain is of paramount importance for the microbiological safety and quality of chilled and frozen foods. Advances in the cold chain to improve food quality and safety upto the final consumer are discussed in detail in James and James (2010). In the following section, the specific stages of the cold chain (chilling, freezing, storage, transport and retail display) are then briefly discussed. Environmental issues and advances in energy reduction are also addressed (Pham, 2014).

In James and James (2010) a very interesting perspective of the interaction between the climate change and the current cold chain is detailed. In this context, the 'contribution' of each stage of the chill chain, from primary chilling to end user refrigerator, to the total energy consumption is presented giving an impressive picture of the current status. Data are also given on the environmental impact of refrigerants used and finally, suggestions are given on improving the energy efficiency of the cold-chain, on each distinct phase of the continuous chill chain. Authors conclude that 'New/alternative refrigeration systems/cycles, such as Trigeneration, Air Cycle, Sorption-Adsorption Systems, Thermoelectric, Stirling Cycle, Thermoacoustic and Magnetic refrigeration, have the potential to save energy in the future if applied to food refrigeration. However, none appear to be likely to produce a step change reduction in refrigeration energy consumption within the food industry within the next decade.'

### 7.3.1 TRANSPORT REFRIGERATION

Tassou et al. (2009) have published a review on the state of the art in food transport refrigeration, assessing also the integrated environmental (including fuel consumption, greenhouse gas emissions, refrigeration engine, etc.) of each methodology. The aim of this study was to identify the carbon foot print, spot weaknesses and suggest alternative refrigeration technologies that could lead to a reduction in energy consumption and environmental impacts. Among their other significant observations they conclude that 'the capacity, size and environmental impact of diesel engine driven systems can be reduced through the use of thermal energy storage (eutectics). Furthermore, they claim that the air cycle technology is quite promising for food transport and other technologies that need further investigation and consideration are: stirling cycle powered systems, magnetic refrigeration, solar energy driven systems and hybrid system arrangements.'

Other innovative systems proposed to enhance food transport, from an energy and environmentally sustainable-wise viewpoint, are vacuum insulated panels, liquid nitrogen transport systems and photovoltaics, which however have limited commercial application possibly due to their high cost (James and James 2010). However, when practical implementation is concerned, as stated by Tassou et al. (2010) the main focus should be given on 'the recovery of thermal energy from the engine exhaust and its use to drive sorption systems, ejector systems, thermoacoustic refrigerators and or/for power generation using thermos electrics or turbogenerators.'

Another interesting approach is presented in the review paper of James et al. (2006) where all types of food transportation are described (sea, air, land transportation), models of heat and mass transfer are proposed, where the principal influencing factors are fluctuating ambient conditions, door openings, product removal/loading, etc., using frequently CFD implementation. Authors point out that transport has received little attention regarding modeling, although it is an important stage of cold chain. They also claim that a model that predicts heat transfer, and thus temperatures, in food products in a refrigerated transport container needs to include numerous parameters and properties (e.g., Conduction through the walls, conduction within the food, heat transfer between the container walls and the refrigerated air, etc.). Since some factors can change abruptly with time due to weather conditions, time of day, and itinerary of the vehicle or ship, a model should therefore be able to integrate the dynamic effect of these changes on the temperature of the food being transported and the energy that has to be extracted by the refrigeration system. As authors underline, there is a gap on this point since no single computer program has been developed that combines all these complex aspects.

## 7.3.2  SUPERMARKET REFRIGERATION SYSTEMS

Retail food outlets are responsible for a significant proportion of total elec-trical energy consumption and Green House Gas (GHG) emissions. Tassou et al. (2011) provide energy consumption data of numerous samples of retail food stores from a number of major retail food chains in the UK, including all major store categories from convenience stores to hypermarkets. Data show a wide variability of energy intensity even within stores of the same retail chain. The paper also discusses the major energy consuming processes in retail food stores and identifies opportunities for energy savings, such as investments on $CO_2$ systems, on their own or in a cascade arrangement with hydrocarbon (HC) or ammonia (R717) systems, improvement of the effi-ciency of the compressors, reduction of the pressure ratio in the system, and continuously match of the refrigeration capacity to the load and the design of more efficient cabinets. Authors conclude that 'research and development areas to be addressed are the reduction of the infiltration rate, reduction of fan and lighting energy consumption, the design of more efficient evaporator coils to increase the evaporating temperature, reduce frosting rates and the implementation of defrost on demand.

There are also opportunities for thermal integration of refrigeration and HVAC (heating ventilation and air conditioning systems) and the application of CHP (Combined Heat and Power) and tri-generation tech-nologies (Tassou et al., 2010).

In Mota-Babiloni et al. (2015b), a review on advances of commercial refrigeration is presented, focusing mainly on energy saving issues, on the use of new, environment-friendly refrigerants and on mathematical modeling used.

Cold storage rooms in the retail level used to maintain and preserve chilled and frozen foods, consume considerable amounts of energy, since about 60–70% of the electrical energy may be used for refrigeration. Therefore, a significant issue posed is to come up with some new energy savings methods and test their applicability. In this context, in Evans et al. (2014a, 2014b) the performance of numerous cold stores worldwide is measured, basic parameters that influence final supermarket consump-tion are described and finally, mathematical models to predict energy consumed are developed. It is obvious that before installing refrigeration systems, supermarket energy models can be used to predict the energy consumption and to propose potential energy saving methods. In recent literature, there are published models to describe several factors effect (e.g., heat recovery, building thermal insulation, etc.) on energy requirements (Acha et al., 2016; Braun et al., 2014; Ducoulombier et al., 2006; Ge and Tassou, 2011; Rasmussen et al., 2016) or mathematical equations to predict costs of several refrigeration systems (Glavan et al., 2016).

### 7.3.3 DOMESTIC STORAGE

Belman-Flores et al. (2015) have presented a detailed review of the latest developments on domestic refrigeration, based on vapor compression thermal distribution, control, environmentally friendly refrigerants, thermal isolation, etc. The main finding of this work was that most research in this area includes modeling the entire refrigerator or some compounds, whereas numerous publications refer to specific cabinet improvements to accessories, secondary parts such as seals, or mechanisms (insulation, defrosting systems), etc. An area of increasing interest is the production of new cooling machines, more efficient, environmentally favorable and sustainable, using, if necessary, hybrid systems. Temperature control is another issue that needs to be addressed. James and James (2010) propose some specific enhancements for evaporators and compressors, such as 'lower viscosity oils; reduction of temperature level inside the compressor; variable speed motors; linear compressors and improved insulation.'

### 7.3.4 MONITORING CURRENT CHILL CHAIN

An overview of the state-of-the-art in monitoring and control of the chill chain is provided by Taoukis et al. (2006). We certainly agree that keeping a steady and adequately cold logistic path for product acceptance, in terms of both safety and quality is of paramount importance for food industry and consumers. In this context, monitoring and control of the cold chain is a prerequisite for reliable quality management and optimization. Continuous and reliable temperature control is essential in all sections of the cool food chain and can be obtained through improved equipment systems, quality assurance systems application, and awareness of all interfering parts. The trend in this area, however is to introduce temperature monitoring in an integrated, structured quality assurance system, based mostly on prevention, through the entire lifecycle of the product.

Recent advances in sensing technology, miniaturization and remote communications have resulted in powerful and accurate temperature loggers that provide continuous recordings of product temperatures during the whole cold chain (Pham, 2014). Besides conventional sensors, readout and recording temperature devices, there is a possibility using radio frequency identification (RFID) technology, to transmit data together with data on geographical location, using mobile wireless systems and the Internet, allowing for intervention to correct problems when deemed necessary (Ruiz-Garcia et al., 2009). The main drawbacks of RFID reported are huge volumes of data, high cost, difficulties in calculating the rate of return and the need to improve reliability and accuracy (Pham, 2014).

An alternative, cost-effective way to individually monitor the temperature conditions of food products throughout the cold chain is the use of time–temperature integrators (TTIs) (Taoukis et al., 2006). The principle of TTI application is based on the use of a physicochemical mechanism and a measurable change to display: (a) the current temperature, (b) the crossover of a preset temperature, or (c) the integrated time–temperature history of the frozen or chilled food. They are systems with an easily measurable response that reflect the accumulated time–temperature history of the product on which there are attached. Their operation is based on irreversible reactions that are initiated at the time of their activation and proceed with an increasing rate, as temperature rises, in a manner that is similar to the temperature dependence of most quality-loss reactions of foods (Lee & Rahman, 2014). These devices are attached on the food itself or outside the packaging and actually follow the food during its whole distribution way from the producer to the chilling equipment of the final consumer. The essence of their application lies on the 'translation' of their response, through the appropriate mathematical formulas, to the actual quality status of the food. Such TTI devices are MonitorMark™, FreezeWatch™ (3M, St. Paul, MN, USA), Fresh-Check® TTI (Temptime Corp., Morriston, NJ, USA), ColdMark™ (Cold Chain Technologies, MA, USA), CheckPoint® TTI (VITSAB A.B., Malmö, Sweden), and OnVu™ TTI (Ciba Specialty Chemicals & Freshpoint, Basel, Switzerland).

Searching briefly current literature, in Wang et al. (2015) a review on the latest research of TTI is presented, its status and the main problems in their application in the food supply chain (e.g., the inaccuracy of temperature monitoring, the high cost of commercial application) are elaborated and the prospect of its development is envisaged, aiming to provide reference and support for researchers in this area. Wan and Knoll (2016) proposed a new type of TTI based on an electrochemical pseudo-transistor that can be effectively combined with RFID technique to check and reassure the safety of perishable food. In a recent work, Kim et al. (2016) a prototype isopropyl palmitate (IPP) diffusion-based TTI system was characterized and evaluated for monitoring microbial quality of non-pasteurized angelica (NPA) juice at various iso-thermal and dynamic temperatures and a mathematical model based on relationships between diffusion and time temperature was establishEd. Results showed that the proposed TTI was only verified for indicating temperature abuse above 13.5°C, and thus the TTI system has the potential for monitoring the microbial quality of perishable food products during distribution and storage. Wu et al. (2015) proposed an enzymatic system, including a lipase type TTI prototype based on the reaction between *Aspergillus niger* lipase and glycerol tributyrate, which time-temperature dependence was mathematically described using the appropriate models. Brizio et al. (2015) evaluated the applicability of a photochromic TTI to monitor the time and temperature history during the period of validity of the whole fish of the cobia specimen stored in ice and the results showed that the smart indicator followed a similar rate of deterioration of the analyzed product visual response, proving to be an efficient shelf life indicator that can assure consumers for the product's quality status easily, cheaply and accurately. A totally different system, namely Gelatin-Templated Gold Nanoparticles was tested as novel TTI by Lim et al. (2012).

Refrigerated and frozen foods have been extensively studied regarding the potential TTI application in order to monitor their temperature throughout the problematic chill chain. However, mainly due to difficulties related to kinetic modeling of frozen foods and TTIs in the subfreezing range, the main focus was given on chilled foods (Brizio and Prentice, 2014; Giannakourou et al., 2005a; Taoukis et al., 1998; Vaikousi et al., 2009), and only a few studies tested TTI at the temperature range of interest for frozen storage. A more complex scheme, using Monte Carlo simulation technique was applied in assessing the performance of Time Temperature Integrators in the real distribution chain of frozen (Giannakourou and Taoukis 2002, 2003), or chilled foods (Giannakourou et al., 2005a; Qi et al., 2014), in a product management system coded Least Shelf Life First Out (LSFO). Its structure is based on validated shelf life modeling of the controlled food product, specification of the initial and the final accepted value of the selected quality parameter, as well as accurate and continuous temperature monitoring in the distribution chain with the appropriate TTI. LSFO aims at reducing the rejected products at the consumer end, by promoting, at selected decision making points of the product life cycle, those product units with the shorter shelf life, according to the response of the attached TTI. A further improvement of the LSFO approach is a chill chain management system coded Shelf-Life Decision System (SLDS) (Giannakourou et al., 2001a, 2001b, 2005b; Koutsoumanis et al., 2002). Compared to LSFO, SLDS policy takes additionally into account the realistic variability of the initial quality state of the product.

## 7.4 MATHEMATICAL MODELING AND SIMULATION OF REFRIGERATION AND FREEZING PROCESSES

Conventional freezing and refrigerated technologies are being optimized through the increase use of modeling techniques. Early works on food refrigeration models concentrated on the prediction of freezing or chilling time based on the mean or center temperature of the food. Plank's simplified equations for the freezing time of foods were presented in 1913, and more recent efforts have concentrated on removing his rather limiting assumptions on the properties and geometry of the product. For a long time, processing time and moisture loss were the major concern of researchers. With the advent of computers, numerical methods were introduced into food research and the whole temperature field inside the product could be modeled. It was quickly realized that this would enable the prediction of many other quantities of interest related to the quality of the food and the shelf-life. At the same time, the use of computers in plant design stimulated research into the prediction of dynamic heat load from food, and integrated models are being developed to take into account the interactions of the food with the processing equipment (Pham, 2014). Further research needs to be carried out on 'weak links' (heat transfer coefficient estimation, food properties estimation, modeling complex product shape, modeling of mass transfer controlled freezing, handling of statistical variations) in the modeling of food refrigeration processes.

### 7.4.1  FREEZING TIME CALCULATION

It is important to accurately predict the freezing time of foods to assess the quality, processing requirements, and economical aspects of food freezing.

In the past, an extensive amount of work has been done to develop mathematical models for the prediction of food freezing time. However, the freezing process is a moving boundary problem. Foods, undergoing freezing, release latent heat over a range of temperatures. Freezing does not occur at a unique temperature. In addition, foods do not have constant thermal properties during freezing. As a result, no exact mathematical model exists for predicting the freezing of foods. Researchers, who have found a solution, have either used numerical finite difference or finite element methods. So, models for predicting freezing times range from approximate analytical solutions to more complex numerical methods. The accuracy of such models is dependent on how closely the corresponding assumptions approach reality. Methods to predict freezing time are of two main types, numerical methods and simple formulas, and an overview is presented in Table 7.4.

A more recent approach involves the use of artificial neural networks (ANN) (Mittal, 2013) or genetic algorithms (Goñi et al., 2008) in order to assess the freezing time or physical properties of frozen foods (Ahmad et al., 2014). Their principle is based on a batch of experimental data, used for "training" of the system in order to select the most appropriate mathematical model. The main advantage of this method is the speed, the rapid calculations achieved and the easiness, that allow for immediate intervention in case some parameters need to be corrected.

As far as future trends in the field of modeling of cool preservation (including refrigeration, freezing, superchilling and chill chain transport/ storage), progress is expected in the area of alternative mathematical tools, rather than the use of analytical solutions. In order to better understand and improve the cooling equipment design, developed mathematical models are necessary that could be integrated in computational techniques. A lot of recent publications describe the evolution recorded in this area, since there are numerous user-friendly tools in computer technology that are related to the refrigeration process (Pham, 2014). It is realized that numerical methods, being very versatile, would enable the prediction of many other quantities of interest related to the quality and shelf life of the food. Heat and mass transfer in solids can be adequately modeled with finite element method (FEM) software such as Ansys, Comsol or Abaqus.

Processes involving fluid flow (i.e., air blast freezing and immersion freezing) can be described by computational fluid dynamics (CFD) software such as Fluent and CFX, allowing for accurate prediction of product temperatures and heat loads during chilling and freezing, the design and

**TABLE 7.4**   Mathematical Models Developed for the Prediction of Food Freezing Time*

| **Mathematical models/approaches-an overview.** |
|---|

1. *Simple prediction formulas*

Plank's equation

It is an analytical solution for a simplified version of the unique phase-change model. The limitations to Plank's equation (1913) for estimation of freezing times for foods are numerous and have been widely discussed.

Modified Plank's equations

- Nagaoka et al., equation (Nagaoka et al., 1956), Levy equation (Levy, 1958).

- Charm and Slavin (1962), Gustschmidt (1964), Mott (1964), Tao (1967), Tien and Geiger (1967, 1968), Tien and Kuomo (1968, 1969), Joshi and Tao (1974): Modifications and alternatives provide improvements in the predictions, but limitations of various types still exist.

- Cleland and Earle equation (Cleland and Earle, 1977, 1979, 1982, 1984): Systematic approach taking into account the sensible heat effects, backed up by carefully collected data covering most of the range of interest to industry.

- de Michelis and Calvelo (1982), Pham method (Pham, 1984, 1986): Correlation experimental data with a minimum of empirical parameters; among these precooling, phase change and post-cooling times.

- Pham (1986): Extension to the freezing of foodstuffs with variations in environmental conditions and to the asymmetric freezing of slabs (Pham, 1987a, b, c).

- The most thorough validation of various methods against data was performed by Cleland (1987a, 1987b, 1990), Pham and Willix (1990) and Hossain et al. (1992). Accurate results for complex shapes when combined with appropriate shape factors.

2. *Numerical equations* (Finite difference (FDM), finite element (FEM) and finite volume (FVM)

- Opportunities to solve complex partial differential equations with temperature-dependent properties (Cleland, 1990, 1991).

- They can be applied to a wide range of conditions including a variety of modes of surface heat transfer, time-variable conditions, and complex object geometries (Mannapperuma and Singh, 1994a, b).

- They also predict full temperature-time profiles not just freezing time.

- Complexity and high implementation costs (particularly for computer-program development and testing and data preparation, but to a lesser extent for computation time).

- For engineering design purposes, a full numerical solution may not be warranted, especially in view of the likely uncertainty in thermal property data, and the imprecise control of freezing conditions that occurs in actual freezers. In these cases the overall accuracy of the predictions is determined less by the calculation precision than the data uncertainty.

*Mittal (2006), Indicative.

optimization of airflow in chillers and freezers, cold stores, refrigerated vehicles, retail display cases, etc. (Sun, 2007). Numerical methods such as finite differences (FD) and finite elements (FE) have been developed and implemented in food cooling calculations. These techniques have the advantage of being applicable for nonstandard geometries, for the heterogeneous food structure, taking into account the variable thermophysical properties of foods, avoiding simplified assumptions.

CFD have been applied, not only to calculate the heat transfer (Davey and Pham, 1997), but also to study different cooling conditions and package designs in order to improve refrigeration processes (Ambaw et al., 2013; Defraeye et al., 2013; Nahor et al., 2005). Another challenging area involves the mathematical modeling of novel technologies applied to freezing and cooling of foods. This will help to better understand and scale-up the equipment to make them efficient and appealing for further industrial applications (Xin et al., 2015).

## 7.5   CONCLUSIONS

This chapter has discussed many innovative refrigeration and freezing technologies. Some of them are essentially improvements of existing methods, such as Impingement and Hydrofluidization freezing that aim at producing significantly higher surface heat transfer rates than previous systems of air blast and immersion freezing, respectively, in order to improve product quality through rapid freezing (James et al., 2015). Other processes described, both for refrigeration and freezing of foods can be implemented as complimentary techniques to existing systems (e.g., pressure shift, magnetic resonance, electrostatic, ultrasound, etc.) with the purpose either to improve product quality (e.g., when ice formation is controlled) or to provide a combined cooling technique with both economic and environmental advantages over the conventional vapor compression Finally, another category of freezing methods is based on altering food properties (such as in dehydrofreezing and the use of antifreeze and ice nucleation proteins).

Many of the innovative technologies described in this chapter are still in the research and development stage, and need to be assessed compared to the traditional ones, not only for their cost, but also for the possible energy savings, their overall carbon foot print, but also for their effect on food quality and preservation while for others the biggest obstacle to adoption by the food industry is high capital cost. In any way, it is a challenge to continue studying these promising alternatives and come up to a conclusion about their efficiency and industrial use potential mainly based on establishing how important quality versus cost is to the consumer and food processor when it comes to refrigerated and frozen foods.

## KEYWORDS

- chilling
- cold chain
- combined techniques
- freezing
- mathematical modelling
- novel techniques and equipment
- superchilling

## REFERENCES

Abdulateef, J. M., Sopian, K., Alghoul, M. A., & Sulaiman, M. Y. (2009). Review on solar-driven ejector refrigeration technologies. *Renewable and Sustainable Energy Reviews, 13*, 1338–1349.

Acha, S., Du, Y., & Shah, N. (2016). Enhancing energy efficiency in supermarket refrigeration systems through a robust energy performance indicator. *International Journal of Refrigeration, 64*, 40–50.

Ahmad, I., Jeenanunta, C., Chanvarasuth, P., & Komolavanij, S. (2014). Prediction of physical quality parameters of frozen shrimp (litopenaeus vannamei): An Artificial Neural Networks and Genetic Algorithm Approach. *Food and Bioprocess Technology, 7*, 1433–1444.

Alawi, O. A., Sidik, N. A. C., & Kherbeet, A. S. (2015). Nanorefrigerant effects in heat transfer performance and energy consumption reduction: A review, *International Communications in Heat and Mass Transfer, 69*, 76–83.

Alizadeh, E., Chapleau, N., de-Lamballerie, M., & Le-Bail, A. (2009). Impact of freezing process on salt diffusivity of seafood: Application to salmon (salmo salar) using conventional and pressure shift freezing.' *Food and Bioprocess Technology, 2*, 257–262.

Allouhi, A., Kousksou, T., Jamil, A., & Zeraouli, Y. (2014). Modelling of a thermal adsorber powered by solar energy for refrigeration applications. *Energy, 75*, 589–596.

Ambaw, A., Delele, M. A., Defraeye, T., Ho, Q. T., Opara, L. U., Nicolaï, B. M., & Verboven, P. (2013). The use of CFD to characterize and design post-harvest storage facilities: Past, present and future. *Computers and Electronics in Agriculture, 93*, 184–194.

Anese, M., Manzocco, L., Panozzo, A., Beraldo, P., Foschia, M., & Nicoli, M. C. (2012). Effect of radiofrequency assisted freezing on meat microstructure and quality.' *Food Research International, 46*, 50–54.

Anon. (2013). Bonduelle signs commercial royalty-bearing license with enwave. Refrigerated & Frozen Foods. http://www.refrigeratedfrozenfood.com/articles/87491. Accessed 6 February 2014.

Arai, S., & Watanabe, M. (1986). Freeze texturing of food materials by ice-nucleation with the bacterium erwinia ananas. *Agricultural and Biological Chemistry, 50*, 169–175.

Astrain, D., Aranguren, P., Martínez, A., Rodríguez, A., & Pérez, M. G. (2016). A comparative study of different heat exchange systems in a thermoelectric refrigerator and their influence on the efficiency. *Applied Thermal Engineering, 103*, 1289–1298.

Awad, T. S., Moharram, H. A., Shaltout, O. E., Asker, D., & Youssef, M. M. (2012). Applications of ultrasound in analysis, processing and quality control of food: a review. *Food Research International, 48*, 410–427.

Barba, F. J., Grimi, N., & Vorobiev, E. (2014). New approaches for the use of non-conventional cell disruption technologies to extract potential food additives and nutraceuticals from microalgae. *Food Engineering Reviews, 7*, 45–62.

Barba, F. J., Parniakov, O., Pereira, S. A., Wiktor, A., Grimi, N., Boussetta, N., et al. (2015). Current applications and new opportunities for the use of pulsed electric fields in food science and industry. *Food Research International, 77*, 773–798.

Barbosa, J. R., Ribeiro, G. B., & De Oliveira, P. A. (2012). A state-of-the-art review of compact vapor compression refrigeration systems and their applications. *Heat Transfer Engineering, 33*, 356–374.

Bassols, J., Kuckelkorn, B., Langreck, J., Schneider, R., & Veelken, H. (2002). Trigeneration in the food industry. *Applied Thermal Engineering, 22*, 595–602.

Belis, E. E., Zorrilla, S. E & Peralta, J. M. (2015). Effect of the number of orifices and operative variables on the heat and mass transfer in a hydrofluidization system with static spheres. *Journal of Food Engineering, 153*, 96–107.

Bellas, I., & Tassou, S. A. (2005). Present and future applications of ice slurries. *International Journal of Refrigeration, 28*, 115–121.

Belman-Flores, J. M., Barroso-Maldonado, J. M., Rodríguez-Muñoz, A. P., & Camacho-Vázquez, G. (2015). Enhancements in domestic refrigeration, approaching a sustainable refrigerator: a review. *Renewable and Sustainable Energy Reviews, 51*, 955–968.

Besagni, G., Mereu, R., & Inzoli, F. (2016). Ejector refrigeration: a comprehensive review. *Renewable and Sustainable Energy Reviews, 53*, 373–407.

Best B. R, Aceves H, J. M., Islas S, J. M, Manzini P, F. L., Pilatowsky F, I., Scoccia, R., & Motta, M. (2013). Solar cooling in the food industry in Mexico: A case study. *Applied Thermal Engineering, 50*, 1447–1452.

Bhaskaracharya, R. K., Kentish, S., & Ashokkumar, M. (2009). Selected applications of ultrasonics in food processing. *Food Engineering Reviews, 1*, 31–49.

Blanda, G., Cerretani, L., Cardinali, A., Barbieri, S., Bendini, A., & Lercker, G. (2009). Osmotic dehydrofreezing of strawberries: Polyphenolic content, volatile profile and consumer acceptance. *LWT-Food Science and Technology, 42*, 30–36.

Braun, M. R., Altan, H., & Beck, S. B. M. (2014). Using regression analysis to predict the future energy consumption of a supermarket in the UK. *Applied Energy, 130*, 305–313.

Brizio, A. P. B. R., Gonzaga Junior, M. A., dos Santos Fogaça, F. H., & Prentice, C. (2015). Dynamic monitoring of the shelf life of Cobia (*Rachycentron canadum*): A study on the applicability of a smart photochromic indicator. *International Journal of Food Science and Technology, 50*, 790–796.

Brizio, A. P. D. R., & Prentice, C. (2014). Use of smart photochromic indicator for dynamic monitoring of the shelf life of chilled chicken based products. *Meat Science, 96*, 1219–1226.

Buckley, S. L., & Lillford, P. J. (2009). Chapter 3-Antifreeze Proteins: Their structure, binding and use A2-kasapis, stefan. In:Norton, I. T., & Ubbink, J. B. (Eds.), Modern Biopolymer Science. San Diego: Academic Press.

Calm, J. M. (2008). The next generation of refrigerants-Historical review, considerations, and outlook. *International Journal of Refrigeration, 31*, 1123–1133.

Carvajal-Rondanelli, P. A., Marshall, S. H., & Guzman, F. (2011). Antifreeze glycoprotein agents: structural requirements for activity. *Journal of the Science of Food and Agriculture, 91*, 2507–2510.

Charm, S. E., & Slavin, J. (1962). A method for calculating freezing time of rectangular packages of food. *Annex Bull. Inst. Int. Froid, 567*–558.

Cheftel, J. C., Lévy, J., & Dumay, E. (2000). Pressure-assisted freezing and thawing: Principles and potential applications.' *Food Reviews International, 16*, 453–483.

Cheftel, J. C., Thiebaud, M., & Dumay, E. (2002). Pressure-assisted freezing and thawing of foods: A review of recent studies.' *High Pressure Research, 22*, 601–611.

Chen, Y., Han, W., & Jin, H. (2015). An absorption compression refrigeration system driven by a mid-temperature heat source for low-temperature applications. *Energy, 91*, 215–225.

Cheng, X. F., Zhang, M., & Adhikari, B. (2014). Effect of ultrasonically induced nucleation on the drying kinetics and physical properties of freeze-dried strawberry. *Drying Technology, 32*, 1857–1864.

Cheng, X., Zhang, M., Xu, B., Adhikari, B., & Sun, J. (2015). The principles of ultrasound and its application in freezing related processes of food materials: A review. *Ultrasonics Sonochemistry, 27*, 576–585.

Choi, M. J., Min, S. G., & Hong, G. P. (2016). Effects of pressure-shift freezing conditions on the quality characteristics and histological changes of pork. *LWT-Food Science and Technology, 67*, 194–199.

Cleland, A. C. (1990). 'Food Refrigeration Processes' Analysis, Design and Simulation, Elsevier, London.

Cleland, A. C., & Earle, R. L. (1977). A comparison of analytical and numerical methods of predicting the Freezing times of foods. *Journal of Food Science, 42*, 1390–1395.

Cleland, A. C., & Earle, R. L. (1979). Predicting freezing times of food in rectangular packages. *Journal of Food Science, 44*, 964–970.

Cleland, A. C., & Earle, R. L. (1984). Freezing time prediction for different final product temperatures. *Journal of Food Science, 49*, 1230–1232.

Cleland, D. J. (1991).' A generally applicable simple method for prediction of food freezing and thawing times, '*Proceedings: XVIII International Congress on Refrigeration, 4*, 1874–1877.

Cleland, D. J., Cleland, A. C., & Earle, R. L. (1987a). Prediction of freezing and thawing times for multidimensional shapes by simple formulae. I. Regular shapes. *International Journal of Refrigeration, 10*, 156–164.

Cleland, D. J., Cleland, A. C., & Earle, R. L. (1987b). Prediction of freezing and thawing times for multidimensional shapes by simple formulae. II. Irregular shapes. *International Journal of Refrigeration, 10*, 234–240.

Crevel, R. W. R., Fedyk, J. K., & Spurgeon, M. J. (2002). 'Antifreeze proteins: Characteristics, occurrence and human exposure.' *Food and Chemical Toxicology, 40*, 899–903.

Cruz, R. M. S., Vieira, M. C., & Silva, C. L. M. (2009).' The response of watercress (*Nasturtium officinale*) to vacuum impregnation: Effect of an antifreeze protein type I.' *Journal of Food Engineering, 95*, 339–345.

Davey, L. M., & Pham, Q. T. (1997). 'Predicting the dynamic product heat load and weight loss during beef chilling using a multi-region finite difference approach,' *International Journal of Refrigeration, 20*, 470–482.

Davies, P. L., & Hew, C. L. (1990). 'Biochemistry of fish antifreeze proteins.' *FASEB Journal, 4*, 2460–2468.

De Michelis, A., & Calvelo, A. (1982).' Mathematical models for non-symmetric freezing of beef.' *Journal of Food Science, 47*, 1211–1217.

Defraeye, T., Lambrecht, R., Tsige, A. A., Delele, M. A., Opara, U. L., Cronjé, P., et al. (2013). 'Forced-convective cooling of citrus fruit: Package design.' *Journal of Food Engineering, 118*, 8–18.

Degner, B. M., Chung, C., Schlegel, V., Hutkins, R., & McClements, D. J. (2014). 'Factors influencing the freeze-thaw stability of emulsion-based foods.' *Comprehensive Reviews in Food Science and Food Safety, 13*, 98–113.

Delgado, A., & Sun, D. W. (2012). 'Ultrasound assisted freezing.' In: D. W. Sun (ed.), Handbook of Frozen Food Processing and Packaging, 2nd edition, CRC Press, Taylor and Francis Group, pp. 645–666.

Deng, J., Wang, R. Z., & Han, G. Y. (2011). 'A review of thermally activated cooling technologies for combined cooling, heating and power systems.' *Progress in Energy and Combustion Science, 37*, 172–203.

Deng, Q., Zinoviadou, K. G., Galanakis, C. M., Orlien, V., Grimi, N., Vorobiev, E., et al. (2014). 'The effects of conventional and non-conventional processing on glucosinolates and its derived forms, isothiocyanates: Extraction, degradation, and applications.' *Food Engineering Reviews, 7*, 357–381.

Dermesonlouoglou, E. K., Boulekou, S., Giannakourou, M. C., & Taoukis, P. S. (2007b). 'Osmodehydro freezing of tomato: from production to consumption.' *Acta Horticulturae.*

Dermesonlouoglou, E. K., Giannakourou, M. C., & Taoukis, P. S. (2007c).' Kinetic modeling of the degradation of quality of osmo-dehydrofrozen tomatoes during storage.' *Food Chemistry, 103*, 985–993.

Dermesonlouoglou, E. K., Giannou, V., & Tzia, K. (2016).' Chilling.' In:K. Tzia & Varzakas, T. (eds.), *Handbook of Food Processing and Engineering*, Vol. I: Food Preservation. CRC Press, Boca Raton, Florida, USA, pp. 224–258.

Dermesonlouoglou, E. K., Pourgouri, S., & Taoukis, P. S. (2008). 'Kinetic study of the effect of the osmotic dehydration pre-treatment to the shelf life of frozen cucumber.' *Innovative Food Science and Emerging Technologies, 9*, 542–549.

Dermesonlouoglou, E., Giannakourou, M. C., & Taoukis, P. (2007a). 'Kinetic modeling of the quality degradation of frozen watermelon tissue: Effect of the osmotic dehydration as a pre-treatment.' *International Journal of Food Science and Technology, 42*, 790–798.

Dermesonlouoglou, E., Giannou, V., & Tzia, C. (2016).' Freezing.' In: T. Varzakas & C. Tzia (Eds.), Handbook of Food Processing Food Preservation. CRC Press, Taylor & Francis Group, USA, p. 223.

Dermesonluoglu, E., Katsaros, G., Tsevdou, M., Giannakourou, M., & Taoukis, P. (2015).' Kinetic study of quality indices and shelf life modeling of frozen spinach under dynamic conditions of the cold chain.' *Journal of Food Engineering, 148*, 13–23.

Ding, X., Zhang, H., Liu, W., Wang, L., Qian, H., & Qi, X. (2014).' Extraction of carrot (*Daucus carota*) antifreeze proteins and evaluation of their effects on frozen white salted noodles.' *Food and Bioprocess Technology, 7*, 842–852.

Ding, X., Zhang, H., Wang, L., Qian, H., Qi, X., & Xiao, J. (2015).' Effect of barley antifreeze protein on thermal properties and water state of dough during freezing and freeze-thaw cycles.' *Food Hydrocolloids, 47*, 32–40.

Ducoulombier, M., Teyssedou, A., & Sorin, M. (2006).' A model for energy analysis in supermarkets.' *Energy and Buildings, 38*, 349–356.

Durance, T. (2013). EnWave signs commercial royalty-bearing license with Bonduelle. EnWave Corporation. http://www. enwave. net/news. php?id=478. Accessed 5 February 2014.

Dymek, K., Dejmek, P., Galindo, F. G., & Wisniewski, M. (2015). 'Influence of vacuum impregnation and pulsed electric field on the freezing temperature and ice propagation rates of spinach leaves.' *LWT-Food Science and Technology, 64*, 497–502.

Erdogdu, F., Sarkar, A., & Singh, R. P. (2005). 'Mathematical modeling of air-impingement cooling of finite slab shaped objects and effect of spatial variation of heat transfer coefficient.' *Journal of Food Engineering, 71*, 287–294.

Estrada-Flores, S. (2012). 'Chilling and freezing by cryogenic gases and liquids (Static and Continuous Equipment), ' in R. H. Mascheroni (ed.), Operation in Food Refrigeration. CRC Press. Taylor & Francis Group. pp. 233.

Evans, J. (2012). 'Chilling and freezing by air (Static and Continuous Equipment), ' in R. H. Mascheroni (ed.), Operation in Food Refrigeration. CRC Press. Taylor & Francis Group, p. 215.

Evans, J. A., Foster, A. M., Huet, J. M., Reinholdt, L., Fikiin, K., Zilio, C., et al. (2014a). 'Specific energy consumption values for various refrigerated food cold stores.' *Energy and Buildings, 74*, 141–151.

Evans, J. A., Hammond, E. C., Gigiel, A. J., Reinholdt, L., Fikiin, K., & Zilio, C. (2014b). 'Assessment of methods to reduce the energy consumption of food cold stores.' *Applied Thermal Engineering, 62*, 697–705.

Fellows, P. J. (2000), 'Freezing.' In: Food Processing Technology, Woodhead Publishing Limited and CRC Press LLC. pp. 418–439

Fellows, P. J. (2009).' Freezing.' In: Food Processing Technology: Principles and Practice, Woodhead Publishing Limited, Cambridge, UK.

Fernandes, M. S., Brites, G. J. V. N., Costa, J. J., Gaspar, A. R., & Costa, V. A. F. (2014). 'Review and future trends of solar adsorption refrigeration systems.' *Renewable and Sustainable Energy Reviews, 39*, 102–123.

Fernández, P. P., Otero, L., Guignon, B., & Sanz, P. D. (2006). 'High-pressure shift freezing versus high-pressure assisted freezing: Effects on the microstructure of a food model.' *Food Hydrocolloids, 20*, 510–522.

Fernández, P. P., Otero, L., Martino, M. M., Molina-García, A. D., & Sanz, P. D. (2008). 'High-pressure shift freezing: Recrystallization during storage.' *European Food Research and Technology, 227*, 1367–1377.

Fikiin, A. G. (1992). 'New method and fluidized water system for intensive chilling and freezing of fish.' *Food Control, 3*, 153–160.

Fikiin, K. (2009). 'Emerging and novel freezing processes.' in J. A. Evans (ed.), *Frozen Food Science and Technology. Blackwell Publishing,* 101–123.

Fong, K. F., Lee, C. K., & Chow, T. T. (2011).' Improvement of solar-electric compression refrigeration system through ejector-assisted vapor compression chiller for space conditioning in subtropical climate.' *Energy and Buildings, 43*, 3383–3390.

Gao, X. Q., Shen, J., He, X. N., Tang, C. C., Li, K., Dai, W., et al., (2016). 'Improvements of a room-temperature magnetic refrigerator combined with Stirling cycle refrigeration effect.' *International Journal of Refrigeration, 67*, 330–335.

Garcia-Perez, J. V., Carcel, J. A., Riera, E., Rosselló, C., & Mulet, A. (2012).' Intensification of low-temperature drying by using ultrasound.' *Drying Technology, 30*, 1199–1208.

Ge, Y. T., & Tassou, S. A. (2011).' Thermodynamic analysis of transcritical $CO_2$ booster refrigeration systems in supermarket.' *Energy Conversion and Management, 52*, 1868–1875.

Ge, Y. T., Tassou, S. A., & Suamir, I. N. (2013).' Prediction and analysis of the seasonal performance of tri-generation and CO2 refrigeration systems in supermarkets.' *Applied Energy, 112*, 898–906.

Giannakourou, M. (2016).' Freezing, ' in T. Varzakas & C. Tzia (eds.), Handbook of Food Processing Food Preservation. CRC Press, Taylor & Francis Group. p. 259

Giannakourou, M. C., & Taoukis, P. S. (2002).' Systematic application of time temperature integrators as tools for control of frozen vegetable quality.' *Journal of Food Science, 67*, 2221–2228.

Giannakourou, M. C., & Taoukis, P. S. (2003).' Application of a TTI-based distribution management system for quality optimization of frozen vegetables at the consumer end.' *Journal of Food Science, 68*, 201–209.

Giannakourou, M. C., Koutsoumanis, K., Dermesonlouoglou, E., & Taoukis, P. S. (2001a).' Applicability of the shelf life decision system (slds) for control of nutritional quality of frozen vegetables, ' *Acta Horticulturae.*

Giannakourou, M. C., Koutsoumanis, K., Nychas, G. J. E., & Taoukis, P. S. (2001b). 'Development and assessment of an intelligent shelf life decision system for quality optimization of the food chill chain.' *Journal of Food Protection, 64*, 1051–1057.

Giannakourou, M. C., Koutsoumanis, K., Nychas, G. J. E., & Taoukis, P. S. (2005a).' Field evaluation of the application of time temperature integrators for monitoring fish quality in the chill chain.' *International Journal of Food Microbiology, 102*, 323–336.

Giannakourou, M. C., Koutsoumanis, K., Nychas, G. J. E., & Taoukis, P. S. (2005b). 'Modelling and reduction of risk of fresh pork products with SMAS: A TTI based chill chain management system, ' *Acta Horticulturae.*

Giannakourou, M., & Giannou, V. (2015). 'Chilling and freezing' in: T. Varzakas & C. Tzia (eds.), Handbook of Food Processing Food Preservation. CRC Press, Taylor & Francis Group. p. 319.

Glavan, M., Gradišar, D., Invitto, S., Humar, I., Juričić, Đ., Pianese, C., & Vrančić, D. (2016). 'Cost optimisation of supermarket refrigeration system with hybrid model.' *Applied Thermal Engineering, 103*, 56–66.

Goñi, S. M., Oddone, S., Segura, J. A., Mascheroni, R. H., & Salvadori, V. O. (2008). 'Prediction of foods freezing and thawing times: Artificial neural networks and genetic algorithm approach.' *Journal of Food Engineering, 84*, 164–178.

Góral, D., & Kluza, F. (2009). 'Cutting test application to general assessment of vegetable texture changes caused by freezing.' *Journal of Food Engineering, 95*, 346–351.

Góral, D., & Kluza, F. (2012).' Heat transfer coefficient in impingement fluidization freezing of vegetables and its prediction.' *International Journal of Refrigeration, 35*, 871–879.

Griffith, M., & Ewart, K. V. (1995).' Antifreeze proteins and their potential use in frozen foods.' *Biotechnology Advances, 13*, 375–402.

Gschneidner Jr., K. A., & Pecharsky, V. K. (2008). 'Thirty years of near room temperature magnetic cooling: Where we are today and future prospects.' *International Journal of Refrigeration, 31*, 945–961.

Gutschmidt, J. (1964). In Lascu, G., Bercescu, V., & Niculescu, L., (eds.), Cooling Technology in the Food Industry, Abacus Press, Turnbridge Wales, Kent.

Harman, T. C., Taylor, P. J., Walsh, M. P., & Laforge, B. E. (2002).' Quantum dot superlattice thermoelectric materials and devices.' *Science, 297*, 2229–2232.

Hassas-Roudsari, M., & Goff, H. D. (2012). 'Ice structuring proteins from plants: Mechanism of action and food application.' *Food Research International, 46*, 425–436.

Hong, G. P., & Choi, M. J. (2015).' Comparison of the quality characteristics of abalone processed by high-pressure sub-zero temperature and pressure-shift freezing.' *Innovative Food Science and Emerging Technologies.*

Hossain, M. D. M., Cleland, D. J., & Cleland, A. C. (1992). 'Prediction of freezing and thawing times for foods of three-dimensional irregular shape by using a semi-analytical geometric factor.' *International Journal of Refrigeration, 15*, 241–246.

Hu, F., Sun, D. W., Gao, W., Zhang, Z., Zeng, X., & Han, Z. (2013). 'Effects of pre-existing bubbles on ice nucleation and crystallization during ultrasound-assisted freezing of water and sucrose solution.' *Innovative Food Science and Emerging Technologies, 20*, 161–166.

Hui, Y. H., Miang-Hoog, Wai-Kit Nip, Scott Smith J., & Yu, P. H. F. (2004). 'Principles of food processing (Part 1. 1)' in J. Scott Smith & Y. H. Hui (eds.), Food Processing Principles and Applications. Blackwell Publishing, pp. 3–31.

Hung, Y. C., & Thompson, D. R. (1983).' Freezing time prediction for slab shape foodstuffs by an improved analytical method.' *Journal of Food Science, 48*, 555–560.

Infante Ferreira, C., & Kim, D. S. (2014).' Techno-economic review of solar cooling technologies based on location-specific data.' *International Journal of Refrigeration, 39*, 23–37.

Iodice, P., Dentice d'Accadia, M., Abagnale, C., & Cardone, M. (2016). 'Energy, economic and environmental performance appraisal of a trigeneration power plant for a new district: Advantages of using a renewable fuel.' *Applied Thermal Engineering, 95*, 330–338.

Islam, M. N., Zhang, M., Adhikari, B., Xinfeng, C., & Xu, B. G. (2014). 'The effect of ultrasound-assisted immersion freezing on selected physicochemical properties of mushrooms.' *International Journal of Refrigeration, 42*, 121–133.

Islam, M. N., Zhang, M., Fang, Z., & Sun, J. (2015). 'Direct contact ultrasound assisted freezing of mushroom (Agaricus bisporus): Growth and size distribution of ice crystals.' *International Journal of Refrigeration, 57*, 46–53.

Iwasaka, M., & Ueno, S. (1998).' Structure of water molecules under 14 T magnetic field.' *Journal of Applied Physics, 83*, 6459–6461.

Iwasaka, M., Onishi, M., Kurita, S., & Owada, N. (2011). 'Effects of pulsed magnetic fields on the light scattering property of the freezing process of aqueous solutions.' *Journal of Applied Physics, 109*, 07E320.

Jackson, T. H., Ungan, A., Critser, J. K., &Gao, D. (1997). 'Novel microwave technology for cryopreservation of biomaterials by suppression of apparent ice formation.' *Cryobiology, 34*, 363–372.

Jacobs, S., Auringer, J., Boeder, A., Chell, J., Komorowski, L., Leonard, J., et al. (2014).' The performance of a large-scale rotary magnetic refrigerator.' *International Journal of Refrigeration, 37*, 84–91.

Jaczynski, J., Tahergorabi, R., Hunt, A. L., & Park, J. W. N. (2012). 'Safety and quality of frozen aquatic food products.' in D. W. Sun (ed.), Handbook of Frozen Food Processing and Packaging (2nd edition), CRC Press, Taylor and Francis Group, USA, pp. 344–385.

Jambrak, A. R., & Herceg, Z. (2014). Application of ultrasonics in food preservation and processing. *Conventional and Advanced Food Processing Technologies.*

James, C., & James, S. (2003). 'Freezing: Cryogenic freezing.' *Encyclopedia of Food Science and Nutrition.* 2725–2732.

James, C., Purnell, G., & James, S. J. (2014a).' A critical review of dehydrofreezing of fruits and vegetables.' *Food and Bioprocess Technology, 7,* 1219–1234.

James, C., Purnell, G., & James, S. J. (2015).' A review of novel and innovative food freezing technologies.' *Food and Bioprocess Technology, 8,* 1616–1634.

James, C., Reitz, B., & James, S. J. (2014b). 'The freezing characteristics of garlic bulbs (allium sativum l.) frozen conventionally or with the assistance of an oscillating Weak magnetic field.' *Food and Bioprocess Technology, 8,* 702–708.

James, S. J., & James, C. (2010). 'Advances in the cold chain to improve food safety, food quality and the food supply chain.' in C. Mena & G. Stevens (eds.), Delivering performance in food supply chains, Woodhead Publishing Limited, USA, pp. 366–386.

James, S. J., & James, C. (2010). 'The food cold-chain and climate change.' *Food Research International, 43,* 1944–1956.

James, S. J., James, C., & Evans, J. A. (2006).' Modeling of food transportation systems a review.' *International Journal of Refrigeration, 29,* 947–957.

Jianfeng, S., Congzhi, Z., Jianlou, M., Huiyong, J., Zhixing, S., & Jie, W. (2013). 'Tentative application of stirling cooler technology in butter churning process.' *European Food Research and Technology, 237,* 223–228.

Joshi, C., & Tao, L. C. (1974).' A numerical method of simulating the axisymmetrical freezing of food systems.' *Journal of Food Science, 39,* 623.

Kaale, L. D., & Eikevik, T. M. (2013). 'A histological study of the microstructure sizes of the red and white muscles of Atlantic salmon (Salmo salar) filets during superchilling process and storage.' *Journal of Food Engineering, 114,* 242–248.

Kaale, L. D., & Eikevik, T. M. (2014).' The development of ice crystals in food products during the superchilling process and following storage, a review.' *Trends in Food Science and Technology, 39,* 91–103.

Kaale, L. D., & Eikevik, T. M. (2015).' The influence of superchilling storage methods on the location/distribution of ice crystals during storage of Atlantic salmon (Salmo salar).' *Food Control, 52,* 19–26.

Kaale, L. D., Eikevik, T. M., Bardal, T., Kjorsvik, E., & Nordtvedt, T. S. (2013a). 'The effect of cooling rates on the ice crystal growth in air-packed salmon filets during superchilling and superchilled storage.' *International Journal of Refrigeration, 36,* 110–119.

Kaale, L. D., Eikevik, T. M., Kolsaker, K., & Stevik, A. M. (2014).' Modelling and simulation of food products in superchilling technology.' *Journal of Aquatic Food Product Technology, 23,* 409–420.

Kaale, L. D., Eikevik, T. M., Rustad, T., & Kolsaker, K. (2011). 'Superchilling of food: A review.' *Journal of Food Engineering, 107,* 141–146.

Kaale, L. D., Eikevik, T. M., Rustad, T., Nordtvedt, T. S., Bardal, T., & Kjørsvik, E. (2013b). 'Ice crystal development in pre-rigor Atlantic salmon filets during superchilling process and following storage.' *Food Control, 31,* 491–498.

Karel, M., & Lund, D. B. (2003).' Freezing, ' in M. Karel & D. B. Lund (eds.), *Physical principles of Food Preservati*on, 2nd edition. New York, Marcel Dekker, Inc.

Karellas, S., & Braimakis, K. (2016). 'Energy–exergy analysis and economic investigation of a cogeneration and trigeneration ORC–VCC hybrid system utilizing biomass fuel and solar power.' *Energy Conversion and Management, 107,* 103–113.

Kauffeld, M., Wang, M. J., Goldstein, V., & Kasza, K. E. (2010).' Ice slurry applications.' *International Journal of Refrigeration, 33*, 1491–1505.

Kawahara, H. (2002). 'The structures and functions of ice crystal-controlling proteins from bacteria.' *Journal of Bioscience and Bioengineering, 94*, 492–496.

Kentish, S., & Feng, H. (2014).' Applications of power ultrasound in food processing.' *Annual Review of Food Science and Technology, 5*, 263–284.

Kiani, H., & Sun, D. W. (2011). 'Water crystallization and its importance to freezing of foods: A review.' *Trends in Food Science and Technology, 22*, 407–426.

Kiani, H., Sun, D. W., & Zhang, Z. (2013a). 'Effects of processing parameters on the convective heat transfer rate during ultrasound assisted low temperature immersion treatment of a stationary sphere.' *Journal of Food Engineering, 115*, 384–390.

Kiani, H., Zhang, Z., & Sun, D. W. (2013b). 'Effect of ultrasound irradiation on ice crystal size distribution in frozen agar gel samples.' *Innovative Food Science and Emerging Technologies, 18*, 126–131.

Kiani, H., Zhang, Z., & Sun, D. W. (2015). 'Experimental analysis and modeling of ultrasound assisted freezing of potato spheres.' *Ultrasonics Sonochemistry, 26*, 321–331.

Kiani, H., Zheng, L., & Sun, D. W. (2014). 'Ultrasonic assistance of food freezing' in D. W. Sun (ed.), Emerging Technologies for Food Processing. San Diego, CA, USA: Elsevier Academic Press, pp. 495–514.

Kim, D. S., & Infante Ferreira, C. A. (2008). 'Solar refrigeration options-a state-of-the-art review.' *International Journal of Refrigeration, 31*, 3–15.

Kim, J. U., Ghafoor, K., Ahn, J., Shin, S., Lee, S. H., Shahbaz, H. M., et al. (2016).' Kinetic modeling and characterization of a diffusion-based time-temperature indicator (TTI) for monitoring microbial quality of non-pasteurized angelica juice.' *LWT-Food Science and Technology, 67*, 143–150.

Knorr, D., Schlueter, O., & Heinz, V. (1998).' Impact of high hydrostatic pressure on phase transitions of foods, ' *Food Technology, 52*.

Koutsoumanis, K., Giannakourou, M. C., Taoukis, P. S., & Nychas, G. J. E. (2002).' Application of shelf life decision system (SLDS) to marine cultured fish quality.' *International, Journal of Food Microbiology, 73*, 375–382.

Lagzian, M., Latifi, A. M., Bassami, M. R., & Mirzaei, M. (2014). 'An ice nucleation protein from *Fusarium acuminatum*: Cloning, expression, biochemical characterization and computational modeling.' *Biotechnology Letters, 36*, 2043–2051.

Le Bail A, Orlowska, M., & Havet, M. (2012). 'Electrostatic field assisted food freezing.' in D. W. Sun (ed.), Handbook of Frozen Food Processing and Packaging (2nd edition), CRC Press, Taylor and Francis Group, USA, pp. 685–91.

Le Bail, A., Tzia, C., & Giannou, V. (2012).' Quality and Safety of Frozen Bakery Products. ' In: D. W. Sun (ed.), Handbook of Frozen Food Processing and Packaging (2nd ed.), CRC Press, Taylor and Francis Group, USA, pp. 501–528.

LeBail, A., Chevalier, D., Mussa, D. M., & Ghoul, M. (2002a). 'High pressure freezing and thawing of foods: A review.' *International Journal of Refrigeration, 25*, 504–513.

LeBail, A., Mussa, D., Rouillé, J., Ramaswamy, H. S., Chapleau, N., Anton, M., et al. (2002b).' High pressure thawing. Application to selected sea-foods.' in H. Rikimaru (ed.) *Progress in Biotechnology*. Elsevier.

Lee, S. J., & Rahman, A. T. M. M. (2014).' Chapter 8-Intelligent Packaging for Food Products, ' in A. Han & H. Jung (eds.), *Innovations in Food Packaging* (Second Edition). San Diego, Academic Press.

Levy, F. L. (1958).' Calculating freezing time of fish in air blast freezers.' *Journal of Refrigeration*, *1*, 55–58.

Li, J., & Lee, T. C. (1995).' Bacterial ice nucleation and its potential application in the food industry.' *Trends in Food Science & Technology*, *6*, 259–265.

Li, Q., Yan, Q., Chen, J., He, Y., Wang, J., Zhang, H., et al. (2012).' Molecular characterization of an ice nucleation protein variant (InaQ) from Pseudomonas syringae and the analysis of its transmembrane transport activity in Escherichia coli.' *International Journal of Biological Sciences*, *8*, 1097–1108.

Lim, S., Gunasekaran, S., & Imm, J. Y. (2012).' Gelatin-Templated Gold Nanoparticles as Novel Time-Temperature Indicator.' *Journal of Food Science*, *77*(9), N45–N49.

Lucas, T., & Raoult-Wack, A. L. (1998).' Immersion chilling and freezing in aqueous refrigerating media: review and future trends.' *International Journal of Refrigeration*, *21*(6), 419–429.

Luo, E. C., Dai, W., Zhang, Y., & Ling, H. (2006). 'Experimental investigation of a thermoacoustic-Stirling refrigerator driven by a thermoacoustic-Stirling heat engine.' *Ultrasonics*, *44*, Supplement, e1531–e3.

Lv, H., Wang, X. D., Meng, J. H., Wang, T. H., & Yan, W. M. (2016). 'Enhancement of maximum temperature drop across thermoelectric cooler through two-stage design and transient supercooling effect.' *Applied Energy*, *175*, 285–292.

Magnussen, O. M., Haugland, A., Torstveit Hemmingsen, A. K., Johansen, S., & Nordtvedt, T. S. (2008), 'Advances in superchilling of food–Process A characteristics and product quality.' *Trends in Food Science & Technology*, *19*, 418–424.

Mannapperuma, J. D., Singh, R. P., & Reid, D. S. (1994a). 'Effective heat transfer coefficients encountered in air blast freezing of whole chicken and chicken parts, individually and in packages, ' *Int. J. Refrig*, *17*, 263–272.

Mannapperuma, J. D., Singh, R. P., & Reid, D. S., (1994b). Effective heat transfer coefficients encountered in air blast freezing of single plastic wrapped whole turkey, *Int. J. Refrig*. *17*, 273–280.

Marra, F., Bedanel, T. F., Uyar, R., Erdogdu, F., & Lyng, J. G. (2015).' Application of radiowave frequency technology in food preservation and processing.' In: S. Bhattacharya (ed.) Conventional and Advanced Food Processing Technologies, 1st Edition, Chapter 20, Wiley Blackwell Ltd, 501–513.

Martinez, A., Astrain, D., Rodriguez, A., & Aranguren, P. (2016). 'Advanced computational model for Peltier effect based refrigerators.' *Applied Thermal Engineering*, *95*, 339–347.

Mascheroni, R. H., & Calvelo, A. (1982).' A simplified model for freezing time calculation in foods.' *Journal of Food Science*. *47*, 1201–1207.

Mihara., & Nakagawa. (2012). 'A method for freezing an object such as foods and organs in a non-destructive state in the presence of a magnetic field, wherein frequency of the magnetic field is 200 Hz or higher.' EP 2 499 924 A1.

Min, G., & Rowe, D. M. (2006).' Experimental evaluation of prototype thermoelectric domestic-refrigerators.' *Applied Energy*, *83*, 133–152.

Mittal, G. S. (2006).' Freezing loads and freezing time calculation.' Handbook of Frozen Food Processing and Packaging. Taylor & Francis Group, LLC, USA.

Mittal, G. S. (2013). Artificial Neural Network (ANN) based process modeling. Handbook of Farm, Dairy and Food Machinery Engineering: Second Edition.

Mohanraj, M., Jayaraj, S., & Muraleedharan, C. (2009).' Environment friendly alternatives to halogenated refrigerants-A review.' *International Journal of Greenhouse Gas Control*, *3*, 108–119.

Mok, J. H., Choi, W., Park, S. H., Lee, S. H., & Jun, S. (2015). 'Emerging pulsed electric field (PEF) and static magnetic field (SMF) combination technology for food freezing.' *International Journal of Refrigeration, 50*, 137–145.

Morais, A. Rd. V., Alencar, Éd. N., Xavier Júnior, F. H., Oliveira, C. Md., Marcelino, H. R., Barratt, G., et al. (2016). 'Freeze-drying of emulsified systems: A review.' *International Journal of Pharmaceutics, 503*, 102–114.

Mota-Babiloni, A., Navarro-Esbrí, J., Barragán-Cervera, Á., Molés, F., & Peris, B. (2015a). 'Analysis based on EU Regulation No 517/2014 of new HFC/HFO mixtures as alternatives of high GWP refrigerants in refrigeration and HVAC systems.' *International Journal of Refrigeration, 52*, 21–31.

Mota-Babiloni, A., Navarro-Esbrí, J., Barragán-Cervera, Á., Molés, F., Peris, B., & Verdú, G. (2015b). 'Commercial refrigeration: An overview of current status.' *International Journal of Refrigeration, 57*, 186–196.

Mota-Babiloni, A., Navarro-Esbrí, J., Barragán, Á., Molés, F., & Peris, B. (2014). 'Drop in energy performance evaluation of R1234yf and R1234ze(E) in a vapor compression system as R134a replacements.' *Applied Thermal Engineering, 71*, 259–265.

Mota-Babiloni, A., Navarro-Esbrí, J., Molés, F., Cervera, Á. B., Peris, B., & Verdú, G. (2016). 'A review of refrigerant R1234ze(E) recent investigations.' *Applied Thermal Engineering, 95*, 211–222.

Mott, L. F. (1964). 'The prediction of product freezing time.' *Aust. Refrig. Air Cond. Heat, 18*, 16.

Nagaoka, J., Takagi, S., & Hotani, S. (1956). 'Experiments on the freezing of fish by air blast freezer.' *Journal of Tokyo University of Fisheries, 42*(1), 65–73.

Nahor, H. B., Hoang, M. L., Verboven, P., Baelmans, M., & Nicolaï, B. M. (2005).' CFD model of the airflow, heat and mass transfer in cool stores.' *International Journal of Refrigeration, 28*, 368–380.

Nair, V., Tailor, P. R., & Parekh, A. D. (2016). 'Nanorefrigerants: A comprehensive review on its past, present and future.' *International Journal of Refrigeration, 67*, 290–307.

Nkwetta, D. N., & Sandercock, J. (2016). 'A state of the-art review of solar air-conditioning systems.' *Renewable and Sustainable Energy Reviews, 60*, 1351–1366.

Norio, O., & Satoru, K., Super-quick freezing method and apparatus therefore. Patent no. US 6250087 B1, 2001 (US Patent).

North, M. F., & Lovatt, S. J. (2012).' Freezing methods and equipment (Chapter 8).' In: D. W. Sun (ed.), Handbook of frozen food processing and packaging. CRC Press, Taylor & Francis Group, USA, pp. 187–200.

Norton, T., & Sun, D. W. (2008). 'Recent advances in the use of high pressure as an effective processing technique in the food industry.' *Food and Bioprocess Technology, 1*, 2–34.

Otero, L., & Sanz, P. D. (2006). 'High-pressure-shift freezing: Main factors implied in the phase transition time.' *Journal of Food Engineering, 72*, 354–363.

Otero, L., & Sanz, P. D. (2012), 'High Pressure Shift Freezing.' In D. W. Sun (Ed.) Handbook of Frozen Food Processing and Packaging, CRC Press, Taylor and Francis Group, pp. USA, 667–683.

Otero, L., Rodríguez, A. C., Pérez-Mateos, M., & Sanz, P. D. (2016). 'Effects of magnetic fields on freezing: application to biological products.' *Comprehensive Reviews in Food Science and Food Safety, 15*, 646–667.

Otero, L., Sanz, P. D., Guignon, B., & Aparicio, C. (2009). 'Experimental determination of the amount of ice instantaneously formed in high-pressure shift freezing.' *Journal of Food Engineering, 95*, 670–676.

Owada N., (2007). Highly-efficient freezing apparatus and high-efficient freezing method. US Patent 7237400 B2.

Owada, N., & Kurita, S., (2001). Super-quick freezing method and apparatus therefore. US Patent 6250087 B1.

Owada, N., & Saito, S., (2010). Quick freezing apparatus and quick freezing method. US Patent 7810340B2.

Palm, B. (2008). 'Hydrocarbons as refrigerants in small heat pump and refrigeration systems: A review.' *International Journal of Refrigeration, 31*, 552–563.

Panarese, V., Herremans, E., Cantre, D., Demir, E., Vicente, A., Gãmez Galindo, F., et al. (2016). 'X-ray microtomography provides new insights into vacuum impregnation of spinach leaves.' *Journal of Food Engineering, 188*, 50–57.

Panarese, V., Rocculi, P., Baldi, E., Wadsö, L., Rasmusson, A. G., & Gómez Galindo, F. (2014). 'Vacuum impregnation modulates the metabolic activity of spinach leaves.' *Innovative Food Science and Emerging Technologies, 26*, 286–293.

Pang, X. F., Deng, B., & Tang, B. (2012).' Influences of magnetic field on macroscopic properties of water.' *Modern Physics Letters B, 26*.

Pang, X., & Deng, B. (2008).' Investigation of changes in properties of water under the action of a magnetic field.' *Science in China, Series G: Physics, Mechanics and Astronomy, 51*, 1621–1632.

Pang, X., & Deng, B., (2009). Investigation of magnetic-field effects on water. 2009 International Conference on Applied Superconductivity and Electromagnetic Devices, ASEMD 2009, pp. 278–283.

Parniakov, O., Bals, O., Lebovka, N., & Vorobiev, E. (2016). 'Effects of pulsed electric fields assisted osmotic dehydration on freezing-thawing and texture of apple tissue.' *Journal of Food Engineering, 183*, 32–38.

Parniakov, O., Lebovka, N. I., Bals, O., & Vorobiev, E. (2015).' Effect of electric field and osmotic pre-treatments on quality of apples after freezing-thawing.' *Innovative Food Science and Emerging Technologies, 29*, 23–30.

Paul, M., & Zhongjie, H. (2015).' A review of carbon dioxide as a refrigerant in refrigeration technology.' South African Journal of Science, 111(9/10), 1–10.

Pearson, A. (2008). 'Refrigeration with ammonia.' *International Journal of Refrigeration, 31*, 545–551.

Peralta, J. M., Rubiolo, A. C., & Zorrilla, S. E. (2009).' Design and construction of a hydrofluidization system. Study of the heat transfer on a stationary sphere.' *Journal of Food Engineering, 90*, 358–364.

Peralta, J. M., Rubiolo, A. C., & Zorrilla, S. E. (2012).' Mathematical modeling of the heat and mass transfer in a stationary potato sphere impinged by a single round liquid jet in a hydrofluidization system.' *Journal of Food Engineering, 109*, 501–512.

Pham, Q. T. (1984). 'Extension to Plank's equation for predicting freezing times of foodstuffs of simple shapes.' *International Journal of Refrigeration, 7*, 377–383.

Pham, Q. T. (1986). 'Simplified equations for predicting the freezing times of foodstuffs.' *Journal of Food Technology, 21*, 209–219.

Pham, Q. T. (1987a). 'Calculation of bound water in frozen food, '*Journal of Food Science, 52*, 210–212.

Pham, Q. T. (1987b). 'Moisture transfer due to temperature changes or fluctuations, '*Journal of Food Engineering, 6*, 33–49.

Pham, Q. T. (1987c). 'A converging front model for the asymmetric freezing of slab-shaped food, '*Journal of Food Science. 52*, 795–800.

Pham, Q. T. (2014). Refrigeration in food preservation and processing. *Conventional and Advanced Food Processing Technologies.*

Pham, Q. T., & Willix, J. (1990).' Effect of biot number and freezing rate on the accuracy of some food freezing time prediction methods.' *Journal of Food Science, 55*, 1429–1434.

Phoon, P. Y., Galindo, F. G., Vicente, A., & Dejmek, P. (2008). 'Pulsed electric field in combination with vacuum impregnation with trehalose improves the freezing tolerance of spinach leaves.' *Journal of Food Engineering, 88*, 144–148.

Picart, L., Dumay, E., Guiraud, J. P., & Cheftel, J. C. (2004). 'Microbial inactivation by pressure-shift freezing: Effects on smoked salmon mince inoculated with *Pseudomonas fluorescens, Micrococcus luteus* and *Listeria innocua.*' *LWT-Food Science and Technology, 37*, 227–238.

Picart, L., Dumay, E., Guiraud, J. P., & Cheftel, J. C. (2005). 'Combined high pressure-subzero temperature processing of smoked salmon mince: Phase transition phenomena and inactivation of *Listeria innocua.*' *Journal of Food Engineering, 68*, 43–56.

Plank, R. (1913). 'DieGefrierdauer von Eisblocken (Freezing time of ice)' Zeitschriftfur die gesamteKalteIndustrie (The Journal of Refrigeration Industry), *20*(6), 109–114.

Qi, L., Xu, M., Fu, Z., Mira, T., & Zhang, X. (2014). 'C2SLDS: A WSN-based perishable food shelf-life prediction and LSFO strategy decision support system in cold chain logistics.' *Food Control, 38*, 19–29.

Qian, S., Nasuta, D., Rhoads, A., Wang, Y., Geng, Y., Hwang, Y., Radermacher, R., & Takeuchi, I. (2016). 'Not-in-kind cooling technologies: A quantitative comparison of refrigerants and system performance.' *International Journal of Refrigeration, 62*, 177–192.

Qureshi, B. A., & Zubair, S. M. (2012). 'The impact of fouling on performance of a vapor compression refrigeration system with integrated mechanical sub-cooling system.' *Applied Energy, 92*, 750–762.

Qureshi, B. A., & Zubair, S. M. (2013a). 'Cost optimization of heat exchanger inventory for mechanical subcooling refrigeration cycles.' *International Journal of Refrigeration, 36*, 1243–1253.

Qureshi, B. A., & Zubair, S. M. (2013b). 'Mechanical sub-cooling vapor compression systems: Current status and future directions.' *International Journal of Refrigeration, 36*, 2097–2110.

Qureshi, B. A., Inam, M., Antar, M. A., & Zubair, S. M. (2013). 'Experimental energetic analysis of a VAPOUR compression refrigeration system with dedicated mechanical sub-cooling.' *Applied Energy, 102*, 1035–1041.

Rahman, M. S., & Velez-Ruiz, J. F. (2007).' Food Preservation by freezing' in M. S. Rahman. (ed.), *Handbook of Food Preservation.* CRC Press, Taylor & Francis Group, LLC. USA, pp. 636–665.

Ramasamy, H. M., & Marcotte, M. (2006).' Low-temperature preservation.' in Food Processing, Principles and Applications. Boca Raton, FL, Taylor & Francis Group, LLC.

Ramløv, H., & Johnsen, J. L. (2014), 'Controlling the Freezing Process with Antifreeze Proteins.' in: D-W. Sun, (ed.), Emerging Technologies for Food Processing (2nd edition), San Diego, CA, USA: Elsevier Academic Press, pp. 539–562.

Rasmussen, L. B., Bacher, P., Madsen, H., Nielsen, H. A., Heerup, C., & Green, T. (2016). 'Load forecasting of supermarket refrigeration.' *Applied Energy, 163*, 32–40.

Reddy, N. J. M. (2016). 'A low power, eco-friendly multipurpose thermoelectric refrigerator.' *Frontiers in Energy, 10*, 79–87.

Regand, A., & Goff, H. D. (2006).' Ice recrystallization inhibition in ice cream as affected by ice structuring proteins from winter wheat grass.' *Journal of Dairy Science, 89*, 49–57.

Reid, D. S. (2000). 'Factors which influence the freezing process: an examination of new sights.' *Bulletin of the International Institute of Refrigeration 2000–2003*, 5–15.

Rezk, A., Elsayed, A., Mahmoud, S., & Al-Dadah, R. (2014).' Adsorption refrigeration.' in V. M. Petrova (ed.)*Advances in Engineering Research. Nova Science Publishers, Vol. 8*, pp. 21-64.

Riffat, S. B., & Ma, X. (2003).' Thermoelectrics: a review of present and potential applications.' *Applied Thermal Engineering, 23*, 913–935.

Ruiz-Garcia, L., Lunadei, L., Barreiro, P., & Robla, J. I. (2009). 'A review of wireless sensor technologies and applications in agriculture and food industry: State of the art and current trends.' *Sensors (Switzerland), 9*, 4728–4750.

Salvadori, V. O., & Mascheroni, R. H. (2002).' Analysis of impingement freezers performance.' *Journal of Food Engineering, 54*, 133–140.

Sánchez-Vega, R., Elez-Martínez, P., & Martín-Belloso, O. (2015). 'Influence of high-intensity pulsed electric field processing parameters on antioxidant compounds of broccoli juice.' *Innovative Food Science and Emerging Technologies, 29*, 70–77.

Santacatalina, J. V., Fissore, D., Cárcel, J. A., Mulet, A., & García-Pérez, J. V. (2015).' Model-based investigation into atmospheric freeze drying assisted by power ultrasound.' *Journal of Food Engineering, 151*, 7–15.

Sarbu, I., & Sebarchievici, C. (2015). 'General review of solar-powered closed sorption refrigeration systems.' *Energy Conversion and Management, 105*, 403–422.

Sari, O., & Balli, M. (2014).' From conventional to magnetic refrigerator technology.' *International Journal of Refrigeration, 37*, 8–15.

Sarkar, A., & Singh, R. P. (2003). 'Spatial variation of convective heat transfer coefficient in air impingement applications.' *Journal of Food Science, 68*, 910–916.

Sarkar, A., & Singh, R. P. (2004a). 'Air impingement technology for food processing: Visualization studies.' *LWT-Food Science and Technology, 37*, 873–879.

Sarkar, A., & Singh, R. P. (2004b). 'Modelling flow and heat transfer during freezing of foods in forced airstreams.' *Journal of Food Science, 69*, E488–E96.

Sarkar, A., Nitin, N., Karwe, M. V., & Singh, R. P. (2004). 'Fluid flow and heat transfer in air jet impingement in food processing.' *Journal of Food Science, 69*, CRH113–CRH22.

Sarkar, J. (2012). 'Ejector enhanced vapor compression refrigeration and heat pump systems-A review.' *Renewable and Sustainable Energy Reviews, 16*, 6647–6659.

Schössler, K., Jäger, H., & Knorr, D. (2012).' Novel contact ultrasound system for the accelerated freeze-drying of vegetables.' *Innovative Food Science & Emerging Technologies, 16*, 113–120.

Shayanfar, S., Chauhan, O. P., Toepfl, S., & Heinz, V. (2013). 'The interaction of pulsed electric fields and texturizing-antifreezing agents in quality retention of defrosted potato strips.' *International Journal of Food Science & Technology, 48*, 1289–1295.

Shayanfar, S., Chauhan, O. P., Toepfl, S., & Heinz, V. (2014). 'Pulsed electric field treatment prior to freezing carrot discs significantly maintains their initial quality parameters after thawing.' *International Journal of Food Science & Technology, 49*, 1224–1230.

Shmroukh, A. N., Ali, A. H. H., & Ookawara, S. (2015). 'Adsorption working pairs for adsorption cooling chillers: A review based on adsorption capacity and environmental impact.' *Renewable and Sustainable Energy Reviews, 50*, 445–456.

Siemer, C., Aganovic, K., Toepfl, S., & Heinz, V. (2015). 'Application of Pulsed Electric Fields in Food.' in S. Bhattacharya (ed.), *Conventional and Advanced Food Processing Technologies,* 1st Edition. Chapter 26, Wiley Blackwell Ltd., 645–672.

Smith, P. G. (2011). 'Low-temperature Preservation (Chapter 11).' In: D. R. Heldman (ed.), *Introduction to Food Process Engineering*, 2ndEd. 275–296.

Soukoulis, C., & Ian, F., (2014). 'To be submitted in Critical Reviews in Food Science and Nutrition Innovative ingredients and emerging technologies for controlling ice recrystallization, texture and structure stability in frozen dairy desserts: A review.' *Critical Reviews in Food Science and Nutrition*.

Steven Brown, J., & Domanski, P. A. (2014). 'Review of alternative cooling technologies.' *Applied Thermal Engineering, 64*, 252–262.

Su, G., Ramaswamy, H. S., Zhu, S., Yu, Y., Hu, F., & Xu, M. (2014). 'Thermal characterization and ice crystal analysis in pressure shift freezing of different muscle (shrimp and porcine liver) versus conventional freezing method.' *Innovative Food Science and Emerging Technologies, 26*, 40–50.

Suamir, I. N. (2014).' Solar driven absorption chiller for medium temperature food refrigeration, a study for application in Indonesia, ' *Applied Mechanics and Materials.*

Suamir, I. N., Tassou, S. A., & Marriott, D. (2012). 'Integration of CO 2 refrigeration and trigeneration systems for energy and GHG emission savings in supermarkets.' *International Journal of Refrigeration, 35*, 407–417.

Suamir, I., & Tassou, S. A. (2013).' Performance evaluation of integrated trigeneration and CO₂ refrigeration systems.' *Applied Thermal Engineering, 50*, 1487–1495.

Sugiartha, N., Tassou, S. A., Chaer, I., & Marriott, D. (2009). 'Trigeneration in food retail: An energetic, economic and environmental evaluation for a supermarket application.' *Applied Thermal Engineering, 29*, 2624–2632.

Sun, D. W. (2007). Computational Fluid Dynamics in Food Processing. CRC Press, Boca Raton, Florida, USA.

Sun, D. W., & Zheng, L. (2006).' Innovations in freezing process.' In: D. W. Sun (ed.), Handbook of Frozen Food Processing and Packaging, Chapter 8, CRC Press, Taylor and Francis Group, pp. 175–195.

Suzuki, T., Takeuchi, Y., Masuda, K., Watanabe, M., Shirakashi, R., Fukuda, Y., et al. (2009).' Experimental investigation of effectiveness of magnetic field on food freezing process.' Transactions of the Japan Society of Refrigerating and Air Conditioning Engineers, No 4, 371–386.

Tao, L. C. (1967). 'Generalized numerical solutions of freezing a saturated liquid in cylinders and spheres.' *AIChEJ. 13*, 165.

Tao, Y., & Sun, D. W. (2015). 'Enhancement of food processes by ultrasound: A review.' *Critical Reviews in Food Science and Nutrition, 55*, 570–594.

Taoukis, P. S., Bili, M., & Giannakourou, M. (1998). 'Application of shelf life modeling of chilled salad products to a TTI based distribution and stock rotation system, ' *Acta Horticulturae.*

Taoukis, P. S., Giannakourou, M. C., & Tsironi, T. N. (2006). 'Monitoring and control of the cold chain.' in D. W. Sun (ed.), *Handbook of Frozen Food Processing and Packaging*, Chapter 14, CRC Press, Taylor and Francis Group, USA, pp. 278–307.

Tassou, S. A., De-Lille, G., & Ge, Y. T. (2009). 'Food transport refrigeration–Approaches to reduce energy consumption and environmental impacts of road transport.' *Applied Thermal Engineering, 29*, 1467–1477.

Tassou, S. A., Ge, Y., Hadawey, A., & Marriott, D. (2011).' Energy consumption and conservation in food retailing.' *Applied Thermal Engineering, 31*, 147–156.

Tassou, S. A., Lewis, J. S., Ge, Y. T., Hadawey, A., & Chaer, I. (2010).' A review of emerging technologies for food refrigeration applications.' *Applied Thermal Engineering, 30*, 263–276.

Tien, R. H., & Geiger, G. E. (1967).' A heat transfer analysis of the solidification of binary eutectic system, ' *Journal of Heat Transfer, 9*, 230.

Tien, R. H., & Geiger, G. E. (1968).' The unidimensional solidification of a binary eutectic system with a time-dependent surface temperature.' *Journal of Heat Transfer, 9C*, 27.

Tien, R. H., & Koumo, V. (1968).' Unidimensional solidification of a subvariable surface temperature. *Trans. Metall. Soc. AIME, 242*, 283.

Tien, R. H., & Koumo, V. (1969).' Effect of density change on the solidification of alloys.' *Am. Soc, Mech. Eng.* (Paper) 69-HT-45.

Tironi, V., De Lamballerie, M., & Le-Bail, A. (2010). 'Quality changes during the frozen storage of sea bass (*Dicentrarchus labrax*) muscle after pressure shift freezing and pressure assisted thawing.' *Innovative Food Science and Emerging Technologies, 11*, 565–573.

Torres-de María, G., Abril, J., & Casp, A. (2005). 'Coefficients d'échanges superficiels pour la réfrigération et la congélation d'aliments immergés dans un coulis de glace (Surface Coefficients for Refrigeration and Freezing of Food Submerged in an Ice Slurry), '*International Journal of Refrigeration, 28*, 1040–1047.

Tu, J., Zhang, M., Xu, B., & Liu, H. (2015). 'Effects of different freezing methods on the quality and microstructure of lotus (Nelumbo nucifera) root.' *International Journal of Refrigeration, 52*, 59–65.

Urrutia, G., Arabas, J., Autio, K., Brul, S., Hendrickx, M., Kakolewski, A., Knorr, D., et al., (2007). 'SAFE ICE: Low-temperature pressure processing of foods: Safety and quality aspects, process parameters and consumer acceptance.' *Journal of Food Engineering, 83*, 293–315.

Ustun, N. S., & Turhan, S., (2015). 'Antifreeze Proteins: Characteristics, Function, Mechanism of Action, Sources and Application to Foods.' *Journal of Food Processing and Preservation, 39*, 3189–3197.

Vaikousi, H., Biliaderis, C. G., & Koutsoumanis, K. P. (2009). 'Applicability of a microbial Time Temperature Indicator (TTI) for monitoring spoilage of modified atmosphere packed minced meat.' *International Journal of Food Microbiology, 133*, 272–278.

Van Buggenhout, S., Grauwet, T., Van Loey, A., & Hendrickx, M. (2007). 'Effect of high-pressure induced ice I/ice III-transition on the texture and microstructure of fresh and pretreated carrots and strawberries.' *Food Research International, 40*, 1276–1285.

Van Buggenhout, S., Messagie, I., Van Der Plancken, I., & Hendrickx, M. (2006). 'Influence of high-pressure-low-temperature treatments on fruit and vegetable quality related enzymes.' *European Food Research and Technology, 223*, 475–485.

Vanderveer, T. L., Choi, J., Miao, D., & Walker, V. K. (2014).' Expression and localization of an ice nucleating protein from a soil bacterium, Pseudomonas borealis.' *Cryobiology, 69*, 110–118.

Velickova, E., Tylewicz, U., Dalla Rosa, M., Winkelhausen, E., Kuzmanova, S., & Gómez Galindo, F. (2013). 'Effect of vacuum infused cryoprotectants on the freezing tolerance of strawberry tissues.' *LWT-Food Science and Technology, 52*, 146–150.

Venketesh, S., & Dayananda, C. (2008).' Properties, potentials, and prospects of antifreeze proteins.' *Critical Reviews in Biotechnology, 28*, 57–82.

Verboven, P., Scheerlinck, N., & Nicol, B. M. (2003).' Surface heat transfer coefficients to stationary spherical particles in an experimental unit for hydrofluidization freezing of individual foods.' *International Journal of Refrigeration, 26*, 328–336.

Volkert, M., Puaud, M., Wille, H. J., & Knorr, D. (2012). 'Effects of high pressure-low temperature treatment on freezing behavior, sensorial properties and air cell distribution in sugar rich dairy based frozen food foam and emulsions.' *Innovative Food Science and Emerging Technologies, 13*, 75–85.

Wan, X., & Knoll, M. (2016). 'A new type of TTI based on an electrochemical pseudo transistor.' *Journal of Food Engineering, 168*, 79–83.

Wang S., & Sun D. W. (2012). 'Antifreeze proteins.' In: D. W. Sun (ed.), Handbook of Frozen Food Processing and Packaging (2nd edition), CRC Press, Taylor and Francis Group, USA, pp. 693–708.

Wang, F., Li, D. Y., & Zhou, Y. (2016). 'Analysis for the ejector used as expansion valve in vapor compression refrigeration cycle.' *Applied Thermal Engineering, 96*, 576–582.

Wang, S., Liu, X., Yang, M., Zhang, Y., Xiang, K., & Tang, R. (2015).' Review of time temperature indicators as quality monitors in food packaging.' *Packaging Technology and Science,* 2i8, 839–867.

Wang, Y., Zhang, M., Adhikari, B., Mujumdar, A. S., & Zhou, B. (2013). 'The Application of Ultrasound Pretreatment and Pulse-Spouted Bed Microwave Freeze Drying to Produce Desalted Duck Egg White Powders.' *Drying Technology, 31*, 1826–1836.

Wäppling RaaholWatanabe, M., & Arai, S. (1987).' Freezing of water in the presence of the ice nucleation active bacterium, erwinia ananas, and its application for efficient freeze-drying of foods.' *Agricultural and Biological Chemistry, 51*, 557–563.

Watanabe, M., Kanesaka, N., Masuda, K., & Suzuki, T. (2011).' Effect of oscillating magnetic field on supercooling in food freezing.' *Proceedings of the 23rd IIR International Congress of Refrigeration; refrigeration for sustainable development, August 21–26,* Prague, Czech Republic. *1*, 2892–2899.

Wiktor, A., Schulz, M., Voigt, E., Witrowa-Rajchert, D., & Knorr, D. (2015).' The effect of pulsed electric field treatment on immersion freezing, thawing and selected properties of apple tissue.' *Journal of Food Engineering, 146*, 8–16.

Woo, M. W., & Mujumdar, A. S. (2010). 'Effects of electric and magnetic field on freezing and possible relevance in freeze drying.' *Drying Technology, 28*, 433–443.

Wowk, B. (2012). 'Electric and magnetic fields in cryopreservation.' *Cryobiology, 64*, 301–303.

Wu, D., Hou, S., Chen, J., Sun, Y., Ye, X., Liu, D., et al., (2015). 'Development and characterization of an enzymatic time-temperature indicator (TTI) based on *Aspergillus niger* lipase.' *LWT-Food Science and Technology, 60*, 1100–1104.

Wu, L., Orikasa, T., Tokuyasu, K., Shiina, T., & Tagawa, A. (2009). 'Applicability of vacuum-dehydrofreezing technique for the long-term preservation of fresh-cut eggplant: Effects of process conditions on the quality attributes of the samples.' *Journal of Food Engineering, 91*, 560–565.

Xanthakis E., Le Bail A., & Havet, M. (2014a). 'Freezing combined with electrical and magnetic disturbances' (2nd edition). in D. W. Sun (ed.), Emerging Technologies for Food Processing. San Diego, CA, USA: Elsevier Academic Press, pp. 563–582.

Xanthakis, E., Havet, M., Chevallier, S., Abadie, J., & Le-Bail, A. (2013).' Effect of static electric field on ice crystal size reduction during freezing of pork meat.' *Innovative Food Science and Emerging Technologies, 20*, 115–120.

Xanthakis, E., Le-Bail, A., & Havet, M. (2014b). Chapter 30-Freezing Combined with Electrical and Magnetic Disturbances A2-Sun, Da-Wen. *Emerging Technologies for Food Processing (Second Edition).* San Diego: Academic Press.

Xanthakis, E., Le-Bail, A., & Ramaswamy, H. (2014c). 'Development of an innovative microwave assisted food freezing process.' *Innovative Food Science and Emerging Technologies, 26*, 176–181.

Xanthakis, E., Le-Bail, A., & Ramaswamy, H. (2014d). 'Development of an innovative microwave assisted food freezing process.' *Innovative Food Science & Emerging Technologies, 26*, 176–181.

Xin, Y., Zhang, M., & Adhikari, B. (2014).' Freezing characteristics and storage stability of broccoli (*Brassica oleracea* L. var. *Botrytis* L.)Under Osmodehydrofreezing and Ultrasound-Assisted Osmodehydro freezing Treatments.' *Food and Bioprocess Technology, 7*, 1736–1744.

Xin, Y., Zhang, M., Xu, B., Adhikari, B., & Sun, J. (2015). 'Research trends in selected blanching pretreatments and quick freezing technologies as applied in fruits and vegetables: A review.' *International Journal of Refrigeration, 57*, 11–25.

Xu, B. G., Zhang, M., Bhandari, B., Cheng, X. f., & Sun, J. (2015).' Effect of ultrasound immersion freezing on the quality attributes and water distributions of wrapped red radish.' *Food and Bioprocess Technology, 8*, 1366–1376.

Xu, B., Zhang, M., Bhandari, B., & Cheng, X. (2014). 'Influence of Ultrasound-Assisted Osmotic Dehydration and Freezing on the Water State, Cell Structure, and Quality of Radish (*Raphanus sativus* L.) Cylinders.' *Drying Technology, 32*, 1803–1811.

Yu, B., Liu, M., Egolf, P. W., & Kitanovski, A. (2010). 'A review of magnetic refrigerator and heat pump prototypes built before the year 2010.' *International Journal of Refrigeration, 33*, 1029–1060.

Zaritzky, N. (2012) 'Physical–Chemical Principles in Freezing.' In: D. W. Sun (Ed.), *Handbook of Frozen Food Processing and Packaging* (2nd edition), CRC Press, Taylor and Francis Group, USA, pp. 3–37.

Zeyghami, M., Goswami, D. Y., & Stefanakos, E. (2015).' A review of solar thermo-mechanical refrigeration and cooling methods.' *Renewable and Sustainable Energy Reviews, 51*, 1428–1445.

Zhang, C., Zhang, H., Wang, L., & Guo, X. (2008).' Effect of carrot (*Daucus carota*) antifreeze proteins on texture properties of frozen dough and volatile compounds of crumb.' *LWT-Food Science and Technology, 41*, 1029–1036.

Zhang, Y., Zhang, H., Wang, L., Qian, H., & Qi, X. (2015).' Extraction of Oat (*Avena sativa* L.)Antifreeze Proteins and Evaluation of Their Effects on Frozen Dough and Steamed Bread.' *Food and Bioprocess Technology, 8*, 2066–2075.

Zhao, J. H., Hu, R., Xiao, H. W., Yang, Y., Liu, F., Gan, Z. L., et al. (2014). 'Osmotic dehydration pretreatment for improving the quality attributes of frozen mango: Effects of different osmotic solutes and concentrations on the samples.' *International Journal of Food Science and Technology, 49*, 960–968.

Zhao, J. H., Liu, F., Pang, X. L., Xiao, H. W., Wen, X., & Ni, Y. Y. (2016). 'Effects of different osmo-dehydrofreezing treatments on the volatile compounds, phenolic compounds and physicochemical properties in mango (*Mangifera indica* L.).' *International Journal of Food Science and Technology, 51*, 1441–1448.

Zheng, L., & Sun, D. W. (2006). 'Innovative applications of power ultrasound during food freezing processes-A review.' *Trends in Food Science and Technology, 17*, 16–23.

Zhou, Z. S., Wang, L. W., Jiang, L., Gao, P., & Wang, R. Z. (2016). 'Non-equilibrium sorption performances for composite sorbents of chlorides–ammonia working pairs for refrigeration.' *International Journal of Refrigeration, 65*, 60–68.

Zimm, C., Jastrab, A., Sternberg, A., Pecharsky, V., Gschneidner, K., Osborne, M., & Anderson, I. (1998). Description and Performance of a Near-Room Temperature Magnetic Refrigerator. In: P. Kittel (ed.), Advances in Cryogenic Engineering, 43, 1759-1760, Boston, MA: Springer, USA.

## CHAPTER 8

# SIMPLE AND MULTIPLE EMULSIONS EMPHASIZING ON INDUSTRIAL APPLICATIONS AND STABILITY ASSESSMENT

REVATHI RAVIADARAN[a,b], KASTURI MUTHOOSAMY[a], and SIVAKUMAR MANICKAM[a*]

[a]Malaysian Palm Oil Board (MPOB), 6 Persiaran Institusi, Bandar Baru Bangi, 43000 Kajang, Selangor, Malaysia

[b]Department of Chemical and Environmental Engineering, Faculty of Engineering, University of Nottingham, Jalan Broga, 43500 Semenyih, Selangor Darul Ehsan, Malaysia, Tel: +6(03) 8924 8156; Fax: +6(03) 8924 8017; E-mail: Sivakumar.Manickam@nottingham.edu.my

## CONTENTS

## 8.1 INTRODUCTION

In recent times, emulsions have become a topic of research and also gained interest in industries such as food, nutraceuticals and pharmaceuticals. In the food industries emulsions could be used to improve the flavor, taste and texture. Also, the nutraceutical industries strive to supply consumers with functional food fortified with nutraceuticals. In this case, emulsions could be used to improve the solubility, physiological activity as well as stability of the nutraceutical compounds. The pharmaceutical industries are rapidly growing and promoting healthcare for the consumers. Research is vastly conducted in pharmaceutical area and emulsions seem to be a suitable tool especially in providing the targeted and improved treatment for various diseases (Putheti, 2015). Amongst the food, nutraceutical and pharmaceutical industries, the pharmaceutically active compounds are the most actively researched for industrial use. Shakeel and Faisal (2010) indicated that nanoemulsion is a promising vehicle for solubility and dissolution enhancement of active compounds in comparison to solid lipid nanoparticle and solid dispersion in their study.

There are two types of emulsions, that is, simple and multiple emulsions. Simple emulsion is also commonly referred to as nanoemulsion existing as oil-in-water (O/W) and water-in-oil (W/O) emulsion. O/W emulsion consists of an oil phase dispersed in a continuous water phase, where each oil globule is surrounded by a thin interfacial layer consisting of emulsifier molecules as shown in Figure 8.1 (Acosta, 2009; Tadros et al., 2004).

Multiple emulsions are also known as double emulsion or emulsion of emulsion (Tang et al., 2013; Muschiolik, 2007). There are two types of

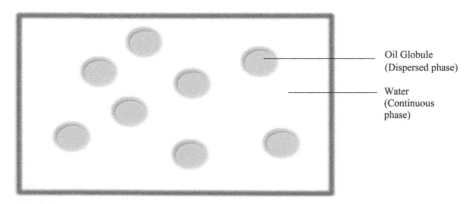

**FIGURE 8.1**  Simple emulsion droplet (O/W).

multiple emulsions, that is, oil-in-water-in-oil (O/W/O) as well as water-in-oil-in-water (W/O/W) emulsion (Kumar et al., 2012; Sahu et al., 2013; Florence and Whitehill, 1982). In a W/O/W system, the oil globules contain smaller internal water droplets, which are dispersed in an external continuous water phase as shown in Figure 8.2. It can be simply explained as an organic phase (hydrophobic) which separates the internal and external aqueous phases (Sahu et al., 2013; Kumar et al., 2012). Typically, O/W and W/O/W are the most commonly researched owing to their evident advantages for transport in human body (Huang et al., 2010). Figure 8.3 shows the photomicrograph of a simple W/O emulsion and a multiple W/O/W emulsion under an optical microscopy adapted from (Bonnet et al., 2009).

This chapter focus is on demonstrating the recent applications in utilizing both the simple and multiple emulsions for the encapsulation of various industrially important active compounds. Emulsion system has the most applications in pharmaceutical industry in comparison to other industries such as food and nutraceutical industries. For pharmaceutical emulsions, various delivery routes such as topical, parenteral and oral are researched. However, it is to be highlighted that only the oral delivery of active compounds has been provided here as the oral route for delivery is regarded as the optimal route for achieving the therapeutic benefits with increased patient compliance. It is the preferred delivery route as it is non-invasive and possesses convenience for self-administration (Shantha Kumar et al., 2006). Stability of both the emulsions and the mechanisms have also been described.

Various destabilization routes such as gravitational separation, flocculation, and coalescence leading to phase separation are discussed in greater

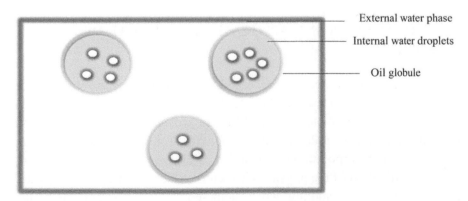

External water phase

Internal water droplets

Oil globule

**FIGURE 8.2**   Multiple emulsion droplet (W/O/W).

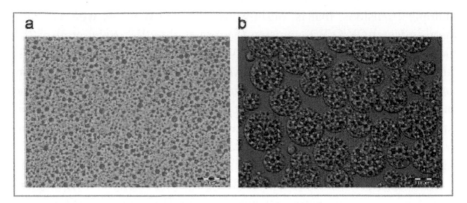

**FIGURE 8.3**  Photomicrograph of (a) W/O emulsion and (b) W/O/W emulsion under an optical microscope (adopted from (Bonnet et al., 2009).

detail below. Besides, the composition of emulsion, that is, types of oils, emulsifiers and stabilizing agents and their ability in improving the stability of an emulsion system are discussed. Also, the common production methods of both simple and multiple emulsions are described and their influence on the stability is given. Additionally, the destabilization of multiple emulsions due to the osmotic pressure difference between the internal and external water phase are broadly shown. The current methods used to overcome this osmotic pressure difference in stabilizing the multiple emulsion and their successes are discussed. Lastly, the common characterization carried out both for the simple and multiple emulsions are briefly given. The current analytical measures in assessing the stability of emulsions are discussed and further suggestions on utilizing the appropriate analytical measures in assessing the stability of emulsions are provided.

## 8.2  ADVANTAGES AND APPLICATIONS OF SIMPLE AND MULTIPLE EMULSIONS

The main advantages and the successful use of emulsions are due to their encapsulation ability. Desai and Park (2005) have broadly discussed the benefits of encapsulation such as the ability to protect the active compounds from degradation by reducing their reactivity from outside environment, reduction in the evaporation of the active compounds to the outer environment, modification of the physical characteristics of the encapsulated material to allow easier handling and to help separate the components of the mixture that would otherwise react with each another. Lu and Bowles (2013) discussed the importance of encapsulation including enhanced stability, protection against oxidation, retention of volatile ingredients, taste masking, moisture-triggered controlled release, pH-triggered controlled release, consecutive delivery of multiple active ingredients and enhanced bioavailability and efficacy. Table 8.1 demonstrates the currently researched industrially active compounds and their potential uses and advantages for the food, nutraceutical and pharmaceutical industries.

Multiple emulsions have also been proven for their advantages in various applications for encapsulating industrially active compounds for the food, nutraceutical and pharmaceutical industries. The details of the reported works associated with the applications as listed above are shown in Table 8.2. For pharmaceutically active compounds, multiple emulsions vastly looked into owing to their ability to provide sustained delivery of the active compounds contained in the innermost phase which requires passing through several phases prior to release at the absorption site (Sahu et al., 2013). The sustained release of entrapped active compounds reduces the transient active compound overload and its side effects (Tang et al., 2012). Corresponding to the sustained release, a reduction in the required dosage could also be achieved. Yanagie et al. (2011) and Takahashi (1973) demonstrated that multiple emulsion is more effective than simple conventional emulsion for cancer treatment. Most importantly, multiple emulsion has the ability to encapsulate both hydrophilic and lipophilic components in a single tool. However, typically there has been no significant research done to explore the potential of encapsulating both hydrophilic and lipophilic components in a multiple emulsion till present.

## 8.3   FACTORS INFLUENCING THE STABILITY OF EMULSIONS

### 8.3.1   DESTABILIZATION MECHANISMS IN THE EMULSIONS

Both simple and multiple emulsions are kinetically stable. This ensures stability against gravitational separation as the small droplet size ensures domination of Brownian motion against gravitational forces (Tadros et al., 2004). The smaller size of droplets in the nanoemulsions provides greater stability to gravitational separation and droplet aggregation compared to the conventional emulsions (McClements and Rao, 2011). This could be

**TABLE 8.1**  Bibliographic Studies with Applications of Simple Emulsions

| Simple emulsions | Applications | References |
|---|---|---|
| **Food** | | |
| Plant essential oils of clove bud (*Syzygium aromaticum*) and oregano (*Origanum vulgare*) are natural compounds with remarkable antimicrobial properties that can serve as natural preservative by forming antimicrobial films. | Both essential oils in nanoemulsion reduced the counts of yeasts and molds in sliced bread during the investigation period of 15 days. | Otoni et al. (2014) |
| | Droplet size reduction provided a further improvement in antimicrobial properties | |
| | Due to increased bioavailability, less preservative is used and still delivered with the same antimicrobial efficiency. | |
| Paprika oleoresin carotenoids as food colorant. | The stability of carotenoids in the nanoemulsions increased as the droplet size reduced. | Pascual-Pineda et al. (2015) |
| | Encapsulation of lipophilic carotenoids was possible with nanoemulsion delivery system. | |
| Coenzyme Q10 (CoQ10) effect of CoQ10 on the physico-chemical stability of emulsions was compared to emulsions without CoQ10. | The emulsion with CoQ10 was used as a functional cream in the cheese making process. | Stratulat et al. (2013) |
| | Quantification of CoQ10 by HPLC showed its retention into cheese matrix by 93%. | |
| | Protein retention and cheese yield were not affected by the addition of the functional CoQ10 cream in the development of functional cheeses. | |
| The effects of oleoresin capsicum (OC) and nanoemulsion OC (NOC) on obesity in obese rats fed with a high-fat diet was tested. | NOC reduced the body weight and adipose tissue mass, whereas OC did not. | Kim et al. (2014b) |
| | NOC could be suggested as a potential anti-obesity agent in obese rats fed with a high fat diet. The effects of NOC on obesity were associated with changes of multiple gene expression, activation of AMPK, and inhibition of glycerol-3-phosphate dehydrogenase in white adipose tissue. | |

| Simple emulsions | Applications | References |
|---|---|---|
| The influence of nanoemulsion encapsulating streptomycin (AUSN-4) on the microbiological, proximal, chemical, and sensory qualities of Indo-Pacific king mackerel (*Scomberomorus guttatus*) steaks stored at 20°C was studied for 72 h. | AUSN-4 nanoemulsion treatment showed an initial reduction in the heterotrophic, $H_2S$ and lactic acid bacterial populations in 12 h. AUSN-4 nanoemulsion treatment significantly decreased the values of chemical indicators of spoilage throughout the storage period with an extension of shelf-life of 48 h. | Joe et al. (2012) |
| Grape berries (*Vitis labruscana* Bailey) are vulnerable to contamination by foodborne pathogens. Nanoemulsions of lemongrass oil (LO) were developed to coat grape berries to improve the shelf-life and microbiological safety. | Antimicrobial effects against *S. Typhimurium* and *E. coli* were exhibited by the LO-nanoemulsion coatings. Shelf-life of grape berries were prolonged. | Kim et al. (2014a) |
| **Nutraceuticals** | | |
| Vitamin E (α-tocopherol) is a nutraceutical compound, which has been shown to possess potent antioxidant and anticancer activity. However, its biological activity may be limited by its poor bioavailability. | Nanoemulsion formulation increased the oral bioavailability when compared to the conventional emulsion. *In vivo* oral bioavailability study revealed that vitamin E in a nanoemulsion form showed a 3-fold increase in the bioavailability when compared to the conventional emulsion. | Parthasarathi et al. (2016) |

**TABLE 8.1** (Continued)

| Simple emulsions | Applications | References |
| --- | --- | --- |
| Curcumin nanoemulsions were formulated due to their high lipophilicity and poor bioavailability. | In a single oral dose study, the rats given 125 mg of curcumin nanoemulsion showed the presence of curcumin in lymph reaching a maximum concentration of 19.16 mg/ml compared to curcumin without emulsification at only 4.58 mg/ml. | Dhavamani and Lokesh (2016) |
| | Pharmacokinetics showed higher levels of curcumin nanoemulsion in serum and liver of rats compared to curcumin without emulsion formation. | Wang et al. (2008) |
| | Enhanced anti-inflammatory activity of curcumin encapsulated in O/W emulsions was observed in the mouse ear inflammation model. | |
| | 43% or 85% inhibition effect was found on 12-O-tetradecanoylphorbol-13-acetate (TPA)-induced edema of mouse ear model. | |
| Piplartine (PL) is an alkaloid found in black-pepper and known for its anticancer activity, however, due to poor solubility its use for oral administration is a challenge. | PL loaded nanoemulsions exhibited enhanced dissolution, cellular permeability and cytotoxic effects as compared to pure PL. | Fofaria et al. (2016) |
| | PL loaded nanoemulsions did not exhibit toxicity in mice upon daily oral administration for 60 days. | |
| | PL loaded nanoemulsions showed 1.5-fold increase in oral bioavailability as compared to free PL. | |
| | PL loaded nanoemulsions showed marked anti-tumor activity at a dose of 10 mg/kg in melanoma tumor bearing mice. | |
| | Administration of PL nanoemulsion enhanced its solubility, oral bioavailability and anti-tumor efficacy. | |
| 5-Demethyltangeretin (5DT) is a unique citrus flavonoid which has been shown to have anti-cancer effects, but its low water-solubility and poor oral bioavailability limits its application. | Emulsion-based delivery systems increased the bioavailability of 5DT and its uptake by intestinal cancer cells. | Zheng et al. (2014) |
| | A cytotoxicity assay showed that 5DT encapsulated in smaller droplets produced stronger growth inhibition than those in larger droplets. | |

| Simple emulsions | Applications | References |
|---|---|---|
| Quercetin loaded nanoemulsions | SDT encapsulated within emulsions with small droplet sizes produced higher cellular absorption and lower cancer cell viability. | Jain et al. (2013) |
| | DPPH free radical scavenging assay showed a comparable antioxidant activity of quercetin nanoemulsion to free quercetin. | |
| | The *in vitro* digestion model demonstrated that the bioaccessibility of quercetin in simulated small intestinal conditions was improved by nanoencapsulation. | |
| | MTT cell viability assay showed that quercetin nanoemulsion did not have significant toxicity on cervix adenocarcinoma. | |
| Antioxidant activities of green tea catechins are shown to exert protective effects against chronic diseases. However, the poor chemical stability and low oral bioavailability limit its usage. | There was 2.78 fold increase in the bioaccessibility of catechins nanoemulsion as compared to unencapsulated catechins. | Bhushani et al. (2016) |
| | There is a significant increase in the intestinal permeability of catechins as assessed by Caco-2 cell model. | |
| Thymoquinone-rich fraction nanoemulsion (TQRFNE) was prepared to investigate the potential acute toxicity of this nanoemulsion in Sprague Dawley rats. | TQRFNE containing 44.5 mg TQ/kg administered orally to rats for 14 days resulted in all the animals appearing normal and healthy throughout the study. | Tubesha et al. (2013) |
| | There were no observed mortality or any signs of toxicity during the experimental period. | |
| Silymarin was used for hepatoprotection with poor oral bioavailability. | The nanoemulsion-treated group showed a significant decrease in the glutamate oxaloacetate transaminase, pyruvate transaminase, alkaline phosphatase, total bilirubin and tissue lipid peroxides and increased the total protein, albumin, globulin and tissue glutathione. | Parveen et al. (2011) |
| | The results indicated an excellent potential of the nanoemulsion formulation for the reversal of tetrachloromethane-induced liver toxicity in rats as compared to standard silymarin. | |

**TABLE 8.1** (Continued)

| Simple emulsions | Applications | References |
|---|---|---|
| **Pharmaceutical** | | |
| Camptothecin is a topoisomerase I inhibitor that acts against cancers. | Camptothecin in nanoemulsion systems showed retarded drug release. | Fang et al. (2009) |
| Clinical application is limited by its insolubility, instability, and toxicity. | Camptothecin in nanoemulsion exhibited cytotoxicity against melanomas and ovarian cancer cells. | |
| | It also increased the release of camptothecin from the nanoemulsion system illustrating a drug-targeting effect. | |
| Nanoemulsion of amlodipine besilate (AB) was prepared to enhance the solubility and oral bioavailability of AB and to achieve localized delivery of drug at the target site. | The release of drug from the nanoemulsion was significantly higher than the marketed tablet formulation. | Chhabra et al. (2011) |
| | The pharmacokinetics and biodistribution studies of nanoemulsion in mice demonstrated a bioavailability of 475% against AB suspension. | |
| | In almost all the tested organs, the uptake of AB from nanoemulsion was significantly higher than AB suspension especially in heart with a drug targeting index of 44.1%, confirming the efficacy of nanosized formulation at therapeutic site. | |
| | A three times increase in the overall residence time of nanoemulsion further enhanced the bioavailability of AB. | |
| Paclitaxel is a hydrophobic drug with poor bioavailability. | A significantly higher concentration in the systemic circulation was observed with paclitaxel nanoemulsion. | Tiwari and Amiji (2006) |
| | The absorbed drug was found to be distributed in the liver, kidneys, and lungs. | |

| Simple emulsions | Applications | References |
|---|---|---|
| Nanoemulsion of efavirenz was developed for pediatric HIV therapy. | *In vitro* release profile showed more than 80% release within 6 h and pharmacokinetic studies also proved a promising *in vivo* absorption profile when compared to efavirenz suspension.<br><br>The developed nanoemulsion proved to be an effective dose for pediatric use. | Kotta et al. (2014) |
| Primaquine (PQ) is an antimalarial drug to combat relapsing form of malaria especially in case of *Plasmodium vivax* and *Plasmodium ovale*. However, application of PQ in higher doses is limited by severe tissue toxicity including hematological and gastrointestinal related side effects. | Primaquine nanoemulsion showed effective antimalarial activity against Plasmodium infection in swiss albino mice at a 25% lower dose level as compared to conventional oral dose.<br><br>It also exhibited improved oral bioavailability and was taken up preferentially by the liver with drug concentration higher at least by 45% as compared with the normal drug suspension. | Singh and Vingkar (2008) |
| Nanoemulsion formulation of Ramipril improved its oral bioavailability. | *In vitro* drug release of the nanoemulsion formulations was highly significant as compared to marketed capsule formulation and drug suspension.<br><br>The bioavailability of Ramipril nanoemulsion in comparison to the conventional capsule form was 229.62% and to drug suspension was 539.49%. | Shafiq et al. (2007) |
| Pterostilbene prevents cancer, diabetes, and cardiovascular diseases but it has poor solubility. | Solubility and *in vitro* release of pterostilbene was significantly improved (96.5% in pH 3.6 buffer; 13.2% in pH 7.4 buffer) in comparison to the pterostilbene suspension (lower than 21.4% in pH 3.6 buffer; 2.6% in pH 7.4 buffer). | Zhang et al. (2014) |
| Nanoemulsion formulation to enhance the bioavailability of the poorly water-soluble drug talinolol. | Nanoemulsion formulation improved the drug release, permeability, and *in vivo* bioavailability as compared to drug suspension due to increased solubility and enhanced permeability of the drug from a nanosized emulsion. | Ghai and Sinha (2012) |

**TABLE 8.1** (Continued)

| Simple emulsions | Applications | References |
|---|---|---|
| Carvedilol has antioxidant and anti-proliferative properties which makes it suitable to combat heart failure. However, it undergoes extensive first pass metabolism and its systemic bioavailability is only 25 to 35%. | Pharmacokinetic studies of carvedilol nanoemulsion in rats revealed a significant increase in oral bioavailability. | Poluri et al. (2011) |
| Simvastatin is poorly bioavailable as it is practically insoluble in water. | *In vitro* release studies showed an increased dissolution rate of nanoemulsion compared with plain drug. | Chavhan et al. (2013) |
| | Pharmacokinetic studies showed relative bioavailability of simvastatin nanoemulsion was 369.0% with respect to plain drug suspension. | |
| | Pharmacodynamic studies conducted in hyperlipidemic rats showed a significant decrease in the total cholesterol and triglyceride levels for nanoemulsion as compared to plain drug proving an improvement in the bioavailability. | |
| Nanoemulsions containing Saquinavir (SQV), an anti-HIV protease inhibitor, for enhanced oral bioavailability and brain disposition. | SQV nanoemulsion concentrations in the systemic circulation were three-fold higher as compared to the control aqueous suspension. | Vyas et al. (2008) |
| | The oral bioavailability and distribution to the brain were significantly enhanced with SQV delivered in nanoemulsion. | |
| | An enhanced rate of SQV absorption following oral administration of nanoemulsions was observed. | |
| Docetaxel has been used for many malignancies. It has high toxicity and non-selective distribution. | Docetaxel nanoemulsion presented antiproliferation effects on both U87 cells and bEnd.3 cells where it's *in vivo* toxicity was found to be significantly low. | Gaoe et al. (2012) |
| | Docetaxel nanoemulsion is a new, less toxic, drug formulation that is effective for brain glioma therapy. | |

| Simple emulsions | Applications | References |
|---|---|---|
| Glipizide is a second generation Sulphonyl urea drug used in the treatment of noninsulin dependent diabetes mellitus. It has less solubility in water and has a short half-life of 2–4 h. | Glipizide nanoemulsions were successful in sustaining the drug release for 12 h.<br><br>Optimized formulation reduced the blood glucose levels up to 12 h. | Srilatha et al. (2013) |
| Clotrimazole as an antimalarial drug. | In four-day suppressive test, mice treated with 10 mg/kg clotrimazole nanoemulsion showed the highest suppression of parasitemia than that of 10 mg/kg clotrimazole suspension.<br><br>The percent reduction of parasitemia was significantly higher in 10 and 15 mg/kg clotrimazole nanoemulsion groups compared to 15 mg/kg suspension group.<br><br>In both murine models, survival of mice treated with nanoemulsion was significantly prolonged compared to suspension at equivalent doses. | Borhade et al. (2012) |
| *Candesartan cilexetil* (CC) exhibited an incomplete intestinal absorption with low oral bioavailability due to its poor aqueous solubility. | Absorption of CC loaded nanoemulsion (CCN) was significantly improved in intestinal tract compared with free CC solution.<br><br>Concentration of active candesartan in rat plasma improved over 10-fold after CC was incorporated into CCN. | Gao et al. (2011) |
| *Benzyl isothiocyanate* (BITC), a compound found in cruciferous vegetables, is an effective chemopreventive agent. | These formulations markedly increased the apical to basolateral transport of BITC in Caco-2 cell monolayers.<br><br>The nanoemulsions were easily taken up by human cancer cells A549 and SKOV-3 and inhibited tumor growth *in vitro*.<br><br>Nanoemulsions enhanced the absorption and bioavailability. | Qhattal et al. (2011) |

**TABLE 8.2** Bibliographic Studies of Multiple Emulsion Applications

| Multiple emulsion | Applications | References |
|---|---|---|
| **Food** | | |
| Flavor | Sensory tests indicated that there is a significant taste difference between multiple emulsion and simple emulsion containing the same ingredients. | Madene et al. (2006) |
| | Delayed release of flavor in multiple emulsion compared to simple emulsion. | Dickinson (2011) |
| Food | Production of reduced fat contents in food by controlling the volume occupied in the internal water droplets. | Lobato-Calleros et al. (2006 (2009) |
| | | Nehir El and Simsek (2012) |
| | | Freire et al. (2015) |
| Healthy lipid cooked meat systems (MS) in which pork back fat was replaced by a W/O/W emulsion prepared with hydroxytyrosol (HXT) within an inner aqueous phase. | Presence of HXT improved oxidative stability and increased DPPH free radical scavenging. | Cofrades et al. (2014) |
| | MS formulated with HXT W/O/W emulsion displayed oxidative stability assuring successful development of potential healthier meat products. | |
| W/O/W emulsion for protection of Lactobacillus acidophilus from gastric juice for utilizing the advantages of the probiotics. | The relative viability of the bacteria included in the W/O/W emulsion was 49% at 2 h in the model gastric juice, whereas the viability of the bacteria directly dispersed in the juice declined to 1.3% at 0.67 h. | Shima et al. (2006) |
| Low-fat food production where beef emulsion systems in which beef fat was replaced by W/O/W emulsion. | Incorporation of W/O/W emulsion resulted in reduced lipid, increased protein content, and modified fatty acid composition. | Serdaroğlu et al. (2016) |
| | W/O/W emulsion treatments had lower jelly and fat separation, higher water-holding capacity and higher emulsion stability than control samples with beef fat. | |
| | W/O/W emulsions had promising impacts on modifying fatty acid composition and developing both oxidatively stable beef emulsion systems. | |

| Multiple emulsion | Applications | References |
|---|---|---|
| Oaxaca cheese is a potential vehicle for delivering probiotic cells. Their processing conditions are harsh and the probiotic cells, *Lactobacillus plantarum* need to be protected. | The exposure of *Lactobacillus plantarum* W/O/W to simulated gastrointestinal conditions did not affect its cells viability. The inclusion of cells in W/O/W emulsion is an effective method for protecting their viability against harsh processing. | Rodríguez-Huezo et al. (2014) |
| *Lactobacillus rhamnosus* was entrapped in the inner aqueous phase of a W/O/W emulsion. | The viability of the entrapped L. rhamnosus in the W/O/W emulsion was compared to that of non-entrapped control cells exposed to low pH and bile salt conditions. The survival of the entrapped cells increased significantly under low pH and bile salt conditions, and their survival was 108% and 128%, respectively. It is concluded that the W/O/W emulsion protected L. rhamnosus against simulated gastrointestinal tract conditions. | Pimentel-González et al. (2009) |
| **Nutraceuticals** | | |
| A W/O/W emulsion was prepared to encapsulate betalain which is an antioxidant, anti-inflammatory, and detoxifying agent. | W/O/W emulsion showed sustained release. | Kaimainen et al. (2015) |
| Anthocyanins are unstable hydrophilic plant pigments for use in functional foods. | It could be shown that it is possible to stabilize anthocyanins in the inner phase of W/O/W. | Frank et al. (2012) |
| The stability of vitamin A was studied in three different emulsions: oil-in-water (O/W), water-in-oil (W/O), and oil-in-water-in-oil (O/W/O). | The stability of vitamin A in the O/W/O emulsion was the highest among the three types of emulsions; remaining percentages at 50°C after 4 weeks in the O/W/O, W/O, and O/W emulsions were 56.9, 45.7, and 32.3, respectively. O/W/O emulsion is the most useful formulation to stabilize vitamin A. | Yoshida et al. (1999) |

**TABLE 8.2** (Continued)

| Multiple emulsion | Applications | References |
|---|---|---|
| Encapsulation of vitamin B12 in W/O/W emulsions was optimized to produce functional cream for cheese milk standardization. | Encapsulation of vitamin B12 in W/O/W emulsions exhibited greater than 96% efficiency and prevented vitamin losses during *in vitro* gastric digestion.\ <br><br> Compared with non-encapsulated vitamin B12, encapsulation in double emulsions reduced vitamin B12 losses and increased retention in cheese from 6.3 to more than 90%. | Giroux et al. (2013) |
| Apigenin-loaded emulsions | The *in vivo* pharmacokinetics revealed that apigenin–loaded emulsions had a higher oral bioavailability than did the orally administrated apigenin suspensions. <br><br> W/O/W multiple emulsions have great potential as targeted delivery systems for apigenin, and may enhance *in vitro* and *in vivo* bioavailability when they pass through the digestive tract. | Kim et al. (2016) |
| Multiple emulsions containing rifampicin were prepared and evaluated for *in vitro* and *in vivo* performance. | The plasma level observed implied that the multiple emulsion gave prolonged drug release. | Nakhare and Vyas (1997) |
| **Pharmaceuticals** | | |
| Urease has wide clinical applications. Urease enzyme serves as a virulence factor and is responsible for pathogenesis in humans. | Urease activity of microbial sources has contributed to the development of many diseases and urease from plant sources is used as vaccine against microbial infection on the basis of its inhibitory activity. <br><br> Multiple emulsion can be used for enzyme immobilization. | May and Li (1974) |
| Chlorpheniramine maleate multiple emulsion | Evaluation of this system showed a gradual and consistent drug release from the delivery system. | Ghosh et al. (1997); |
| Indomethacin as an anti-inflammatory drug | Prolonged oral release in both *in vitro* and *in vivo* studies in rabbit. | Roy and Gupta (1993) |

| Multiple emulsion | Applications | References |
|---|---|---|
| Nattokinase-loaded self-double-emulsifying drug delivery system (SDEDDS). | Encapsulation of nattokinase was up to 86.8 ± 8.2%. The cumulative release of nattokinase within 8 h was about 30%, exhibiting a sustained release effect. | Wang et al. (2015) |
| | The pharmacodynamics study indicated that nattokinase-loaded SDEDDS could significantly prolong the whole blood clotting time in mouse and effectively improve the carrageenan-induced tail thrombosis compared with nattokinase solution. | |
| Chloroquine – antimalarial agent | Taste masking. | Vaziri and Warburton (1994) |
| Chlorpromazine – antipsychotic drug | Ability to mask the bitter taste efficiently. | Khan (2007) |
| Barbiturates | Drug overdose treatment. | Chiang et al. (1978) |
| | Entrapping excess drug in multiple emulsions cures over dosage treatment by utilizing the difference in the pH. | |
| Insulin | A multiple emulsion for oral administration of insulin showed hypoglycemic activity for a long period after oral administration to rats. | Toorisaka et al. (2003) |
| Diclofenac sodium | Inverse targeting. | Talegaonkar and Vyas (2005) |
| | Multiple emulsion retards the reticuloendothelial system uptake mainly to liver, brain and targeting to non-reticuloendothelial system tissues such as lungs and inflammatory tissue. | |
| Vaccine adjuvant of surface antigen hemagglutinin. | Multiple emulsion formulations were more effective in eliciting an immune response against influenza virus in rats than the conventional vaccine due to their adjuvant properties. | Bozkir and Hayta (2004) |
| | | Bozkir et al. (2004) |
| | | Shah et al. (2015) |

**TABLE 8.2** (Continued)

| Multiple emulsion | Applications | References |
|---|---|---|
| Boron-entrapped W/O/W emulsion applied for boron delivery carrier in hepatocellular carcinoma treatment. | Multiple emulsion was superior to those by conventional emulsion. | Yanagie et al. (2011) |
| Anticancer agent 5-fluorouracil (5-FU) | W/O/W multiple emulsion of anticancer agent 5-FU showed better radioactivity in the regional lymph nodes after intratesticular injection compared to conventional emulsion. | Takahashi (1973) |
| Water soluble Epirubicin | Benefits of small droplet size. | Higashi et al. (1995) |
| | W/O/W emulsion in treating hepatocellular carcinoma accumulates in the small vessels of the tumor when injected to the liver via the hepatic artery. | Higashi and Setoguchi (2000) |
| | Clinical studies showed that the size of microdroplets influenced the anti-tumor effect of the therapy. | |
| Aqueous oxygen | Oxygen substitute. | Zheng et al. (1993, 1991) |
| | Benefits of small droplet size. | |
| | A hemoglobin multiple emulsion has small droplet size to provide oxygen flow through blood vessels to desired body tissues or organs thereby providing a blood substitute by maintaining high oxygen exchange activity. | |
| Insulin | Protection against enzymatic degradation caused by pepsin, trypsin and a-chymotrypsin. | Cunha et al. (1997) |
| | Insulin encapsulated in multiple emulsion has promoted intestinal membrane permeability. | Onuki et al. (2004) |
| Elemenum – anti-cancer drug for inhibition on the growth and proliferation of the human lung adenocarcinoma cell line. | Reduced side effects. | Saraf (2010 (Li et al. (2005) |
| | No marrow inhibition and no harm to the heart and liver. | |

| Multiple emulsion | Applications | References |
|---|---|---|
| W/O/W emulsion encapsulating Tacrolimus – immuno suppressive drug used for patients undergoing liver transplantation. | Suppressed side effects of nephrotoxic and neurotoxic effects.<br><br>Reduced dosage.<br><br>Enhanced immunosuppressive efficacy.<br><br>Delivery to targeted organ site. | Onuki et al. (2004) |
| Aspirin | Sustained delivery.<br><br>Reduced side effects in gastrointestinal tract, kidney and platelets. | Tang et al. (2012) |
| Doxorubicin hydrochloride | Sustained delivery by multiple emulsion in Sprague Dawley rats.<br><br>Reduced dosage. | Lin et al. (1992) |
| W/O/W emulsion of Penclomedine – novel cytotoxic agent for antitumor activity in mice by both intraperitoneal and intravenous administration. | Preliminary results suggest that this emulsion preparation shows improved cytotoxic activity compared to a suspension of the solid drug in a comparable dosage range. | Prankered et al. (1988) |
| W/O/W emulsion of Etoposide. | Sustained release.<br><br>Reduced side effect. | Ma et al. (1993) |
| W/O/W emulsion of Vinblastine sulfate | Sustained delivery.<br><br>Reduced dosage.<br><br>Multiple emulsion formulation as a single aqueous injection caused a gradual increase in the number of bone marrow cells arrested in metaphase for up to 4 h after administration.<br><br>The increase in the number of arrested metaphases was still continuing after 48 h. | Elson et al. (1970)<br><br>Herbert (1965) |
| W/O/W emulsion of 6-mercaptopurine-antitumor activity on the murine leukemia cell line *in vitro*. | Sustained release with extended blood circulating carriers.<br><br>Reduced side effects due to decreased uptake and cytotoxicity. | Khopade and Jain (1999) |

**TABLE 8.2** (Continued)

| Multiple emulsion | Applications | References |
|---|---|---|
| Taxol | Improved solubility of Taxol. Formulation of ethanol-in-oil-in-water multiple emulsions which are capable to encapsulate functional components that have a low solubility with respect to water and oil but soluble in ethanol. | Nakajima et al. (2003) |
| Methotrexate is effective antineoplastic drug but for effective cancer chemotherapy | Sustained delivery Parenteral administration of multiple emulsion enhances its therapeutic activity against the mouse lymphoma and rat Walker tumor. | Omotosho et al. (1989) |
| Pentazocine – narcotic drug of the benzomorphan class of opioids used to treat moderate to moderately severe pain. | Sustained drug release. Formulation of pentazocine multiple emulsion showed good tolerance to various pH range of the gastrointestinal tract during *in vitro* studies. *In vivo* studies carried out in mice showed prolonged drug levels in various tissues after oral administration. | Mishra and Pandit (1990) |
| Vancomycin – antibiotic used to treat a number of bacterial infections. | Sustained drug release Vancomycin multiple emulsion administered intravenously to rats showed prolonged drug concentrations in plasma compared to the Vancomycin solution. | Okochi and Nakano (2000) |
| Cytarabine (cytosine arabinoside) – antimetabolite for the treatment of acute lymphocytic and granulocytic leukemia. | Reduced dosage. A chemotherapeutic effect similar to that of five daily doses of cytosine arabinoside in aqueous solution could also be obtained against the lymphoma by a single dose of this drug in a W/O/W emulsion. | Benoy et al. (1974) |

described by the Stokes' law as in Eq. (1), where the velocity of a droplet moving upward is influenced by gravity (g), droplet radius (R), the density difference between the continuous and dispersed phases ($\Delta\rho$), and the dynamic viscosity of the continuous phase ($\mu$). An increase in the droplet radius directly results in the increase in velocity of the droplet movement. In the case of higher density of dispersed phase than the continuous phase the droplet moves upwards and vice versa.

$$V_{stokes} = \frac{2}{9}\frac{(\Delta P)}{\mu}gR^2 \qquad (1)$$

However, emulsions are thermodynamically unstable as with given sufficient time phase separation occurs (Tadros et al., 2004, Gutiérrez et al., 2008, Wooster et al., 2008, Solans et al., 2003). Thermodynamic instability in both the simple and multiple emulsion could lead to breakdown through various physicochemical mechanisms such as flocculation, coalescence, Ostwald ripening and creaming. Figure 8.4 below summarizes the destabilization mechanisms leading to phase separation in the emulsions. Ostwald ripening is a process whereby large droplets grow at the expense of smaller ones due to the high Laplace pressure in the small droplets (Binks and Lumsdon, 2000). Flocculation is a process where two or more droplets come together to form aggregations in emulsions although at the same time droplets retain their individual integrity (McClements, 2005). Flocculation occurs when the kinetic energy released during droplet collision brings the droplets over the repulsive force barrier into a region where attractive forces cause the droplets to attach to each other (Claesson et al., 1990). Coalescence is a process where two or more droplets merge together forming a single large droplet (McClements, 2005). Coalescence can only occur when the droplets are close to each other and the interfacial membranes are disrupted.

Gupta et al. (2016) indicated that Ostwald ripening is the dominant destabilization mechanism in the emulsion. For instance in an O/W simple emulsion, breakdown could occur through flocculation and coalescence of the oil globules within the dispersed phase. Whereas, in a W/O/W multiple emulsion, breakdown could occur through flocculation, coalescence within the internal water droplets, coalescence of the oil globules and coalescence of the internal and external water droplets with the oil globules (Qian and McClements, 2011; Dickinson, 2011; Pawlik et al., 2010). In addition, loss of the internal water droplets from the oil globules through diffusion,

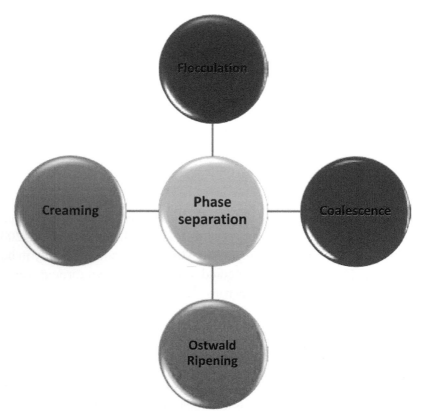

**FIGURE 8.4**   Destabilization mechanisms leading to phase separation in an emulsion

aggregation, or expulsion also causes breakdown in the multiple emulsions. These phenomenon in both simple and multiple emulsion result in creaming followed by phase separation. Jiao and Burgess (2003) stated that flocculation may or may not cause creaming, and thus it is not considered as a mark of instability for emulsions. Coalescence represents a potential step towards creaming. This leads to phase separation into oil phase, unseparated emulsion and water phase (Jiao and Burgess, 2003). Figure 8.5 adapted from Gupta et al. (2016) shows the summary of mechanisms resulting in the phase separation of emulsion.

### 8.3.2   COMPOSITION OF EMULSION

The composition of emulsion formulation is very important to ensure its stability. Emulsion is composed of oil, water, emulsifiers, co-emulsifiers,

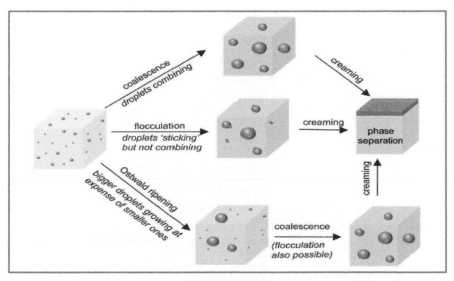

**FIGURE 8.5**   Destabilization mechanisms in the emulsions (adapted from Gupta et al., 2016).

co-surfactants and stabilizing agents as summarized in Figure 8.6. Many evidence from conducted researches shows the need for using co-emulsifiers (blends of emulsifiers) as a single chain of emulsifier could be insufficient to reduce the interfacial tension in an emulsion system. Co-solvents for the oral delivery such as the organic solvents, for example, polyethylene glycol, propylene glycol and ethanol are suitable for addition in the emulsion system. They ensure good solubilization of a large quantity of active compounds within the system. Other than emulsifiers, stabilizing components are also added to the multiple emulsion. It includes gelling or viscosity-increasing as well as thickening agents added to internal and/or external aqueous phase in multiple emulsion system such as Xanthan gum, Arabic gum, and Hydroxypropyl Methylcellulose.

## 8.3.2.1   Oil Phase and Its Influence in the Stability of Emulsion

The oil phase is an important component in the formulation due to its ability to solubilize the desired quantity of the lipophilic active compounds for transport through the intestinal lymphatic system. The oil phase employed for an oral based emulsion formulation could be of vegetable origin (soybean, sesame, olive, peanut, safflower, palm, sunflower, corn, castor, canola

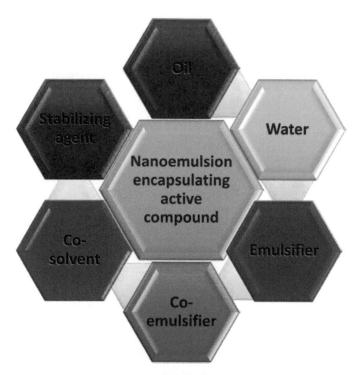

**FIGURE 8.6**  Composition of an emulsion system encapsulating active compounds.

and rapeseed oil), mineral oil (liquid paraffin, medium chain triglyceride) and specialty oils such as Moringa oleifera oil. Oils derived from vegetable sources are biodegradable, whereas those based on mineral oils are only removed from the body very slowly (Kumar et al., 2012). As a general rule, mineral oil produces more stable emulsions than vegetable oils (Florence and Whitehill, 1981). Medium chain triglyceride can also be used since it could more easily be digested than long chain triglyceride of vegetable oil (Bach and Babayan, 1982). Hence, gastrointestinal tract absorption depends upon the type of triglyceride molecule. In addition, the viscosity of oil phase is important for stabilization as it controls the transfer rates of solutes across the oil membrane (Bhattacharya et al., 2016). (Omotosho et al., 1986) revealed that oil phase with greater polarity could result in a greater tendency of the water passage from the internal phase to the external phase in a multiple emulsion causing the shrinkage of droplets. The chemical structure of the oil such as the chain length of fatty acids and the number of unsaturated bonds are essential for the stability of the emulsion (Ushikubo and Cunha, 2014).

## 8.3.2.2 Emulsifiers and Their Influence in the Stability of Emulsion

As nanoemulsions are nanosized, they have a large surface area with abundant surface free energy (Jantzen and Robinson, 2002). The interfacial free energy is associated with the interfacial tension which can be mathematically represented as in Eq. (2), where G is the interfacial energy, $\gamma$ is the interfacial tension and A is the total interfacial area.

$$\Delta G = \gamma \, \Delta A \tag{2}$$

As the interfacial area increases, there is a direct proportional increase in the interfacial tension and interfacial free energy. This renders the emulsion system to be thermodynamically unstable. Thus, an increase in the interfacial tension could be reduced by adding emulsifiers to the system. Emulsifiers adsorb at the interfacial layer thus preventing the coalescence of droplets. Long term stability is dependent on the strength of this film (Jiao and Burgess, 2008). Opawale and Burgess (1998) stated that for the film to be an effective barrier, it should remain intact at all the points between two droplets. It should have surface elasticity to reform rapidly once broken. This also shows that interfacial strength and interfacial elasticity can predict the stability in an emulsion system (Opawale and Burgess, 1998).

The surfactants that are used to stabilize the emulsion system could be non-ionic, zwitterionic, and ionic (cationic/anionic). The hydrophilic head of non-ionic or ionic surfactants influences the stabilizing interactions in the emulsion. The stabilization mechanism of ionic surfactants in an emulsion system is by the formation of an electrical double layer. For a non-ionic surfactant, dipole-dipole interaction and hydrogen bonding between the hydrophilic head of the surfactant with the water phase governs the stabilization. Zwitterionic surfactants may be more effective when used in combination with ionic surfactants. However, due to the toxicity issues the ionic surfactants are hardly used in the oral based emulsion formulations (Strickley, 2004). The non-ionic surfactants are usually preferable for use in oral formulations (Narang et al., 2007). Florence and Whitehill (1982) stated that non-ionic surfactants are preferred for pharmaceutical use not only because of their lower toxicity but also due to their less likely interactions with ionic compounds. Matsumoto et al. (1976) found that non-ionic surfactants gave better yield of emulsion droplets than ionic emulsifiers.

Non-ionic emulsifiers are available as monomeric and polymeric emulsifiers. There have been several investigations on the suitability of monomeric and polymeric emulsifiers for use in a multiple emulsion. Garti (1997) indicated that monomeric emulsifiers have some limitations in terms of emulsion stabilization. Morris and Newcombe (1993) and Benichou et al. (2002) suggested to employ naturally occurring macromolecular substances such as polysaccharides and proteins as they play a crucial role in stabilizing the emulsion system through steric and electrostatic stabilization. Macromolecules can act as polymeric emulsifiers at the interface in a slow process by forming an interfacial layer comprised of loops and tails that can be well anchored into the oil and water phases (Harzallah et al., 1998; Schuch et al., 2015). Also, it was suggested to use high molecular weight polymers rather than short-chain polymers (Tadros and Vincent, 1983; Abd-Elbary et al., 1984). Garti (1997) reinforced that the polymeric emulsifier used needs to have an effective adsorbing properties and good solubility in the continuous and dispersed medium. This was also seen in an investigation by Yuan et al. (2008) on beta carotene emulsification, where it was concluded that the nature of emulsifiers had considerable influence on the physicochemical stability. The degradation rate of beta-carotene was appreciably faster in emulsion stabilized with small molecule emulsifiers in comparison to large molecule emulsifiers (Sameh et al., 2012).

Stabilization of the emulsion system by monomeric emulsifiers could be explained by the Gibbs-Marangoni effect which states that during any stretching at the interface there is a decrease in surfactant excess concentration and an increase in the surface tension at the interface. Thus, the surface tension gradient causes emulsifier to flow toward the stretched region, thus providing a "healing" force and resists against further thinning. In contrast, polymeric natural emulsifiers, for instance, protein, the adsorption at the oil-water interface occurs due to the hydrophobic groups present within their structure. Once adsorbed, the molecules can unfold to maximize the number of hydrophobic groups that are in contact with the surface enabling the hydrophilic groups to rearrange and protrude away from the surface into the water phase (Evans et al., 2013). Interaction can sometimes occur between adjacent adsorbed protein molecules through hydrophobic bonding or disulfide bond formation leading to the formation of a viscoelastic layer at the oil-water interface (Evans et al., 2013). This prevents the breakage of the film thereby stabilizing the emulsion system (Bhattacharya et al., 2016).

### 8.3.2.3    mulsifiers for Use in the Formation of Simple Emulsion

For a simple emulsion, it is easier to select an appropriate emulsifier for stabilization. For instance, to form a W/O emulsion, it is required to utilize a hydrophobic emulsifier with Hydrophilic-Lipophilic Balance (HLB) value of 2–7. Whereas, for an O/W emulsion hydrophilic emulsifiers with a HLB value of 6–16 is utilized. Figure 8.7 shows the droplets with emulsifier used in a simple O/W emulsion.

### 8.3.2.4    Emulsifiers for Use in the Formation of Multiple Emulsion

However, for multiple emulsion due to its complex nature utilization of emulsifier must be carefully considered to ensure a long-term stability. Multiple emulsion requires at least two emulsifiers to be present in the system using the HLB values (Morris, 1965; Griffin, 1949). Hydrophobic emulsifier with a low HLB value of 2–7 is used to stabilize the primary emulsion, while a hydrophilic emulsifier with a high HLB value of 6–16 is used to stabilize the secondary emulsion (Sahu et al., 2013; Yaqoob Khan et al., 2006). Moreover, it has been found out in majority of the cases that the most stable emulsions are formed when both the emulsifying agents have the same hydrocarbon chain length, for instance, Span 80 as hydrophobic emulsifier and Tween

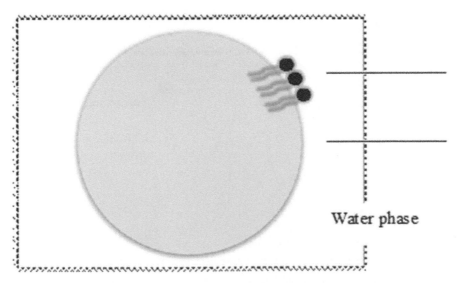

**FIGURE 8.7**    Single droplet of simple O/W emulsion with emulsifiers.

80 as hydrophilic emulsifier. Figure 8.8 shows the droplets with emulsifiers used in a multiple W/O/W emulsion system.

The concentration and ratio of these two emulsifiers are important to obtain a stable W/O/W emulsion. Insufficient and excess amount of emulsifier may result in the destabilization of multiple emulsion. Besides, an excess amount of emulsifier may lead to toxic effects upon consumption (Greene et al., 1988, McClements and Rao, 2011). An excess amount of hydrophobic emulsifier can cause the inversion of W/O/W multiple emulsion to simple O/W emulsion. Jiao and Burgess (2003) stated that an excess amount of hydrophilic surfactant in the external water phase beyond its critical micelle concentration may lead to the solubilization of the hydrophobic emulsifier in the primary W/O. As the hydrophilic/hydrophobic nature between the external water phase and the internal water phase of a multiple emulsions is similar, the internal interphase tends to be attracted towards the external interphase. This could lead to the rupture of the oil layer and the loss of internal water droplets due to the decrease in the concentration of the hydrophobic emulsifier. Therefore, the optimal concentration of hydrophobic emulsifier needs to be determined during the formulation.

Kita et al. (1978) reported that the weight ratio of the hydrophobic to the hydrophilic emulsifiers in the system should be between 2 to 20 to obtain a sufficient volume of W/O emulsion encapsulated within W/O/W multiple

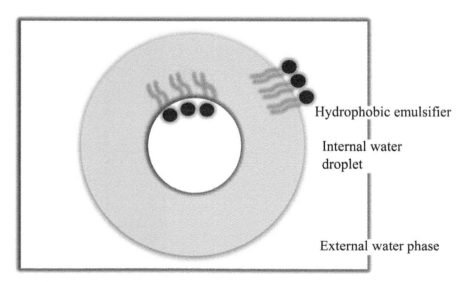

**FIGURE 8.8**    Single droplet of multiple W/O/W emulsion with emulsifiers.

emulsions. Hence, the widespread practice is to use ten-fold ratio of hydrophobic emulsifier to hydrophilic emulsifier to obtain 90% or higher yields of the W/O/W emulsion. This is because the hydrophilic emulsifier causes a negative effect on the movement of water molecules from the inner phase to the external phase. This shows that there is a major influence of the hydrophilic emulsifier on the water movement. At a higher concentration of hydrophilic emulsifier, higher solubility of this emulsifier in the external water phase is present ensuring transport in and out of both the internal and external water phase to be controlled by the osmotic pressure gradient (Kita et al., 1977; Garti et al., 1985). Therefore, it is imperative to keep the concentration of hydrophilic emulsifier as low as possible. However, concentrations lower than 0.5% would invert W/O/W into an O/W emulsion. Garti et al. (1985) demonstrated that lower concentrations of hydrophilic emulsifier at 1 g in 100 g of external water phase gave higher yields of multiple emulsion at any given HLB, while at higher concentrations of hydrophilic emulsifier, the yields were low and the emulsifiers with lower HLB were preferred. Emulsifiers with lower HLB slowed the water transport between both the internal and external water phases in a multiple emulsion. For instance, Tween 80 with the high HLB of 15 is capable of boosting water transport more than Tween 85 with the lower HLB of 11. It is also known that certain mixtures of emulsifiers provide better performance than pure emulsifier. Trotta et al. (2003) observed that the mixed surfactant films are usually more flexible and thus more suitable to form emulsified spherical droplets. The synergism of blended emulsifiers provides minimum energy than a single emulsifier (Turabi et al., 2008).

In summary, altering the composition can be accomplished by using different strategies for a multiple emulsion which ultimately improves its stabilization. These could be categorized as follows:

The inner water phase:
    (i) Stabilizing the inner W/O emulsion in the presence of better emulsifiers.
    (ii) Increasing the viscosity of the inner water by using thickening agents.

The oil phase:
    (i) Modifying the nature of oil phase by increasing its viscosity.
    (ii) Adding complexing agents such as thickeners to the oil.

The interfaces:
  (i)  Stabilizing the inner and/or the outer emulsion by using poly-
       meric emulsifiers, macromolecular amphiphiles such as proteins
       and polysaccharides to form strong and more rigid film at the
       interface.

The external water phase:
  (i)  Adding viscosity-increasing/gelling/thickening agents to control
       the movement of water between the internal and external water
       phases.

### 8.3.3 PHASE VOLUME RATIO AND ITS INFLUENCE IN THE STABILITY OF SIMPLE AND MULTIPLE EMULSION

Phase volume ratio is the ratio between the dispersed and continuous phase
in a simple emulsion. Whereas, in a multiple emulsion it is the ratio of W/O
internal emulsion in a W/O/W multiple emulsion. This ratio can have an
influence on the stability of emulsions. Lovelyn and Attama (2011) stated
that very high phase volume ratios may result in coalescence during emulsification for simple emulsion. Also, Binks and Lumsdon (2000) concluded
that the volume of W/O emulsion in W/O/W emulsion can have an influence
on the stability of the system. Cameron (2005) found that the droplet size
of the emulsion increases significantly with an increase in the volume ratio
of the W/O to W/O/W emulsion. Therefore, an optimal (22–40%) internal
phase volume ratio can be utilized for the emulsion formulation to ensure
good stability upon storage.

### 8.3.4 METHODS TO PRODUCE EMULSION

As explained above, the size of droplets both in the simple and multiple
emulsion is essential in ensuring the long-term stability in the system.
Smaller sized droplets are resistant to destabilization mechanisms such
as gravitational separation, coalescence, flocculation and creaming in an
emulsion system. Both simple and multiple emulsions can be produced by
low and high energy methods. Low energy method used to produce emulsions are phase inversion methods, that is, phase inversion temperature
(PIT), phase inversion composition (PIC) and emulsion inversion point,

spontaneous emulsification and membrane emulsification (Sanguansri and Augustin, 2006; Anton et al., 2008). The PIT or PIC method forms emulsion through phase transition and conversion by changing the emulsion composition or environment conditions such as temperature, pH and ionic strength. Whereas, in the EIP method, the only change involved is the volume fraction of the oil and water phase instead of the emulsifier (McClements, 2011, Fernandez et al., 2004). Spontaneous emulsification occurs when immiscible oil and water are in contact. This phenomenon triggered by gradients of chemical potential between both the phases, which under certain conditions leads to negative values of free energy of emulsification (Solans et al., 2016). Membrane emulsification is also a low energy consumption method where the dispersed phase is pressed through a porous membrane while the continuous phase flows along the membrane surface (Laouini et al., 2012).

High energy approach can be classified into high-pressure homogenization (Quintanilla-Carvajal et al., 2010) ultrasound (Sanguansri and Augustin, 2006) and high-speed homogenization (Anton et al., 2008). For a high energy method, high shear is applied to overcome the Laplace pressure and break-up droplets into smaller size (Sivakumar et al., 2014). Ultrasound utilizes the formation, growth and implosive collapse of bubbles which are responsible for the droplet break-up and formation of fine droplets (Tang et al., 2012). The common dispersing equipments used include the high-pressure valve, colloidal mill (rotor-stator principle), static mixers (plate type design), high shear mixers, hydrodynamic cavitation reactor, ultrasound cavitation reactor and microfluidizer. Figure 8.9 summarizes the methods used to generate emulsion.

To form nanoemulsion, the free energy required can be given by the following Eq. (3), where A is the interfacial area, $\gamma$ is the interfacial tension, T is the temperature and S is the entropy.

$$\Delta G = \Delta A \gamma - T \Delta S \tag{3}$$

where $\Delta A \gamma$ is the free energy required to increase the oil-water interface and $T \Delta S$ is the free energy associated with increasing the number of droplet formation. Sufficient free energy in the system is required to overcome the interfacial tension and increase the oil-water interface for the formation of nanoemulsion droplets. The change in entropy of a nanoemulsion is insufficient to overcome the free energy required to expand the oil-water interface. Therefore, an additional free energy source is required to expand the

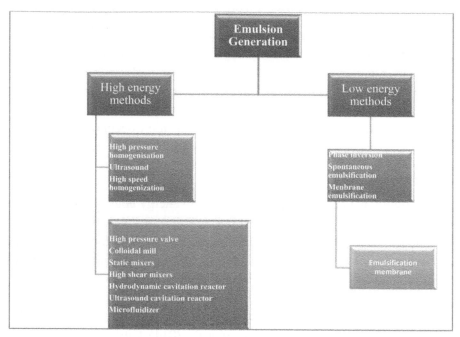

**FIGURE 8.9**   Different methods used for the generation of emulsion.

oil-water interface to cause droplet formation (Tadros et al., 2004). This free energy can be provided by mechanical devices or by the chemical potential of the system (Solans et al., 2005). In high-energy methods, this free energy originates from the mechanical forces such as shear, turbulence or cavitation applied to the system. Whereas, in low-energy methods, the free energy for emulsion formation comes from the physicochemical processes instead of the application of mechanical forces. However, low energy methods possess the limitations such as using large quantity of emulsifier and instability of the droplets in a long-term storage. Therefore, it is more advisable to use high energy input techniques in comparison to the low energy input techniques.

The benefit of using high energy input techniques to obtain a small droplet size of emulsion could be seen in Eq. (4). From this equation, it could be observed that as the droplets get smaller, Laplace pressure becomes higher in the droplets and thus requiring high energy source for droplet break-up. $\Delta P$ is the Laplace pressure, $\gamma$ is the oil-water interfacial tension and R is the radius of curvature.

$$\Delta P = \frac{2\gamma}{R}$$

(4)

In high energy approaches, intense disruptive forces are applied on the sample to be emulsified. Anandharamakrishnan (2014) stated that during emulsification an increase in energy intensity or duration decreases the interfacial tension. This enhances the emulsifier adsorption rate and consequently the dispersed to continuous phase viscosity ratio, which falls within a range of 0.05–5. This viscosity range aids in producing smaller droplet sizes. Two opposing processes namely droplet disruption and droplet coalescence take place during emulsification. Mahdi Jafari et al. (2006) indicated that the achievement of a balance between these two processes leads to the production of a smaller droplet. For instance, intense energy from a mechanical device such as homogenizer is required to generate disruptive forces greater than the restoring forces for a spherical droplet formation (Schubert and Engel, 2004). For high energy approaches, conditions such as the number of passes/cycles, pressure valve and impingement, surfactant concentration, pressure and amplitude difference can influence the droplet size formed (Kentish et al., 2008).

### 8.3.4.1   Production of Simple and Multiple Emulsions

The production of simple emulsion is straightforward. Using high energy approaches, it just needs to be pre-mixed before subjecting into any high energy equipment to obtain nanoemulsion. Multiple emulsion is a little complicated and can be produced in two stages, that is, two-stage and single stage production (Figures 8.10 and 8.11). The two-stage multiple emulsion production method is more convenient and reliable (Okochi and Nakano, 2000). Multiple emulsion produced with two-stage production has higher stability, more homogeneous droplets, easily reproducible, and gives high yield (Okushima et al., 2004).

Single stage multiple emulsion production is not commonly used as it is difficult to control, has large droplet size, uses higher amount of synthetic surfactant leading to toxic effect and limits its route of administration (Sigward et al., 2013). It also has poor reproducibility and shows no long-term stability (Sigward et al., 2013). It also involves heating which limits the application to temperature resistant drug (Sigward et al., 2013). Furthermore, an excess mixing in single stage utilizing stirrers at high speed with high speed energy output can also lead to breakdown of the inner W/O emulsion in a W/O/W multiple emulsion (Kumar et al., 2012). Therefore,

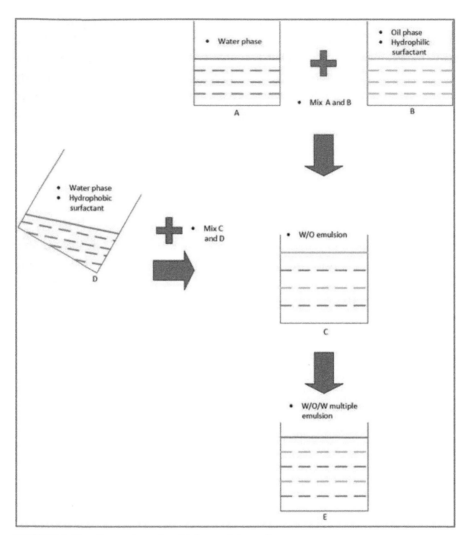

**FIGURE 8.10**   Generation of multiple emulsion using two stages.

it can be noted that two stage production of multiple emulsion has better advantages than the single stage production.

## 8.3.5   OSMOTIC PRESSURE IN MULTIPLE EMULSION

A significant phenomenon that only occurs in a multiple emulsion leading to destabilization is the difference in the osmotic pressure between the inner and external water droplets. The thermodynamic driving force induces the

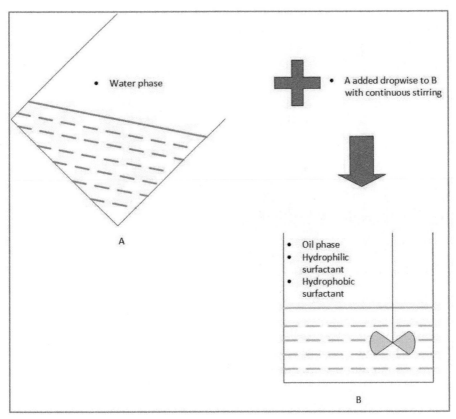

**FIGURE 8.11**   Generation of multiple emulsion using single stage.

transport of water molecules from the internal to external phase and vice versa due to the difference in the solute concentration in both the phases (Mezzenga et al., 2004). Solute here refers to the encapsulated water soluble active compounds in the internal water phase. Consequently, the internal water droplets swell, thinning the oil layer and eventually results in the rupture of oil layer (Matsumoto et al., 1978; Bibette et al., 1999; Florence and Whitehill, 1982). Alternatively, the internal water droplets can also shrink due to the transport of water from the internal to external droplets forming simple O/W emulsion. This phenomenon is also illustrated in Figure 8.12.

Apart from the osmotic pressure difference due to the solute concentration, the movement of water in and out of both the phases can also occur due to the presence of hydrophobic emulsifiers in the oil phase (Benichou et al., 2004). The internal water phase could travel through the oil layer of the multiple emulsion droplets as the hydrophobic fatty acid tails of emulsifiers

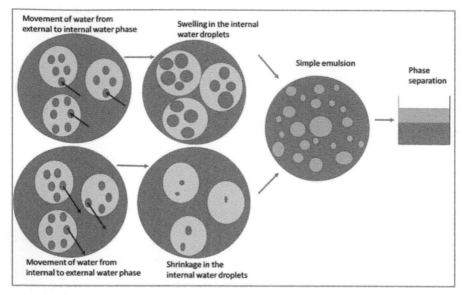

**FIGURE 8.12**   Influence of osmotic pressure on the multiple emulsion droplets.

act as water-permeable membrane between both the water phases (Khokra et al., 2012). Jiao and Burgess (2008) showed that the Laplace pressure due to the curvature of the dispersed internal water droplets could lead to Ostwald ripening as they coalesce. They also stated that for a W/O emulsion with a stable interfacial film, the addition of a small quantity of oil insoluble osmotic agent to the dispersed internal water phase would have a stabilizing effect by counteracting the effect of Laplace pressure. Thus, it can be concluded that the long-term stability of the multiple emulsion requires a balance between the Laplace and osmotic pressure. Addition of an osmotic agent also forms a rigid interfacial layer between the oil and internal aqueous phases. Landfester (2006) indicated that the effect of concentration of emulsifier on the droplet size depends on the amount of osmotic agent added to the multiple emulsion. Therefore, the growth and shrinkage of internal water droplets can be hindered by altering the osmotic pressure between both the water phases (Pawlik et al., 2010; Mezzenga et al., 2004; Leal-Calderon et al., 2012).

Generally, osmotic agents such as salt, glucose or glycerol are utilized in the multiple emulsion. The presence of salt is crucial for the formation of emulsion and the stability of primary W/O emulsions (Aronson and Petko, 1993). Davis (1976) investigated the effects of salt on the size of multiple emulsion droplets and found that little or no change in the size occurs over some time. Salt is a co-stabilizer, a lipophobe that builds-up the osmotic pressure to counter balance the Laplace pressure, and consequently stabilizes the emulsions against diffusional degradation known as Ostwald ripening (Landfester, 2006; Capek, 2010; Colmán et al., 2014; Pawlik et al., 2010). Lutz et al. (2009) stated that the addition of salt in the inner water phase strengthens the interaction between the salt and emulsifier, thus providing a better packing of the emulsifier at the interfacial layer. Therefore, increasing the elasticity of the layer decreases the interfacial mobility leading to increasing the stability of emulsion by "stiffening" the interface (Lutz et al., 2009). Aronson and Petko (1993) concluded that the major effect of osmotic agent is to increase the resistance of the oil film to coalescence by increasing the adsorption density and lowering the interfacial tension of the interfacial film.

However, there are also investigations suggested that the water in the internal phase is transported at an enhanced rate even in the absence of an osmotic agent due to the availability of lipophilic emulsifier in the oil phase (Garti and Benichou, 2001). Benichou et al. (2004) demonstrated that even if the osmotic pressure of the two phases was equilibrated, coalescence of internal water droplets still takes place and the water droplets are transported out mostly through a 'reverse micellar mechanism' controlled by the viscosity of the oil phase and the nature of oil membrane. In addition, Davis et al. (1976) also added lesser amounts of salt to the internal aqueous phase but found that this approach could not overcome the inequality in osmotic pressure upon storage. Tang et al. (2012) studied the mean droplet size of the resulting multiple emulsions containing different amounts of glucose both in the inner and outer aqueous phases. It was noted that the changes in the content of glucose both in the inner and outer water phases did not have any considerable influence on the droplet diameter as well as the stability of multiple emulsion. Garti and Benichou (2004) found that the release of encapsulated matter in the presence of monomeric emulsifiers indicated that the water transport takes place even if the droplets are very stable to coalescence and even when the osmotic pressure of the two phases has been equilibrated. Aston et al. (1992) concluded that even with close-packed molecular areas of emulsifier and salt in the internal water phase, transport of water between the internal and external phase was still observed. Jiao and Burgess (2003) stated that the viscosity of W/O/W multiple emulsions was dependent on the volume fraction of the primary W/O emulsion in the system. Water transport from the inner water phase to the external water phase or vice versa affects the volume fraction and therefore changes the viscosity of W/O/W emulsion. As such, they indicated that the incorporation of salt in the internal water phase increases the viscosity of W/O/W multiple emulsions because of increased droplet swelling and volume fraction of W/O emulsion. The swollen droplets may eventually rupture, which destabilizes the emulsions. Therefore, they concluded that a significant increase in the viscosity due to salt addition to the internal water phase is an indication of poor long-term stability of W/O/W emulsions (Jiao and Burgess, 2003). Hence, regardless of the salt concentration, some degree of creaming of the W/O/W emulsion would occur. It is suggested to reassess the requirement of osmotic agents to improve the osmotic gradient and transport of water both in the external and internal water phases due to the mixed observations noted in the previous investigations.

## 8.4   CHARACTERIZATION OF SIMPLE AND MULTIPLE EMULSIONS IN ASSESSING THEIR STABILITY

There are many common characterization methods carried out on the formed simple and multiple emulsions. These methods are also used in evaluating the stability in the emulsion system. The organoleptic properties are physical characteristics where the color and smell are noted and any change in color or smell of the emulsion upon storage could be used as an indication for the changes in the stability (Mahmood and Uzair, 2010). Liquefaction is also an organoleptic property that can be attributed to the passage of water from the internal phase to external phase in multiple emulsion which could result in phase separation and is also a method to identify the change in stability (Mahmood and Uzair, 2010). The mean droplet size, polydispersity index and zeta potential determined by a Zetasizer are popularly used to characterize the formed emulsion (Schwarz et al., 2012; Amid and Mirhosseini, 2014). Any changes in the above parameters upon storage have been widely used to evaluate the stability of an emulsion system.

However, the above-mentioned characterization of emulsions is not sufficient to conclude the stability of emulsions. In the recent times, analyzing the rheology of emulsions have become a crucial factor in assessing its stability. Rheological measurements are important for the characterization of emulsions as the rheological properties can affect the rate of creaming and the shelf-life of products which is also an indication of stability (Pereda et al.,

2007). Rheological measurements include shear stress, shear rate and apparent viscosity analysis which can be performed using a dynamic shear rheometer or viscometer (Mahmood et al., 2013). Also, interfacial rheology is also an important analysis to monitor the stability of the surfactant film. The interfacial strength and interfacial elasticity of the surfactant film could be related to the stability of the emulsion system and can be measured by oscillatory surface rheometer (Opawale and Burgess, 1998, Sahu et al., 2013). Issa et al. (2005) noted that monitoring the pH is crucial for determining the emulsions' stability. In fact, pH changes indicate the occurrence of chemical reactions that can give an idea on the quality of the final product (Mahmood and Uzair, 2010). pH determination can be carried out using a pH meter. For a multiple emulsion, it is important to analyze the morphology of the emulsion for stability evaluation. The morphology of emulsions would enable to identify coalescence of internal droplets resulting in the loss and swell of internal droplet. This can be done using transmission electron microscopy (Clausse et al., 2005), scanning electron microscopy (Yang et al., 2001), scanning transmission electron microscopy (Tang et al., 2013) as well as an optical (Cournarie et al., 2004) and confocal microscopy (Mahmood and Uzair, 2010). Therefore, to assess the stability of emulsions, it is important to note that droplet size and zeta potential alone are not sufficient. To obtain a reliable stability assessment of the emulsion system, morphology, pH and the rheological characteristics are important.

Due to the presence of osmotic pressure difference between the internal and external water phase in a multiple emulsion, the change in this pressure can act as a stability indicator. Osmotic pressure or osmolality in a multiple emulsion can be determined using a vapor pressure osmometer (Bozkir and Hayta, 2004). Another method of quantifying the change in the osmotic pressure is by incorporating a conductometric tracer in the inner water phase of the multiple emulsion to detect any leakage from the internal to the external water phase. Conductimetric analysis of the tracer in undiluted samples could be done by a conductivity meter (Mahmood et al., 2013).

Solubility, encapsulation efficiency and *in vitro* release of the encapsulated active compound in an emulsion system are essential for the evaluation of the formulation efficacy. These properties are important for ensuring the efficacy of the emulsion system for food, nutraceutical and pharmaceutical active compounds. The active compound to be encapsulated can be qualitatively tested for their solubility in various solvents such as water, methanol, ethanol and acetonitrile. The encapsulation efficiency of the active compound in emulsions is generally determined using dialysis (Fu et al., 2005), centrifugation (Neeraja et al., 2014) and filtration (Yang et al., 2001) method. Commonly, the *in vitro* release can be estimated using the conventional dialysis technique using the dialysis bag. Also, to evaluate the encapsulation efficiency, an internal tracer/marker can be entrapped within the emulsion. The evaluation of this tracer/ marker contained in the emulsion could be correlated to the encapsulation efficiency (Sahu et al., 2013). The quantification of the loss of encapsulated active compound due to exposure to environmental stress such as heat and light within the emulsion could be analyzed using High Performance Liquid Chromatography (Schwarz et al., 2012). Measurement of change in conductivity also provides the information on the *in vitro* release of the encapsulated active compounds in the emulsion. The amount of release is directly proportional to the amount of active compound that is available in the external water phase. The conductivity of the emulsions can be directly measured using a digital conductivity meter.

Stability tests can be performed at different storage conditions for both the primary and multiple emulsions by manipulating the humidity, temperature and light. Freeze- thawing stability analysis have become common due to the freeze-thaw cycles an emulsion undergoes during processing and storage. Centrifugation could be done as an accelerated stability measurement to quantify the phase separation utilizing the centrifuge (Lerche, 2002). A turbiscan could be utilized to evaluate the demulsification efficiency and to analyze the destabilization process. It measures the change in the optical characteristics of an emulsion to evaluate the stability of an emulsion (Celia et al., 2009).

## 8.5  CONCLUSIONS

Both simple and multiple emulsions are largely advantageous to be formulated for human consumption. Various beneficial active compounds especially for use in the pharmaceutical, nutraceutical and food industry were successfully formulated by the simple and multiple emulsion formulation. Natural constituents such as the protein, polysaccharides, starch, xanthan gum, agar gum are becoming widely acceptable for the formulation owing to the increasing health concerns. They also enhance the stability of emulsion formulations. Thus, the future formulations of simple and multiple emulsion should utilize the natural constituents for the formulation. Stability of the emulsion system especially the multiple emulsion if it is well controlled, would enhance its marketability. In overcoming the osmotic pressure which results in instability, the addition of osmotic agents to the emulsion system such as salt, glycerol, and glucose needs to be re-assessed due to the varying views obtained in the past. Also, the droplet size alone is not sufficient to judge and conclude on the stability of emulsion. It is best to also use morphological, pH and the rheological analysis to evaluate the stability of emulsions.

## ACKNOWLEDGEMENTS

The authors would like to acknowledge the financial and facilities support provided by MPOB and the Faculty of engineering, University of Nottingham Malaysia Campus for this research work.

## KEYWORDS

- **nanoemulsion**
- **multiple emulsion**
- **stability**
- **utrition**
- **drug delivery**
- **analytical methods**

## REFERENCES

Abd-Elbary, A., Nour, S., & Mansour, F. (1984). Efficacy of different emulsifying agents in preparing O/W/O or W/O/W multiple emulsions. *Pharmazeutische Industrie, 46,* 964–969.

Acosta, E. (2009). Bioavailability of nanoparticles in nutrient and nutraceutical delivery. *Current Opinion in Colloid & Interface Science, 14,* 3–15.

Amid, B. T., & Mirhosseini, H. (2014). Stabilization of water in oil in water (W/O/W) emulsion using whey protein isolate-conjugated durian seed gum: Enhancement of interfacial activity through conjugation process. *Colloids and Surfaces B: Biointerfaces, 113,* 107–114.

Anandharamakrishnan, C. (2014). Techniques for formation of nanoemulsions. *In Techniques for Nanoencapsulation of Food Ingredients. New York, Springer*, 7–16.

Anton, N., Benoit, J. P., & Saulnier, P. (2008). Design and production of nanoparticles formulated from nano-emulsion templates a review. *Journal of Controlled Release, 128*, 185–199.

Aronson, M. P., & Petko, M. F. (1993). Highly concentrated water-in-oil emulsions: Influence of electrolyte on their properties and stability. *Journal of Colloid Interface Science, 159*, 134–149.

Aston, M. S., Herrington, T. M., & Tadros, T. F. (1992). Interfacial properties of mixtures of sorbitan sesquioleate with a block copolymer surfactant at the water/air and water/oil interfaces. *Colloids and Surfaces, 62*, 31–39.

Bach, A. C., & Babayan, V. K. (1982). Medium-chain triglycerides: an update. *The American Journal of Clinical Nutrition, 36*, 950–962.

Benichou, A., Aserin, A., & Garti, N. (2002). Protein-polysaccharide interactions for stabilization of food emulsions. *Journal of Dispersion Science and Technology, 23*, 93–123.

Benichou, A., Aserin, A., & Garti, N. (2004). Double emulsions stabilized with hybrids of natural polymers for entrapment and slow release of active matters. *Advances in Colloid and Interface Science, 108*, 29–41.

Benoy, C. J., Schneider, R., Elson, L., & Jones, M. (1974). Enhancement of the cancer chemotherapeutic effect of the cell cycle phase specific agents methotrexate and cytosine arabinoside when given as a water-oil-water emulsion. *European Journal of Cancer, 10*, 27–33.

Bhattacharya, R., Mukhopadhyay, S., & Kothiyal, P. (2016). Review on microemulsion-as a potential novel drug delivery system. *World Journal of Pharmacy and Pharmaceutical Sciences, 5*, 700–729.

Bhushani, J. A., Karthik, P., & Anandharamakrishnan, C. (2016). Nanoemulsion based delivery system for improved bioaccessibility and Caco-2 cell monolayer permeability of green tea catechins. *Food Hydrocolloids, 56*, 372–382.

Bibette, J., Calderon, F. L., & Poulin, P. (1999). Emulsions: basic principles. *Reports on Progress in Physics, 62*, 969–1033.

Binks, B., & Lumsdon, S. (2000). Catastrophic phase inversion of water-in-oil emulsions stabilized by hydrophobic silica. *Langmuir, 16*, 2539–2547.

Bonnet, M., Cansell, M., Berkaoui, A., Ropers, M., Anton, M., & Leal-Calderon, F. (2009). Release rate profiles of magnesium from multiple W/O/W emulsions. *Food Hydrocolloids, 23*, 92–101.

Borhade, V., Pathak, S., Sharma, S., & Patravale, V. (2012). Clotrimazole nanoemulsion for malaria chemotherapy. Part II: Stability assessment, in vivo pharmacodynamic evaluations and toxicological studies. *International Journal of Pharmaceutics, 431*, 149–160.

Bozkir, A., & Hayta, G. (2004). Preparation and evaluation of multiple emulsions water-in-oil-in-water (w/o/w) as delivery system for influenza virus antigens. *Journal of Drug Targeting, 12*, 157–164.

Bozkir, A., Hayta, G., & Saka, O. (2004). Comparison of biodegradable nanoparticles and multiple emulsions (water-in-oil-in-water) containing influenza virus antigen on the in vivo immune response in rats. *Die Pharmazie: An International Journal of Pharmaceutical Sciences, 59*, 723–725.

Cameron, N. R. (2005). High internal phase emulsion templating as a route to well-defined porous polymers. *Polymer, 46*, 1439–1449.

Capek, I. (2010). On inverse miniemulsion polymerization of conventional water-soluble monomers. *Advances in Colloid and Interface Science, 156*, 35–61.

Celia, C., Trapasso, E., Cosco, D., Paolino, D., & Fresta, M. (2009). Turbiscan Lab® Expert analysis of the stability of ethosomes® and ultradeformable liposomes containing a bilayer fluidizing agent. *Colloids and Surfaces B: Biointerfaces, 72*, 155–160.

Chavhan, S. S., Petkar, K. C., & Sawant, K. K. (2013). Simvastatin nanoemulsion for improved oral delivery: design, characterisation, in vitro and in vivo studies. *Journal of Microencapsulation, 30*, 771–779.

Chhabra, G., Chuttani, K., Mishra, A. K., & Pathak, K. (2011). Design and development of nanoemulsion drug delivery system of amlodipine besilate for improvement of oral bioavailability. *Drug Development and Industrial Pharmacy, 37*, 907–916.

Chiang, C. W., Fuller, G. C., Frankenfeld, J. W., & Rhodes, C. (1978). Potential of liquid membranes for drug overdose treatment: in vitro studies. *Journal of Pharmaceutical Sciences, 67*, 63–66.

Claesson, P. M., Eriksson, J. C., Herder, C., Bergenståhl, B. A., Pezron, E., Pezron, I., et al. (1990). Forces between non-ionic surfactant layers. *Faraday Discussions of the Chemical Society, 90*, 129–142.

Clausse, D., Gomez, F., Pezron, I., Komunjer, L., & Dalmazzone, C. (2005). Morphology characterization of emulsions by differential scanning calorimetry. *Advances in Colloid and Interface Science, 117*, 59–74.

Cofrades, S., Santos-López, J., Freire, M., Benedí, J., Sánchez-Muniz, F., & Jiménez-Colmenero, F. (2014). Oxidative stability of meat systems made with W 1/O/W 2 emulsions prepared with hydroxytyrosol and chia oil as lipid phase. *LWT-Food Science and Technology, 59*, 941–947.

Colmán, M., Chicoma, D., Giudici, R., Araújo, P., & Sayer, C. (2014). Acrylamide inverse miniemulsion polymerization: In situ, real-time monitoring using nir spectroscopy. *Brazilian Journal of Chemical Engineering, 31*, 925–933.

Cournarie, F., Rosilio, V., Chéron, M., Vauthier, C., Lacour, B., Grossiord, J. L., et al. (2004). Improved formulation of W/O/W multiple emulsion for insulin encapsulation. Influence of the chemical structure of insulin. *Colloid and Polymer Science, 282*, 562–568.

Cunha, A. S., Grossior, J., Puisieux, F., & Seiller, M. (1997). Insulin in w/o/w multiple emulsions: preparation, characterization and determination of stability towards proteases in vitro. *Journal of Microencapsulation, 14*, 311–319.

Davis, S. (1976). The emulsion obsolete dosage form or novel drug delivery system and therapeutic agent? *Journal of Clinical Pharmacy and Therapeutics, 1*, 11–27.

Davis, S., Purewal, T., & Burbage, A. (1976). The particle size analysis of multiple emulsions [proceedings]. *The Journal of pharmacy and pharmacology, 28*, 60P–60P.

Desai, K. G. H., & Jin Park, H. (2005). Recent developments in microencapsulation of food ingredients. *Drying Technology, 23*, 1361–1394.

Dhavamani, S., & Lokesh, B. (2016). ID: 88: Phospholipid based lipoid™ nanoemulsion of curcumin in linseed oil showed increased bioavailability and assists in elevating the levels of docosahexaenoic acid in serum and liver lipids of rats. *Journal of Investigative Medicine, 64*, 930–930.

Dickinson, E. (2011). Double emulsions stabilized by food biopolymers. *Food Biophysics, 6*, 1–11.

Elson, L., Mitchlev, B., Collings, A., & Schneider, R. (1970). Chemotherapeutic effect of a water-oil-water emulsion of methotrexate on the mouse L1210 leukemia. *European Journal of Clinical and Biological Research, 15*, 87–90.

Evans, M., Ratcliffe, I., & Williams, P. (2013). Emulsion stabilization using polysaccharide protein complexes. *Current Opinion in Colloid & Interface Science, 18*, 272–282.

Fang, J. Y., Hung, C. F., Hua, S. C., & Hwang, T. L. (2009). Acoustically active perfluorocarbon nanoemulsions as drug delivery carriers for camptothecin: drug release and cytotoxicity against cancer cells. *Ultrasonics, 49*, 39–46.

Fernandez, P., André, V., Rieger, J., & Kühnle, A. (2004). Nano-emulsion formation by emulsion phase inversion. *Colloids and Surfaces A: Physicochemical and Engineering Aspects, 251*, 53–58.

Florence, A., & Whitehill, D. (1981). Some features of breakdown in water-in-oil-in-water multiple emulsions. *Journal of Colloid and Interface Science, 79*, 243–256.

Florence, A., & Whitehill, D. (1982). The formulation and stability of multiple emulsions. *International Journal of Pharmaceutics, 11*, 277–308.

Fofaria, N. M., Qhattal, H. S. S., Liu, X., & Srivastava, S. K. (2016). Nanoemulsion formulations for anti-cancer agent piplartine Characterization, toxicological, pharmacokinetics and efficacy studies. *International Journal of Pharmaceutics, 498*, 12–22.

Frank, K., Walz, E., Gräf, V., Greiner, R., Köhler, K., & Schuchmann, H. P. (2012). Stability of Anthocyanin-Rich W/O/W-Emulsions Designed for Intestinal Release in Gastrointestinal Environment. *Journal of Food Science, 77*, N50–N57.

Freire, M., Bou, R., Cofrades, S., Solas, M. T., & Jiménez-Colmenero, F. (2015). Double emulsions to improve frankfurter lipid content: impact of perilla oil and pork backfat. *Journal of the Science of Food and Agriculture, 96*, 900–908.

Fu, X., Ping, Q., & Gao, Y. (2005). Effects of formulation factors on encapsulation efficiency and release behavior *in vitro* of huperzine A-PLGA microspheres. *Journal of microencapsulation, 22*, 705–714.

Gao, F., Zhang, Z., Bu, H., Huang, Y., Gao, Z., Shen, J., Zhao, C., & Li, Y. (2011). Nanoemulsion improves the oral absorption of candesartan cilexetil in rats: performance and mechanism. *Journal of Controlled Release, 149*, 168–174.

Gaoe, H., Pang, Z., Pan, S., Cao, S., Yang, Z., Chen, C., et al. (2012). Anti-glioma effect and safety of docetaxel-loaded nanoemulsion. *Archives of Pharmacal Research, 35*, 333–341.

Garti, N. (1997). Progress in stabilization and transport phenomena of double emulsions in food applications. *LWT-Food Science and Technology, 30*, 222–235.

Garti, N., & Benichou, A. (2001). Double emulsions for controlled-release applications: progress and trends. *Encyclopedic Handbook of Emulsion Technology. New York: Marcel Dekker. 377*–407.

Garti, N., & Benichou, A. (2004). Recent developments in double emulsions for food applications. *Food emulsions. New York: Marcel Dekker, 353*–370.

Garti, N., Magdassi, S., & Whitehill, D. (1985). Transfer phenomena across the oil phase in water-oil-water multiple emulsions evaluated by Coulter counter: Effect of emulsifier 1 on water permeability. *Journal of Colloid and Interface Science, 104*, 587–591.

Ghai, D., & Sinha, V. R. (2012). Nanoemulsions as self-emulsified drug delivery carriers for enhanced permeability of the poorly water-soluble selective β 1-adrenoreceptor blocker Talinolol. *Nanomedicine: Nanotechnology, Biology and Medicine, 8*, 618–626.

Ghosh, L., Ghosh, N., Thakur, R., Pal, M., & Gupta, B. (1997). Design and evaluation of controlled-release W/O/W multiple-emulsion oral liquid delivery system of chlorpheniramine maleate. *Drug Development and Industrial Pharmacy, 23*, 1131–1134.

Giroux, H. J., Constantineau, S., Fustier, P., Champagne, C. P., St-Gelais, D., Lacroix, M., et al. (2013). Cheese fortification using water in oil in water double emulsions as carrier for water soluble nutrients. *International Dairy Journal, 29*, 107–114.

Greene, H. L., Hambidge, K., Schanler, R., & Tsang, R. C. (1988). Guidelines for the use of vitamins, trace elements, calcium, magnesium, and phosphorus in infants and children receiving total parenteral nutrition: report of the Subcommittee on Pediatric Parenteral Nutrient Requirements from the Committee on Clinical Practice Issues of the American Society for Clinical Nutrition. *The American Journal of Clinical Nutrition, 48*, 1324–1342.

Griffin, W. R. (1949). Residual gravity in theory and practice. *Geophysics, 14*, 39–56.

Gupta, A., Eral, H. B., Hatton, T. A., & Doyle, P. S. (2016). Nanoemulsions: formation, properties and applications. *Soft Matter, 12*, 2826–2841.

Gutiérrez, J., González, C., Maestro, A., Sole, I., Pey, C., & Nolla, J. (2008). Nano-emulsions: New applications and optimization of their preparation. *Current Opinion in Colloid & Interface Science, 13*, 245–251.

Harzallah, B., Aguié-Béghin, V., Douillard, R., & Bosio, L. (1998). A structural study of β-casein adsorbed layers at the air water interface using X-ray and neutron reflectivity. *International Journal of Biological Macromolecules, 23*, 73–84.

Herbert, W. (1965). Multiple emulsions: a new form of mineral-oil antigen adjuvant. *The Lancet, 286*, 771.

Higashi, S., & Setoguchi, T, (2000). Hepatic arterial injection chemotherapy for hepatocellular carcinoma with epirubicin aqueous solution as numerous vesicles in iodinated poppy-seed oil microdroplets: clinical application of water-in-oil-in-water emulsion prepared using a membrane emulsification technique. *Advanced Drug Delivery Reviews, 45*, 57–64.

Higashi, S., Shimizu, M., Nakashima, T., Iwata, K., Uchiyama, F., Tateno, S., et al. (1995). Arterial injection chemotherapy for hepatocellular carcinoma using monodispersed poppy seed oil microdroplets containing fine aqueous vesicles of epirubicin. Initial medical application of a membrane-emulsification technique. *Cancer, 75*, 1245–1254.

Huang, Q., Yu, H., & Ru, Q. (2010). Bioavailability and delivery of nutraceuticals using nanotechnology. *Journal of Food Science, 75*, R50–R57.

Issa, M. M., Köping-Höggård, M., & Artursson, P. (2005). Chitosan and the mucosal delivery of biotechnology drugs. *Drug Discovery Today: Technologies, 2*, 1–6.

Jain, S., Jain, A. K., Pohekar, M., & Thanki, K. (2013). Novel self-emulsifying formulation of quercetin for improved in vivo antioxidant potential: Implications for drug-induced cardiotoxicity and nephrotoxicity. *Free Radical Biology and Medicine, 65*, 117–130.

Jantzen, G. M., & Robinson, J. R. (2002). Sustained-and controlled-release drug delivery systems. *Drugs and The Pharmaceutical Sciences, 121*, 501–528.

Jiao, J., & Burgess, D. J. (2003). Rheology and stability of water-in-oil-in-water multiple emulsions containing Span 83 and Tween 80. *AAPS Pharmsci, 5*, 62–73.

Jiao, J., & Burgess, D. J. (2008). Multiple emulsion stability: pressure balance and interfacial film strength. *Multiple Emulsions: Technology and Applications. Hoboken: Wiley-Interscience*, 1–28.

Joe, M. M., Chauhan, P. S., Bradeeba, K., Shagol, C., Sivakumaar, P., K. & SA, T. (2012). Influence of sunflower oil based nanoemulsion (AUSN-4) on the shelf life and quality of Indo-Pacific king mackerel (Scomberomorus guttatus) steaks stored at 20°C. *Food Control, 23*, 564–570.

Kaimainen, M., Marze, S., Järvenpää, E., Anton, M., & Huopalahti, R. (2015). Encapsulation of betalain into w/o/w double emulsion and release during in vitro intestinal lipid digestion. *LWT-Food Science and Technology, 60*, 899–904.

Kentish, S., Wooster, T., Ashokkumar, M., Balachandran, S., Mawson, R., & Simons, L. (2008). The use of ultrasonics for nanoemulsion preparation. *Innovative Food Science & Emerging Technologies, 9,* 170–175.

Khan, A. Y. (2007). Potentials of liquid membrane system: an overview. *Pharmainfo.net, 5,* Issue 6.

Khokra, S. L., Parashar, B., Dhamija, H. K., Kumar, K., & Arora, S. (2012). Formulation and evaluation of novel sustained release multiple emulsion containing chemotherapeutic agents. *International Journal of PharmTech Research, 4,* 866–872.

Khopade, A., & Jain, N. (1999). Fine multiple emulsions bearing 6-mercaptopurine: *In Vitro* and *In Vivo* antitumor studies. *Drug Delivery, 6,* 181–185.

Kim, B. K., Cho, A. R., & Park, D. J. (2016). Enhancing oral bioavailability using preparations of apigenin-loaded W/O/W emulsions: *In Vitro* and *In Vivo* evaluations. *Food Chemistry, 206,* 85–91.

Kim, I. H., Oh, Y. A., Lee, H., Song, K. B., & Min, S. C. (2014a.). Grape berry coatings of lemongrass oil-incorporating nanoemulsion. *LWT-Food Science and Technology, 58,* 1–10.

Kim, J. Y., Lee, M. S., Jung, S., Joo, H., Kim, C. T., Kim, I. H., Seo, S., Oh, S., & Kim, Y. (2014b). Anti-obesity efficacy of nanoemulsion oleoresin capsicum in obese rats fed a high-fat diet. *International Journal of Nanomedicine, 9,* 301–310.

Kita, Y., Matsumoto, S., & Yonezawa, D. (1977). Viscometric method for estimating the stability of W/O/W-type multiple-phase emulsions. *Journal of Colloid and Interface Science, 62,* 87–94.

Kita, Y., Matsumoto, S., & Yonezawa, D. (1978). Permeation of water through oil layer in WOW type multiple-phase emulsions. *Nippon Kagaku Kaishi,1,* 11–14.

Kotta, S., Khan, A. W., Ansari, S. H., Sharma, R. K., & Ali, J. (2014). Anti HIV nanoemulsion formulation: optimization and *In Vitro–In Vivo* evaluation. *International Journal of Pharmaceutics, 462,* 129–134.

Kumar, R., Kumar, M. S., & Mahadevan, N. (2012). Multiple emulsions: a review. *International Journal of Recent Advances in Pharmaceutical Research, 2,* 9–19.

Landfester, K. (2006). Synthesis of colloidal particles in miniemulsions. *Annual Review of Material Research, 36,* 231–279.

Laouini, A., Fessi, H., & Charcosset, C. (2012). Membrane emulsification: A promising alternative for vitamin E encapsulation within nano–emulsion. *Journal of Membrane Science, 423,* 85–96.

Leal-Calderon, F., Homer, S., Goh, A., & Lundin, L. (2012). W/O/W emulsions with high internal droplet volume fraction. *Food Hydrocolloids, 27,* 30–41.

Lerche, D. (2002). Dispersion stability and particle characterization by sedimentation kinetics in a centrifugal field. *Journal of Dispersion Science and Technology, 23,* 699–709.

Li, X., Wang, G., Zhao, J., Ding, H., Cunningham, C., Chen, F., Flynn, D., Reed, E., & Li, Q. (2005). Antiproliferative effect of β-elemene in chemoresistant ovarian carcinoma cells is mediated through arrest of the cell cycle at the G2-M phase. *Cellular and Molecular Life Sciences CMLS, 62,* 894–904.

LIN, S. Y., WU, W., & LUI, W. (1992). *In vitro* release, pharmacokinetic and tissue distribution studies of doxorubicin hydrochloride (Adriamycin HCl) encapsulated in lipiodolized w/o emulsions and w/o/w multiple emulsions. *Die Pharmazie, 47,* 439–443.

Lobato-Calleros, C., Rodriguez, E., Sandoval-Castilla, O., Vernon-Carter, E., & Alvarez-Ramirez, J. (2006). Reduced-fat white fresh cheese-like products obtained from W

1/O/W 2 multiple emulsions: Viscoelastic and high-resolution image analyzes. *Food Research International, 39*, 678–685.

Lovelyn, C., & Attama, A. A. (2011). Current state of nanoemulsions in drug delivery. *Journal of Biomaterials and Nanobiotechnology, 2*, 626–639.

Lu, J., & Bowles, M. (2013). How will nanotechnology affect agricultural supply chains? *International Food and Agribusiness Management Review, 16*, 21–42.

Lutz, R., Aserin, A., Wicker, L., & Garti, N. (2009). Double emulsions stabilized by a charged complex of modified pectin and whey protein isolate. *Colloids and Surfaces B: Biointerfaces, 72*, 121–127.

Ma, J. L., Xiong, Q. M., & Tao, T. (1993). Physico-chemical Properties and its Release *In Vitro* of Multiple Emulsion Containing Etoposide. *Chinese Journal of Pharmaceuticals, 24*, 357–357.

Madene, A., Jacquot, M., Scher, J., & Desobry, S. (2006). Flavour encapsulation and controlled release–a review. *International Journal of Food Science & Technology, 41*, 1–21.

Mahdi Jafari, S., He, Y., & Bhandari, B. (2006). Nano-emulsion production by sonication and microfluidization a comparison. *International Journal of Food Properties, 9*, 475–485.

Mahmood, A., & Uzair, M. (2010). Formulation and characterization of a multiple emulsion containing 1% L-ascorbic acid. *Bulletin of the Chemical Society of Ethiopia, 24*, 1–10.

Mahmood, T., Akhtar, N., Khan, B. A., Rasul, A., & Khan, H. M. S. (2013). Fabrication, physicochemical characterization and preliminary efficacy evaluation of a W/O/W multiple emulsion loaded with 5% green tea extract. *Brazilian Journal of Pharmaceutical Sciences, 49*, 341–349.

Matsumoto, S., Kita, Y., & Yonezawa, D. (1976). An attempt at preparing water in oil in water multiple phase emulsions. *Journal of Colloid and Interface Science, 57*, 353–361.

Matsumoto, S., Ueda, Y., Kita, Y., & Yonezawa, D. (1978). Preparation of water in olive oil in water multiple phase emulsions in an eatable form. *Agricultural and Biological Chemistry, 42*, 739–743.

May, S. W., & Li, N. N. (1974). Encapsulation of enzymes in liquid membrane emulsions. *In Enzyme Engineering Volume, 2*, 77–82, Springer US.

McClements, D. J. (2005). Theoretical analysis of factors affecting the formation and stability of multilayered colloidal dispersions. *Langmuir, 21*, 9777–9785.

McClements, D. J. (2011). Edible nanoemulsions: fabrication, properties, and functional performance. *Soft Matter, 7*, 2297–2316.

McClements, D. J., & Rao, J. (2011). Food grade nanoemulsions: Formulation, fabrication, properties, performance, biological fate, and potential toxicity. *Critical Reviews in Food Science and Nutrition, 51*, 285–330.

Mezzenga, R., Folmer, B. M., & Hughes, E. (2004). Design of double emulsions by osmotic pressure tailoring. *Langmuir, 20*, 3574–3582.

Mishra, B., & Pandit, J. (1990). Multiple water-oil-water emulsions as prolonged release formulations of pentazocine. *Journal of controlled Release, 14*, 53–60.

Morris, G., & Newcombe, G. (1993). Granular activated carbon: the variation of surface properties with the adsorption of humic substances. *Journal of Colloid and Interface Science, 159*, 413–420.

Morris, H. P. (1965). Studies on the development, biochemistry, and biology of experimental hepatomas. *Advances in Cancer Research, 9*, 227–302.

Muschiolik, G. (2007). Multiple emulsions for food use. *Current Opinion in Colloid & Interface Science, 12*, 213–220.

Nakajima, M., Nabetani, H., Ichikawa, S., & Xu, Q. Y. (2003). U.S. Patent No. 6,538,019. Washington, DC: U.S. Patent and Trademark Office.

Nakhare, S., & Vyas, S. (1997). Multiple emulsion based systems for prolonged delivery of rifampicin: *In Vitro* and *In vivo* characterization. *Die Pharmazie, 52*, 224–226.

Narang, A. S., Delmarre, D., & Gao, D. (2007). Stable drug encapsulation in micelles and microemulsions. *International Journal of Pharmaceutics, 345*, 9–25.

Neeraja, P., Amaleshwari, M., & Ravali, G. (2014). Formulation and evaluation of nifedipine multiple emulsions. *International Journal of Pharmaceutical, Chemical & Biological Sciences, 4*, 673–680.

Nehir E. L, S., & Simsek, S. (2012). Food technological applications for optimal nutrition: An overview of opportunities for the food industry. *Comprehensive Reviews in Food Science and Food Safety, 11*, 2–12.

Okochi, H., & Nakano, M. (2000). Preparation and evaluation of w/o/w type emulsions containing vancomycin. *Advanced Drug Delivery Reviews, 45*, 5–26.

Okushima, S., Nisisako, T., Torii, T., & Higuchi, T. (2004). Controlled production of monodisperse double emulsions by two-step droplet breakup in microfluidic devices. *Langmuir, 20*, 9905–9908.

Omotosho, J. A., Whateley, T. L., Law, T. K., & Florence, A. T. (1986). The nature of the oil phase and the release of solutes from multiple (w/o/w) emulsions. *Journal of Pharmacy and Pharmacology, 38*, 865–870.

Omotosho, J., Whateley, T., & Florence, A. (1989). Methotrexate transport from the internal phase of multiple w/o/w emulsions. *Journal of Microencapsulation, 6*, 183–192.

Onuki, Y., Morishita, M., & Takayama, K. (2004). Formulation optimization of water in oil water multiple emulsion for intestinal insulin delivery. *Journal of Controlled Release, 97*, 91–99.

Opawale, F. O., & Burgess, D. J. (1998). Influence of interfacial properties of lipophilic surfactants on water-in-oil emulsion stability. *Journal of Colloid and Interface Science, 197*, 142–150.

Otoni, C. G., Pontes, S. F., Medeiros, E. A., & Soares, N. D. F. (2014). Edible films from methylcellulose and nanoemulsions of clove bud (Syzygium aromaticum) and oregano (Origanum vulgare) essential oils as shelf life extenders for sliced bread. *Journal of Agricultural and Food Chemistry, 62*, 5214–5219.

Parthasarathi, S., Muthukumar, S., & Anandharamakrishnan, C. (2016). The influence of droplet size on the stability, in vivo digestion, and oral bioavailability of vitamin E emulsions. *Food & Function, 7*, 2294–2302.

Parveen, R., Baboota, S., Ali, J., Ahuja, A., Vasudev, S. S., & Ahmad, S. (2011). Effects of silymarin nanoemulsion against carbon tetrachloride-induced hepatic damage. *Archives of Pharmacal Research, 34*, 767–774.

Pascual-Pineda, L. A., Flores-Andrade, E., Jiménez-Fernández, M., & Beristain, C. I. (2015). Kinetic and thermodynamic stability of paprika nanoemulsions. *International Journal of Food Science & Technology, 50*, 1174–1181.

Pawlik, A., Cox, P. W., & Norton, I. T. (2010). Food grade duplex emulsions designed and stabilized with different osmotic pressures. *Journal of Colloid and Interface Science, 352*, 59–67.

Pereda, J., Ferragut, V., Quevedo, J., Guamis, B., & Trujillo, A. (2007). Effects of ultra-high pressure homogenization on microbial and physicochemical shelf life of milk. *Journal of Dairy Science, 90*, 1081–1093.

Pimentel-González, D., Campos-Montiel, R., Lobato-Calleros, C., Pedroza-Islas, R., & Vernon-Carter, E. (2009). Encapsulation of *Lactobacillus rhamnosus* in double emulsions formulated with sweet whey as emulsifier and survival in simulated gastrointestinal conditions. *Food Research International, 42*, 292–297.

Poluri, K., Sistla, R., Veerareddy, P., Narasu, L. M., Raje, A. A., & Hebsiba, S. M. (2011). Formulation, characterization and pharmacokinetic studies of carvedilol nanoemulsions. *Current Trends in Biotechnology and Pharmacy, 5*, 1110–1122.

Prankered, R., Frank, S., & Stella, V. (1988). A novel, practically water insoluble cytotoxic agent. *Journal of Parental Science and Technology, 42*, 76–81.

Putheti, S. (2015). Application of nanotechnology in food nutraceuticals and Pharmaceuticals. *Journal of Science and Technology, 2*, 17–23.

Qhattal, H. S. S., Wang, S., Salihima, T., Srivastava, S. K., & Liu, X. (2011). Nanoemulsions of cancer chemopreventive agent benzyl isothiocyanate display enhanced solubility, dissolution, and permeability. *Journal of Agricultural and Food Chemistry, 59*, 12396–12404.

Qian, C., & McClements, D. J. (2011). Formation of nanoemulsions stabilized by model food-grade emulsifiers using high-pressure homogenization: factors affecting particle size. *Food Hydrocolloids, 25*, 1000–1008.

Quintanilla-Carvajal, M. X., Camacho-Díaz, B. H., Meraz-Torres, L. S., Chanona-Pérez, J. J., Alamilla-Beltrán, L., Jimenéz-Aparicio, A., et al. (2010). Nanoencapsulation: A new trend in food engineering processing. *Food Engineering Reviews, 2*, 39–50.

Rodríguez-Huezo, M., Estrada-Fernández, A., García-Almendárez, B., Ludena-Urquizo, F., Campos-Montiel, R., & Pimentel-González, D. (2014). Viability of *Lactobacillus plantarum* entrapped in double emulsion during Oaxaca cheese manufacture, melting and simulated intestinal conditions. *LWT-Food Science and Technology, 59*, 768–773.

Roy, S., & Gupta, B. K. (1993). *In Vitro-In Vivo* correlation of indomethacin release from prolonged release w/o/w multiple emulsion system. *Drug Development and Industrial Pharmacy, 19*, 1965–1980.

Sahu, D., Sanjay, K., & Piyush, A. (2013). Recent advancement, technology & applications of multiple emulsions. *Innovare Journal of Health Sciences, 1*, 19–23.

Sameh, H., Wafa, E., Sihem, B., & Fernando, L. C. (2012). Influence of diffusive transport on the structural evolution of W/O/W emulsions. *Langmuir, 28*, 17597–17608.

Sanguansri, P., & Augustin, M. A. (2006). Nanoscale materials development–a food industry perspective. *Trends in Food Science & Technology, 17*, 547–556.

Saraf, S. (2010). Applications of novel drug delivery system for herbal formulations. *Fitoterapia, 81*, 680–689.

Schubert, H., & Engel, R. (2004). Product and formulation engineering of emulsions. *Chemical Engineering Research and Design, 82*, 1137–1143.

Schuch, A., Helfenritter, C., Funck, M., & Schuchmann, H. (2015). Observations on the influence of different biopolymers on coalescence of inner water droplets in W/O/W (water-in-oil-in-water) double emulsions. *Colloids and Surfaces A: Physicochemical and Engineering Aspects, 475*, 2–8.

Schwarz, J. C., Klang, V., Karall, S., Mahrhauser, D., Resch, G. P., & Valenta, C. (2012). Optimisation of multiple W/O/W nanoemulsions for dermal delivery of aciclovir. *International Journal of Pharmaceutics, 435*, 69–75.

Serdaroğlu, M., Öztürk, B., & Urgu, M. (2016). Emulsion characteristics, chemical and textural properties of meat systems produced with double emulsions as beef fat replacers. *Meat Science, 117*, 187–195.

Shafiq, S., Shakeel, F., Talegaonkar, S., Ahmad, F. J., Khar, R. K., & Ali, M. (2007). Development and bioavailability assessment of ramipril nanoemulsion formulation. *European Journal of Pharmaceutics and Biopharmaceutics, 66*, 227–243.

Shah, R. R., Brito, L. A., O'hagan, D. T., & Amiji, M. M. (2015). Emulsions as vaccine adjuvants. *In Subunit Vaccine Delivery,* 59–76. Springer New York.

Shakeel, F., & Faisal, M. S. (2010). Nanoemulsion: a promising tool for solubility and dissolution enhancement of celecoxib. *Pharmaceutical Development and Technology, 15,* 53–56.

Shantha Kumar, T., Soppimath, K., & Nachaegari, S. (2006). Novel delivery technologies for protein and peptide therapeutics. *Current Pharmaceutical Biotechnology, 7,* 261–276.

Shima, M., Morita, Y., Yamashita, M., & Adachi, S. (2006). Protection of Lactobacillus acidophilus from the low pH of a model gastric juice by incorporation in a W/O/W emulsion. *Food Hydrocolloids, 20,* 1164–1169.

Sigward, E., Mignet, N., Rat, P., Dutot, M., Muhamed, S., Guigner, J. M., Scherman, D., et al. (2013). Formulation and cytotoxicity evaluation of new self-emulsifying multiple W/O/W nanoemulsions. *International Journal of Nanomedicine, 8,* 611–625.

Singh, K. K., & Vingkar, S. K. (2008). Formulation, antimalarial activity and biodistribution of oral lipid nanoemulsion of primaquine. *International Journal of Pharmaceutics, 347,* 136–143.

Sivakumar, M., Tang, S., Y., & Tan, K., W. (2014). Cavitation technology–a greener processing technique for the generation of pharmaceutical nanoemulsions. *Ultrasonics Sonochemistry, 21,* 2069–2083.

Solans, C., Esquena, J., Forgiarini, A. M., Uson, N., Morales, D., Izquierdo, P., et al. (2003). Nano-emulsions: formation, properties, and applications. *Surfactant Science Series,* 525–554.

Solans, C., Izquierdo, P., Nolla, J., Azemar, N., & Garcia-Celma, M. (2005). Nano-emulsions. *Current Opinion in Colloid & Interface Science, 10,* 102–110.

Solans, C., Morales, D., & Homs, M. (2016). Spontaneous emulsification. *Current Opinion in Colloid & Interface Science, 22,* 88–93.

Srilatha, R., Aparna, C., Srinivas D. R., & Prathima, S. M. (2013). Formulation evaluation and characterization of glipizide nanoemulsion. *Asian Journal of Pharmaceutical and Clinical Research, 6,* 66–71.

Stratulat, I., Britten, M., Salmieri, S., St-Gelais, D., Champagne, C. P., Fustier, P., et al. (2013). Encapsulation of coenzyme Q 10 in a simple emulsion-based nutraceutical formulation and application in cheese manufacturing. *Food Chemistry, 141,* 2707–2712.

Strickley, R. G. (2004). Solubilizing excipients in oral and injectable formulations. *Pharmaceutical Research, 21,* 201–230.

Tadros, T. F., & Vincent, B. (1983). Emulsion stability. *Encyclopedia of Emulsion Technology, 1,* 129–285.

Tadros, T., Izquierdo, P., Esquena, J., & Solans, C. (2004). Formation and stability of nanoemulsions. *Advances in Colloid and Interface Science, 108,* 303–318.

Takahashi, T., Mizuno, M., Fujita, Y., Nishioka, B., & Majima, S. (1973). Increased concentration of anticancer agents in regional lymph nodes by fat emulsions, with special reference to chemotherapy of metastasis. *GANN Japanese Journal of Cancer Research,* 64, 345-350.

Talegaonkar, S., & Vyas, S. (2005). Inverse targeting of diclofenac sodium to reticuloendothelial system-rich organs by sphere-in-oil-in-water (s/o/w) multiple emulsion containing poloxamer 403. *Journal of Drug Targeting, 13,* 173–178.

Tang, S. Y., Sivakumar, M., & Nashiru, B. (2013). Impact of osmotic pressure and gelling in the generation of highly stable single core water-in-oil-in-water (W/O/W) nano multiple emulsions of aspirin assisted by two-stage ultrasonic cavitational emulsification. *Colloids and Surfaces B: Biointerfaces, 102,* 653–658.

Tang, S. Y., Sivakumar, M., Ng, A. M. H., & Shridharan, P. (2012). Anti-inflammatory and analgesic activity of novel oral aspirin-loaded nanoemulsion and nano multiple emulsion formulations generated using ultrasound cavitation. *International Journal of Pharmaceutics, 430,* 299–306.

Tiwari, S. B., & Amiji, M. M. (2006). Improved oral delivery of paclitaxel following administration in nanoemulsion formulations. *Journal of Nanoscience and Nanotechnology, 6,* 3215–3221.

Toorisaka, E., Ono, H., Arimori, K., Kamiya, N., & Goto, M. (2003). Hypoglycemic effect of surfactant-coated insulin solubilized in a novel solid-in-oil-in-water (S/O/W) emulsion. *International Journal of Pharmaceutics, 252,* 271–274.

Trotta, M., Debernardi, F., & Caputo, O. (2003). Preparation of solid lipid nanoparticles by a solvent emulsification–diffusion technique. *International Journal of Pharmaceutics, 257,* 153–160.

Tubesha, Z., Imam, M. U., Mahmud, R., & Ismail, M. (2013). Study on the potential toxicity of a thymoquinone-rich fraction nanoemulsion in sprague dawley rats. *Molecules, 18,* 7460–7472.

Turabi, E., Sumnu, G., & Sahin, S. (2008). Rheological properties and quality of rice cakes formulated with different gums and an emulsifier blend. *Food Hydrocolloids, 22,* 305–312.

Ushikubo, F., & Cunha, R. (2014). Stability mechanisms of liquid water-in-oil emulsions. *Food Hydrocolloids, 34,* 145–153.

Vaziri, A., & Warburton, B. (1994). Slow release of chloroquine phosphate from multiple taste-masked W/O/W multiple emulsions. *Journal of Microencapsulation, 11,* 641–648.

Vyas, T. K., Shahiwala, A., & Amiji, M. M. (2008). Improved oral bioavailability and brain transport of Saquinavir upon administration in novel nanoemulsion formulations. *International Journal of Pharmaceutics, 347,* 93–101.

Wang, X., Jiang, S., Wang, X., Liao, J., & Yin, Z. (2015). Preparation and evaluation of nattokinase-loaded self-double-emulsifying drug delivery system. *Asian Journal of Pharmaceutical Sciences, 10,* 386–395.

Wang, X., Jiang, Y., Wang, Y. W., Huang, M. T., HO, C. T., & Huang, Q. (2008). Enhancing anti-inflammation activity of curcumin through O/W nanoemulsions. *Food Chemistry, 108,* 419–424.

Wooster, T. J., Golding, M., & Sanguansri, P. (2008). Impact of oil type on nanoemulsion formation and Ostwald ripening stability. *Langmuir, 24,* 12758–12765.

Yanagie, H., Kumada, H., Nakamura, T., Higashi, S., Ikushima, I., Morishita, Y., et al. (2011). Feasibility evaluation of neutron capture therapy for hepatocellular carcinoma using selective enhancement of boron accumulation in tumor with intra-arterial administration of boron-entrapped water-in-oil-in-water emulsion. *Applied Radiation and Isotopes, 69,* 1854–1857.

Yang, Y. Y., Chung, T. S., & Ng, N. P. (2001). Morphology, drug distribution, and in vitro release profiles of biodegradable polymeric microspheres containing protein fabricated by double-emulsion solvent extraction/evaporation method. *Biomaterials, 22,* 231–241.

Yaqoob Khan, A., Talegaonkar, S., Iqbal, Z., Jalees Ahmed, F., & Krishan Khar, R. (2006). Multiple emulsions: an overview. *Current Drug Delivery, 3,* 429–443.

Yoshida, K., Sekine, T., Matsuzaki, F., Yanaki, T., & Yamaguchi, M. (1999). Stability of vitamin A in oil-in-water-in-oil-type multiple emulsions. *Journal of the American Oil Chemists' Society, 76,* 1–6.

Yuan, Y., Gao, Y., Zhao, J., & Mao, L. (2008). Characterization and stability evaluation of β-carotene nanoemulsions prepared by high pressure homogenization under various emulsifying conditions. *Food Research International, 41*, 61–68.

Zhang, Y., Shang, Z., Gao, C., Du, M., Xu, S., Song, H., & Liu, T. (2014). Nanoemulsion for solubilization, stabilization, and in vitro release of pterostilbene for oral delivery. *AAPS PharmSciTech, 15*, 1000–1008.

Zheng, J., Li, Y., Song, M., Fang, X., Cao, Y., McClements, D. J., & Xiao, H. (2014). Improving intracellular uptake of 5-demethyltangeretin by food grade nanoemulsions. *Food Research International, 62*, 98–103.

Zheng, S., Beissinger, R., & Wasan, D. (1991). The stabilization of hemoglobin multiple emulsion for use as a red blood cell substitute. *Journal of Colloid and Interface Science, 144*, 72–85.

Zheng, S., Zheng, Y., Beissinger, R. L., Wasan, D. T., & Mccormick, D. L. (1993). Hemoglobin multiple emulsion as an oxygen delivery system. *Biochimica et Biophysica Acta (BBA)-General Subjects, 1158*, 65–74.

# CHAPTER 9

# RAPID DETECTION OF PATHOGENS AND TOXINS

CHRISTINA G. SIONTOROU,[1] VASILIOS N. PSYCHOYIOS,[2] GEORGIA-PARASKEVI NIKOLELI,[2] DIMITRIOS P. NIKOLELIS,[3] STEPHANOS KARAPETIS,[2] SPYRIDOULA BRATAKOU,[2] and KONSTANTINOS N. GEORGOPOULOS[1]

[1]Laboratory of Simulation of Industrial Processes, Department of Industrial Management and Technology, School of Maritime and Industry, University of Piraeus, Greece

[2]Laboratory of Inorganic and Analytical Chemistry, School of Chemical Engineering, Dept 1, Chemical Sciences, National Technical University of Athens, 9 Iroon Polytechniou St., Athens 157 80, Greece

[3]Laboratory of Environmental Chemistry, Department of Chemistry, University of Athens, Panepistimiopolis-Kouponia, GR-15771 Athens, Greece

## CONTENTS

## ABSTRACT

Food safety relies heavily on timely and reliable pathogen and/or toxin detection. Although culture-based analysis seems to remain indispensable for the near future, novel nano-enabled technologies have been developed and optimized, aiming at counterbalancing comparable reliability with much reduced assay times. The current state-of-the-art in food analysis is presented herein in two main sections, referring to pathogen detection strategies for reducing assay times and toxin detection schemes, respectively. Approaches based on DNA/RNA, nucleic acid sequencing, immunology, and biosensing systems are reviewed, putting emphasis on the developed bench-scale sensors and commercial systems. Technology barriers, unresolved issues and methodological uncertainties are, also, discussed, along with future trends and emerging technologies.

## 9.1   INTRODUCTION

Food safety is a critical health parameter with serious ramifications and repercussions that may extend well beyond geographical origin or timeline. As food chain becomes more globalized and more preserved, a range of opportunities and risks are emerging. Improved hygiene management systems and strict regulations are not sufficient to eliminate the threat of foodborne pathogens and the disease outbreaks worldwide. Health threats originate from bacteria, viruses, fungi, toxins, and parasites (Dwivedi and Jaykus, 2011). According to the Centers for Diseases Control and Prevention (CDC, 2011), 44% of food-derived illnesses are caused by 31 pathogens, the majority of which involve noroviruses, *salmonella*, clostridia, campylobacter, toxoplasma, *E.coli*, listeria and staphylococci. Novel strategies, such as biocontrol with, for example, the use of prebiotic and probiotic additives at farming (Jordan et al, 2014), could potentially reduce the primary production risks but their effectiveness during storage, transport, preparation and distribution of food is doubtful.

Pathogens and toxins penetrate the food chain through water, air, contact with soil, and the processing/storage environment. Ready-to-eat food is subjected to stricter safety control during preparation than food consumed raw (such as fruits and vegetables) although their forward logistics risk (e.g., at market display) is still considerable. Undoubtedly, testing is indispensable in order to ensure safety and minimize the occurrence of foodborne illness.

Detection methods should be fast, easy to use, even by minimally trained or untrained personnel, reliable, selective, and largely automated, yielding a straightforward and comprehensive result output at a reasonable cost. The subjectivity of assay interpretation seems to be a critical issue in detection reliability (Mandal et al., 2011); the higher the specificity of the method, the easier, and more reliable, the interpretation of the result.

Culture-based techniques are well-established and commonly used, offering low-cost sensitivity and accuracy. The procedure is simple yet tiresome and, most importantly, time consuming, including sample homogenization, cultivation in nutrient media and species identification using biochemical testing or microscopy. Mean analysis times are in the order of several days; technological advances in certain steps of the procedure, for example, in polymerase chain reaction (PCR) or in cultural enrichment, reduce assay time to a few days or, at the most, to 30 h (Lopez-Campos et al., 2012). At any case, the time required for risk assessment does not allow for timely risk management. Toxin detection, on the other hand, may seem more manageable regarding assay time, given the present state-of-the art of diagnostics, but the huge number of metabolites along with their structural similarities present challenges that add up to the complexity of the sample matrix and the very low detection limits required. For example, shiga toxin produced by *Shigella dysenteriae* differs by only one amino acid from Shiga-like toxin 1 (Stx-1) produced by some *E. coli* strains; shiga-like toxin 2 (Stx-2), produced by other stains of *E. coli*, differs from shiga toxin by a different amino acid sequence, and is 400 times as toxic than Stx-1 (Sandvig et al., 2015). The toxins are usually present in food matrices at concentration levels of pg/ml or less (Verhaegen et al., 2016), whereas most sensing methods offer ng/ml detection limits (Rasooly et al., 2015). Considering that food samples can be rarely tested on an 'as is' basis, sample processing and toxin extraction raise detectable levels in sample manifold.

Although culture-based detection still remains necessary (Taskila et al., 2012), quicker culture-independent assessments are developing, capitalizing on biotechnology and nanotechnology tools (Figure 9.1). Significant reductions in assay times yield, inevitably, compromises in selectivity and sensitivity, unless careful designed and validated devices are developed. Thus, the research domain is reluctant to adapt early a completely different detection rationale, concentrating vastly on optimizing the established laboratory (*ex situ*) testing. Although recent trends are focusing towards on site, less-than-an-hour assays, a number of challenges pertaining to both, availability of technology and quality assurance, need to be studied and overcome.

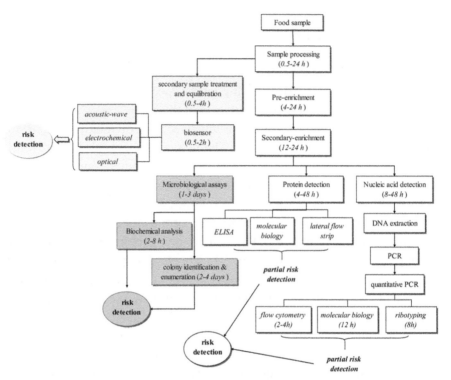

**FIGURE 9.1**   Overview of the assay protocol involved in conventional (main stream) food sample analysis. Microbiological assays (grayed stages) provide the highest accuracy level but they are lengthy; nucleic acid based techniques could be simultaneously performed but a set of tests are required to yield a comprehensive result. New approaches, such as biosensors, are more rapid and straightforward. Entries in italics represent alternative technology options.

## 9.2   PATHOGEN DETECTION STRATEGIES FOR REDUCING ASSAY TIMES

Strategies proposed (Figure 9.2) are mostly nucleic acid, immunological or biosensor based (Zhao et al., 2014). Molecular techniques have been developed to optimize various assay stages and analytical parameters (sensitivity, selectivity and discrimination of viable pathogens) with a distant (but distinct) prospect for engineering field (on-site) detectors. Antibody-based methods are advantageous on assay handling and sensitivity for a range of pathogens and toxins, with a proven field suitability along with a well-known cross-reactivity. Developed biosensors offer detection times of minutes and comparable sensitivity and selectivity to

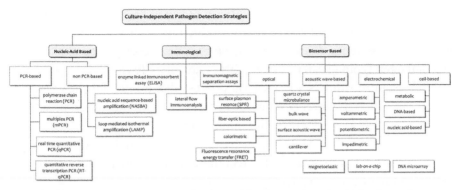

**FIGURE 9.2**   State-of-the art strategies for rapid pathogen detection in food samples.

immunoassays; these devices are just starting to enter mass production and launching, especially for on-site (continuous or intermittent) monitoring. The proposed methods are critically presented and reviewed here and a brief comparison of the methods is shown in Figure 9.3.

## 9.2.1   NUCLEIC ACID BASED METHODS

These methods target specific DNA or RNA sequences of the pathogen by hybridization with a complimentary synthetic probe. The very basis of

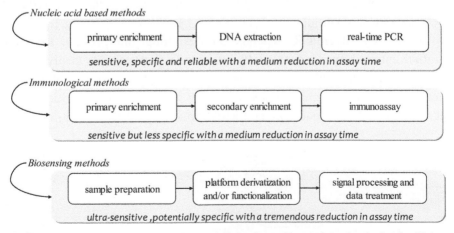

**FIGURE 9.3**   Brief comparison of the workflows for pathogen detection in food utilizing three main strategic approaches: nucleic acid-based, immunological and biosensor technology.

detection ensures high levels of specificity (Zhao et al., 2014). Gene targeting, especially those related to toxin production from *Clostridium botulinum*, *Vibrio cholerae*, *Staphylococcus aureus*, or *E. coli* (Fusco et al., 2011) and pathogen typing (Adzitey et al., 2012) are, also, feasible.

PCR is an *in situ* DNA replication process that allows for the exponential amplification of target DNA in the presence of synthetic oligonucleotide primers and DNA polymerase; the initial number of target sequences doubles with each cycle. Identification based on PCR amplification of target genes is considered to be a reliable technique when properly developed and validated for a certain species.

The detection of a number of pathogens within one analysis requires the simultaneous amplification of more than one locus. Multiplex PCR (mPCR) uses several primer sets, each specific for a certain pathogen, combined into a single PCR assay (Chen and Lin, 2007). Therefore, mPCR is advantageous in that less time and effort is required to produce the same results as simple PCR (Figure 9.4). Multiplex PCR without enrichment has been used to detect *Listeria monocytogens*, *Staphylococcus aureus*, *Streptococcus agalactiae*, *Enterobacter sakazakii*, *Escherichia coli* O157:H7, *Vibrio parahaemolyticus*, *Salmonella spp.* and *Pseudomonas fluorescens* in milk and poultry samples (Chiang et al., 2012); the detection limit of all species ranged between $10^2$ to $10^4$ colony forming units (CFU) per mL of sample homogenate. Lee et al. (2014) developed a similar assay for processed food vegetables and oysters with comparable detection limits. Timmons et al. (2013) designed primers with 5' flaps and reduced detection limit ten-fold.

Reliability of mPCR depends on designing the primers to avoid mutual interactions and to yield amplicons that can be easily distinguished after thermal cycling. The use of mPCR is currently limited to determining the dynamics and structure of microbial communities (Zhao et al., 2014). Mukhopadhyay and Mukhopadhyay (2007) used *fliCh₇* and

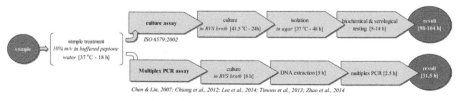

**FIGURE 9.4**  Flowchart of the general protocol in traditional cultural assays (top workflow) and the real-time multiplex PCR assay (bottom workflow).

*iap* gene-specific primers to establish a multiplex-PCR assay for *E. coli* O157:H7 and *Listeria monocytogenes*. Analysis implemented modified enrichment and harvesting methods that enhanced considerably sensitivity and assay time.

Real-time or quantitative PCR (qPCR) monitors the amplicon formation throughout the process (Jung et al., 2005). qPCR allows the quantification of specific microorganisms in food samples, whereas the targeting of specific genes to monitor their expressions provides useful information regarding the susceptibility of the pathogen to environmental factors (temperature, pH, lipid composition of food, etc.) Monitoring the process real-time reduces overall assay time as post-amplification treatment is no longer necessary. Gomez et al. (2010) developed a qPCR to quantify the total aerobic bacteria and fungi on fresh produce, using the centrifugation water produced during sample processing as a reference in order to compensate for the complexity of food matrix; 35% of the natural bacterial population and 64% of inoculated bacteria were recovered in the reference. In effect, the qPCR enumeration of cell number is similar to plate assay but easier and quicker.

PCR-based assays are disadvantageous in discriminating between live and dead cells. Li and Chen (2013) used propidium monoazide pretreatment to inhibit DNA amplification from dead cells; notwithstanding, detectability was only reduced to $10^3$ CFU/g of sample.

The use of RNA instead of DNA introduced the quantitative reverse transcription PCR (RT-qPCR). RNA is reverse transcribed into complementary DNA (cDNA) using a reverse transcriptase and specific primers (Bustin 2000). The cDNA is then used as template for PCR. RT-qPCR can be performed with a single-step protocol to minimize contamination risks. Yet, the method is not rugged as small variations in sample treatment or amplification can impact the results. The use of RNA is generally better in discriminating between live and dead cells due to its rapid degradation outside of cells (Yaron and Matthews, 2002). Furthermore, the mRNA levels of an organism under certain conditions (e.g., food composition, temperature, pH, oxygen levels) can be deduced (Zhao et al., 2014). An RT-qPCR assay for the detection of *L. monocytogenes* and *S. enterica* was developed for dairy products (Lucore et al., 2000); metal hydroxides were used to immobilize bacterial cells before RNA extraction. The detection limits were $10^2$ CFU/mL in milk and 10 CFU/mL for ice cream. Rantsiou et al. (2008) proposed an RT-qPCR method for the detection of *L. monocytogenes* in soft cheese, fermented sausage, cured ham, minced meat, and milk food matrices; detectability ranged between $10^3$ to $10^4$ CFU/mL or g of food sample.

Nucleic acid sequence-based amplification (NASBA) is a non-PCR primer-dependent technology that amplifies nucleic acids in an one-step procedure to detect specific RNA sequences *in vitro*. The tri-enzymatic system, including avian myeloblastosis virus reverse transcriptase (AMV-RT), T7 DNA-dependent RNA polymerase, and RNase H (Dwivedi and Jaykus, 2011), is coupled with two primers: the forward primer contains a 5' promoter region recognized by the T7 RNA polymerase, followed by a complementary sequence to the target RNA; the reverse primer is identical to a sequence on the target RNA. The forward primer binds to its target sequence and is extended by AMV-RT to create an RNA-cDNA hybrid; RNase H hydrolyzes the RNA component and the free cDNA is annealed. T7 RNA polymerase binds to its promoter region on the double-stranded cDNA resulting in the transcription of many target RNA copies. Gel electrophoresis or enzyme-linked gel assays are used to detect the products (Law et al., 2015). O'Grady et al. (2009) used this technique to detect 1–10 CFU/mL of *S. aureus* in milk, whereas Min and Baeumner (2002) detected viable *E. coli* in drinking water with a limit of 40 CFU/mL.

Loop mediated isothermal amplification (LAMP) is a novel nucleic acid amplification method that uses autocycling strand displacement DNA synthesis performed by the Bst DNA polymerase large fragment, amplified under irothermal conditions between 59°C and 65°C; the amplicons are mixtures of many different sizes of stem loop DNAs with several inverted repeats of the target sequence and cauliflower-like structures with multiple loops. The method has been proven to be rapid, low-cost, simple, extremely highly sensitive and specific (Notomi et al., 2000). The advantages over PCR-based methods that use 4 to 6 primers to target 6–8 specific regions of the target gene, are apparent. The method yields usually $10^3$ times or higher more amplification products within 60 min than PCR; the products can be easily visualized with SYBR Green I dye to avoid gel electrophoresis analysis. Chen et al. (2011) developed LAMP-based assay for acidophilic thermophilic bacteria in pure juices with a detection limit of 22.5 CFU/ml of sample. LAMP has been also used in reverse transcription and multiplex assays (Jasson et al., 2010). *In situ* LAMP assay (Ikeda et al., 2007) and real-time reverse transcription LAMP assay (Liu et al., 2009) have been developed and employed for the detection of various foodborne pathogens.

## 9.2.2   IMMUNOLOGICAL METHODS

Antibody-based analysis (Figure 9.5) is perhaps the only technology at present with a proven efficacy for the detection of cells, spores, and viruses (Iqbal et al., 2000). Polyclonal antibodies can be raised quickly and cheaply and do not require the time or expertise associated with the production of monoclonal antibodies; yet, their specificity and availability is limited. Progress achieved in recombinant antibody phage display technology rendered immunological assays with improved sensitivity, specificity, reproducibility and reliability to enable vast commercialization. While nucleic acid-based detection may be more specific and sensitive, immunological-based detection is more robust and has the ability to detect not only contaminating organisms but also their biotoxins that may not be expressed in the organism's genome (Iqbal et al., 2000). Even though both antibody-based and nucleic acid-based detection have greatly decreased assay times compared to traditional culture techniques, they still lack the capability and prospect for real-time detection.

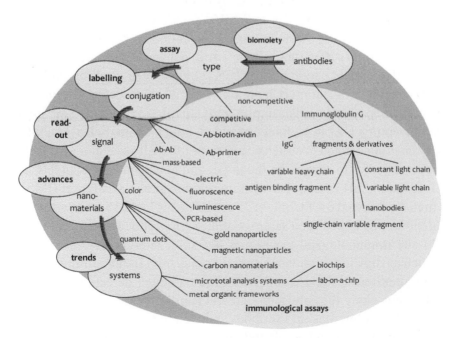

**FIGURE 9.5**   Materials, methods, advancements and trends in immunological food analysis.

ELISA-based assays can be directly applied for the detection of foodborne pathogens, such as *L. monocytogenes* (Tully et al., 2006) and *Salmonella enterica* spp. (Valdivieso-Garcia et al., 2001). A major draw-back lies in analysis time which is usually lengthy; typical assays are multi-step, involving blocking, washing, and incubation prior to sub-strate development. Protein microarrays are excellent candidates for high-throughput analysis of biomolecular interactions in miniaturized assays formats (MacBeath and Schreiber, 2000). The implementation of antibodies on such platforms for pathogen detection offers a flexible approach for the screening of many bacterial isolates from numerous sample matrices.

Array formats typically consist of a panel of pathogen-specific anti-bodies spotted onto individual positions on a microarray slide by dedi-cated robotic handling (printing), with subsequent pathogen detection commonly employing sandwich ELISA formats. Gehring et al. (2006) printed a biotinylated antibody (caprine-derived) for *E. coli* O157:H7 on a streptavidin-coated microarray slide. Captured cells were further probed with a fluorescein-labeled secondary antibody and microarray spots were visualized through the use of fluorescent microscopy; the lin-ear detection range was $3 \times 10^6$ to $9 \times 10^7$ cells/mL.

Nanomaterials have been proposed as antibody binding platforms where their large surface-to-volume ratio enhances antibody density. Single-walled carbon nanotubes (SWCNTs) have been used for antibody and enzyme immobilization (Chunglok et al., 2011) to detect *S. enterica* serovar *Typhimurium* in milk samples with ten times lower detection limit than conventional ELISA. Functionalized gold nanoparticles (AuNPs) have been implemented to structure a network that allowed more enzyme-labeled antibodies to connect to the target and to enhance the signal (Cho and Irudayaraj, 2013); coupled with immunomagnetic separation (IMS), this method detected 3 CFU/mL of *E. coli* O157:H7 and 15 CFU/mL of *S. Typhimurium* in spiked food samples within 2 h of inoculation.

On-site immunological techniques based on lateral flow immunoas-says such as dipstick, immunochromatography, and immunofiltration are gaining attention in the area of foodborne pathogen detection. The test sample flows along the solid substrate via capillary action; the sample is mixed with a colored agent (antibody, labeled antigen or gold nanopar-ticles) and moves through a substrate pretreated with an antibody or anti-gen (Gomez at al., 2010). Most lateral flow assays are basically designed to incorporate a visual response about 2–10 min after the application of the sample. Jung et al. (2005) developed a colloidal immunochromatographic strip for the detection of *Escherichia coli* O157:H7 in enriched samples, reporting that the minimum limit was $1.8 \times 10^5$ CFU/ml without enrichment and 1.8 CFU/ml after enrichment.

Immunomagnetic separation assays (IMS) allow the conjunction of food analysis with advanced detection and imaging technology. The use of immunomagnetic beads as capturing agents provides selectivity as the growth of interfering microorganisms is suppressed. Separation is a two-step process: the target cells are mixed with the beads, incubated for 1 h and magnetically separated (Mandal et al., 2011); magnetic handling is fast and efficient, whereas it only slightly affects the target analytes. Immunoprecipitation, isolation and identification of pathogens can utilize a range of bioreactive agents to improve the resolution of magnetic resonance imaging (Stevens and Jaykus, 2004). An immunomagnetic bead-immunoliposome fluorescence assay has been proposed for *Escherichia coli* O157:H7 (DeCory et al., 2005); overall assay time was 8 min.

### 9.2.3 BIOSENSOR SYSTEMS

A biosensor-based process utilizes compact devices to detect biochemical interactions translated into straightforward signals by a variety of transducers (Figure 9.6). Most approaches provide response times of minutes and overall assay times of <1 h. Nanotechnology offers amply a wide variety of new materials such as nanowires, nanofibers, nanoparticles, nanobelts or nanoribbons and nanotubes that are promptly incorporated into biosensor platforms. Nanotechnology enables the detection of a large number of pathogens within a single, label-free assay, utilizing surface plasmon resonance, amperometric, potentiometric, mass-sensitive, magnetoelastic, microbial, DNA, and impedimetric configurations. Most detectors proposed are antibody- or DNA based, although alternative schemes have been proposed and are currently validated.

#### 9.2.3.1 Optical Systems

Optical illumination of a metal surface is considered as a suitable basis for pathogen detection. The basic principle allows real-time monitoring

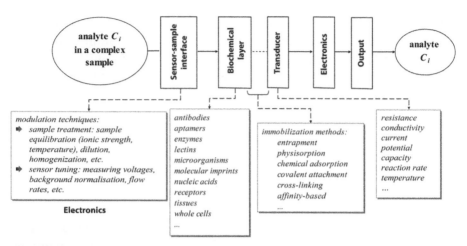

**FIGURE 9.6** Overview of the biosensor architecture and commonly used materials, techniques and methods for food analysis.

of biomolecular interactions. Surface plasmon resonance (SPR) sensors work on this principle. Antibodies are immobilized on thin gold film just over the reflecting surface of waveguide, to capture various pathogens (Arora et al., 2011); at certain wavelengths and near IR region, strong resonance is generated through the interaction of light with the electron cloud in the metal. Commercial SPR systems are available from Biacore and Texas Instruments.

SPR sensors monitor the binding events in very close proximity of the transducer allowing for rinse-free assays and very low detection limits. Wei et al. (2007) developed an SPR biosensor for *C. jejuni* in poultry. A detection limit of $10^3$ CFU/mL of *C. jejuni* was achieved and the overall assay time was 60 min, although matrix effects have been reported. A sandwich assay was employed for the detection of *E. coli* on spinach (Linman et al., 2010). A secondary anti-*E. coli* antibody labeled with horseradish peroxidase was bound to the captured cells; the enzymatic reaction produced an insoluble dark blue product to enhance the SPR measurement that gave $10^4$ CFU/mL detection limit. A miniaturized SPR platform was used to detect *S. Enteritidis* in poultry (Son et al., 2007), while Taylor et al. (2006) used an 8-channel customized SPR platform to detect *E. coli* O157:H7, *S. Typhimurium*, *L. monocytogenes*, and *C. jejuni*, simultaneously in apple juice.

Optical fibers serve as excellent transducers for food safety evaluation. The main advantage of this technology is the capability of remote sensing. Commercial platforms are available such as the RAPTOR (Research International, Inc., Monroe, Wash., USA). Fluorescence resonance energy transfer (FRET) is a phenomenon of non-radiative energy transfer from a fluorescent donor molecule to an acceptor molecule due to dipole–dipole interactions when in close proximity. A FRET immunosensor for *S. Typhimurium* in pork has been developed using a silica fiber core for fluorophore-labeled antibody protein G complexes immobilization and fluorophore excitation by the evanescent wave (Ko and Grant, 2006); the detection limit achieved, however, is quite high ($10^5$ CFU/mL). Other optical methods have been implemented for pathogen detection. Functionalized polymer vesicles engineered to respond to specific bacterial metabolites, were used for the colorimetric detection of *S. aureus* and *E. coli* in apple juice (Pires et al., 2011). You et al. (2011) developed a handheld device based on light scattering in latex immunoagglutination assays for *E.coli* detection directly on fresh vegetables.

### 9.2.3.2 Acoustic Wave Systems

Acoustic wave platforms are based on the detection of mechanical acoustic waves produced by a biochemical reaction. The basis of signal production is an oscillating crystal that resonates at a fundamental frequency; the crystal is coated with a biological agent (e.g., antibody) that, when exposed to the complementary analyte (e.g., antigen), a quantifiable change occurs in the resonant frequency of the crystal, which correlates to mass changes at the crystal surface; nanogram levels of mass changes can be detected, which is usually not low enough for sensitive detection requiring post-amplification. The vast majority of acoustic wave biosensors utilize piezoelectric transducers that generate and transmit acoustic waves in a frequency-dependent manner. The most commonly used piezoelectric materials include quartz ($SiO_2$) and lithium niobate ($LiTaO_3$) (Leonard et al., 2003). In order to acquire an active surface for use in a piezoelectric biosensor the surface must be chemically stable with a high number of the actively immobilized biological elements, whereas the coating surface should be as thin as possible.

Acoustic wave platforms offer label-free, on-line assays for immunological interactions at a cost-effective and straightforward protocols. Salmain et al. (2011) successfully designed a direct, label-free immunosensor for the rapid detection and quantification of staphylococcal enterotoxin A (SEA) in buffered solutions using the quartz crystal microbalance with dissipation; the detection limit was 20 ng/ml and the overall assay time was 15 min.

Disadvantages of these platforms include the necessary long incubation times for surface derivatization and functionalization, problems in crystal surface regeneration for consecutive assays and the number of washing and drying steps required. The problem of surface functionalization is common to most biosensor formats due to issues related to the suitability and efficiency of the biomoieties immobilization methods. A quartz crystal microbalance biosensor using AuNPs capped with a DNA probe for signal amplification has been developed for *E. coli* O157:H7 *eaeA* gene in apple juice, milk, and ground beef (Chen et al., 2008); the detection limit achieved was $10^2$ CFU per mL or g of sample. New materials and immobilization approaches have been, also, considered. Babacan et al. (2000) have shown that the immobilization of anti-*Salmonella* antibodies onto a gold electrode of a piezoelectric quartz crystal through Protein A coupling has proven more reproducible and more stable than coupling of the antibody with polyethylenimine. Yilmaz et al. (2015) proposed the use of molecular imprinting-based synthetic receptors for bacterial detection and evaluated the biosensor in apple juice samples; although the detection limit was not satisfactory, this study demonstrates a certain potential for a new detection strategy for foodborne pathogens.

Cantilever sensors based on piezoelectric materials for mass-based sensing have also been reported for pathogen detection. *E. coli* O157:H7 at a level of 10 cells/mL in beef has been identified within a 10-min assay (Maraldo and Mutharasan, 2007). The method was optimized and adopted for milk samples and fruit extracts (Sharma and Mutharasan, 2013).

The development of a dual channel surface acoustic wave device was also reported for the detection of *Legionella* and *Escherichia coli* (Howe and Harding, 2000); some problems associated with measurements at the liquid phase have been reported. Even when the electrodes were protected, for example, using $SiO_2$ coating, sensor sensitivity did not improve.

### 9.2.3.3 Electrochemical Systems

Electrochemical systems utilize amperometric, voltammetric, potentiometric, or conductometric/impedimetric transducers in order to screen liquid, solid or gaseous samples. Mostly electrode-based, miniaturization of the devices is feasible but regeneration might prove tedious. Compared to SPR in a study for *E. coli* detection using the same antibody immobilization method, an impedimetric biosensor reached a detection limit 40 times lower than SPR (Maalouf et al., 2007).

Enzymatic reactions and electroactive labels, including nanomaterials are commonly used in culture-independent electrochemical platforms to yield disposable sensors. Lin et al. (2008) fabricated a disposable strip, based on an indirect ELISA assay with the primary antibody immobilized on AuNP-modified electrodes. An enzyme-linked sandwich immunoassay was also combined with magnetic separation in a voltammetric biosensor to detect $10^3$ CFU of *E. coli* O157:H7 per ml of pre-treated apple juice (Gehring and Tu, 2005). Multiplex detection of *E. coli* O157:H7, *Campylobacter*, and *Salmonella* in milk has been performed using a multiwalled carbon nanotube-polyallylaminemodified screen-printed electrode with immobilized antibodies; antibody-functionalized nanocrystals (CdS, CuS, and PbS) were used for labeling (Viswanathan et al., 2012). Tang et al. (2010) developed a voltammetric sandwich immunosensor using horseradish peroxidase-nanosilica-doped multiwalled carbon nanotubes for signal amplification to detect staphylococcal enterotoxin B; correlation to ELISA results was quite satisfactory.

Amperometric transducers were used for the measurement of the biochemical reactions taking place during various metabolic processes in bacterial cells. One approach relies on the detection of specific marker enzyme after incubating in a suitable medium. This strategy aids in easy coliforms detection in water samples by the mere presence of enzyme $\beta$-d-glucuronide, glucuronosohydrolase and $\beta$-d-galactosidase. Conventional enzyme-based methods for *E. coli* detection are time-consuming and need spectrophotometric verification. Biosensors employ efficient electro-oxidative methods for enzyme detection (Arora et al., 2011). A rapid viable *E. coli* detection method was developed by Perez et al. (2001) using $\beta$-d-galactosidase to convert 4-aminophenyl-$\beta$-d-galactopyranoside to 4-aminophenol after hydrolysis. Serra et al. (2005) developed a novel and improved tyrosinase composite biosensor based on amperometric detection of $\beta$-galactosidase activity; the enzymatic hydrolysis of phenyl-d-galactopyranoside yields phenol as end product, readily detected by the sensor.

Enzyme-based or supported systems exhibit small dynamic ranges (usually a decade of concentration) due to limitations imposed by the saturation kinetics of the enzymes. Interference from the oxidation/reduction of matrix components is, also, a problem. Despite, real-time measurements can be routinely performed. A real-time aptamer-based potentiometric biosensor was reported for the detection of *E. coli* CECT 675 as a nonpathogenic surrogate for *E. coli* O157:H7 in milk and apple juice (Zelada-Guillen et al., 2010).

Ion-sensitive field-effect transistor (ISFET) and light-addressable potentiometric sensor (LAPS) are commonly proposed approaches for food analysis; still, fabrication of the sensors compromise the functionality of the immobilized biological species and hinders the biochemical reaction (Wand and Salazar, 2016). The commercially available Threshold® Immunoassay System targets *E. coli* O157:H7 in food samples.

Impedance spectroscopy employs a cyclic function of small amplitude and variable frequency to a transducer; the resulting current is used to calculate the impedance at each of the probed frequencies (Zhao et al., 2014). Detection is based on the measurement of changes in the electrical properties of bacterial cells when they are attached to, or associated with, the electrodes. Louie et al. (1998) developed an impedance-based, on-site biosensor system to detect *E. coli* O157:H7 and *Salmonella* spp; stable readings could be obtained within 1 min. Recently, Pal et al. (2008) developed a sandwich conductometric immunosensor for *B. cereus*. The sensitivity of the sensor has been improved by using lateral flow and high-density electrodes. *Listeria innocua*, a surrogate for *L. monocytogenes*, was detected via the recognition of the target by bacteriophage-encoded peptidoglycan hydrolases (endolysin) immobilized on a gold screen printed electrode (Tolba et al., 2012); the biosensor was validated with milk spiked with the target organism but not with real samples. Graphene has become an attractive material for biosensing due to its unique nanostructure and electrical properties. A graphene based conductometric sensors has been developed for bacterial detection at 10 CFU/mL (Huang et al., 2011), although not tested in food samples.

### 9.2.3.4   Advanced Technology and Trends

Analytical performance and operational stability remain an on-going intense struggle for the diagnostics domain. Reproducibility and in-field performance are considered as top priority for rapid safety assessments, while detections limits continue to be reduced with every technology advancement that is rapidly assimilated by the domain to shift research trajectories and alternate strategies. Nanotechnology, offering new tools, procedures and materials, enable new detection strategies to be investigated for designing and validating multiplex detection configurations, microfluidic techniques, integrated sensors and wireless devices.

DNA-based biosensors are considered more suitable and reliable for pathogen detection. Short nucleic acid sequences, probe specific for a particular pathogen, are immobilized on the surface of a transducer. Complementary binding to the bacterial DNA sequences in the probe sequence (hybridization) provides pathogen identification. The extent of hybridization determines the presence or absence of complementary sequences in the sample (Arora et al., 2011). Detection of different pathogenic bacteria can be done with disposable low-density genosensor arrays, fabricated using screen-printed arrays of gold electrodes with immobilized thiol-tethered oligonucletide and biotinylated signaling probes (Farabullini et al., 2007); detection of sequence complementarity is achieved through the identification of toxin produced by the specific bacteria. Wang (2002) has successfully developed novel genosensors for *Cryptosporidium*, *E. coli*, *Giardia* and *Mycobacterium tuberculosis*.

Microarray and microfluidics technology is expected to boost DNA diagnostics. A microarray is a miniature device with short (25–80 bp) single-stranded DNA oligonucleotide probes, complementary to genes or genomic markers of pathogens, attached to glass slides or chips (McLoughlin, 2011). The DNA (or RNA) from a target organism is extracted and fluorescent-tagged; the DNA is then denatured to generate single-stranded molecules which bind to their corresponding complementary probes on the array. The fluorescence signal intensity is proportional to the concentration of the target DNA sequence (Lauri and Mariani, 2009). Wang et al. (2007) used this technique to successfully detect 22 pathogens from broth cultures. Microarray technology is highly sensitive and specific, suitable for multiple analyzes and easy to perform. The production, however, of the probes and the fluorescent tagging remain quite expensive and expertise-intensive. Up to now, few commercially available kits for food analysis are available.

Microfluidic assays (lab-on-a-chip kits) test nano- or pico-volumes of liquid samples through a number of micro-channels, each hosting a different nucleic-acid-based assay; the small sample volumes usually require pre-concentration steps. Microfluidic devices have been applied to the detection of *S. Enteritidis*, *S. aureus*, *E. coli*, and *C. jejuni* (Kim et al., 2014). A microfluidic device has been coupled to mPCR for the detection of *S. enterica*, *E. coli*

O157:H7, and *L. monocytogenes* in hotdog, milk, and banana food matrices (Zhang et al., 2011); the detection limits achieved were on the order of $10^4$ genome copies of each pathogen per 1 µL of purified DNA and the overall assay time was 14 min.

Magnetoelastic sensors, based on amorphous ferromagnetic alloys, present a new trend in biosensor platforms. When excited by an external time-varying magnetic field, the materials exhibit a magnetoelastic resonance which can be detected by using a pickup coil (Ruan et al., 2003). When a target pathogen comes in contact with the alloy sensor surface, the added mass causes a change in resonance frequency and the signal can be remotely detected by the pick-up coil. Magnetoelastic sensing can be easily engineered into remote and wireless detectors for pathogens in fresh produce with minimal or no sample preparation. Filamentous E2 phages were immobilized on a magnetoelastic platform and the developed sensor detected *S. Typhimurium* in tomatoes, without any sample treatment (i.e., 'as is') within 30 min (Li et al., 2010).

## 9.3  TOXIN DETECTION SCHEMES

There exist a large number of microbe-produced compounds with a toxic profile, and a larger number of their variants. Toxin detection in food follows two main strategies: indirectly, via the identification of pathogens and the toxin-producing genes, and directly, by developing assays and sensors to target each specific compound. Certain compounds, such as cholera or shiga toxin, exert high toxicity even at ultra low concentration levels; thus, the former approach becomes advantageous over direct assays that, considering the need of sample treatment prior to analysis, have to demonstrate femtomolar detection limits to become competitive.

Bioassays based on animal testing or phenotypic methods are accurate but not rapid (Wang and Salazar, 2016). Yet, prompt screening of food (fresh or processed) is highly essential to minimize adverse health effect. Similar to pathogens, the existing detection methods include ELISA, biosensors, DNA-based and immunochemical methods, all of which are sensitive and possess multiplexing capability. While DNA-based assays may be more sensitive than immunoassays, the latter has an important advantage in the detection of extracellular toxins. The list of toxins with the higher epidemiological profile includes cholera toxin, Staphylococcal enterotoxin, shiga

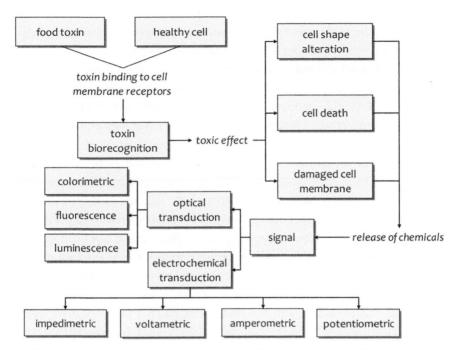

**FIGURE 9.7** General strategy for cell-based toxin biosensing in foodstuff.

toxin, *Salmonella* enterica and brevetoxins. To no surprise, research efforts have concentrated towards those species.

Electrochemical methods, coupled and conjugated to a wide variety of pre- and post- modifications, offer less-than-a-minute response times and low detecatbility. Viswanathan et al. (2006) developed an electrochemical immunosensor for cholera (CT) detection, using liposomic magnification by linking an anti-CT-B subunit monoclonal antibody with poly(3,4-ethyl-enedioxythiophene) coated on Nafion-supported multiwalled carbon nanotube film in a glassy carbon electrode. To improve sensitivity, detection was based on sandwich-type assay involving electronic transducers and ganglioside functionalized liposomes. In the presence of CT, the liposomes release potassium ferrocyanide that can be measured by adsorptive square-wave stripping voltammetry; the limit of detection and linear range were at the pg/ml level. In a colorimetric bioassay developed by Schofield et al. (2007), a thiolated-lactose derivative self-assembled on gold nanoparticles aggregated upon binding to the CT-B subunit, and the detection principle was based on a color change from red to purple; the limit of detection achieved was quite high (3 mg/mL). Nikoleli et al. (2011) developed a thin-film CT biosensor with micromolar detectability based on polymerized lipids incorporated with ganglioside GM1, the natural CT receptor; response time was less than 5 min but the sensor was not tested in real food samples.

Staphylococcal enterotoxins comprise a group of 21 heat-stable toxins produced by *S. aureus*. Although the existing antibody-based detection methods are suitable for high throughput screening, sensitivity and assay time need improvement (Inbaraj and Chen, 2016). Yang et al. (2008) developed an optical immunosensor using carbon nanotubes and immobilized anti-enterotoxin antibodies; a horseradish peroxidase -labeled secondary antibody was used for visualization through fluorescence. The sensor demonstrated significantly enhanced sensitivity over the standard-type immunosensor assay with 0.1 ng/ml detectability; food analysis required purification by carboxymethyl cellulose chromatography. The method has been modified with gold nanoparticles: the antibodies were immobilized through physical adsorption along with a dye conjugate; the product was detected by chemiluminescence (Yang et al., 2009). The limit of detection achieved was 10 times lower than that obtained with the traditional ELISA method; also, overall sensor performance was better than carbon nanotubes-based screening.

More than 200 different serotypes of *E. coli* can produce shiga toxin, 150 of which are potent toxicants (Johnson et al., 2006). Identification of strains is indispensable (using multiplex PCR, for example), but other strategies have been proposed for rapid detection. The B subunit of shiga toxins specifically recognizes the globotriose ($P^k$) blood group antigen, which contains the trisaccharide $\alpha$Gal(1$\rightarrow$4)$\beta$Gal(1$\rightarrow$4)$\beta$Glc, and each of five B subunits has three available binding sites for $P^k$ (Inbaraj and Chen, 2016). Chien et al. (2008) developed a chip-based sensor to host an SPR competition assay using glyconanoparticles obtained by self-assembling two derivatives of $P^k$ onto gold nanoparticles of various sizes; the longer chain length was shown to enhance binding affinity of the $P^k$ moiety, resulting in a greater flexibility of $P^k$ ligand to bind onto more sites on the toxin surface. Nagy et al. (2008) proposed a chromatic shiga sensor with Gal-a1,4-Gal glycopolydiacetylene nanoparticles; the method demonstrated selectivity, speed of response and sensitivity.

Brevetoxins are cyclic polyether neurotoxins produced by the marine dinoflagellate *Karenia brevis* responsible for shellfish poisoning. For rapid screening, Tang et al. (2011) developed a sensitive electrochemical immunosensor on GNPs-decorated amine-terminated polyamidoamine dendrimers; horseradish peroxidase-labeled anti-brevetoxin antibodies were used for the competitive immunoassay to give a detection limit of 0.01 ng/mL.

Aflatoxins are produced by *Aspergillus flavus* and *Aspergillus parasiticus,* fungi abundant in warm and humid areas. The toxin can be found in agricultural crops such as maize, peanuts, cottonseed, and tree nuts. Aflatoxin B1 is a class-1 human carcinogen with maximum permissible residual levels less than 1 fg/ml (Commission Regulation No 1881/2006). Evtugyn et al. (2014) proposed an electrochemical aptasensor based on glassy carbon electrodes modified with electropolymerized neutral red and polycarboxylated macrocyclic ligands onto which the DNA aptamers were covalently attached. The interaction with the toxin resulted in the decrease of the cathodic peak current of the probe measured by cyclic voltammetry; detection limit was found to be 0.1 nM but improved significantly when electrochemical impedance spectroscopy replaced cyclic volatammetry. Recovery studies on peanuts, cashew nuts, white wine and soy sauce spiked samples ranged between 85 and 100%.

Basu et al. (2015) developed a field effect capacitive aflatoxin B1 sensor using electrophoretically deposited reduced graphene oxide films; detection limit was 0.1 femtogram of toxin per ml of sample but the sensor has not been tested in real food samples.

## 9.4   TECHNOLOGICAL BARRIERS TO RAPID FOOD ASSAYS

The goal of any assay method, established or emerged, is to manage to detect within a few hours a small number of pathogens (or toxin molecules) amid a large number of normal microflora (or biochemical molecules) within a complex sample matrix. Low detectability at excellent selectivity and minimal cross-reactivity is a challenging task for most techniques even today. The accuracy of DNA-based techniques comes with step-by-step protocols to follow that increase assay time; enzyme-based techniques are usually more rugged and time-effective but the sample matrix may (and usually does) contain enzyme inhibitors that increase uncertainty of the results (Enserink, 2001).

Time reductions in DNA-based detection are feasible using, for example, rapid PCR and nucleic acid sequence-based amplification (NASBA) instead of DNA hybridization, with best detection limits of $10^2$–$10^3$ CFU/gr gram of sample to target 1–10 CFU/g in the food product. Real-time NASBA, commercially available as TaqMan® (Applied Biosystems, CA, US) and NucliSens® (bioMerieux, Durham, NC, US) can further reduce analysis time with fluorescence resonance energy transfer (FRET). Conventional post-amplification electrophoresis gives results and confirmation within 12–72 hours; substituting with FRET, overall assay times reduce to 4–11 hours (Jaykus, 2003).

The TaqMan® assay is based on the endogeonous 5'→3' exonuclease activity of Taq DNA polymerase by including a dual fluorophore-labeled oligonucleotide probe during the PCR amplification cycle. The probe hybridizes to its complementary amplicon during the PCR reactions and the 5' nuclease activity of Taq polymerase lops off the fluorescent 5' dye molecule, which fluoresces freely without the need of 3' quencher dye. The NucliSens® Basic Kit, on the other hand, is based on a transcription-driven isothermal RNA amplification method, combining NASBA with molecular beacons (dual fluorophore-labeled oligonucleotide probes). The beacon probe sequence is flanked by a stem-and-loop structure that masks fluorescence; when hybridization to specific amplicons occurs during amplification, this structure is disrupted and the fluorophore agent fluoresces. The advantage of fluorescence lies in the ability to detect and confirm the amplicon during amplification. The disadvantages include larger sample volumes and detection limits, trained analysts, higher analysis costs and the regulatory problem of nonviable pathogens detected.

Notwithstanding, matrix effects affect reliability in pathogen and toxin detection in a multifaceted manner. Food is a heterogeneous multi-component matrix; difference in viscosity due to fats may interfere with mixing and sample homogenization (Mandal et al., 2011). Fat and conjugates interfere with antibody-binding, whereas carbohydrates inhibit nucleic acid amplification (Dwivedi and Jaykus, 2011). Sharma and Mutharasan (2013) developed a piezoelectric cantilever sensor for *L. monocytogenes* in milk; results indicated significant matrix effects due to the large amounts of dissolved sugars, proteins and fats. Native microflora, such as *Enterococcus* or *Lactobacillus* spp., also interfere with *L. monocytogenes* detection (Nero et al., 2009). Another study using surface plasmon reference (SPR) showed that the lipid and protein components of broiler meat inhibited the detection of *Campylobacter jejuni* (Wei et al., 2007). *E. coli* O157:H7 SPR detection in cucumber revealed significant nonspecific adsorption of carbohydrates, vitamins, and dietary fiber (Wang et al., 2013). Moreover, significant interference in this study was caused by large numbers of nonpathogenic *E. coli* cells, which blocked the transport of the target pathogen to the surface of the sensor.

Sodium chloride, sucrose, and lysine can act either as DNA adducts or as polymerase inhibitors (Wilson, 1997). Polyphenolics in fruit extracts (Siebert et al., 1996) and milk (Ongol et al., 2009) exert mild inhibitory effects on DNA extraction due to protein-protein interactions. Fruit extracts pose also pH problems: Taylor et al. (2006) had to neutralize samples before analysis with an SPR sensor in order to optimize antibody–antigen binding. Pre-treatment of meat and poultry samples may cause the release of enzymes or antimicrobial agents that interfere with analysis and detection (Bhunia, 2014).

Toxin load is, also, influenced by the complexity of the matrix. *E.coli* toxins detected in milk, spinach, and ground beef matrices using a capacitive immunosensor (Li et al., 2011), showed that adherence of toxins differed significantly between the food items, requiring matrix-dependent variability measurements. Other studies demonstrated significant interference of unknown origin (D'Souza and Jaykus, 2003).

## 9.5 FUTURE TRENDS

Food safety management currently relies on pre-market screening of processed products and the implementation of good agricultural practices for fresh produce and processed food alike. While this strategy reduces the release of contaminated finished product, it cannot prevent failure in the forward logistics chain.

Today, more effective food safety efforts strive to eliminate pathogens by focusing on the entire logistics chain from pre-processed raw-material to at-market-display products. Following certain severe outbreaks, for example, the EHEC/STEC outbreak related to sprouts in Germany, risk management focuses towards more proactive measures; this shift, however, might require a significant revision of microbiological criteria and testing.

Leaving aside the problems related to sampling and statistics, mainly due to sample representativeness issues (Zwietering et al., 2016), risk management is challenging due to both, the nature of the object and the severity of the risk. Most well-established control strategies cannot guarantee food safety. Cost-effective and efficient environmental monitoring, for example, relies on alarm sensors, placed *in situ*, that provide a rough, but much needed, estimate of below-threshold or above-threshold pollution levels; in case of a suspected increase in pollution inflow, more elaborate analysis with sensitive and selective detectors provides reliable results to guide counter measures and anti-stress activities. This strategy might be applicable and efficient in processed food, providing that the environmental conditions (temperature, light, air/water-tightness, etc.) during product distribution are strictly implemented and controlled; still, this management framework does not provide the necessary ruggedness or versatility to ensure the required safety level. Despite, fresh produce is not easy to control and, further, food contamination incidences are not easy to constrain.

The risks associated with food storage and distribution make on-site rapid pathogen and toxin testing a decisive factor for disease prevention, while, considering the costs related to disease outbreaks, economically advantageous. Faster results would mean that harmful products are withdrawn timely and quality products reach the consumers earlier. The need for alternative testing strategy is eminent; a 2000 study estimated the number of weekly tests in a dairy plant to be 636, followed closely by 444 weekly microbiological tests from processed-food plants (Alocilja and Radke, 2003).

Technology is mature for supplying a range of field instruments and *in situ* applied methods. Trajectories clearly indicate that biosensors have the potential to offer a wide diversity of hand-held devices or remote sensors (Siontorou and Batzias, 2014). Nanomaterial and nanoengineering tools succeeded in developing devices with rapid response times and ultra-trace detection limits; most importantly, new detection strategies emerged that gave an 'out-of-the-box' view in pathogen and toxin detection. Still, lab-based and culture-based verification of results or phenotyping is indispensable. Besides, *in situ* or on-site analysis needs to solve sample treatment; most fresh produce needs some degree of treatment to become samples, which includes time consuming and laborious homogenization, filtration and extraction.

The post-genomic era might provide new tools for rapid detection. Epigenomics study the mechanisms of gene expression and the changes induced, for example, by toxins or bioactive food components (Cifuentes, 2012). Two different analytical approaches are emerging for quantifying changes: microarray technology and rapid DNA sequencing, both coupled with real-time PCR (Herrero et al., 2012). Other interesting approaches that could handle the wide natural variability of food include metabolic fingerprinting and profiling that are still bound to computer power and spectrometric or nuclear magnetic resonance equipment. At the huge technology pace, they might gain a more routine character in the near future.

## 9.6 CONCLUSIONS

The detection and identification of foodborne pathogens and toxins currently rely on very elaborate, time-consuming and expensive conventional culturing techniques in order to minimize uncertainty and risk. These techniques, based on sampling, sample handling and laboratory analyzes, cannot be engineered (minimized and adapted) into on-site rapid assays. As a result, food industry needs real-time, portable pathogen and toxin detection sensors with high sensitivity and selectivity, simple operation and reliability.

Advances in immunology, molecular biology, engineering and bioinformatics continue to support the development of fast testing. These trends, especially those implicating metabolism-linked strategies, can gradually replace conventional schemes. There are still many problems to be solved related to both, food matrix and technology. Coordinated research must be intensified around alternative detection strategies, sensor optimization and validation in real-world applications with a view to rapidly moving proof-of-concept set-ups for rapid screening to prototyping and commercialization.

## KEYWORDS

- **biosensor-based detection**
- **food safety**
- **immunoarrays**
- **immunoassays**
- **molecular techniques**
- **nucleic acid-based assays**

## REFERENCES

Adzitey, F., Rusul, G., Huda, N., Cogan, T., & Corry, J. (2012). Prevalence, antibiotic resistance and RAPD typing of *Campylobacter* species isolated from ducks, their rearing and processing environments in Penang, Malaysia. *Int. J. Food Microbiol*, *154*, 197–205.

Alocilja E. C., & Radke S. M. (2003). Market analysis of biosensors for food safety. *Biosens. Bioelectron.* *18*, 841–846.

Arora P., Sindhu A., Dilbaghi N., & Chaudhury A. (2011). Biosensors as innovative tools for the detection of food borne pathogens. *Biosens. Bioelectr, 28*, 1–12.

Babacan S., Pivarnik P., Letcher S., & Rand A. G., (2000). Evaluation of antibody immobilization methods for piezoelectric biosensor application. *Biosens. Bioelectron., 15,* 615–621.

Basu J., Datta S., & RoyChaudhuri C. (2015). A graphene field effect capacitive immunosensor for sub-femtomolar food toxin detection. *Biosens. Bioelectron, 68,* 544–549.

Bhunia A. K. (2014). One day to one hour: how quickly can foodborne pathogens be detected? *Future Microbiol, 9,* 935–946.

Bustin S. A. (2000). Absolute quantification of mRNA using real-time reverse transcription polymerase chain reaction assays. *J. Mol. Endocrinol, 25,* 169–193.

CDC. (2011). Estimates of Foodborne Illness in the United States. Available online at http://www.cdc.gov/foodborneburden/2011-foodborneestimates.html. (accessed 28 March 2016).

Chen S., Wu, V. C. H., Chuang Y., & Lin, C. (2008). Using oligonucleotide-functionalized Au nanoparticles to rapidly detect foodborne pathogens on a piezoelectric biosensor. *J. Microbiol. Methods, 73,* 7–17.

Chen, H. M., & Lin, C. W. (2007). Hydrogel-coated streptavidin piezoelectric biosensors and applications to selective detection of Strep-Tag displaying cells. *Biotechnol. Prog, 23,* 741–748.

Chen, S. Y., Wang, F., Beaulieu, J. C., Stein, R. E., & Ge, B. L. (2011). Rapid detection of viable *salmonella*e in produce by coupling propidium monoazide with loop-mediated isothermal amplification. *Appl. Environ. Microbiol, 77,* 4008–4016.

Chiang, Y. C., Tsen, H. Y., Chen, H. Y., Chang, Y. H., Lin, C. K., Chen, C. Y., et al. (2012). Multiplex PCR and a chromogenic DNA macroarray for the detection of *Listeria monocytogens, Staphylococcus aureus, Streptococcus agalactiae, Enterobacter sakazakii, Escherichia coli* O157:H7, *Vibrio parahaemolyticus, Salmonella* spp. and *Pseudomonas fluorescens* in milk and meat samples. *J. Microbiol. Methods, 88,* 110–116.

Chien, Y. Y., Jan, M. D., Adak, A. K., Tzeng, H. C., Lin, Y. P., Chen, Y. J., et al. (2008). Globotriose-functionalized gold nanoparticles as multivalent probes for Shiga-like toxin. *Chembiochem, 9,* 1100–1109.

Cho, I., & Irudayaraj, J. (2013). In-situ immuno-gold nanoparticle network ELISA biosensors for pathogen detection. *Int. J. Food Microbiol, 164,* 70–75.

Chunglok, W., Wuragil, D. K., Oaew, S., Somasundrum, M., & Surareungchai, W. (2011). Immunoassay based on carbon nanotubes-enhanced ELISA for *Salmonella enterica* serovar *Typhimurium. Biosens. Bioelectron, 26,* 3584–3589.

Cifuentes, A. (2012). Food analysis: present, future, and foodomics, *ISRN Anal. Chem.* ID 801607, http://dx.doi.org/10.5402/2012/801607.

D'Souza D. H., & Jaykus L. A. (2003). Nucleic acid sequence based amplification for the rapid and sensitive detection of *Salmonella enterica* from foods. *J. Appl. Microbiol. 95,* 1343–1350.

DeCory, T. R., Durst, R. A., Zimmerman, S. J., Garringer, L. A., Paluca, G., DeCory, H. H., et al. (2005). Development of an immunomagnetic bead-immunoliposome fluorescence assay for rapid detection of *Escherichia coli* O157:H7 in aqueous samples and comparison of the assay with a standard microbiological method. *Appl. Environ. Microbiol, 71,* 1856–1864.

Dwivedi, H. P., & Jaykus, L. (2011). Detection of pathogens in foods: The current state-of-the-art and future directions. *Crit. Rev. Microbiol, 37,* 40–63.

Enserink, M. (2001). News focus: Biodefense hampered by inadequate tests. *Science, 294,* 1266–1267.

Evtugyn, G., Porfireva, A., Stepanova, V., Sitdikov, R., Stoikov, I., Nikolelis, D., et al. (2014). Electrochemical aptasensor based on polycarboxylic macrocycle modified with neutral red for aflatoxin B1 detection. *Electroanalysis, 26*, 2100–2109.

Farabullini, F., Lucarelli, F., Palchetti, I., Marrazza, G., & Mascini, M. (2007). Disposable electrochemical genosensor for the simultaneous analysis of different bacterial food contaminants. *Biosens. Bioelectron, 22*, 1544–1549.

Fusco, V., Quero, G. M., Morea, M., Blaiotta, G., & Visconti, A. (2011). Rapid and reliable identification of *Staphylococcus aureus* harboring the enterotoxin gene cluster (egc). and quantitative detection in raw milk by real time PCR. *Int. J. Food Microbiol, 144*, 528–537.

Gehring, A. G., Albin, D. M., Bhunia, A. K., Reed, S. A., Tu, S. I., & Uknalis, J. (2006). Antibody microarray detection of *Escherichia coli* O157:H7: Quantification, assay limitations, and capture efficiency. *Anal. Chem. 78*, 6601–6607.

Gehring, A., & Tu, S. (2005). Enzyme-linked immunomagnetic electrochemical detection of live *Escherichia coli* O157: H7 in apple juice. *J. Food Prot, 68*, 146–149.

Gomez P., Pagnon M., Egea-Cortines M., Artes F., & Weiss J. (2010). A fast molecular nondestructive protocol for evaluating aerobic bacterial load on fresh-cut lettuce. *Food Sci. Technol. Int, 16*, 409–415.

Herrero M., Simó C., García-Cañas V., Ibáñez E., & Cifuentes A. (2012). Foodomics: MS-based strategies in modern Food Science and Nutrition, *Mass Spectr. Rev, 31*, 49–69.

Howe E., & Harding G. (2000). A comparison of protocols for the optimisation of detection of bacteria using a surface acoustic wave (SAW) biosensor. *Biosens. Bioelectron, 15*, 641–649.

Huang, Y., Dong X., Liu, Y., Li, L., & Chen, P. (2011). Graphene-based biosensors for detection of bacteria and their metabolic activities. *J. Mater. Chem, 21*, 12358–12362.

Ikeda, S., Takabe, K., Inagaki, M., Funakoshi, N., & Suzuki, K. (2007). Detection of gene point mutation in paraffin sections using in situ loop-mediated isothermal amplification. *Pathol. Int, 57*, 594–599.

Inbaraj, B. S., & Chen, B. H. (2016). Nanomaterial-based sensors for detection of foodborne bacterial pathogens and toxins as well as pork adulteration in meat products. *J. Food Drug Anal. 24*, 15–28.

Iqbal, S. S., Mayo, M. W., Bruno, J. G., Bronk, B. V., Batt, C. A., & Chambers, P. (2000). A review of molecular recognition technologies for detection of biological threat agents. *Biosens. Bioelectron, 15*, 549–578.

Jasson, V., Jacxsens, L., Luning, P., Rajkovic, A., & Uyttendaele, M. (2010). Alternative microbial methods: an overview and selection criteria. *Food Microbiol, 27*, 710–730.

Jaykus, L. A. (2003). Challenges to developing real-time methods to detect pathogens in foods. *ASM News, 69*, 341–347.

Johnson, K. E., Thorpe, C. M., & Sears, C. L. (2006). The emerging clinical importance of non-O157 Shiga toxin-producing *Escherichia coli. Clin. Infect. Dis, 43*, 1587–1595.

Jordan, K., Dalmasso, M., Zentek, J., Mader, A., Bruggeman, G., Wallace, J., et al. (2013). Microbes versus microbes: control of pathogens in the food chain. *J. Sci. Food Agric. 94*, 3079–3089.

Jung, B. Y., Jung, S. C., & Kweon, C. H. (2005). Development of a rapid immunochromatographic strip for detection of *Escherichia coli* O157. *J. Food Prot, 68*, 2140–2143.

Kim, T. H., Park, J., Kim, C. J., & Cho, Y. K. (2014). Fully integrated lab-on-a-disc for nucleic acid analysis of food-borne pathogens. *Anal. Chem, 86*, 3841–3848.

Ko, S., & Grant, S. A. (2006). A novel FRET-based optical fiber biosensor for rapid detection of *Salmonella typhimurium. Biosens. Bioelectron, 21,* 1283–1290.

Lauri, A., & Mariani, P. O. (2009). Potentials and limitations of molecular diagnostic methods in food safety. *Genes Nutr, 4,* 1–12.

Law, J. W., Ab, Mutalib N. S., Chan, K. G., & Lee, L. H. (2015). Rapid methods for the detection of foodborne bacterial pathogens: Principles, applications, advantages and limitations. *Front Microbiol, 5,* 770, 1–19.

Lee, N., Kwon, K. Y., Oh, S. K., Chang, H. J., Chun, H. S., & Choi, S. W. (2014). A multiplex PCR assay for simultaneous detection of *Escherichia coli* O157:H7, Bacillus cereus, *Vibrio parahaemolyticus, Salmonella* spp., *Listeria monocytogenes,* and *Staphylococcus aureus* in Korean ready-to-eat food. *Foodborne Pathog. Dis, 11,* 574–580.

Leonard, P., Hearty, S., Brennan, J., Dunne, L., Quinn, J., Chakraborty, T., et al. (2003). Advances in biosensors for detection of pathogens in food and water. *Enzyme Microb. Technol, 32,* 3–13.

Li, B., & Chen, J. Q. (2013). Development of a sensitive and specific qPCR assay in conjunction with propidium monoazide for enhanced detection of live *Salmonella* spp. in food. *BMC Microbiol. 13,* 273, 1–13.

Li, D., Feng, Y., Zhou, L., Ye, Z., Wang, J., Ying, Y., et al. (2011). Label-free capacitive immunosensor based on quartz crystal Au electrode for rapid and sensitive detection of *Escherichia coli* O157:H7. *Anal. Chim. Acta, 687,* 89–96.

Li, S., Li, Y., Chen, H., Horikawa, S., Shen, W., Simonian, A., & Chin, B. A. (2010). Direct detection of *Salmonella typhimurium* on fresh produce using phage-based magnetoelastic biosensors. *Biosens. Bioelectron, 26,* 1313–1319.

Lin, Y., Chen, S., Chuang, Y., Lu, Y., Shen, T. Y., Chang, C. A., & Lin, C. (2008). Disposable amperometric immunosensing strips fabricated by Au nanoparticles-modified screen-printed carbon electrodes for the detection of foodborne pathogen *Escherichia coli* O157:H7. *Biosens. Bioelectron, 23,* 1832–1837.

Linman, M. J., Sugerman, K., & Cheng, Q. (2010). Detection of low levels of *Escherichia coli* in fresh spinach by surface plasmon resonance spectroscopy with a TMB-based enzymatic signal enhancement method. *Sensors Actuators B: Chem, 145,* 613–619.

Liu, Y., Chuang, C. K., & Chen, W. J. (2009). *In situ* reverse transcription loop-mediated isothermal amplification (*in situ* RT-LAMP) for detection of Japanese encephalitis viral RNA in host cells. *J. Clin. Virol, 46,* 49–54.

Lopez-Campos, G., Martinez-Suarez, J. V., Aguado-Urda, M., & Lopez-Alonso, V. (2012). Detection, identification, and analysis of foodborne pathogens. Springer US. New York, NY 13–32.

Louie, A. S., Marenchic, I. G., & Whelan, R. H. (1998). A fieldable modular biosensor for use in detection of foodborne pathogens. *Field Anal. Chem. Technol, 2,* 371–377.

Lucore, L. A., Cullison, M. A., & Jaykus, L. A. (2000). Immobilization with metal hydroxides as a means to concentrate food-borne bacteria for detection by cultural and molecular methods. *Appl. Environ. Microb. 66,* 1769–1776.

Maalouf, R., Fournier-Wirth, C., Coste, J., Chebib, H., Saikali, Y., & Vittori, O. (2007). Label-free detection of bacteria by electrochemical impedance spectroscopy: comparison to surface plasmon resonance. *Anal. Chem, 79,* 4879–4886.

MacBeath, G., & Schreiber, S. L. (2000). Printing proteins as microarrays for high-thoughput function determination. *Science, 289,* 1760–1763.

Mandal, P. K., Biswas, A. K., Choi, K., & Pal, U. K. (2011). Methods for rapid detection of food borne pathogens: An overview. *Am. J. Food Technol, 6,* 87–102.

Maraldo, D., & Mutharasan, R. (2007). 10-minute assay for detecting *Escherichia coli* O157: H7 in ground beef samples using piezoelectric-excited millimeter-size cantilever sensors. *J Food Prot, 70*, 1670–1677.

McLoughlin, K. S. (2011). Microarrays for pathogen detection and analysis. *Brief Funct. Genomics* 10, 342–353.

Min, J., & Baeumner, A. J. (2002). Highly sensitive and specific detection of viable *Escherichia coli* in drinking water. *Anal. Biochem, 303*, 186–193.

Mukhopadhyay, A., & Mukhopadhyay, U. K. (2007). Novel multiplex PCR approaches for the simultaneous detection of human pathogens: *Escherichia coli* O157:H7 and *Listeria monocytogenes*. *J. Microbiol. Methods, 68*, 193–200.

Nagy, J. O., Zhang, Y., Yi, W., Liu, X., Motari E., Song, J. C., et al. (2008). Glycopolydiacetylene nanoparticles as a chromatic biosensor to detect Shiga-like toxin producing *Escherichia coli* O157:H7. Bioorg. Med. *Chem. Lett, 18*, 700–703.

Nero, L. A., de, Mattos, M. R., Barros, M. A., Beloti, V., & Franco, B. D. (2009). Interference of raw milk autochthonous microbiota on the performance of conventional methodologies for *Listeria monocytogenes* and *Salmonella* spp. detection. *Microbiol. Res, 164*, 529–535.

Nikoleli, G. P., Nikolelis, D. P., & Tzamtzis, N. (2011). Development of an electrochemical biosensor for the rapid detection of cholera toxin using air stable lipid films with incorporated ganglioside GM1. *Electroanalysis, 23*, 2182–2187.

Notomi, T., Okayama, H., Masubuchi, H., Yonekawa, T., Watanabe, K., Amino, N., et al. (2000). Loop-mediated isothermal amplification of DNA. *Nucleic Acids Res, 28*, E63.

O'Grady, J., Lacey, K., Glynn, B., Smith, T. J., Barry, T., & Maher, M. (2009). TM-RNA– a novel high-copy-number RNA diagnostic target–its application for *Staphylococcus aureus* detection using real-time NASBA. *FEMS Microbiol. Lett, 301*, 218–223.

Ongol, M. P., Tanaka, M., Sone, T., & Asano, K. (2009). A real-time PCR method targeting a gene sequence encoding 16S rRNA processing protein, rimM, for detection and enumeration of Streptococcus thermophilus in dairy products. *Food Res, Int, 42*, 893–898.

Pal, S., Ying, W., Alocilja, E. C., & Downes, F. P. (2008). Sensitivity and specificity performance of a direct-charge transfer biosensor for detecting Bacillus cereus in selected food matrices. *Biosyst. Eng, 99*, 461–468.

Perez, F. G., Tryland, I., Mascini, M., & Fiksdal, L. (2001). Rapid detection of *Escherichia coli* in water by a culture-based amperometric method. *Anal. Chim. Acta, 427*, 149–154.

Pires, A. C. D. S., Soares, N. D. F. F., da Silva, L. H. M., De Almeida, M. V., Le Hyaric, M., Andrade, N. D., et al. (2011). A colorimetric biosensor for the detection of foodborne bacteria. *Sensor Actuat. B-Chem, 153*, 17–23.

Planche, T., Aghaizu, A., Holliman, R., Riley, P., Poloniecki, J., Breathnach, A., et al. (2008). Diagnosis of Clostridium difficile infection by toxin detection kits: a systematic review. *Lancet Infect. Dis. 8*, 777–784.

Rasooly, R., Balsam, J., Hernlem, B. J., & Rasooly, A. (2015). Sensitive detection of active Shiga toxin using low cost CCD based optical detector. *Biosens. Bioelectr, 68*, 705–711.

Salmain, M., Ghasemi, M., Boujday, S., Spadavecchia, J., Techer, C., & Val, F. (2011). Piezoelectric immunosensor for direct and rapid detection of staphylococcal enterotoxin A (SEA) at the ng level. *Biosens. Bioelectron, 29*, 140–144.

Sandvig, K., Lingelem, A. B. D., Skotland, T., & Bergan, J. (2015). Shiga toxins: properties and action on cells. In: Ladant D., Popoff M. R., & Alouf J. E. (Eds.). The Comprehensive Sourcebook of Bacterial Protein Toxins (Fourth Edition), Elsevier Ltd Netherlands, 2015, 267–286.

Schofield, C. L., Field, R. A., & Russell, D. A. (2007). Glyconanoparticles for the colorimetric detection of cholera toxin. *Anal. Chem, 79*, 1356–1361.

Serra, B., Morales, M. D., Zhang, J. B., Reviejo, A. J., Hall, E. H., & Pingarrón, J. M. (2005). In-a-day electrochemical detection of coliforms in drinking water using a tyrosinase composite biosensor. *Anal. Chem, 77*, 8115–8121.

Sharma, H., & Mutharasan, R. (2013). Rapid and sensitive immunodetection of *Listeria monocytogenes* in milk using a novel piezoelectric cantilever sensor. *Biosens. Bioelectron, 45*, 158–162.

Siebert K. J., Troukhanova N. V., & Lynn, P. Y. (1996). Nature of polyphenol-protein interactions. *J. Agric. Food Chem, 44*, 80–85.

Siontorou, C. G., & Batzias, F. A. (2014). A methodological combined framework for road mapping biosensor research: A fault tree analysis approach within a strategic technology evaluation frame. *Crit. Rev. Biotechnol, 34*, 31–55.

Son, J. R., Kim, G., Kothapalli, A., Morgan, M. T., & Ess, D. (2007). Detection of *Salmonella* Enteritidis using a miniature optical surface plasmon resonance biosensor. *J Phys: Conference Series, 61*, 1086–1090.

Stevens, K. A., & Jaykus, L. A. (2004). Bacterial separation and concentration from complex sample matrices: A review. *Crit. Rev. Microbiol, 30*, 7–24.

Tang, D., Tang, J., Su, B., & Chen, G. (2010). Ultrasensitive electrochemical immunoassay of staphylococcal enterotoxin B in food using enzyme-nanosilica-doped carbon nanotubes for signal amplification. *J. Agric. Food Chem, 58*, 10824–10830.

Tang, D., Tang, J., Su, B., & Chen, G. (2011). Gold nanoparticles-decorated amine-terminated poly(amidoamine) dendrimer for sensitive electrochemical immunoassay of brevetoxins in food samples. *Biosens. Bioelectron, 26*, 2090–2096.

Taskila, S., Tuomola, M., & Ojamo, H. (2012). Enrichment cultivation in detection of foodborne *Salmonella*. *Food Control, 26*, 369–377.

Taylor, A. D., Ladd, J., Yu, Q., Chen, S., Homola, J., & Jiang, S. (2006). Quantitative and simultaneous detection of four foodborne bacterial pathogens with a multi-channel SPR sensor. *Biosens. Bioelectron, 22*, 752–758.

Timmons, C., Dobhal, S., Fletcher, J., & Ma, L. M. (2013). Primers with 5' flaps improve the efficiency and sensitivity of multiplex PCR assays for the detection of *Salmonella* and *Escherichia coli* O157:H7. *J. Food Prot, 76*, 668–673.

Tolba, M., Ahmed, M. U., Tlili, C., Eichenseher, F., Loessner, M. J., & Zourob, M. (2012). A bacteriophage endolysin-based electrochemical impedance biosensor for the rapid detection of *Listeria* cells. *Analyst, 137*, 5749–5756.

Tully, E., Hearty, S., Leonard, P., & O'Kennedy, R. (2006). The development of rapid fluorescence-based immunoassays, using quantum dot-labeled antibodies for the detection of *Listeria monocytogenes* cell surface proteins. *Int. J. Biol. Macromol, 39*, 127–134.

Valdivieso-Garcia, A., Riche, E., Abubakar, O., Waddell, T. E., & Brooks, B. W. (2001). A double antibody sandwich enzyme-linked immunosorbent assay for the detection of *Salmonella* using biotinylated monoclonal antibodies. *J. Food Prot, 64*, 1166–1171.

Verhaegen, B., Van Damme, I., Heyndrickx, M., Botteldoorn, N., Elhadidy, M., Verstraete, K., et al. (2016). Evaluation of detection methods for non-O157 Shiga toxin-producing *Escherichia coli* from food. *Int. J. Food Microbiol, 219*, 64–70.

Viswanathan, S., Rani, C., & Ho, J. A. (2012). Electrochemical immunosensor for multiplexed detection of food-borne pathogens using nanocrystal bioconjugates and MWCNT screen-printed electrode. *Talanta, 94*, 315–319.

Viswanathan, S., Wu, L. C., Huang, M. R., & Ho, J. A. (2006). Electrochemical immuno-sensor for cholera toxin using liposomes and poly(3,4-ethylenedioxythiophene)-coated carbon nanotubes. *Anal. Chem, 78*, 1115–1121.

Wang X. W., Zhang L., Jin, L. Q., Jin, M., Shen, Z. Q., An, S., et al. (2007). Development and application of an oligonucleotide microarray for the detection of food-borne bacterial pathogens. *Appl. Microbiol. Biotechnol, 76*, 225–233.

Wang Y., & Salazar J. K. (2016). Culture-independent rapid detection methods for bacterial pathogens and toxins in food matrices. *Compr. Rev. Food Sci. Food Safety, 15*, 183–205.

Wang, J. (2002). Electrochemical nucleic acid biosensors. *Anal. Chim. Acta, 469*, 63–71.

Wang, Y., Ye, Z., Si, C., & Ying, Y. (2013). Monitoring of *Escherichia coli* O157:H7 in food samples using lectin based surface plasmon resonance biosensor. *Food Chem, 136*, 1303–1308.

Wei, D., Oyarzabal, O. A., Huang, T., Balasubramanian, S., Sista, S., & Simonian, A. L. (2007). Development of a surface plasmon resonance biosensor for the identification of *Campylobacter jejuni*. *J. Microbiol. Methods, 69*, 78–85.

Wilson, I. G. (1997). Inhibition and facilitation of nucleic acid amplification. *Appl. Environ. Microb, 63*, 3741–3751.

Yang, M., Kostov, Y., & Rasooly, A. (2008). Carbon nanotubes based optical immunodetec-tion of staphylococcal enterotoxin B (SEB) in food. *Int. J. Food Microbiol, 127*, 78–83.

Yang, M., Kostov, Y., Bruck, H. A., & Rasooly, A. (2009). Gold nanoparticle-based enhanced chemiluminescence immunosensor for detection of staphylococcal enterotoxin B (SEB) in food. *Int. J. Food Microbiol, 133*, 265–271.

Yaron, S., & Matthews, K. R. (2002). A reverse transcriptase-polymerase chain reaction assay for detection of viable *Escherichia coli* O157:H7: investigation of specific target genes. *J. Appl. Microbiol, 92*, 633–640.

Yilmaz, E., Majidi, D., Ozgur, E., & Denizli, A. (2015). Whole cell imprinting based *Escherichia coli* sensors: a study for SPR and QCM. *Sensor Actuat. B-Chem, 209*, 714–721.

You, D. J., Geshell, K. J., & Yoon, J. (2011). Direct and sensitive detection of foodborne pathogens within fresh produce samples using a field-deployable handheld device. *Biosens. Bioelectron, 28*, 399–406.

Zelada-Guillen, G. A., Bhosale, S. V., Riu, J., & Rius, F. X. (2010). Real-time potentiometric detection of bacteria in complex samples. *Anal. Chem, 82*, 9254–9260.

Zhao, X., Lin, C. W., Wang, J., & Oh, D. H. (2014). Advances in rapid detection methods for foodborne pathogens. *J. Microbiol. Biotechn, 24*, 297–312.

Zwietering, M. H., Jacxsens, L., Membréc, J. M., Nauta, M., & Peterz, M. (2016). Relevance of microbial finished product testing in food safety management. *Food Control, 60*, 31–43.

# CHAPTER 10

# FOOD IRRADIATION TECHNOLOGY: A NOVEL ASPECT FOR THE FUTURE

DEBASHIS DUTTA and MIRA DEBNATH DAS

*School of Biochemical Engineering, Indian Institute of Technology (Banaras Hindu University), Varanasi – 221005, India*

## CONTENTS

## 10.1  INTRODUCTION

### 10.1.1  BACKGROUND

Radiation processing of food involves deployment short wave (near $10^{-12}$ m) energy on food materials to achieve a specific purpose together with extension of shelf life, insect, disinfestations, elimination of food borne pathogens such as *E.coli* 0157:H7, *Camplyobacter, Salmonellae, Shigella, toxoplasma, Listeria* and parasites. In comparison with heat or chemical treatment, irradiation is considered a greater powerful and appropriate technology to destroy food borne pathogens. It offers a number of advantages from producer, processor, retailers and consumers. In 1980 the joint FAO/ IAEA/WHO Expert Committee convened on the wholesomeness of radiation processed food in Geneva concluded that any food irradiation up to 10 KGy (kilo gray) is secure for human intake. Independent technical and scientific committees constituted by numerous countries including Canada, Denmark, France, India, Sweden and Europe Economic Community (United Kingdom) and United States of America have declared that radiation process is secure for human consumption. Today some 50 countries globally including India have approved the exploitation of irradiation for over 100 food items and about 30 of these are applying the technology on a limited commercial scale. The USA, Canada, China, The Netherlands, Belgium, Brazil, Thailand and Australia are among the leaders in adopting the technology whereas countries, such as Denmark, Germany and Luxembourg are adversarial. Interest in the commercial application of the process is thus emerging for many reasons. High food losses caused by the insect infestations microbial contamination and spoilage, mounting challenges over food borne diseases harmful residue of chemical fumigants and impact of these chemicals on the environment this stiff standard of best and quarantine restrictions in international trade are some of the reasons. Though irradiations alone cannot solve the problems of food preservations altogether, it may play vital role in reducing post harvest losses and undue use of chemical fumigants.

### 10.1.2   OVERVIEW

Radiation processing of food has been very well and considerably studied so one can ensure its toxicological, nutritional and microbiological safety. In 1980, FAO/IAEA/WHO collectively with Joint Expert Committee on Food Irradiation (JECFI) reviewed the substantial records on wholesomeness gathered up to that point and concluded that irradiation of any commodity up to an overall dose of 10 KGy leaves no toxicological hazards and introduces no unique dietary or microbiological issues (WHO, 1981). An professional Expert Group constituted by WHO in 1994 yet again reviewed the wholesomeness facts available till then and validated the earlier conclusion of JECFI. In 1998 another professional Expert Group constituted by WHO/FAO/IAEA affirmed the safety of food irradiated to doses above 10 kGy. In view of this advice the Codex Committee on Food Standards of The Codex Alimentarius Commission has also revised additionally in 2003 the Codex General Standard for Irradiated Foods that sets standards for processed foods.

The Agreements on Sanitary and Phytosanitary (SPS) Practices and Technical Barriers to Trade (TBT) under the World Trade Organization (WTO) has furnished an awesome incentive to the adoption of irradiation as an SPS measure in international trade under the precept of equivalence. Thus, irradiation can be applied to overcome quarantine limitation, and to hygienise merchandise for international trade. These agreements are administered under the standards, guidelines, and recommendations of the international organizations such as Codex Alimentarius Commission, International Plant Protection Convention (IPPC). This should encourage application of radiation for improving international trade in agro-horticultural foods among the WTO member states. In 2003 IPPC included irradiation as a plant quarantine measure. Plant and Animal Health Inspection Service (APHIS) of the USDA has issued a very last Rule on 'Irradiation as Phytosanitary Treatment for Imported Fruits and Vegetables' (Hall, 1989).

This regulation has opened the US Market to irradiated fruits and vegetables, handled to satisfy quarantine requirements from other countries. More than 40 countries are irradiating food for processing industries and institutional catering. These radiation processed food items are labeled to indicate the treatment and its purpose. The quantity and range of radiation processed food products entering trade has grown steadily in latest years, particularly in China, Belgium, France, The Netherlands, South Africa and the United States. Over 600,000 tons of food materials had been irradiated in the year 2011.

Food Standards Australia New Zealand (FSANZ) has authorized the irradiation of nine tropical fruits – mango, papaya, breadfruit, carambola, custard apple, litchi, longan, mangosteen and rambutan – for the treatment of pest infestation (FSANZ Standard 1.5.3.). Growing volumes of irradiated tropical fruit are now efficaciously being exported from Australia to New Zealand. In 2010, this trade in irradiated tropical fruit grew to a couple of thousand tons of mangoes, papayas and litchis. Today, New Zealand is an export fulfillment story for Australian horticulture, accounting for more than

25 per cent of total Australian mango exports. According to the Australian Mango Industry Association, New Zealand is the quickest growing market for Australian mangoes and helped the enterprise acquired considered one of its first rate export seasons in 2010. The food safety and biosecurity approval of irradiated tropical fruits in New Zealand is steady with global treaties and regulatory frameworks. For example, Codex Alimentarius, the arena body accountable for food safety, has issued a General Standard for Irradiated Foods and the International Plant Protection Convention recognizes irradiation as a phytosanitary measure (International Standards for Phytosanitary Measures No.18).

The Government of India have set up a country wide monitoring business enterprise National Monitoring Agency (NMA) in 1987 to recall the various aspects of radiation processing of foods. It approved radiation processing of potato, onion, spices and frozen seafood for export, and thereafter the notion become notified as draft under Prevention of Food Adulteration Act (PFA), guidelines in 1992. The very last notification of acclaim for both export and domestic consumption underneath PFA amended and published in 1994 vide GSR No. 614(E) dated August 9, 1994. In April 1998 PFA approved radiation processing of additional items and some more items of food had been accepted vide GSR No. 320(E) dated May 2, 2001.

In India, the Ministry of Health and Family Welfare amended the Prevention of Food Adulteration Rules (1954) via a Gazette notification dated August 9, 1994, allowing irradiation of onion, potato and spices for internal marketing and consumption. In 1998 and 2001 a number of other food items were approved for radiation processing. Approval for additional items including fish and shrimp for shelf-life extension and pathogen control and disinfestations of dry fish and pulses is anticipated soon.

In 1994 Government of India amended Prevention of Food Adulteration Act (1954) Rules and approved irradiation of onion, potato and spices for domestic market. Additional items have been accredited in April 1998 and in May 2001. In 2004, Ministry of Agriculture & Co-operation, Government of India, amended plant protection and quarantine regulations to include irradiation as a quarantine degree. Legal guidelines and rules enacted under the Atomic Energy Act, enforced by the Atomic Energy Regulatory Board, an independent body, govern operations of irradiators used to process non-food products, such as medical supplies as well as food. Many clinical product irradiators are running in India and round the world. The plants that need be approved by the government earlier construction and operation are subjected to regular inspection, safety audits, and other opinions to make sure that they are adequately well operated. Only those foods accredited under the Prevention of Food Adulteration Act rules may be irradiated and offered in domestic market.

### 10.1.3  INTERNATIONAL CONSULTATIVE GROUP ON FOOD IRRADIATION

International Project on Food Irradiation (IFIP) changed into accountable for examining the wholesomeness of foods irradiated as much as the dose of 10 kGy after which terminated in 1982, to form global platform "International Consultative Group on Food Irradiation" (ICGFI) in 1983 convened by the UN agencies FAO, IAEA and WHO (Derr, 1993). The 3 UN agencies and 19 founding member governments' representatives signed a assertion, which mounted the ICGFI in 1984.

The International Consultative Group on Food Irradiation (ICGFI) turned into conceived at a 1983 meeting convened through the UN groups FAO, IAEA and WHO. The three UN agencies and 19 founding member governments' representatives signed a statement, which set up the ICGFI in 1984.

(IFIP) had effectively furnished its mission of examining the wholesomeness of foods irradiated up to the dose of 10 kGy and was terminated in 1982, the governments of participating nations and the international groups FAO/IAEA/WHO felt that the international platform furnished via IFIP considering that 1970 turned into very useful and have to be renewed.

The foremost goal of ICGFI is to evaluate international developments and to provide a focus of advice on the application of food irradiation to member states. The best precedence is assigned to its program of work to promote public facts on food irradiation, discussing the process in an objective manner. It offers guides at the safety, the effectiveness and commercialization of the process, legislative aspects and control of irradiation facilities and additionally organizes training courses for plant technical employees, food inspectors, journalists and others. ICGFI membership has grown to 44 member states in 1995.

### 10.2  CHRONOLOGICAL DEVELOPMENTS

Reports of successful experiments in the United States inspired similar efforts in other countries. Quickly countrywide research application were underway in Belgium, Canada, France, The Netherlands, Poland, Russia, Germany and United Kingdom. However, health authorities in these countries nevertheless hesitated to grant permissions to market irradiated foods. Hot debates about the safety of irradiated foods for human consumption had been identified because essential obstacle to commercialization of the process. As a result of this recognition, under the sponsorship of the International Atomic Energy Agency (IAEA) in Vienna and the Food and Agriculture Organization (FAO) in Rome, a group of 19 countries which promptly grew to 24 formed the International Project on Food Irradiation (IFIP), in 1970, with headquarters in Karlsruhe, Germany. The World Health Organization (WHO) in Geneva showed major concern on high food losses caused by the insect infestations microbial contamination and spoilage, mounting concern over food borne diseases harmful residue of chemical fumigants and impact of these chemicals on the environment and quarantine regulation in international trade. Resources of the member countries were pooled to carry out chemical analyzes and animal feeding studies on a extensive variety of irradiated foods, such as meat, fish, fruit, spices, wheat and rice. The Joint FAO/IAEA/WHO Expert Committee on Food Irradiation (JECFI) convened meetings in 1970, 1976 and 1980. On the 1980 meeting, the JECFI decisively stated:

a)  The Committee concluded that the irradiation of any food commodity up to an overall average dose of 10 kGy provides no toxicological danger; therefore, toxicological trying out of foods so dealt with is no longer required.
b)  The Committee considered that the irradiation of food up to an overall average dose of 10 kGy introduces no special nutritional or microbiological troubles.

Based on JECFI findings, the World Health Organization published a report titled "Wholesomeness of Irradiated Foods," in Geneva, in 1981. The report concluded that no further toxicological or nutritional research is wanted on foods irradiated as much as an universal dose of 10 kGy (Thayer, 1994c).

However, global studies in food irradiation continues. To date, food irradiation has been studied greater than any other food process. All proof gathered from almost a century of medical and technical research ends in the realization that food irradiation is a secure, useful and sensible technique.

**TABLE 10.1** Chronology of Food Irradiation

| Year | Milestone |
|------|-----------|
| 1895 | X-rays discovered by Von Roentgen. |
| 1896 | Radioactivity discovered by Becquerel. Proposal to use ionizing radiation to preserve food by destroying microorganisms was published by Minsch. |
| 1904 | Studies on bactericidal effects of ionizing radiation was published by Prescott at MIT. |
| 1905 | U.S. and British patents were issued for use of ionizing radiation to kill bacteria in foods. |
| 1905 to 1920 | Much research conducted on the physical, chemical, and biological effects of ionizing radiation. |
| 1921 | USDA researcher Schwartz published studies on the lethal effect of x-rays on *Trichinella spiralis* in raw pork. |
| 1923 | First published results of animal feeding studies to evaluate the wholesomeness of irradiated foods. |
| 1930 | French patent issued for the use of ionizing radiation to preserve foods. |
| 1943 | MIT group, under U.S. Army contract, demonstrates the feasibility of preserving ground beef by x-rays. |
| Late 1940s and early 1950s | Beginning of era of food irradiation development by U.S. Government (under Atomic Energy Commission, industry, universities, and private institutions) including long-term animal feeding studies by U.S. Army and Swift Company. |
| 1950 | Beginning of food irradiation program by England and numerous other countries. |
| 1953 | The U.S. Army develops the National Food Irradiation Program and begins experiments with fruits, vegetables, dairy products, fish and meats. |
| 1960–61 | Canada's approval of potato irradiation marks one of the very first approvals of irradiated food for general consumption by Canadian consumers. A team led by Frank Warland builds a mobile food irradiator which becomes known as the Mobile Irradiator Program. It demonstrates the utility of irradiation to inhibit sprouting and extend produce shelf life. |
| 1963–64 | The U.S. Food and Drug Administration (FDA) approves the irradiation of potatoes, wheat and flour. |
| 1964 | The Newfield Products irradiation facility is established in St. Hilaire, Quebec, Canada by John Masefield. |
| 1964 | Nordion designs and manufactures the first C-188 type Cobalt-60 source used to irradiate food. |
| 1970 | The National Aeronautics and Space Administration (NASA) adopts irradiation to sterilize meat for astronauts to eat while in space. |

**TABLE 10.1**   (Continued)

| Year | Milestone |
|------|-----------|
| 1980 | U.S. Department of Agriculture (USDA) inherits the food irradiation program from the U.S. Army. |
| 1981 | An international expert committee publishes *Wholesomeness of Irradiated Food*. |
| 1983 | Canada and the U.S. approve irradiation of spices. |
| 1983 | The worldwide standard for the application of irradiation up to 10 kGy is adopted. |
| 1984 | Screwworm flies are eradicated from the U.S. and Mexico using the Sterile Insect Technique (SIT). |
| 1986 | The U.S. approves the irradiation of fruits and vegetables up to 1 kGy. |
| 1986 | As part of their approval, the U.S. FDA and other international bodies require that irradiated foods be labeled with a RADURA symbol and a statement "treated with radiation." |
| 1987 | The Canadian Irradiation Centre (CIC)—now Nordion's Gamma Centre of Excellence (GCE)—is established in Laval, Quebec, Canada. |
| 1988 | The FAO, WHO, IAEA, ITC and WTO adopt an international food irradiation trade agreement. |
| 1989 | Thai Irradiation Center is established. |
| 1990 | The U.S. FDA approves the irradiation of poultry. |
| 1992 | Food Technology Service, Inc. (FTSI) begins commercial operations in Mulberry, Florida. |
| 1993 | The American Medical Association (AMA) endorses food irradiation. |
| 1993 | The 'Jack in the Box' *E. coli* outbreak involves Jack in the Box restaurants in California, Idaho, Washington and Nevada. |
| 1997 | The U.S. FDA approves the irradiation of meat products. |
| 1998 | A consumer acceptance of irradiated food study is published. |
| 2000 | Ion Beam Applications (IBA) and Nordion open a joint venture irradiation facility in Mexico. |
| 2000 | Omaha Steaks begins a program to irradiate all ground beef products. |
| 2000 | Hawaii Pride irradiation facility is established. |
| 2002 | Wegmans introduces irradiated ground beef. |
| 2005 | The U.S. FDA approves irradiation of live mollusks. |
| 2005 | Sadex Corporation is established in Sioux City, Iowa, U.S. |
| 2006 | USDA-APHIS publishes a pioneering rule providing generic low-dose radiation quarantine treatments to control insects. |
| 2006 | *E. coli* in uncooked spinach infects 199 people in 26 U.S. states. |
| 2007 | Researchers study the effect of education on consumer acceptance of irradiated foods. |

| Year | Milestone |
|------|-----------|
| 2007 | First legal shipment of Indian mangoes lands at Kennedy Airport, U.S. |
| 2008 | U.S. FDA approves irradiation of spinach and leafy greens for pathogen reduction. |
| 2008 | First importation of Mexican guava is allowed into the U.S. |
| 2011 | USDA Animal and Plant Health Inspection Service (APHIS) propose a rule to allow irradiation facilities to be built in the Southern U.S. |
| 2011 | Benebion irradiation facility opens in Guadalajara, Mexico. |
| 2012 | Gateway America establishes an irradiation facility in Gulfport, Mississippi, U.S. |
| 2013 | Tomato and capsicum irradiation is approved for importation to Australia/New Zealand. |
| 2014 | U.S. opens market to South African persimmons. |
| 2014 | U.S. FDA approves irradiation of shellfish. |
| 2014 | Canada's contribution to food irradiation showcased at a museum exhibition. |
| 2015 | Approval of Steritech Queensland facility by United States Department of Agriculture (USDA), for irradiating Australian mangoes and lychees for export into the USA. |

Adapted from American Council on Science and Health. Irradiated Food.

## 10.3   HISTORY OF FOOD IRRADIATION

### 10.3.1   DEVELOPMENT OF IRRADIATION TECHNOLOGY

Initiation of food irradiation constituted with the discovery of x-rays by W.K. Roentgen in 1895 and the discovery of radioactive substances by H. Becquerel in 1896 (Brynjolfsson, 1989). Because of lack of techno-financial feasibility of the procedure of producing ionizing radiation no commercial development changed into observed till 1940 (Urbain, 1989). Within the context of nuclear safety and human health, the maximum relevant types of radiation are alpha and beta particles and gamma rays (Frazer and Westhoff, 1988).

The usage of radiation in food processing is a novel technique on account that its inception. Meats, fish, fruits and vegetables have been preserved for centuries through the sun's energy. These days, infrared and microwave radiation has been brought to the list of radiant energies in food processing.

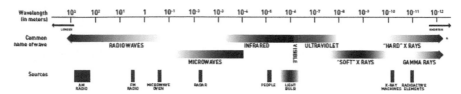

**FIGURE 10.1**  The Electromagnetic Spectrum.

The concept of using ionizing radiation in food preservation almost right away accompanied Henri Becquerel's discovery of radioactivity in 1895. E. coli and Salmonella have usually been the most causative food borne pathogens and nevertheless being a countrywide threat to human health (Dutta and Debnath, 2014). Use of ionizing energy can wreck pathogenic and spoilage microorganisms in food were mentioned in a German medical journal in the same year (Stewart, 2004a). The biological effects of ionizing radiation are generally measured in units called millisieverts (mSv).

Food irradiation became conceived as novel method of food preservation upkeep seeing that thermal canning and pasteurization of wine, beer, and milk within the 19th century (Josephson and Dymsza, 1999). The contemporary era of food irradiation applications research started out while the United States Atomic Energy Commission (USAEC) initiated a coordinated research program in the use of ionizing radiation for food preservation in 1950 (Proctor and Goldblith, 1951) and began to provide spent fuel rods from nuclear reactors (Thayer, 1986). Most of these experiments took place at the National Laboratory in Lemont, Illinois. Already in the early stages of this method, the limitation of spent fuel rods have became increasingly obvious, in particular with regard to exact dosimetry. Cobalt-60 (Co-60), a deliberately produced radioisotope was turned into observed substantially greater suitable for this motive. Cobalt-60 sources have been made to be had with the aid of the USAEC to several U.S. academic institutions, such as the Massachusetts Institute of Technology (MIT), University of California at Davis, University of Washington at Seattle and University of Florida at Gainesville, in the early 1960's. Afterwards, the Marine Products Development Irradiator, with 235 kCi (kilocuries) of Co-60 turned into built with the aid of National Marine Fisheries Services at Gloucester, Massachusetts, observed through a Grain Products Irradiator with 35 kCi of Co-60 on the USDA's Entomological Research Centre in Savannah, Georgia.

The U.S. Armed Forces performed an important role in the early years of food irradiation studies. The U.S. Army Natick Laboratories at Natick,

Massachusetts acquired a 1.3 MCi (megacurie) cobalt source and an 18 kW (kilowatt) electron linear accelerator (Hannan, 1955). Food irradiation studies started out in early 1950's. After 1960, the U.S. Army focused on high dose applications, to develop sterile meat products, to replace for canned or frozen military rations. The U.S. Army continues to be an active member of the global community of researchers in the field of food irradiation (Skala, 1986).

## 10.3.2  HISTORY OF COMMERCIALIZATION

The first industrial use of food irradiation occurred in 1957 in Stuttgart, Germany, whilst a spice producer determined to enhance the hygienic quality of his product by treating it with accelerated electrons produced by a Van de Graaff electron accelerator. The machine becomes dismantled later in 1959.

In Canada, irradiation of potatoes to inhibit sprouting become first permitted in 1960. Rapidly afterwards, an irradiation company named Newfield Products Ltd. Turned into formed at Mont St. Hillaire, near Montreal. The plant was designed to process some 15,000 tons of potatoes per month, using a Co-60 source. After the first year of operation, Newfield Products bumped into economic problems and ceased operation.

A significant event took place in December 1988. A number of UN agencies, specifically the FAO, WHO, IAEA, ITC and GATT (now renamed WTO) sponsored the International Conference on the Acceptance, Control of, and Trade in Irradiated Food in Geneva. Official delegations from 57 countries brought 220 participants together, comprising government officials at the senior policy-making level, experts in international law, health, energy, and food, and representatives of consumer unions. An International Document on Food Irradiation became adopted by means of consensus, which made recommendations on inter-governmental and governmental activities, process control and trade.

The IAEA/WHO/FAO Joint Division publishes reviews on volumes of commercially irradiated food products. Nonetheless, it is hard to obtain reliable statistics on quantities of commercially irradiated products because such information, which comes from irradiation companies, is often considered commercially confidential. However, each year, about 500,000 tons of food products are commercially irradiated in some 26 countries, notably in The Netherlands, France, Belgium, South Africa and Ukraine. It is reported that the grain irradiator in the port of Odessa, Ukraine, radiation disinfests about 300,000 tons of grain per annum. Other countries where foods are commercially irradiated are Canada, Hungary, Japan, Thailand and USA.

In terms of commercial developments in North America, perhaps the most important milestone in commercialization was the establishment of the first North American dedicated food irradiation facility. Food Technology Services, Inc. (FTSI, formerly Vindicator) changed into commissioned in the last quarter of 1991, in the town of Mulberry, near Tampa, Florida. This state-of-the-art pallet irradiation facility has started its commercial activities by irradiating strawberries and citrus for sale on the market in Miami and Chicago. Currently, a variety of irradiated fruits and vegetables are regularly distributed in retail outlets in Florida, Illinois, Ohio and Indiana. Some quantities of poultry are being processed for institutional customers and a number of articles are processed under military food research programs. As well as, the facility supplies irradiated food to NASA which is used in the space program.

### 10.3.3 HISTORY OF LABELING

Labeling of foods treated with ionizing energy has been one of the most controversial issues related to commercial production. The Joint FAO/IAEA/WHO Expert Committee concluded that for irradiated foods which had been approved as safe to eat, there was no valid scientific reason for identifying the products with a label at the retail level when similar labeling is not required for the other commonly used processing methods (WHO, 1981).

The United Nation's Codex Alimentarius Commission, after receiving the recommendations of the Joint FAO/IAEA/WHO Expert Committee, referred the labeling issue to its Committee on Labeling. This committee, which meets every 2 years, usually in Ottawa, Canada, is concerned with uniformity in labeling among the approximately 130 Codex member countries, including Canada and the United States, to facilitate international trade. The committee agreed to recommend that the use of a logo or symbol be optional, but that the label of an irradiated food should carry a written statement indicating that it had been irradiated.

In both the United States and Canada, wholly irradiated foods, which are sold either in pre-packaged or bulk form, must be identified as having been irradiated, by using the international irradiation symbol. Additionally, the statement "Treated with Radiation," "Treated by Irradiation" or "Irradiated" is required. Other statements that explain the reason for irradiation, or the benefits, may be used on the same label. The main purpose of the label is to advise consumers of the choice, rather than to warn. Indeed, in some countries, the irradiation label has become a symbol of high quality. Irradiated ingredients representing 10% or more of a finished product are to be described as "irradiated" on the list of ingredients. Ingredients in processed foods (i.e., spices), which represent less than 10% of a finished product have no labeling requirements. Foods that have been subjected to irradiation treatment are to be identified as such in any advertisements.

### 10.3.4 LABELING REQUIREMENTS OF IRRADIATED FOOD

The current version (ACSH, 1988) of the FDA low-dose irradiation rule requires that the retail label consist of the internationally agreed-upon symbol, and the phrases "treated with radiation" or "treated by irradiation."

All foods shall be labeled in accordance with the provisions of PFA Rules. The label shall bear a logo of Radura symbol and include following information. Irradiated commodities like potato and onion have only food value and should not be used for cultivation (Schutz et al., 1989). This fact should be displayed on the food package.

| PROCESSED BY IRRADIATION METHOD |
| DATE OF IRRADIATION _____ |
| LICENCE NO. _____ |
| PURPOSE OF IRRADIATION _____ |

#### 10.3.4.1   Re-Irradiation

Food once processed by radiation shall not be processed by radiation again unless specifically permitted under the PFA rules.

### 10.4   COMMERCIAL ASPECTS

### 10.4.1 CONTROLLING ECONOMIC LOSS

- Improve food security and safety;

- Enhance exports;
- Prevent use of chemicals harmful to human health and environment;
- Facilitate distribution from production centers to consumption centers;
- Cut down cycles of glut and scarcity;
- Better returns to farmers and price stabilization.

## 10.4.2 NUTRITIONAL VIABILITY OF IRRADIATED FOOD

Irradiated foods are entirely wholesome and nutritious (Brynjolfsson, 1985). All known methods of food processing and even storing food at room temperature for a few hours after harvesting can lower the content of some nutrients, such as vitamins. At low doses of radiation, nutrient losses are either not measurable or, if they can be measured, are not significant. At the higher doses used to extend shelf-life or control harmful bacteria, nutritional losses are less than or about the same as cooking and freezing. For this reason irradiation is often coined as cold sterilization or non-thermal technology. On the basis of the dose of radiation the application of radiation of food processing is generally divided into three main categories:

- **Low Dose Irradiation up to 1 KGy**: practiced for sprout inhibition; delay of ripening; insect disinfestations; parasite inactivation
- **Medium Dose Irradiation, 1–10 KGy**: applied for reduction in numbers of spoilage microorganisms; reduction in numbers or elimination of non-spore-forming pathogens, that is, disease causing microorganisms.
- **High Dose Irradiation above 10 KGy**: employed for reduction in numbers of microorganisms to the point of sterility.

The killing effect of irradiation on microbes is measured in D-values. One D-value is the amount of irradiation needed to kill 90% of that organism. For *E. coli* O157, if 0.3 KGy is sufficient to kill 90% of the population, D-value of *E. coli* will be 0.3 kGy and D-value will be forwarded exponentially to kill 99% (2D) of the organisms present and to kill 99.9% (3D) and so on. Therefore, D value plays significant role to quantify time of irradiation and hence design of irradiation treatment. The D-values are different for each organism, and need to be measured for each organism. They can even vary by temperature, and by the specific food.

At higher irradiation doses (10 KGy), some vitamins such as A, B1, C, E, and K can degrade to some extent. Irradiation can also cause some changes in the sensory characteristics of food and the functional properties of food components. Irradiation initiates the auto-oxidation of fats, which gives rise to rancid off flavors. The extent of irradiation-induced lipid oxidation depends on factors associated with oxidation such as temperature, oxygen availability, fat composition, and pro-oxidants. Specific comparison is given in Tables 10.2 and 10.3.

Gamma irradiation of the chicken resulted in a decrease in alpha tocopherol at 3 kGy and 2°C, by 6% without any significant changes in gamma tocopherol (Lakritz and Thayer, 1994).

Ascorbate, cysteine, and quinoid reductants were also demonstrated to be naturally present in sufficient quantities to account for the lower rates of loss of thiamin and riboflavin observed during irradiation of pork meat, as compared to irradiation in buffered solution (Fox et al., 1992).

The rate of loss of thiamin upon irradiation was found to be about 3 times as fast in skeletal muscle as well as in liver and the loss decreasing with increasing reductant titer. For the same amount of thiamin loss, liver could be irradiated to 3 times the dose as could muscle (Fox et al., 1993).

On the other hand, the niacin content of bread from irradiated flour was 17% more than the non-irradiated control flour (Diehl, 1991).

Milk develops an off-flavor at relatively low doses but various cheese show good tolerance at doses up to 3 kGy (Diehl, 1983).

In an editorial commentary in the Journal of the American Medical Association, Dupont (1992), it was recommended that the microbiological safety of poultry food in the United States may be improved through the use of irradiation against transmission of bacterial enteropathogens like *Salmonella, Shigella, Campylobacter,* and *Vibrios.*

Food borne illness is one of the largest preventable public health problems in the United States (Lee, 1994), as shouted by The Assistant Secretary for Health, Director, U.S. Public Health Service.

Agriculture scientists and the American Medical Association (AMA, 1993) agreed that food irradiation as "a safe and effective process that

**TABLE 10.2** Thiamin Retention Comparison

| Meat | % in irradiated sample | % in canned sample |
|------|------------------------|--------------------|
| Beef | 21 | 44 |
| Chicken | 22 | 66 |
| Pork | 12 | 57 |

Thomas, et al. (1981).

**TABLE 10.3**  Vitamin Content Comparison of Cooked Chicken

| Vitamin | Non irradiated sample | Irradiated sample |
|---|---|---|
| Vit-A (IU) | 2200 | 2450 |
| Vit-E (mg) | 3.3 | 2.15 |
| Thiamin (mg) | 0.58 | 0.42 |
| Riboflavin (mg) | 2.1 | 2.25 |
| Niacin (mg) | 58.0 | 55.5 |
| Vit-B6 (mg) | 1.22 | 1.35 |
| Vit-B12 (mg) | 21 | 28 |
| Pantothenic Acid (mg) | 13 | 17 |
| Folacin (mg) | 0.23 | 0.18 |

Amounts are for 2.2 pounds (1 kilogram) cooked chicken (Josephson, et al., 1978).

increases the safety of food when applied according to governing regulations (Marsden, 1994).

### 10.4.3  COMMERCIAL PROSPECTS IN INDIA

In India radiation processing of food can be undertaken both for export and domestic markets. India is well known for its export of rice (Basmati Rice exported about 2,016,775.0 Mt with a value of 1,088,913.34 lacs, Non-Basmati Rice exported about 139,540.76 Mt with a value of 36,529.60 lacs), seafood (exported of 467,297 metric tons valued at Rs. 68,810 million or US\$1.43 billion. The U.S. emerged as the largest market for Indian marine products during 2002–03 relegating Japan to second place), spices (exported about 317,800 tons 368,525.12 lacs), poultry (exported about 170,211.11 Mt with a value of 82.66 Crore), meat (exported about 107984.96 Mt with a value of 1272.17 Crore), onion (exported about 1,664,922.37 Mt with a value of 231,942.97 lacs), mango (exported about 74,460.62 Mt with a value of 20,053.96 lacs), grapes (exported about 131,153.64 Mt with a value of 54,533.86 lacs). The export market of other fruit is about 260,675.41 Mt with a value of 52,283.34 lacs, and other vegetable is about 419,241.31 Mt with a value of 73,185.87 lacs. Besides, Minor Forest Products (Medicinal and Aromatic Plants) exported about a value of 1872.90 Crore (APEDA, Govt. of India).

The Marine Products Export Development Authority (MPEDA) is the government agency for export promotion as well as a primary source of information for social, economics, legal and regulatory environments in the global marine product market. Radiation processing can also be used for restructuring cost of bulk commodities in export markets and for selling value added packaged commodities directly in retail markets. India has one of the world's largest domestic markets with huge quantities of cereals, pulses & their products, fruits and vegetables, and seafood which are distributed throughout the country. During storage, agriculture commodities worth approx. 60,000 crore are wasted due to insect infestation and microbial spoilage. Though, the interest of the entrepreneurs in this technology has steadily increased. In the next few years, the volume of food and allied products processed by irradiation is thus expected to significantly improve in the country. A study reported that 20 grays for 0.25, 0.5, 1.0, or 100 minutes reduced adult emergence of Mexican fruit flies from larvae by more than 99% (Lester and Wolfenbarger, 1990). Pasteurizing doses of irradiation can kill or reduce the populations of both food spoilage and pathogenic microorganisms in food. For example, *Salmonella* spp. and *Campylobacter*

**TABLE 10.4**  Food Items Approved for Radiation Preservation by the Ministry of Health and Family Welfare Under Preservation of Food Adulteration Act, 1955 [GSR 614 (E) Dated 09.0.94 and GSR 172 E, Dated 06.04.98]

| Food items | Purpose | Minimum Dose (KGy) | Maximum Dose (KGy) |
|---|---|---|---|
| Onion | Sprout inhibition | 0.03 | 0.09 |
| Potato | | 0.06 | 0.15 |
| Ginger | | 0.03 | 0.15 |
| Garlic | | 0.03 | 0.15 |
| Shallots (Small onion) | | 0.03 | 0.15 |
| Mango | Disinfestations (Quarantine) | 0.25 | 0.75 |
| Rice, semolina (sooji, rawa), wheat atta, maida Raisin, figs, dried dates | Disinfestation | 0.25 | 1.0 |
| | | 0.25 | 1.0 |
| | | 0.25 | 0.75 |
| | | 0.25 | 0.75 |
| Meat and meat products including chicken | Shelf-life extension and pathogen control | 2.5 | 4.0 |
| Spices | Microbial decontamination | 6.0 | 14.0 |

**TABLE 10.5**    Additional Food Items Recommended by the Central Committee for Food Standards

| Food items | Purpose | Minimum Dose (KGy) | Maximum Dose (KGy) |
|---|---|---|---|
| Pulses, dried | Disinfestations | 0.25 | 1.0 |
| sea-foods | | 0.25 | 1.0 |
| Fresh sea-foods | Shelf-life enhancement | 1.0 | 3.0 |
| Frozen sea-foods | Pathogen control | 4.0 | 6.0 |

*jejuni* can be eliminated from poultry, and trichinae from pork (ACSH, 1988). *Escherichia coli* O157:H7, *Salmonella, C. jejuni, Listeria monocytogenes,* and *Staphylococcus aureus* can be eliminated in uncooked ground beef (Beuchat et al., 1993). Irradiated foods are used by astronauts during space travel. In a review of the use of irradiated foods, Karel (1989) stated that, "On earth, food irradiation will most likely be used in combination with other preservation techniques. In space, how irradiation will be used will depend on the length of the voyage."

## 10.4.4    COMMERCIAL VIABILITY OF FOOD IRRADIATION

| Method of Processing | Energy required (KJ/Kg) |
|---|---|
| Irradiation of rice, pulse, wheat, sea food within 2.5 KGy | 21 |
| Irradiation with 30 KGy for sterilization of meat products | 157 |
| Heat sterilization (marine products) | 157 |
| Blast freezing | 918 |
| Storing the products at –250c for 3.5 weeks | 7552 |
| Refrigerated storage for 5.5 days at 0°C | 318 |
| Refrigerated storage for 10.5 days at 0°C | 396 |

## 10.4.5    EFFECTS ON MICROORGANISMS

Lethality due to ionizing radiation, as proposed by the target theory results from the ionization of the water molecules, to yield free radicals. This effect is reduced if food is irradiated in the frozen state. And can cause mutations in the organisms present (Frazier, and Westhoff, 1988).

Bacterial spores are more resistant (due to presence of calcium salt of dip-
icolinic acid) to ionizing radiation than are vegetative cells. Gram-positive
bacteria are more resistant than gram-negative bacteria. The resistance of
yeasts and molds varies considerably, but some are more resistant (due to
have manan, glucan and chitin in their cell wall) than most bacteria.

Irradiation effect thus depends upon the kind and species of the organism, the numbers of organisms (or spores) originally present, the composition of the food. Some constituents, for example, proteins, catalase, and
reducing substances (nitrites, sulfites, and sulfhydryl compounds) are protective. Compounds that combine with the SH groups would be sensitizing,
the presence or absence of oxygen, as it varies with the organism, ranging
from no effect to sensitization of the organism. Undesirable "side reactions"
are likely to be intensified in the presence of oxygen and to be less frequent
in a vacuum, the physical state of the food during irradiation. Both moisture content and temperature affect different organisms in different ways,
the condition of the organisms, age, temperature of growth and sporulation,
and state (vegetative or spore) may affect the sensitivity of the organisms.

Other mechanisms involved in irradiation microbial inactivation are cell
membrane alteration, denaturation of enzymes, alterations in ribonucleic
acid (RNA) synthesis, effects on phosphorylation, and DNA compositional
changes. According to the dose used and the goal of the treatment, food irradiation can be classified into three categories:

- Radurization is a process comparable with thermal pasteurization.
  The goal of radurization is to reduce the number of spoilage microorganisms, using doses generally below 10 kGy.
- Radicidation is a process in which the irradiation dose is enough to
  reduce specific non-spore-forming microbial pathogens. Doses generally range from 2.5 to 10 kGy, depending on the food being treated.
- Radappertization is a process designed to inactivate spore-forming
  pathogenic bacteria, similar to thermal sterilization. Irradiation doses
  must be between 10 and 50 kGy.

Molds and Gram-positive vegetative bacteria are more tolerant than
Gram-negative bacteria.

Resistance of *Salmonella typhimurium* increased at reduced irradiation
temperatures. Chemical compounds with nutritional or flavor functions can
also be affected by ionizing irradiation; At higher irradiation doses, such as
those required for food sterilization, some vitamins such as A, B1, C, E, and
K can degrade to some extent. Irradiation may cause some changes in the

**TABLE 10.6**   Effects of Ionizing Radiation on Microorganisms

| Organism | Approximate lethal dose (kGy) |
|---|---|
| Insects | 0.22 to 0.93 |
| Viruses | 10 to 40 |
| Yeasts (fermentative) | 4 to 9 |
| Yeasts (film) | 3.7 to 18 |
| Molds (with spores) | 1.3 to 11 |
| Bacteria (cells of pathogens): | |
| *Mycobacterium tuberculosis* | 1.4 |
| *Staphylococcus aureus* | 1.4 to 7.0 |
| *Cornybacterium diphtheriae* | 4.2 |
| *Salmonella* spp. | 3.7 to 4.8 |
| Bacteria (cells of saprophytes): | |
| Gram-negative: | |
| *Escherichia coli* | 1.0 to 2.3 |
| *Pseudomonas aeruginosa* | 1.6 to 2.3 |
| *Pseudomonas fluorescens* | 1.2 to 2.3 |
| *Enterobacter aerogenes* | 1.4 to 1.8 |
| Gram-positive | |
| *Lactobacillus* spp. | 0.23 to 0.38 |
| *Streptococcus fecalis* | 1.7 to 8.8 |
| *Leuconostoc dextranicum* | 0.9 |
| *Sarcina lutea* | 3.7 |
| Bacterial spores: | |
| *Bacillus subtillus* | 12 to 18 |
| *Bacillus coagulans* | 10 |
| *Clostridium botulinum* (A) | 19 to 37 |
| *Clostridium botulinum* (E) | 15 to 18 |
| *Clostridium perfringens* | 3.1 |
| *Putrefactive anaerobe* 3679 | 23 to 50 |
| *Bacillus stearothermophilus* | 10 to 17 |

Adapted from Frazier and Westhoff (1988).

sensory characteristics of food and the functional properties of food components. Irradiation initiates the autoxidation of fats, which gives rise to rancid off flavors. The extent of irradiation-induced lipid oxidation depends on

factors associated with oxidation such as temperature, oxygen availability, fat composition, and pro-oxidants.

## 10.5 INTERNATIONAL PROSPECTS

### 10.5.1 ASPECTS AND DEMAND

India has a strong raw material base, but has been unable to tap the potential for processing and value addition in perishables like fruits and vegetables. Only about 6.5% of the fruits and vegetables in India are being processed, which is much lower when compared to countries like China (23%), Thailand (30%), USA (65%), Brazil (70%) Philippines (78%) South Africa (80%) Malaysia (83%). Even, within the country, share of fruits and vegetables processed is much less when compared to other agricultural products such as milk (35%) and marine products (26%).

In 1959, Russia started irradiation of fruits, vegetables, spices, cereals, meats and poultry while many other countries commenced in the 1970s. Thailand initiated irradiating onions to delay sprouting in 1971 followed by fermented pork sausage. Also in 1971, South Africa set out the irradiation of potatoes, onion, fruits, spices, meat, fish, and chicken. Israel approved the irradiation of animal feed in 1973 while Japan took off for marketing irradiated potatoes in 1974. China is currently the biggest user of irradiation. Currently, among 65 countries worldwide, Argentina, Australia, Bangladesh, Thailand, the USA and Vietnam have legislation allowing phytosanitary uses of irradiation. National legislation in India, Australia and New Zealand have recently approved irradiation facilitated to sanitize agricultural produce after harvesting. In 2008, the US FDA approved the use of irradiation to control food-borne pathogens in iceberg lettuce and spinach, in addition to existing US approvals for a variety of products, including herbs and spices, grains, poultry, ground beef, and seafood. In India, significant volumes of food for sanitary purposes, including spices, seafood, vegetables, grains, potatoes, sterilized meals and meats are being continued.

World trade in agricultural, fisheries and forestry products have been crossed totals about $600 billion each year. Now the IAEA have decided to strengthen the national capacities of FAO and IAEA Member States by applying irradiation to control insect pests in exported fruits and vegetables (phytosanitary treatment) and for the control of bacterial contamination as well as extending food's shelf-life.

**TABLE 10.7**   International Demand

| Country | Food Products |
| --- | --- |
| Argentina | Potatoes, strawberries, onions, garlic. |
| Australia | Spices, Herbs, Vegetable Seasoning, and 9 tropical fruit include mango and papaya. |
| Bangladesh | Mangoes of Himsagar variety, fresh water fish, Prawn, beef casing, flour, turtle meat, macaroni and others. |
| Belgium | Potatoes, strawberries, onions, garlic, shallots, paprika, pepper, gum arabic, 78 spices. |
| Bulgaria | Potatoes, onions, garlic, grain, dry food concentrates, dried fruits, fresh fruits. |
| Canada | Potatoes, onions, wheat flour, poultry, cod and haddock filets, spices and certain dried vegetables. |
| Finland | Spices, herbs, hospital meals. |
| Chile | Potatoes, papaya, wheat, chicken, onions, rice, fish products, spices. |
| China | Vegetables and fruits, grains and beans, meats and poultry (fresh, chilled or frozen), cooked meat, spices and dehydrated vegetables and dried fruits and nuts. Major products included rice, garlic, spices, dehydrated vegetables, and health food. |
| France | Potatoes, onions, garlic, shallots, spices, dried fruits and vegetables. |
| Germany | Hospital meals. |
| Israel | Potatoes, onions, poultry, 36 spices, fresh fruits and vegetables. |
| Korea | Mushrooms, spices and their products, dried meat products, dried fish and shellfish products, soybean paste and hot pepper paste powder, starch for condiments, dried vegetables, yeast and enzyme products, aloe products and ginseng products. |
| Czechoslovakia | Potatoes, onions, mushrooms. |
| Japan | Mangoes, Shrimp. |
| Malaysia | Papaya, Basmati Rice. |
| Netherlands | Asparagus, cocoa beans, strawberries, mushrooms, hospital meals, potatoes, shrimp, onions, poultry, soup greens, fish filets, frozen frog legs, rice and ground rice products, rye bread, spices, endive, powdered batter mix. |
| Philippines | Potatoes, onions, garlic. |
| South Africa | Potatoes, onions, garlic, chicken, papaya, mangoes, strawberries, dried bananas, avocados, beans. |
| Spain | Potatoes, onions |
| Sri Lanka | Rice, the enzyme bromelin and other products, herbs, fermented pork sausages or 'nham'—295 tons, bromelin enzyme, pet products. |
| Thailand | Potatoes, onions, garlic, dates, wheat, rice, fish, chicken. |

**TABLE 10.7** (Continued)

| Country | Food Products |
|---------|---------------|
| USSR | Potatoes, grain, fresh and dried fruits and vegetables, dry food concentrates, poultry, onions, prepared meat products. |
| UK | Hospital meals. |
| USA | Mango, Spices, Vegetables and Sea Food (mainly Shrimp/Prawn), Wheat and wheat flour, potatoes, spices, pork, fresh fruits and vegetables. |
| Vietnam | Rice, litchi and mushrooms. |
| Yugoslavia | Cereals, potatoes, onions, garlic, poultry, dried fruits and vegetables. |

## 10.5.2  INTERNATIONAL APPROVALS

To date, clearances are in place in the USA for spices and dry aromatic ingredients, fresh fruits and vegetables ("fresh foods"), pork, poultry, red meats, shell eggs and food enzymes. Canadian legislation has remained unchanged since the 1989 reclassification of irradiation as a process, rather than an additive. Foods cleared to date include potatoes, onions, wheat and wheat flour, spices and dry aromatic ingredients.

Globally, national legislation is still very divergent. This lack of international harmonization is seen as a major impediment to international trade as it constitutes a non-tariff barrier. For instance, the European Union has still not reached agreement on a guideline for the regulation of food irradiation, due to resistance from Germany. In contrast to Germany, The Netherlands, Belgium and France routinely irradiate many foods. Regulatory developments in the Republic of South Africa deserve a separate mention as, in addition to being one of the pioneers in commercialization, it is also the only country where precooked, shelf-stable meat products irradiated at 45 kGy are allowed for retail sale. To date, 40 countries have collectively approved irradiation of more than 50 different foods.

Canada and USA were the first to enlist their irradiation plant/facilities under their respective legislation regulating additives. Needless to say, their initiatives caused quite some confusion as it is well known that irradiation is a process using electromagnetic energy, rather than an additive. The rationale of the agencies is that irradiation is administered as if it were an additive, to be able to take advantage of stricter controls in the legislation governing additives. The USFDA continues to administer irradiation in this

**TABLE 10.8**   Food Irradiation: Regulation and Approval

| Year | Country |
|------|---------|
| 1958 | The Food, Drug and Cosmetic Act is amended, directing that food irradiation be evaluated as a food additive, not as a physical process. All new food additives, including radiation, must be approved by the FDA before they can be used. The U.S. Congress passed legislation, which President Eisenhower signed in 1958. This legislation is still the law of the land. |
| 1958 to 1959 | U.S.S.R. approves potato and grain irradiation for sprout inhibition and insect disinfestations. |
| 1960 | Canada approves potato irradiation for sprout inhibition at a maximum dose of 0.1 kilogray (KGy) which was increased in 1963 to 0.15 kGy. In 1965, a clearance to irradiate onions up to the same dose was added to the list. |
| 1963 to 1964 | U.S. FDA approves irradiated bacon, potatoes and as a result of a petition to "Process Wheat and Wheat Products for the Control of Insect Infestation." |
| 1964 to 1967 | U.S. FDA approves flexible packaging material for food containers during irradiation processing. |
| 1976 | Joint FAO/IAEA/WHO Expert Committee on (safety/wholesomeness of) Food Irradiation (JECFI) approves several irradiated foods and recommends that food irradiation be classified as a physical process. |
| 1979 | U.S. FDA Bureau of Foods (Center for Food Safety and Applied Nutrition) forms internal Irradiated Foods Committee. |
| 1980 | Joint FAO/IAEA/WHO Expert Committee on (safety/wholesomeness of) Food Irradiation (JECFI) approves all irradiated foods treated with a maximum average dose of 10 kGy**. |
| 1983 | U.S. FDA and Canadian Health and Welfare Department approve sterilization of spices with irradiation. |
| 1985 | U.S. FDA approves irradiation pasteurization of pork to control trichinosis (with a minimum dose of 0.3 kGy and a maximum of 1.0 kGy of ionizing radiation). |
| 1986 | U.S. FDA approves irradiation of fruits and vegetables and other foods up to 1 kGy. |
| 1990 to 1992 | The U.S. government announced approval of ionizing radiation treatments of poultry to eliminate foodborne pathogens. The regulation for irradiation of poultry products from the USDA Food Safety and Inspection Service requires minimum and maximum doses of 1.5 and 3.0 kGy, respectively. |
| 1994 | The Food Safety and Inspection Service of the USDA has indicated dosages of ionizing radiation treatments for red meat products to the FDA. A maximum level of 4.5 kGy is proposed for unfrozen red meat, and 7.5 kGy for frozen red meat. Approval of radiation treatment for red meat products is expected early in 1995. |

manner, whereas the Health Protection Branch of Health Canada reclassified irradiation as a food process in March 1989 (Loaharanu, 1989).

## 10.6 PURPOSE OF IRRADIATION OF FOOD AND AGRO-PRODUCTS

### 10.6.1 PURPOSES OF IRRADIATION

1. Phytosanitation to overcome quarantine barriers in fruits like mangoes and vegetables.
2. Inhibition of Sprouting in potatoes and onions.
3. Delay in ripening and senescence of fruits.
4. Enhancement in shelf-life by destruction of spoilage microbes.
5. Overcome quarantine barriers in flowers (Rajanigandha, lotus, rose, lilly, etc.) by preventing insects.

### 10.6.2 ADVANTAGES OF IRRADIATION

1. No rise in temperature, products remain in their original states.
2. Not dependent on factors such as air humidity, penetration of gas, temparature, etc.
3. Exposure time in the processing chamber determines the dose received; only one parameter.
4. Gamma processed products don't contain residues.
5. Gamma processed products are ready for immediate use.
6. Radiation processing is a good process ad therefore unlike heat it can be used on agriculture commodities without changing their fresh-like characters.
7. Radiation processing does not alter significant nutrition value, flavor texture and appearance of food.
8. Radiation using Co-60 cannot induce any radioactivity in food and does not leave any harmful or toxic radioactive residues on foods as is the case with chemical fumigants.
9. Due to the highly penetrating nature of the radiation energy it is a very effective method.
10. Pre-packaged foods can be treated for hygienization and improving self-life.

11. The radiation processing facility are quite environment friendly and are safe to handlers and public around.

## 10.7  DETAILS OF TECHNOLOGY

Three different techniques of irradiation exists that use different kind of rays viz; gamma rays, electron beams and x-rays (Jones, 1992).

### 10.7.1  GAMMA CHAMBER 5000 (LAB SCALE)

- It is a compact self shielded Cobalt-60 gamma irradiator providing an irradiation volume of approximately 5000 cc. The material for irradiation is placed in an irradiation chamber located in the vertical drawer inside the lead flask. This drawer can be moved up and down with the help of a system of motorized drive which enables precise positioning of the irradiation chamber at the center of the radiation field.
- Radiation field is provided by a set of stationary cobalt-60 sources placed in a cylindrical cage. The sources are doubly encapsulated in corrosion resistant stainless steel pencils and are tested in accordance with international standards. Two access holes of 8 mm diameter are provided in the vertical drawer for introduction of service sleeves for gases, thermocouple, etc. A mechanism for rotating/stirring samples during irradiation is also incorporated. The lead shield provided around the source is adequate to keep the external radiation field well within permissible limits. The Gamma Chamber 5000 unit can be installed in a room measuring 4 meters X 4 meters X 4 meters.
- Specifications of Gamma Chamber 5000:

| Maximum Co-60 source capacity | 518 TBq (14000 Ci) |
|---|---|
| Dose rate at maximum capacity | ~ 9 kGy/hr (0.9 Mega Rad/hr) at the center of sample chamber |
| Dose rate uniformity | +25% or better radially; –25% or better axially |
| Irradiation volume | 5000cc approx |
| Size of sample chamber | 17.2 cm (dia) x 20.5 cm (ht) |
| Shielding material | Lead and stainless steel |
| Weight of the unit | 5600 kg. approx. |
| Size of unit | 125 cm (l) x 106.5 cm (w) x 150 cm (ht) |
| Timer range | 6 seconds onwards |

## 10.7.2   DOSIMETRY AND ASSESSMENT

Dosimetry is an important aspect of radiation processing. Food irradiation processing involves wide dose ranges and variable dose-rates. Many accurate, reliable dosimetry systems are now available. Dosimetry helps in preservation of quality of product, fulfillment of legal aspects and economic operation of the plants. With globalization, the need for reliable process control for irradiated food/agro-products for the international trade is ever increasing. In all the various guidelines and standards/amendments developed for food irradiation, the activities of principal concern are process validation and process control. The objective of such formalized procedures is to establish documentary evidence that the irradiation process has achieved the satisfied results. The key role of such activities is inevitably a well characterized reliable dosimetry system that is traceable to recognized national and international dosimetry standards. Only such dosimetry systems can help establish the required documentary evidence. In addition, industrial radiation processing such as irradiation of foodstuffs and sterilization of health care products are both highly regulated, in particular with regard to dose. Besides, dosimetry is necessary for scaling up processes from the research level to the industrial level. Thus, accurate dosimetry is indispensable.

## 10.7.3   IMPORTANCE OF DOSIMETRY STANDARDS

- Research & Development:
    Transfer of experience and data → Quantitative baseline
- Scaling up of a new radiation process (Radiation induced effects tied to dosimetry):
    Laboratory → Pilot scale → Commercial plant
- Service irradiation facility (Quality assurance through documentation):
  Customer products ⇔ Irradiation facility (e.g., ISOMED, BRIT)
- Independent control:
    Commercial operation → financial success

## 10.8   FOOD IRRADIATION ASPECTS ON HEALTH

Radioactivity in foods may be happened either contact of foods with radioactive substances directly or by excitation into the nuclei of the atoms in foods.

However, the food itself never contacts a radioactive substance and the ionizing radiation from irradiators is not strong enough to excite the nucleus of even one atom of a food molecule. USAEC and FDA reviewed several hundred studies on the effects of food irradiation before giving approval about the general safety of the treatment. In order to make recommendations specifically about poultry irradiation, U.S. Food and Drug Administration scientists reviewed separately. Independent scientific committees in Denmark, Sweden, United Kingdom, Canada and India also have reaffirmed the safety of food irradiation.

In addition, food irradiation has received official international endorsement from the World Health Organizations.

Radiation processing is a method for producing chemical, physical and microbiological changes in substances by exposing to ionizing radiation. Unique advantages of ionizing radiation as an industrial tool is that radiation induced effects are reproducible and quantifiable. The dose of radiation is measured in the SI unit known as Gary (Gy). One Gray (1Gy) dose of radiation means of 1 Joule of energy absorbed per kg of food material while the biological effects of ionizing radiation are generally measured in units called millisieverts (mSv). High doses of ionizing radiation, however, can damage healthy tissues and cause serious illness. While a safe level of radiation has not been conclusively established, research shows that radiation doses less than 100 mSv/year resulted no significant measureable health effects in humans.

Natural background radiation worldwide is on average 2.4 mSv/year, while in Ramsar, Iran, natural radiation levels can reach 260 mSv/year – 13 times more than allowed for workers in Canadian nuclear facilities.

## 10.9   CONCLUSION

Food irradiation technology provides a novel approach toward safely preserves food and eradication of pathogens. Chronological development and in depth research have resulted in amendment of regulatory approvals for this process in developing countries. The commercialization, labeling and nutritional viability of food irradiation is also increasing. Retail stores that offer irradiated products for sale are experiencing positive feedback due to have their intra governmental subsidized support. Irradiation research has been extended to other biological process/tissues (e.g., Development

**TABLE 10.8**  Radiation Effects on Human Body/Biological Tissue

| Dose (mSv) | Effect on Human Body/Biological Tissue* |
|---|---|
| 10,000 | Fatal within weeks |
| 6,000 | Average dose to Chernobyl emergency workers who died within a month |
| 5,000 | Single dose which would kill half of those exposed to it within a month |
| 1,000 | Single dose which would cause radiation sickness, nausea, but not death |
| 400 | Maximum hourly radiation levels recorded at Fukushima on March 14, 2011 |
| 350 | Exposure of Chernobyl residents who were relocated |
| 100 | Lowest level linked to increased cancer risk |
| 10 | Full-body CT scan |
| 2.4 | Natural radiation to which we are all exposed, per year |
| 0.1 | Chest x-ray |
| 0.01 | Dental x-ray |
| 0.001 | Annual public dose due to nuclear power reactors in Canada |

*Canadian Nuclear safety Commission report; International Atomic Energy Agency (IAEA).

of blood irradiator). Important UN agencies such as the World Health Organization (WHO) and the Food and Agriculture Organization (FAO) now recognize irradiation as another important method of controlling pathogens and food spoilage in global issue. Consumers and food processing companies both will benefit from the commercialization of this process.

## KEYWORDS

- **dosimetry**
- **hygienise**
- **phytosanitary**
- **sanitary**

## REFERENCES

American Council on Science and Health (ACSH). *Irradiated Food.* Third edition. American Council on Science and Health.

Anon. (1990). Dept. of Health and Human Services Food and Drug Administration. 21 CFR, part 179 final rule; irradiation in the production processing and handling of food. *Fed. Regist, 55*, 18538–18544.

Anon. (1992). Dept. of Agriculture Food Safety and Inspection Service 9 CFR part 381 final rule: irradiation of poultry products. Fed. Regist, *57*, 435,888–43,600.

Beuchat, L. R., Doyle, M. P., & Brackett, R. E. (1993). Irradiation inactivation of bacterial pathogens in ground beef; Final Report. American Meat Institute Foundation.

Blumenthal, D. (1990). Food Irradiation: Toxic to Bacteria, Safe for Humans; FDA Consumer. November.

Brynjolfsson, A. (1985). Wholesomeness of irradiated foods: A review. *J. Food Safety, 7,* 107–126.

Brynjolfsson, A. (1989). Future radiation sources and identification of irradiated foods. *Food Technol, 43*(7), 84–89, 97.

Center for Disease Control, Frequently Asked Questions about Food Irradiation (last reviewed Sept. 29, 1999). http://www.cdc.gov/ncidod/dbmd/diseaseinfo/foorirradiation.htm

Derr, D. (1993). Radiation Processing. ASTM Standardization News. *21*(7), 25–27.

Diehl, J. F. (1983). Radiolytic effects on foods, in: *Preservation of Foods by Ionising Radiation*, CRC Press Inc. *Boca Raton, Florida, 1,* pp. 279–357.

Diehl, J. F. (1991). Nutritional effects of combining irradiation with other treatments. *Food Control.* 2 (1), 20–25.

Dutta, D., & Debnath, M. E. (2014). Coli And *Salmonella*-The Most Causative Food Borne Pathogens-National Threat To Human Health. Scientific India, Vol. 2., Issue 2., ISSN-2349–1418

Economic Research Service, ERS Estimates Foodborne Disease Costs at 6.9 Billion Per Year (updated March 12, 2001) http://www.ers.usda.gov/Emphases/SafeFood/features.htm.

Elias, P. S. (1989). New concepts for assessing the wholesomeness of irradiated foods. *Food Technol, 43*(7), 81–83.

FDA. (1990). Irradiation in the production, processing and handling of food: labeling. Food and Drug Administration. *Fed. Regist, 55*(5), 646.

Fox, J. B., Ackerman, S., & Thayer, D. W., The effect of radiation scavengers on the destruction of thiamin and riboflavin in buffers and pork due to gamma irradiation. *Biotehnol. Rev, 30*(4), 171–175.

Fox, J. B., Lakritz, L., & Thayer, D. W., Effect of reductant level in skeletal muscle and liver on the rate of loss of thiamin due to gamma-irradiation. *Int. J. Radiat. Biol, 64*(3), 305–309.

Frazier, W. C., & Westhoff, D. C. (1988). Chapter 10. Preservation by radiation: In: *Food Microbiology*, 4th ed., McGraw-Hill: New York, NY.

Hall, R. L. (1989). Commercialization of the food irradiation process. *Food Technol, 43* (7), 90–92.

Hannan, R. S. (1955). Scientific and technological problems in using ionizing radiations for the preservation of food; Dept. Sci. Indust. Res. Food Investigation Report No. 61, H.M.S.O.

Institute of Food Technologists' Expert Panel on Food Safety and Nutrition. (1983). Radiation preservation of Foods. A scientific status summary. *Food Technol, 37*(2), 55–60.

International Atomic Energy Agency (IAEA). (1988). Acceptance, Control of and Trade in Irradiated Foods. Proceedings of FAO, IAEA, WHO, ITC-UNCTAD/GATT International Conference. Geneva Switzerland.

Jay B., Fox, L., Lakrit, J., Richardson, K., Ward, D., & Thayer W. (1995). Gamma irradiation effects on thiamin and riboflavin in beef, lamb, pork, and turkey. *Journal of Food Science.* Volume 60, Issue 3, Date: May Pages, 596–598.

Jay, James. (1986). Modern Food Microbiology. Van Nostrand Reinhold Company: New York, NY.

Jones, J. M. (1992). Chapter 12. Food Irradiation in Food Safety; Eagan Press. St. Paul, M. N.

Josephson, E. S., & Dymsza, H. A. (1999). "Food Irradiation," *Technology, 6,* 235–238

Josephson, E. S., Thomas, M. H., & Calhoun, W. K. (1979). Nutritional aspects of food irradiation: An overview. *J. Food Proc. Press, 2,* 299–313.

Karel, M. The future of irradiation applications on earth and in space. *Food Technol, 43*(7), 95–97.

Kim, M., (1998). Morehouse, Food Irradiation: The Treatment of Foods with Ionizing Radiation, Food Testing and Analysis, edition (Vol. 8., No. 3) June/July (1998).

Lakritz, L., & Thayer, D. W. (1994). Effect of gamma radiation of total tocopherols in fresh chicken breast muscle. *Meat Science, 37,* 439–448.

Lee, P. R. (1994). Irradiation to prevent foodborne illness. *J. Am. Med. Assoc, 272*(4), 261.

Lester, G. E., & Wolfenbarger, D. A. (1990). Comparisons of Cobalt-60 gamma irradiation dose rates on grapefruit flavedo tissue and on Mexican fruit fly mortality. *J. Food Protect, 53*(4), 329–331.

Loaharanu, P. (1989). International trade in irradiated foods: Regional status and outlook. *Food Technol, 43*(7), 77–80.

Marsden, J. L. (1994). Irradiation and food safety; American Meat Institute Issues Briefing July, 1–4.

Mead, P. S., Slutsker, L., Dietz, V., McCaig, L. F., Bresee, J.S., Shapiro, C., et al. (1999). (Centers for Disease Control and Prevention, Atlanta, Georgia), Food-Related Illness and Death in the United States, Emerging Infectious Diseases, Sept, Oct. (Vol. 5, No. 5)

Proctor, B. E., & Goldblith, S. A. (1951). Food Processing with ionizing radiations. *Food Technol, 5,* 376.

Schutz, H. G., Bruhn, C. M., & Diaz-Knauf, K. V. (1989). Consumer attitudes toward irradiated foods: Effects of labeling and benefits information; Paper No. 84. In Annual Meeting of Int. of Food Technologists. Chicago, IL, June 25–29.

Skala, J. H., McGown, E. L., & Waring, P. P. (1987). Wholesomeness of irradiated foods. *J. Food Protect, 50*(2), 150–160.

Stewart, Eileen, M. (2004a). Food irradiation: more pros than cons? *Biol, 51*(1), 91–94.

Stewart, Eileen, M. (2004b). Food irradiation: more pros than cons? *Biol, 51*(2), 141–144.

Thayer, D. W. (1993c). Extending shelf life of poultry and red meat by irradiation processing. *J. Food Protect, 56*(10), 831–833, 846.

Thayer, D. W. (1993d). Irradiation for control of foodborne pathogens on meat and poultry. In Safeguarding the Food Supply through Irradiation Processing Techniques (pp. 23–45). Orlando, Florida. Agriculture Research Inst. Bethesda, MD, October 25–31.

Thayer, D. W. (1994c). Wholesomeness of irradiated foods. *Food Technol, 48*(5), *132,* 134. 136.

Thayer, D. W., & Boyd, G. (1992). Gamma ray processing to destroy *Staphylococcus aureus* in mechanically deboned chicken meat. *J. Food Sci, 57*(4), 848–851.

Thayer, D. W., & Boyd, G. (1993b). Elimination of *Escherichia coli* O157:H7 in meats by gamma irradiation. *Appl. Environ. Microbiol,* (4), 1030–1034.

Thayer, D. W., Boyd, G., & Jenkins, R. K., Low-dose gamma irradiation and refrigerated storage *in vacuo* affect microbial flora of fresh pork. *J. Food Sci, 58* (4), 717–719.

Thayer, D. W., Fox, J. B., & Lakritz, L. (1993a). Chapter 23. Effects of ionizing radiation treatments on the microbiological, nutritional, and structural quality of meats. In *Food Flavor and Safety*. Spanier, A. M. Okai, H., & Tamura, M. (Eds.). *Am. Chemical Society,* 294–302.

Thayer, D. W., Lachica, R. V., Huhtanen, C. N., & Wierbicki, E. (1986). Use of irradiation to ensure the microbiological safety of processed meats. *Food Technol, 40*(4), 159–162.

Thayer, D. W., Lachica, R. V., Huhtanen, C. N., & Wierbicki, E. (1986). Use of irradiation to ensure the microbiological safety of processed meats. *Food Technol, 40*(4), 159–162.

The report of the 1964 meeting in Rome was published in WHO Technical Report Series, No. 316 (Geneva: World Health Organization, 1966). The report of the 1969 meeting in Geneva was published in WHO Technical Report Series, No. 451 (Geneva: World Health Organization, 1970). The report of the 1976 meeting in Geneva was published in WHO Technical Report Series, No. 604 (Geneva: World Health Organization, 1977). The report of the 1980 meeting in Geneva was published in WHO Technical Report Series, No. 659 (Geneva: World Health Organization, 1981). The report of the 1997 meeting in Geneva was published in WHO Technical Report Series, No. 890 (Geneva: World Health Organization, 1998).

Urbain, W. M. (1989). Food Irradiation: The past fifty years as prolog to tomorrow. *Food Technol, 43* (7), 76, 92

WHO. (1981). Technical Report Series 659. Report of a Joint FOA/IAEA/Who Expert Committee. Wholesomeness of irradiated food. Geneva, Switzerland.

WHO. (1987). Food Irradiation. In Point of Fact. No. 40. Geneva, Switzerland.

# CHAPTER 11

# PROSPECTS OF ANTARCTIC KRILL UTILIZATION BY INDIA

C. N. RAVISHANKAR,[1] C. O. MOHAN,[2]
SATYEN KUMAR PANDA,[3] and B. MEENAKUMARI[4]

[1]*ICAR-Central Institute of Fisheries Technology, Matsyapuri P.O., Willingdon Island, Cochin – 682029, Kerala, India*

[2]*Fish Processing Division, ICAR-Central Institute of Fisheries Technology, Matsyapuri P.O., Willingdon Island, Cochin – 682029, Kerala, India, E-mail: comohan@gmail.com*

[3]*Quality Assurance and Management Division, ICAR-Central Institute of Fisheries Technology, Matsyapuri P.O., Willingdon Island, Cochin – 682029, Kerala, India*

[4]*Indian Council of Agricultural Research, Krishi Anusandhan Bhawan II, New Delhi – 110012, India*

## CONTENTS

## ABSTRACT

Krill forms one of the most abundantly available animal sources of marine origin. It has very good quantity of useful omega 3 fatty acids and good quality proteins and minerals. In spite of this, it is being underutilized mainly due to its high fluoride content and high protease content of muscle, which degrades the muscle protein rapidly. Use of krill meat for human consumption can be increased by overcoming these problems. At present it is mainly used as fish feed for aquaculture and as bait for sport fishing. A variety of medicinally important high value products can be prepared from krill which can find extensive utilization of this underutilized resources. India is expanding its aquaculture activities and at present ranks second in aquaculture production. Huge quantity of feed will be required to meet the ever-expanding aquaculture industry. In this paper, possibility of utilizing krill as source of aqua feed is explored.

## 11.1 INTRODUCTION

Krill are open ocean pelagic crustaceans commonly encountered in Southern Ocean. Out of the 85 species of Krill reported in the world oceans, *Eupahusia superba* (Antarctic Krill) and *Euphausia pacifica* (North Pacific Krill) contribute significantly to the Krill fishery of the world. Morphologically similar to shrimp, krill are miniature organisms with size of 0.01–2.00 g weight and 0.8–6.0 cm length (Nicol and Endo, 1997). The species of krills ranging in size from under 0.5 inch (1 cm) up to 5.5 inches (14 cm) long. The dominant krill in the southern polar oceans is the Antarctic krill (*Euphausia superba*) (Figure 11.1), which is up to 2.3 inches (6 cm) long and weighs about 0.035 ounces (1 g). Antarctic krill have a life span of about 5 to 10 years. Antarctic Krill is considered to be a keystone species, an organism upon which very many Antarctic predators depend. Krill eat phytoplankton, single-celled plants that float in the seas near the surface. Some tropical krill also eat zooplankton. Krill spend their days in the dark depths of the ocean (about 100 m deep), safe from their major predators (like baleen whales and sea birds). They swim to the surface each night to eat phytoplankton. They can fast for up to 200 days and shrinks and becomes very thin during this time. Krill have a hard exoskeleton, many legs and a segmented body. Females produce almost 1,000 eggs each summer; the eggs are laid at the

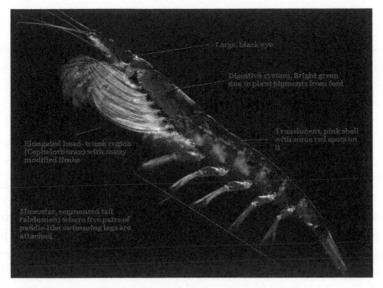

**FIGURE 11.1**    Different parts of Antarctic krill (*Euphausia superba*).

surface, but fall to great depths. The hatchlings swim back to the surface to feed. Like all crustaceans, krill molt their exoskeleton as they grow.

High concentration of Antarctic Krill exists in the Southern Atlantic and in some regions of Indian Ocean, in close proximity to Antarctic continent. The surface distribution of Antarctic Krill extends to 36 million square kilometer, which makes it the most abundant and largest multi-cellular animal species on the planet. The Krill Fisheries in the Southern Ocean is currently managed by Commission for the Conservation of Antarctic Marine Living Resources (CCAMLR). As per CCAMLR, Antarctic krill (*Euphausia superba*) may be fished in the Southern Ocean in Subareas 48.1 to 48.4, Subarea 48.6 and Divisions 58.4.1 and 58.4.2. The fisheries in Subareas 48.1 to 48.4, and Divisions 58.4.1 and 58.4.2 are established fisheries, and the fishery in Subarea 48.6 is an exploratory fishery. However, fishing is currently conducted only in Subareas 48.1 to 48.4. The combined catch limit for krill is currently fixed at 620,000 tons. The total catch of Krill during the year 2013 was 217,357 tons. Midwater trawls and beam trawls operating up to 250m depths are used by fishing vessels to harvest krill. Continuous fishing system or fish pumps are used to handle large volume of catch.

## 11.2 NUTRITIONAL PROFILE OF ANTARCTIC KRILL

Considerable attention is given to Krill due to its nutritional significance as it contains good quality protein, lipid and minerals. Krill resembles other seafood in being rich source of high-quality protein, with the advantage of being low in fat and good source of omega-3 fatty acids and antioxidants (Tou et al., 2007). The moisture content of whole krill ranges between 77.9 to 83%, whereas the crude protein and lipid varies in the range of 10.75 to 15.4% and 0.5 to 3.6%, respectively (Grantham, 1977, Ravishankar et al., 2013). In terms of essential amino acid content, they are rich in leucine, lysine, isoleucine, arginine and glutamic acid. The essential amino acid content is more than 40% of total amino acid. The level of indispensable amino acid in krill protein complies with FAO/WHO/UNU amino acid requirement for adults. Compared to other animal foods, the lipid content of Krill is quite low, making it ideal for human nutrition. The fatty acid profile of Krill indicates low content of SFA (26.1%) and MUFA (24.2%), but predominantly higher content of PUFA (48.5%) compared to other seafood. The n–3 fatty acids, especially EPA and DHA are also abundant, accounting up to 19% of total fatty acids (Kolakowska et al., 1994), due to which many industries are developing krill oil in capsule form (Figure 11.2). Krill lipids on the whole have fatty acid compositions similar to fish lipids, except for the near absence of C18:0 (Ravishankar et al., 2013). Another typical feature of Krill is that most of the fatty acids are incorporated as phospholipids. The cholesterol content of Krill is comparatively higher than fish, but lower than shrimp, which is reported to be 62.1–71.6 mg/100 g by Watanabe (1979) and 101.7 mg/100 g by Ravishankar et al. (2013).

Antarctic krill provides good source of minerals, especially Selenium and iron compared to other seafood (Chi et al., 2013). The primary carotenoid in krill is astaxanthin, which is reported to be present at a level of 15–20 mg/Kg in the krill tissue (Nicol et al., 2000). Nutritional evaluation of Krill tail meat indicates protein efficiency ratio (PER) of 2.2 and 82.3% protein digestibility (Ravishankar et al., 2013).

## 11.3 UTILIZATION OF KRILL RESOURCES

Krill is the single largest under-utilized commercial marine resource as the present level of harvest is far below the regulatory catch quota (Naylor et al., 2009). There are vigorous commercial interest to utilize Krill in

**FIGURE 11.2** Omega 3 fatty acid rich Krill oil in capsule form.

development of pharmaceutical and nutraceutical products as well as in and aquaculture feeds. Use of krill meal and krill oil for development of aquaculture feed was forecasted as the driving force behind growth of krill fishing industry (Nicol and Foster, 2003). There is a high stake involved in utilization of Krill resources for human use and other commercial applications. Empirical estimate shows that out of the total global krill catch, about 45% is used for sport fishing bait, 43% for aquaculture feed, and 12% for human consumption, primarily in the form of supplements. In Japan, nearly 43% of Antarctic krill is used for direct human consumption (Nicol and Endo, 1997). The nutritional value and potential health benefits of Krill from the perspective of human consumption has been reviewed by Tou et al. (2007). Currently krill is consumed as frozen raw krill, frozen boiled krill, peeled krill meat and dried krill (Figure 11.3) in many parts of the world, which is expected to diversify further with technological advancements. Fermented krill products could also be used as a potential source of nutrients and natural antioxidants (Faithong et al., 2010).

Freeze dried products are used as fishing bait in UK and Japan. The organic amines extracted from Krill are used in cosmetic products for screening UV radiation. Krills are good source of Taurine (0.29%), which can be used as an emerging nutritional supplement.

**FIGURE 11.3**    Fresh, frozen, and dried Antarctic krill (Top right: Photo from Uwe Kils; https://creativecommons.org/licenses/by-sa/3.0/)

In spite of its potential as a high quality lipid and protein source, use of krill as human food has been limited. Krill is mainly used in reduction fisheries for manufacture of krill meal due to its high astaxanthin content. Additionally, encapsulated krill oil is used as dietary supplement for protection against cardiovascular disease. Krill oil is comparable to fish oil in terms of PUFA content. Despite its high PUFA content, krill oil is considered relatively resistant to oxidation which is attributed to the presence of antioxidants (Suzuki and Shibata, 1990).

The krill is reported to undergo quick autolysis due to its very high proteolytic activity. This has raised a lot of interest and a number of studies on the characteristics of these enzymes. To mention few, krill enzymes have been shown to have a higher activity toward fibrin and casein compared to trypsin. Also, a serine proteinase from krill referred to as euphauserase has been sequenced, modified, and produced. Krill enzymes are now being exploited for medical applications, such as debridement of necrotic wounds and removal of dental plaque (Westerhof et al., 1990). The degradation efficiency of the trypsin-like krill enzyme was found to be higher than that of bovine trypsin at 37°C. At lower temperature (1 3°C), the efficiency of krill enzymes is 60 times higher than that of bovine trypsin. It is also reported that krill enzyme possesses both trypsin and carboxypeptidase activity.

Antarctic Krill resources can be harnessed for extraction of chitin as Krill has a higher percentage of chitin than other highly exploited aquatic resources such as crab and shrimp. Chitin, a partly deacetylated (1–4)-2-acetamido-2-deoxy-β-d-glucan, found in crustacean shells, has been proven to be a useful bioactive component and polymer material. Owing to its outstanding biocompatibility, biodegradability, and bioactivity, chitin has been applied in various fields, including functional food, cosmetics, agriculture, medicine, paper industry, immobilization support, and wastewater treatment. Additionally, in recent decades, chitin derivatives have received great attention for their special properties. Given this background, the development of Antarctic krill chitin would provide an effective means to utilize the Antarctic krill shell. Antarctic krill chitin corresponded to the α-polymorph, which constituted small, stable, and uniform micro-crystals. The chitin content in Krill is reported to be in the range of 2.4–2.7% (Nicol and Hosie, 1993). The degree of deacetylation value ranges between 11 to 15%.

In recent years increased attention is given to extraction and characterization of collagen from Antarctic krill (Mizuta et al., 1998). Muscle connective tissues of multicellular animals contain collagen as a fundamental proteinaceous constituents, which plays mechanically and physiologically important functions. As for crustacean species, the occurrence of two genetically distinct types of collagen α 1 (AR-I) and α 2(AR-I) components were shown to be widely distributed in the muscular tissues of other decapod and stomatopod species. The krill collagens are highly insoluble not only in dilute acid solvent probably owing to the existence of acid- or heat-resistant cross-links or of stabilizing materials such as polysaccharides. The solubility of the krill collagen in the limited pepsin digestion is low (approximately 32%) compared with that of the collagen from other prawn muscle (more than 80%). The krill muscle collagen or α 1(Kr) component has similar amino acid composition to that of the a1 (AR-I) component of other crustacean varieties. The krill muscle collagen or α 1(Kr) component has a main localization in the relatively thick connective tissues, epimysium and perimysium, which is almost identical feature to that of the α 1(AR-I) component.

## 11.4   CONCERNS ASSOCIATED WITH UTILIZATION OF ANTARCTIC KRILL

Proper utilization of this valuable resource is of extreme importance at a time when most of the conventional resources are reaching or exceeding optimum sustainable limits. A major hindrance to commercial processing of krill and development of new krill-based food products is the intense postmortem proteolytic and lipolytic activity (Kawamura et al., 1981). The protease and lipases are released immediately upon the death of krill, resulting in autolysis, which leads to a rapid spoilage. The high activity of enzymes combined with its small size makes krill processing for human food a significant challenge.

Lack of advanced krill harvesting systems, onboard handling and processing of large volumes, long distance to the fishing grounds and other problem of logistics pose great challenge for utilizing krill resources by India.

Another limiting factor for utilization of Krill for human consumption as well as for manufacture of aquaculture feed is presence of high content of fluorine. Fluorine is part of the exoskeleton of Krill and the natural fluorine content is reported to vary between 1000 to 6000 ppm (Moren et al., 2007). The European Union directive 2008/76/EC specifies a maximum content of 3000 ppm fluorine in krill to be used as fish feed ingredient and the final manufactured fish feed should not contain more than 350 ppm. Fluoride content in Krill meal is reported as high as 940–1160 ppm (Moren et al., 2007). Further, fluoride content of Salmon diet prepared with replacement of 20–100% fishmeal with krill meal was in the range of 130–900 ppm.

As discussed earlier, the major interest evinced by world nations to harvest krill was to use for aquaculture feed. But, complete replacement of fishmeal with krill meal in fish feed is found to negatively impact growth of cultured aquatic species. It has been shown that dietary fluoride derived from Antarctic krill could inhibit fish growth when krill meal is used to replace fishmeal. In a three year feeding trial of Salmonids with Antarctic krill, cultured in cages in Kiel Fjord, considerable accumulation of fluoride as high as 670 mg/kg was observed in the bony parts (Grave, 1981). Weight gain, feed intake and specific growth rate of rainbow trout (*Oncorhynchus mykiss*) significantly decreased when fed with a diet containing 30% krill meal (Yoshitomi et al., 2006). On the other hand, replacement of fishmeal with low fluoride Krill meal in the diet did not influence any growth parameter, when fed to rainbow trout (*Oncorhynchus mykiss*) cultured in freshwater (Yoshitomi et al., 2007). It was shown that yellowtail (*Seriola quinqueradiata*) fed with 100% whole krill meal showed adverse growth with accumulation of fluoride in bones (Yoshitomi and Nagano, 2012), but use of de-shelled krill meal did not exhibit any deleterious impact on growth. Feeding trials of

Atlantic salmon (*Salmo salar*), Atlantic cod (*Gadus morhua*), rainbow trout (*Oncorhyncus mykiss*) and Atlantic halibut (*Hippoglossus hippoglossus*) with diets partly or fully substituted proteins from krill meal showed no effect on growth and health parameters as well as no increased accumulation of fluorine in hard tissues (Moren et al., 2007). In culture of Atlantic Salmon (*Salmo salar*) it has been reported that fishmeal can be replaced up to 40% krill meal as a source of dietary protein without affecting the product quality (Suontama et al., 2007). In another study it is found that increased salinity of grow-out systems neutralizes the negative impact of krill meal (Julshamn et al., 2004). In Atlantic Salmon reared in seawater, the fluoride concentration in tissues and hard parts were not affected by dietary fluoride concentration, and Krill meal could be included in salmon diets up to 30% to partially substitute for fishmeal without any adverse effect on growth performance or survival.

In Krill, high amount of fluoride is concentrated in exoskeleton (Sands, 1998). In *Euphausia superba*, level of fluoride all exoskeletal regions are reported to be in the range of 1290–4028 µg/g (Sands et al., 1998). It is hypothesized that high level of fluoride is actively taken up for deposition in hardening exoskeleton (Zhang et al., 1993).

A reduction in fluoride concentration in krill meal can be obtained by separating the exoskeleton from the muscle prior to meal production. However, the fluoride leaches out from the exoskeleton to the muscle during cold storage. During frozen storage, fluoride migrates from exoskeleton to muscle tissues and hence fluoride levels as high as $1119 \pm 328$ µg/g is reported in in the muscle tissues of krill (Adelung et al., 1987; Sands et al., 1998).

Natural high level of some of the metallic elements in Krill is also another discouraging feature in its utilization. Krill meal produced from Antarctic krill is reported to contain high amount of Copper (22–81 ppm) and Cadmium (>1.4 ppm), which sometime breaches the regulatory limits stipulated for feed ingredients (Hansen et al., 2010).

## 11.5  POSSIBLE APPROACHES FOR UTILIZATION OF KRILL RESOURCES

### 11.5.1  DEVELOPMENT OF KRILL HARVESTING AND PROCESSING SYSTEMS

Preliminary work has already been done by Central Institute of Fisheries Technology (CIFT) during the First Indian Antarctic Expedition (FIKEX)

(27 December 1995 – 10 March 1996) onboard FORV Sagar Sampada in Fishing Area 58 in the India Ocean Sector of Southern Ocean. During FIKEX, a total catch of 12070 kg was landed by aimed trawling operations. Antarctic krill constituted 47.1% of the total catch while salps constituted 52.5% in the landings of krill midwater trawl (DOD, 1996).

In order to sustainably harvest krill from southern ocean, fishing vessels with onboard processing facility must be made available, either by joint venture or in public private partnership (PPP) mode. The harvested Krill must be processed within 4–6 hours to overcome the problem of autolytic degradation. Removal of exoskeleton shall ensure reduction of fluoride content in krill meal. Another possibility of utilization of Krill for human consumption is by separating the tail meat for development of value added products. Onboard drying and freezing facility must be ensured for subsequent extraction of krill oil and krill meal.

## 11.5.2 POSSIBILITY OF SUBSTITUTING FISHMEAL WITH KRILL MEAL FOR AQUACULTURE SECTOR IN INDIA

A significant, but gradually diminishing proportion of world fisheries production is still processed into fishmeal and fish oil. As per FAO estimates, in 2012, more than 86 percent (136 million tons) of global fish production was utilized for direct human consumption. The residual 14 percent (21.7 million tons) was destined to non-food uses, of which 75 percent (16.3 million tons) was reduced to fishmeal and fish oil (SOFIA, 2014). The average fishmeal production during 2011–13 was 5.189 million tons (OECD, 2014). With aquaculture expansion, more pressure is expected on fishmeal and fish oil. Globally, aquaculture sector uses about 3.06 million tons or 56.0% of the world's fishmeal production. Global aquaculture feed production is reported to be around 40.4 million metric tons (Alltech, 2014).

As per estimates of International Fishmeal and Fish Oil Organization (IFFO), massive changes in the use of both fishmeal and fish oil has been noticed over the years. In 1960, around 98% of fishmeal was used in pig and chicken diets, but by 2010, 73% of global fishmeal production is being used in aquaculture. Present production of fishmeal in India is around 357.94 thousand tons out of which 29.52 thousand tons is exported. India also imports 10.36 thousand tons of fishmeal worth 89.18 crore INR.

**TABLE 11.1**  Estimation of Requirement of Fish Feed, Fishmeal and Krill for India at Current (2012–14) Level of Aquaculture Production

| | Finfish | | Crustaceans | Molluscs | Total | Source |
|---|---|---|---|---|---|---|
| | Inland Aquaculture | Mariculture | | | | |
| Present production (million tons) | 3.81 | 0.08 | 0.30 | 0.01 | 4.21 | SOFIA (2014) |
| % Fed aquaculture | 70.00% | 100.00% | 100 | 0 | | |
| Production from Fed Aquaculture (million tons) | 2.67 | 0.08 | 0.30 | 0.00 | | |
| FCR assumed | 2:1 | 3:1 | 2:1 | - | | |
| Assumed Present Feed Requirement (million tons) | 5.34 | 0.25 | 0.60 | - | 6.19 | |
| Assumed % of protein in feed | 25 | 50 | 45 | - | | |
| Protein requirement | 1.33 | 0.13 | 0.27 | - | 1.73 | |
| Average Percentage of Protein contributed by fishmeal | 7 | 70 | 45 | - | | Tacon A.G.and Metian, M. (2008); Aquaculture 285 (2008) 146–158 |
| Fishmeal requirement | 0.09 | 0.09 | 0.12 | - | 0.3 | |
| Substitution with Krill meal (30%) | 0.03 | 0.027 | 0.036 | - | 0.093 | Hansen et al. (2010). Aquaculture 310 (2010) 164–172 |
| **Krill to be harvested (with 20% yield)** | | | | | **0.465 Million tons** | |

**TABLE 11.2** Predicted Requirement of Fish Feed, Fishmeal and Krill for India by 2023

| | Finfish | | Crustaceans | Molluscs | Total | Source |
|---|---|---|---|---|---|---|
| | Inland Aquaculture | Mariculture | | | | |
| Predicted Production of aquaculture by 2023 (million tons) | 5.22 | 0.12 | 0.41 | 0.02 | 5.77 | OECD (2014) |
| % Fed aquaculture | 70% | 100% | 100% | 0% | | |
| Production from Fed Aquaculture (million tons) | 3.66 | 0.12 | 0.41 | 0.00 | | |
| FCR Assumed | 2:1 | 3:1 | 2:1 | - | | |
| Assumed Present Feed Requirement (million tons) | 7.31 | 0.35 | 0.82 | - | 8.48 | |
| Assumed % of protein in feed | 25 | 50 | 45 | - | | |
| Protein requirement | 1.83 | 0.17 | 0.37 | - | | |
| Average Percentage of Protein contributed by fishmeal | 7 | 70 | 45 | - | | Tacon A.G.and Metian, M. (2008). Aquaculture 285 (2008) 146–158 |
| Fishmeal requirement | 0.13 | 0.12 | 0.17 | - | 0.42 | |
| Substitution with Krill meal (30%) | 0.04 | 0.04 | 0.05 | - | 0.12 | Hansen et al. (2010). Aquaculture 310 (2010) 164–172 |
| **Krill to be harvested (with 20% yield)** | | | | | **0.62 Million tons** | |

As per the 2014 Alltech Global Feed Survey findings, the current aquaculture feed production by India is 1.1 million metric tons (Anon, 2014). As shown in Table 11.1, based on the current level of aquaculture production, India needs to harvest 0.465 million tons of Krill. By 2023, the aquaculture production of India is projected to be 5.77 million tons as per OECD-FAO projections (OECD, 2014). Hence the predicted requirement of krill meal by 2023 would be 0.12 million tons, which translates to 0.62 million tons of krill resource (Table 11.2).

## 11.6 CONCLUSION

It is projected that by 2020 India needs 416 thousand tons of fishmeal for aquaculture sector alone (World Bank, 2013). Hence, there is a possibility of substituting fishmeal with krill meal to an extent of 30–40%, provided detailed risk analysis with regard to deleterious impact of high content of fluoride and heavy metals is carried out for Indian aquacultured species. Further, detailed economic analysis has to be carried out for harvest, transport and commercial scale manufacturing of krill meal and other krill based products.

## KEYWORDS

- fluoride
- harvesting
- krill meal
- mantarctic Krill
- nutritional profile
- utilization patter

## REFERENCES

Adelung, D. F., Buchholz, B., Culik, B., & Keck, A. (1987). Fluoride in tissues of krill *Euphausia superba* Dana and *Meganyctiphanes norvegica* M. Sars in relation to the molt cycle. *Polar Biol, 7*, 43–50.
Alltech. (2014). Alltech Global Feed Survey and Summary (2014). Available at http://www.alltech.com/ sites/default/files/alltechglobalfeedsummary2014.pdf. Accessed on 26th August 2014.

Anon. (2014). Alltech Global Feed Survey findings, AQUA Culture Asia Pacific Magazine, May/June 2014), 40.

Chi, h., Li, X., & Yang, X. (2013). Processing status and utilization strategies of Antarctic Krill (*Euphausia superba*) in China. *World J. Fish Mar. Sci., 5*(3), 275–281.

Faithong, N., Benjakul, S., Phatcharat, S., & Binsan, W. (2010). Chemical composition and antioxidative activity of Thai traditional fermented shrimp and krill products. *Food Chem, 119*, 133–140.

Grave, H. (1981). Fluoride content of salmonids fed on Antarctic krill. *Aquaculture, 24*, 191–196.

Hansen, J. Ø., Penn, M., Øverland, M., Shearer, K. D., Krogdahl, Å., Mydland, L. T., et al., (2010). High inclusion of partially de-shelled and whole krill meals in diets for Atlantic salmon (*Salmo salar*). Aquaculture, *310* (1–2), 164–172.

Kawamura, Y., Nishimura, K., Igarashi, S., Doi, E., & Yonezawa D. (1981). Characteristics of autolysis of Antarctic krill. *Agric. Biol. Chem, 45*, 93–100.

Kolakowska, A., Kolakowski, E., & Szcygielski, M. (1994). Winter season krill (Euphausia superba Dana) as a source of n-3 polyunsaturated fatty acids. *Die Nahrung, 38*, 128–134.

Moren, M., Malde, M. K., Olsen, R. E., Hemre, G. I., Dahl, L., Karlsen, Ø., et al. (2007). Fluorine accumulation in Atlantic salmon (Salmo salar), Atlantic cod (Gadus morhua), rainbow trout (Oncorhyncus mykiss) and Atlantic halibut (Hippoglossus hippoglossus) fed diets with krill or amphipod meals and fish meal based diets with sodium fluoride (NaF) inclusion. Aquaculture, *269*, 525–531.

Naylor, R. L., Hardy, R. W., Bureau, D. P., Chiu, A., Elliott, M., Farrell, A. P., et al. (2009). Feeding aquaculture in an era of finite resources. PNAS, *106* (36), 15103–15110.

Nicol S., & Hosie, G. W. (1993). Chitin production by krill. Biochem. *System. Ecol, 21*, 181–184.

Nicol S., Forster I., & Spence J. (2000). Products derived from krill. In: Krill: Biology, Ecology and Fisheries, by Everson, I. (Ed.), Malden, MA: Blackwell Sciences Ltd., 262–283.

Nicol, S., & Endo, Y. (1997). Krill Fisheries of the World. FAO Fisheries Technical Paper 367.

Nicol, S., & Foster, J. (2003). Recent trends in the fishery for Antarctic krill. Aquat. Living Resour, *16*, 42–45.

OECD. (2014). OECD-FAO Agricultural Outlook 2014. Fish and seafood projections: Production and trade. DOI:10.1787/agr_outlook-2014-table180-en.

Ravishankar, C. N., Srinivasa Gopal, T. K., & Meenakumari, B. (2013). Post harvest technological aspects of Antarctic Krill. In: Scientific and geopolitical interests in Arctic and Antarctic, by Ramesh, R., Sudhakar, M., & Chattopadhyay S., (Eds.), Proceedings of International Conference on Science and Geopolitics of Arctic and Antarctic, (iSaGAA), March 2013, Lights Research Foundation, 183–194.

Sands, M., Nicol, S., & McMinn, A. (1998). Fluoride in Antarctic marine crustaceans. *Mar. Biol, 132*, 591–598.

Suontama, J., Kiessling, A., Melle, W., Waagbø, R., & Olsen, R. E. (2007). Protein from northern krill (Thysanoessa inermis), Antarctic krill (Euphausia superba) and the Arctic amphipod (Themisto libellula) can partially replace fishmeal in diets to Atlantic salmon (Salmo salar) without affecting product quality. *Aquaculture Nutrition, 13*, 50–58.

Suzuki, T., & Shibata, N. (1990). The utilization of Antarctic krill for human food. *Food Reviews International, 6*, 119–147.

Tou, J. C., Jaczynski, J., & Chen, Y. (2007). Krill for Human Consumption: Nutritional Value and Potential Health Benefits. *Nutrition Reviews, 65*(2), 63–67.

Westerhof, W., van Ginkel, C. J., Cohen, E. B., & Mekkes, J. R. (1990). Prospective randomized study comparing the debriding effect of krill enzymes and a non-enzymatic treatment in venous leg ulcers. *Dermatologica, 181*, 293–297.

World, Bank. (2013). Fish to 2030: Prospects for fisheries and aquaculture. Agriculture and environmental services discussion paper 03, World Bank Report Number 83177-GLB. The World Bank, Washington DC 20433, 80.

Yoshitomi, B., & Nagano, I. (2012). Effect of dietary fluoride derived from Antarctic krill (*Euphausia superba*) meal on growth of yellowtail (Seriola quinqueradiata). Chemosphere, *86* (9), 891–897.

Yoshitomi, B., Aoki, M., & Oshima, S. (2007). Effect of total replacement of dietary fish meal by low fluoride krill (*Euphausia superba*) meal on growth performance of rainbow trout (*Oncorhynchus mykiss*) in fresh water.

Yoshitomi, B., Aoki, M., Oshima, S., & Hata, K. (2006). Evaluation of krill (*Euphausia superba*) meal as a partial replacement for fish meal in rainbow trout (*Oncorhynchus mykiss*) diets. Aquaculture, *261*, 440–446.

Zhang H., Jianming P., Xianhao C., & Biying Z. (1993). Biogeochemistry research of Fluoride in Antarctic Ocean. I. The study of fluoride anomaly in krill. *Antarctic Res*, 4, 55–61.

# CHAPTER 12

# PROCESSING AND PACKAGING OF DAIRY-BASED PRODUCTS

LATHA SABIKHI,[1] YOGESH KHETRA,[2] and P. NARENDER RAJU[2]

[1]*Principal Scientist and Head, Dairy Technology Division, ICAR-National Research Institute, Karnal, Haryana, India*

[2]*Scientist, Dairy Technology Division, ICAR-National Research Institute, Karnal, Haryana, India*

## CONTENTS

## 12.1   INTRODUCTION

Milk has existed, ever since evolution, as the most wholesome single food available in nature. It is as ancient as mankind itself, as it is the substance created to feed the mammalian infant, be it the minute mouse or the mammoth whale. Species such as cows, buffalos, sheep, goats and camels continue to be used globally for the producing milk for human use since ancient times. Although intended by nature for the neonate, milk plays an important role in meeting the nutritional requirement of children, adolescents, elderly persons and pregnant and nursing mothers. Milk and dairy products are useful to all categories of population, as it the major source of protein and calcium in the readily available form. The world milk production has increased by more than 50% in the past five decades, increasing from 344 million tons (MT) in 1983 to 769 MT in 2013 (faostat3.fao.org). India, with ~18% of the world's production, is the largest milk producer (135.6 MT) followed by the USA (91.3 MT), China (40.6 MT), Pakistan (39.1 MT), Brazil (34.4 MT), Germany (31.1 MT) and the Russian Federation (30.5 MT) (faostat3.fao.org, 2013 data). South Asia has seen the highest growth rate since the 1970s in milk production. Africa, on the other hand has witnessed slow growth, probably owing to poverty and adverse climatic conditions. The countries with the highest milk deficits are China, Italy, the Russian Federation, Mexico, Algeria and Indonesia, while those with the highest milk surpluses are New Zealand, USA, Germany, France, Australia and Ireland (www.fao.org). Several southeast Asian countries, including China do not have a long tradition of dairying, whereas milk and milk products have an important role in the diet in several other countries. These include countries in the Mediterranean and Near East, the Indian subcontinent, the savannah regions of West Africa, the highlands of East Africa and parts of South and Central America.

Milk is a highly nutritious food, containing most of the essential nutrients in a balanced amount. It is an excellent medium for the growth of micro-organisms that can lead to spoilage as well as diseases. Milk, therefore, is highly perishable and has an extremely short shelf life. Processing of milk can extend its life by days to months and also add value to it. In the developing countries where milk production is scattered and based on marginal and small farm systems, processing opportunities at the farm gives small-scale dairy producers higher returns than selling raw milk and offers better opportunities to reach regional and urban markets. As individual farmers cannot establish such units, farmer producer companies and self-help groups are being formed to pool the milk and process it. The processing of milk into packaged milk and products can benefit entire communities by generating off-farm jobs in milk collection, transportation, processing, engineering and maintenance, accountancy and marketing. Converting raw milk into products with longer shelf life also helps to overcome issues of seasonal variations in milk production. The shelf life of milk can be extended for several days through processing techniques such as chilling (most suitable for raw milk) and fermentation (for yogurt, *dahi,* kefir, koumiss, etc.). Milk can also be converted to value added (butter, cheese, *ghee*), concentrated (condensed milk) and dried (milk powder, dairy creamer/whitener) dairy products with long shelf lives and which are easy to transport. Milk, on arrival at the dairy plant is processed into products by applying several unit operations. Several of these operations are routine compulsory processes common to all dairies, while certain others are undertaken for specialized purposes in selected dairy plants. Some common dairy processing operations are listed in Figure 12.1. The relative share of processing milk to convert it into selected products is listed in Table 12.1.

## 12.2   FILTRATION

Visible foreign particles (dirt, feed, manure, dandruff, hair, flies, dust particles, worn out cells, cell debris etc.) often mar the appearance of milk. The large particles among these are removed by simple filtration through specially woven filter cloth. Milk filtering devices are provided with closely woven cotton cloth, cellulose pads or nylon filters that have very small pore size. These are effective in removing visible sediment from milk but cannot remove cell fragments, leucocytes and micro-organisms. The margins

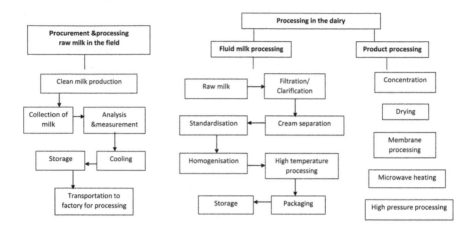

**FIGURE 12.1**    Selected dairy processing operations.

of filter cloth/pad/mesh are compressed in a frame to provide support and covered with metal perforated screen. The entire assembly is fitted in an enclosure with suitable inlet and outlet connections for sanitary piping. The design of the system is such that removing the filters for washing is easy. The frequency of changing the filter cloth depends on the amount of foreign matter present in milk and temperature of filtration. In continuous large scale operations, two or more filters are used in the line with a flow valve in between so that processing goes on uninterrupted during filter removal.

## 12.3   CLARIFICATION

A continuous process of centrifugal clarification removes sediments, cell debris and even bacteria more efficiently than filtration. The clarifier is a mechanical device with a centrifuge bowl with baffles inserted in the form of conical discs. The discs are stacked together, forming a single unit called the disc stack. They rest on each other and are kept apart by radial strips. Milk flows through the channel formed by the strips. The milk enters the equipment at the outer edge of the disc stack and flows radially towards the axis of rotation, being subjected to centrifugal force that acts on all particles. First, the particle settles on the upper disc so that it may be separated. The liquid velocity is so small at this point that the settled particle is no longer carried along with the liquid. Under the influence of centrifugal force, it slides outwards along the underside of the disc, and is pushed off the outer edge

TABLE 12.1 Continent-Wise Production of Selected Milk Products: Global Share (%)*

| Continent | Fresh cream | Yoghurt | Cheese | Butter and Ghee | Evaporated and Condensed Milk | Skimmed milk powder | Whole Milk Powder | Condensed Whey | Whey Powder |
|---|---|---|---|---|---|---|---|---|---|
| Africa | 0.30 | 2.60 | 4.00 | 2.60 | 1.30 | 0.50 | 1.00 | NA | 0.10 |
| Americas | 7.30 | 18.50 | 30.10 | 12.50 | 37.30 | 23.10 | 30.40 | 35.30 | 32.10 |
| Asia | 1.00 | 48.40 | 6.40 | 30.00 | 11.70 | 6.60 | 4.60 | 10.20 | 0.20 |
| Europe | 90.70 | 30.50 | 56.70 | 48.70 | 48.20 | 59.60 | 45.20 | 54.50 | 64.00 |
| Oceania | 0.70 | NA | 2.80 | 6.20 | 1.50 | 10.20 | 18.80 | NA | 3.60 |

* 2013 figures.

NA – Information not available.

Source: http://faostat3.fao.org.

and deposited on the wall of the centrifuge bowl. As the milk passes along the full radial width in the discs, the time of passage allows smaller particles to be separated. Thus, the denser particles are forced towards the periphery and collected as clarifier sludge in the space outside the disc stack. The clarified milk is taken out through the outlet.

The filtration and clarification equipment is generally located in the raw milk line, depending on the type of clarification – cold or warm. Cold stream clarifiers are placed between reception and raw milk storage tanks or between storage tank and pasteurizer. Warm milk clarifiers are located either between regeneration and heating section of the pasteurizer or between heating and holding sections of the pasteurizer. To avoid post contamination milk should not be clarified after pasteurization. Filtration and clarification do not improve the keeping quality of milk. The cream layer decreases after both the processes, the effect increasing with increase in clarifying temperatures. The composition of milk is not affected by the removal of clarifier slime.

## 12.4   CREAM SEPARATION

The recovery of fat is of importance in dairy processing because of its high economic value. Fat content of various dairy products needs to be adjusted to meet the legal standards, before the milk is processed into products. The process of centrifugal cream separation separates milk into cream and skim milk on the basis of differences in the densities of fat (910–980 kg/m$^3$), obtained as cream and solids-not-fat (1027–1032 kg/m$^3$) obtained as skim milk. To enable the process of separation of two phases, the following three criteria must be fulfilled.
 i.   The substance to be treated must be a mixture of two distinct phases, one of which should be continuous phase.
 ii.  The phases to be separated must not be soluble in each other.
 iii. They must also have different densities.

Milk, with serum as continuous and heavier phase and fat as dispersed and lighter phase satisfies the requirements for separation of the two phases.

### 12.4.1   BATCH PROCESS: SEPARATION BY GRAVITY

If milk is left undisturbed, milk fat rises to the surface owing to its low density. The fat globules being lighter tend to move upwards and collect at the surface in the form of a layer of cream, which is skimmed off manually.

The diameter of milk fat globules varies from 2–10 microns and determines the rate of rise of the fat globules. Depending on the size, fat globules take different times to rise to the surface in order to form the cream layer. The sedimentation velocity ($V_g$) of a particle the particle is calculated using the Stoke's Law, represented by the following formula.

$$V_g = (D^2 (d_p - d_c)/ 18\, \eta)\, g$$

where, $V_g$ – sedimentation velocity in m/s; D – particle diameter in m; $d_p$ – particle density in kg/m³; $d_c$ – density of continuous phase (skim milk) in kg/m³; $\eta$ – viscosity of continuous phase in kg/m-s; and g – acceleration due to gravity (9.81 m/s²).

It is evident from the formula that the floatation velocity of the fat globule is influenced by the square of the particle diameter, difference in densities between the phases and viscosity of continuous phase. Hence, the flotation velocity of a larger globule is higher than smaller ones. Although fat globules cluster into larger aggregates and behave as one entity, the process is extremely slow. Let us calculate the floatation velocity of a fat globule with a diameter of 5 microns.

$$D = 5 \times 10^{-6}\, m;\ d_p - d_c = 980 - 1028 = 48;\ \eta = 1.42 \times 10^{-3}\ kg/m\text{-}s;\ g = 9.81\ m/s^2$$

Substituting these values in the Stoke's formula, we arrive at:

$$V_g = (25 \times 10^{-12} \times 48/18 \times 1.42 \times 10^{-3}) \times 9.81$$
$$= 928.68 \times 10^{-9}\, m/s\ \text{or}\ 3343248 \times 10^{-9}\, m/h\ \text{or}\ 3.343 \times 10^{-3}\, m/h\ \text{or}\ 3.343\ mm/h,$$

which $V_g$ is a very slow rate of separation.

The three basic methods of gravity creaming of milk in use were the shallow pan method, deep setting method and the water dilution method. In the first, a circular pan 0.3–0.6 m in diameter and 0.1–0.15 m deep is filled two third with milk and left undisturbed in a cool place to facilitate the fat to rise. In the deep set method, milk is placed in metal containers 0.3–0.6 m deep and 0.2–0.25 m in diameter and kept in cold water. The skim milk is removed through a faucet provided at the bottom of the can after 24 hours. The quality of cream is superior, owing to the cold temperatures. The third method is based on the fact that by diluting the milk, its viscosity reduces, thereby increasing the rate of rise of fat globules. Milk is diluted with equal volume of water and allowed to set for 12 hours. The method reduces the value of skim milk which becomes unsuitable even for animal feeding. It gives incomplete separation of the fat and does not have adequate control on temperature resulting in deterioration in the quality of cream. The fat contents in the skim milk by the three methods are 0.5–1.0%, 0.2–0.35% and 0.3–0.4%, respectively. None of the gravity separation systems is suitable for large quantities of milk.

## 12.4.2  CONTINUOUS PROCESS: CENTRIFUGAL SEPARATION

When a liquid contained in a vessel is rotated around an axis, a centrifugal force is generated, wherein the acceleration is not constant like the gravitational acceleration in the stationary vessel. The centrifugal acceleration increases with increasing distance from the axis of rotation and with speed of rotation. Let us substitute the centrifugal acceleration ($r\omega^2$) for the gravitational acceleration (g) in the Stokes's formula explained in the previous section.

$$V_c = (D^2(d_p - d_c)/ 18\, \eta)\, r\omega^2$$

where, $V_c$ – sedimentation velocity in m/s; D – particle diameter in m; $d_p$ – particle density in kg/m³; $d_c$ – density of continuous phase (skim milk) in kg/m³; $\eta$ – viscosity of continuous phase in kg/m-s; r – distance from the axis of rotation or radius of the rotating bowl in m; and $\omega$ – angular velocity in m/s.

Let us now consider the separation efficiency using centrifugal force, of the same fat globule (5 microns diameter) discussed in the section above. Let us say it is rotating at a speed of n = 5600 and is placed at a distance 0.2 m from the axis of rotation.

$D = 5 \times 10^{-6}\,m$; $d_p - d_c = 980 - 1028 = 48$; $\eta = 1.42 \times 10^{-3}$ kg/m-s; r = 0.2 m;

angular velocity $\omega = 2\,\pi n/60$ radians/s,
where n = 5600 rpm (So, $\omega$ = 586.67 rad/s).
Substituting these values in the Stoke's formula, we get:

$$V_c = (25 \times 10^{-12} \times 48/18 \times 1.42 \times 10^{-3}) \times 0.2 \times 586.67^2$$

$= 6516506.64 \times 10^{-9}$ m/s or $23459423904 \times 10^{-9}$ m/h or 23459.42 mm/h.

Comparing this to the floatation velocity by the gravity separator $V_g$ obtained in the previous section, the centrifugal velocity is 23459.42/3.343 = 7017.47 (~ 7018) times faster.

The construction of a cream separator is similar to that of a clarifier, except for a few features. A greater space is provided outside the discs in the clarifier for gathering heavier material on the wall of the bowl. The clarifier has only one outlet for the milk, whereas in the separator there are two, one for the skim milk and the other for the cream. In the separator, the disc stack is equipped with vertically aligned distribution holes, while these are absent in the clarifier. The milk enters through vertically aligned distribution holes at outer edge while the discs (typically 80–200 in number) rotate around the axis at high speed (typically 10,000–18,000 rpm) and is subjected to centrifugal force. The heavier portion is forced towards the periphery and collected as skim milk to the space outside the disc stack, from where it travels through the channel between the top of the disc stack and conical hood of the separator bowl to a concentric skim milk outlet. The fat globules, being less dense, rise towards the inner edge of the discs towards the axis of rotation and are collected as cream separately.

The solid impurities (leucocytes, hair, udder cells, straw and bacteria), which are the heaviest, settle outwards towards the periphery of the separator and collect in the sediment space. This is called the separator sludge or the separator slime. The sediment space volume depends on the size of the separator and is typically 10–20 liters. The total amount of sediment may vary, but is typically ~ 1 kg for every 10,000 liters of milk. Solid retaining type separators need to be dismantled and cleaned to empty the sediment space at relatively frequent intervals. However, modern separators equipped with self-cleaning or solid-ejecting bowls for automatic ejection of accumulated sediment at 30–60 minutes interval, save time and labor. These separators have a sliding bottom, which is kept pushed up in place during separation, against a seal ring in the bowl hood by hydraulic pressure created by water below it. Draining this water results in the drop of pressure, causing the separator bottom to slide down and result in discharging the collected segment. The sediment is discharged from the frame by gravity either directly into the sewerage or a vessel. A fresh charge of water into the system increases the pressure and pushes the bottom up again. The sediment discharge operation can be carried out manually or by a sensor-operated automatic timer. It takes only about one tenth of a second and thus does not involve stopping the separator for cleaning.

## 12.4    STANDARDIZATION OF MILK

Standardization is the process of adjustment of the solids [fat and/or solid-not-fat (SNF)] contents of milk to pre-determined values. Whole (full fat) milk has to be standardized to meet the legal standards prescribed for fluid milk products such as standard (4.5% fat/8.5% SNF), toned (3% fat/8.5% SNF) and double toned (1.5% fat/ 9% SNF) milk or for the manufacture of various other milk products. Standardization is normally done by three different ways as follow.

1.  Pre-processing standardization in which whole milk is separated and standardized to pre-set composition before it is subjected to thermal processing (Figure 12.2).
2.  Post-processing standardization, where pasteurized whole milk is mixed with pasteurized cream or skim milk to adjust the fat and SNF.
3.  Direct standardization, where a calculated proportion of the cream is directed straight from the separator to the skim milk line using the cream/skim milk screws in order to adjust the fat content desired level.

Raw material available and their fat and SNF contents are assessed before mixing them in the correct proportions. The raw material may include whole milk, skim milk, cream, anhydrous milk fat, skim milk powder (SMP) or whole milk powder (WPM). Once the relative proportions have been determined based on their composition and the composition desired in the final product, the exact amount of each is mixed together to arrive at a certain weight of final product.

### 12.4.1    PEARSON'S SQUARE METHOD

If only a single component (fat or SNF) is to be standardized, the Pearson's Square method is the easiest to achieve this. Let us consider the example

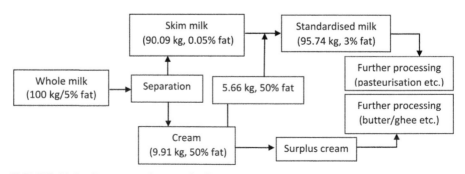

**FIGURE 12.2**    Pre-processing standardization.

where 100 kg whole milk (WM) containing 6% fat is to be standardized to toned milk (TM) with 3% fat by adding skim milk (SM) with 0.05% fat.

The fat percent required in the final product is placed at the center of an imaginary square. The known fat contents of WM (6.0%) and SM (0.05%) are placed at the upper and lower left hand corners. The numeral at the center (3.0) is subtracted from the numerals at the left hand corners (6.0 and 0.05) and the results (3.0 and 2.9) are placed diagonally opposite, ignoring the negative signs. The numerals on the right hand side now represent the proportion of each corresponding raw materials whose fat contents are on the left hand side of the square. Thus in this case, 2.95 is the proportion of WM that should be mixed with 3.0 parts of SM to yield 5.95 (3.0+2.95) parts of the standardized milk containing 3.0% fat.

The example indicates that if 2.95 kg of whole milk containing 6% fat is mixed with 3.0 kg of skim milk containing 0.1% fat, the resultant product will be 5.95 kg of standardized milk containing 3.0% fat. As the given quantity of WM is 100 kg, the proportions can be calculated as follows:

Quantity of SM = 100 x 3/2.95 = 101.7 kg
Total quantity of the product = 100+101.7 = 201.7 kg.

Proof:
*Raw materials:*
100 kg of 6.0% whole milk contains 100 x 6.0/100 = 6 kg fat
101.7 kg of 0.05% skim milk contains 101.7 x 0.05/100 = 0.05085 kg fat
Total fat in raw materials (6 + 0.05085) = 6.05085 kg fat ~ 6.051 kg fat.

*Product:*
201.7 kg of 3.0% standard milk contains 201.7 x 3.0/100 = 6.051 kg fat

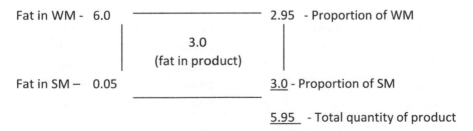

**FIGURE 12.3**   Pearson Square method of standardization.

Thus, the fat content in the input is equal to the fat content of the output, which validates the calculation. This is a batch process of standardization where the original components are first tested and then calculated amounts of these are mixed to obtain the resultant product. The final product so obtained is sent for further processing.

### 12.4.2 ALGEBRAIC METHOD

If both fat and SNF are to be standardized, the method is more complicated and involves algebraic equations. Let us do this with an example.

Whole milk (WM – 4% fat/8.3% SNF) is to be mixed with skim milk (SM – 0.1% fat/9% SNF) to obtain 100 kg toned milk (TM – 3% fat/8.5% SNF). Calculate the quantities of ingredients required.

Solution:

Let a and b kg of WM and SM be the respective quantities required.

Quantity equation: $a + b = 1001$        (1)

Fat equation: $4a/100 + 0.1b/100 = 100 \times 3/100$

$\Leftrightarrow 4a + 0.1b = 300$                    (2)

SNF equation: $8.3a/100 + 9b/100 = 100 \times 8.5/100$

$\Leftrightarrow 8.3a + 9b = 850$                    (3)

Solving Eqs. (2) and (3),$(4a + 0.1b = 300) \times 9$

$(8.3a + 9b = 850) \times 0.1$
$\Leftrightarrow 36a + 0.9b = 2700$
$\underline{0.83a + 0.9b = 85}$
$35.17a = 2615$ or $a = 74.35$
$b = 100 - 74.35 = 25.65$ kg
Proof:

| Ingredient | Quantity kg | Fat % | Fat kg | SNF % | SNF kg |
|------------|-------------|-------|--------|-------|--------|
| WM | 74.35 | 4 | 2.9740 | 8.3 | 6.17105 |
| SM | 25.65 | 0.1 | 0.02565 | 9 | 2.3085 |
| Total (input) | | | 2.99965 | | 8.47955 |
| TM (output) | 100 | 3 | 3.00 | 8.5 | 8.50 |

Hence, as the total of inputs is equal to output, the calculations are correct.

### 12.4.3   IN-LINE STANDARDIZATION

Large dairies handling huge volumes of milk require fast, correct and constant standardization methods irrespective of the fluctuations in the incoming raw milk. In-line standardization is the best method to achieve this. The amount of fat to be removed from a given quantity of milk is calculated and accordingly a known amount of preheated whole milk is separated. The cream having predetermined fat content is collected separately or diverted to cream pasteurizer. When required amount of fat is removed the cream flow is routed to remix with adequate amount of skim milk to give final product of the required fat content. The pressure in the skim milk outlet must be kept constant with the aid of the constant pressure valve located close to skim milk outlet, in order to achieve accurate standardization. Various control valves, flow and density meters with or without computerized control loops are used to adjust fat content of cream and milk to the desired levels. The variable parameters like fat content of incoming milk, milk throughput and preheating temperatures are measured and monitored to achieve precision in the process.

## 12.5   BACTOFUGATION

Bactofugation is the process of separating micro-organisms from milk using a specially designed centrifuge machine called bactofuge. The process is widely used in industries that manufacture fluid milk, cheese, infant foods and dried milk. Heat-resistant bacterial spores are separated from milk by centrifugation on the basis of their higher densities. Thus, the process can remove organisms that can survive thermal processing treatments also. Bactofugation leads to an 86–92% reduction of the total bacterial count and removes 94–98% of aerobic spores. The process is always combined with a thermal process and can be linked to existing pasteurization equipment. It can complement other heat processes like pasteurization, sterilization and thermisation to increase the shelf life of milk and products by two to five days in a simple and economical way.

The equipment comes in two forms. The single phase bactofuge has only one outlet at the top of the bowl for milk from which bacteria have been removed. The bactofugate (bacteria concentrate) is collected in the sludge space and discharged at preset intervals. The two-phase bactofuge has two outlets at the top, one for the bactofugate and the other for the reduced-bacteria-milk. The top disc is specially designed to ensure continuous and simultaneous discharge of bacteria-reduced milk and bactofugate through the separate outlets. The amount of bactofugate from two phase bactofuge is ~3%, while that from the single phase bactofugeis lower at ~0.2% of the feed. Owing to the losses of larger casein micelles during the process, the bactofugate has higher dry matter content than original milk. The optimal temperature of milk for bactofugation should be ~55–60°C, as higher temperatures lead to higher protein discharge in bactofugate.

## 12.6   HOMOGENIZATION

The fat in milk occurs in the form of globules of sizes varying from 0.1 to 20 microns. Owing to this wide variation in globule size and the lower density, the fat in milk rises to the top in the form of cream layer, when the milk is left undisturbed, rendering non-uniformity and unsightly appearance to the milk. Gaulin in 1899 invented the process of homogenization, a mechanical process wherein the fat globules in milk are broken up to smaller sizes that they do not rise to the surface to form cream layer. The process is a part of the standard global industrial protocol today. The milk is forced through a small passage at high velocity and pressure in order to disintegrate the large fat globules in milk and reduce their size to one micron or less in order to decrease their ability to aggregate. The newly formed fat globules, because of the larger surface area, lose their original membrane, which is replaced by which a mixture of proteins adsorbed from the serum phase of milk.

### 12.6.1   TYPES OF HOMOGENIZERS

The mechanical device used for the process is called a homogenizer. The machines available can be classified according to pressure developed or the number of stages employed for homogenization. There are basically three types of machines operating on the basis of pressure. In the 'no pressure' type, the homogenization effect is achieved by the creation of high velocity frequency by sonic vibration. The 'low pressure' machines used where the pressure required is relatively less, are of rotary type and provide partial homogenization. Milk is partially homogenized at a pressure of around 5–7 MPa in a rotary disc centrifuge, which is a good example of low pressure homogenization. The 'high pressure' machines are fitted with pistons that are connected through the crank shaft to the transmission rod. The piston runs in a high pressure block cylinder with a back pressure device. All modern high pressure homogenizers are examples of this. They operate at high pressures in the range of 15–40 MPa and homogenize the feed material completely.

Homogenizers can operate single, double or multiple stages, most universal being the single or two stage processes. Single stage process is used for homogenization of products with high fat content or products demanding high viscosity. On the other hand, the two stage machines are used primarily to break-up fat clusters in products with high fat content, where low viscosity is desired. Here, owing to shearing and cavitation forces created during the first stage, the fat globules break to small size. The second stage breaks the fat clusters into individual globules. The homogenization pressure $(P_1)$ is achieved at the first stage in both the processes. The back pressure $(P_2)$ is created by the process in single stage homogenization, while in two-stage homogenization the back pressure $(P_2)$ is created by the second stage. Maintaining $P_1/P_2$ at about five results in the maximum efficiency in modern homogenizers.

Two stage homogenizers are normally placed after the first regenerative section of the pasteurizer. During the ultra high temperature (UHT) treatment, however, they are always located downstream, that is, in the aseptic chain after the UHT. The homogenizers used in the UHT plant are of special aseptic design to avoid any post-processing contamination. In an indirect UHT plant, the homogenizer is generally located upstream. Downstream location is also ideal for processing high fat and high protein products because the machine helps to break the aggregated protein and fat molecules that form during the UHT process.

## 12.6.2  EFFICIENCY OF HOMOGENIZATION

The efficiency of the homogenization process is maximum, when the fat phase is in liquid state and when the concentration is normal. The homogenization temperatures and pressures applied depend on the product and usually range between 55 and 80°C and 10 to 25 MPa (100–250 bar) respectively. As fat is mostly in the solid phase and incompletely dispersed in cold milk (below 40°C), homogenization is ineffective at low and ambient temperatures. Excessive clumping makes it difficult to homogenize products with high fat content. One gram fat needs nearly 0.2 g of casein is needed as membrane material for good homogenization effect. For this reason, cream with more than 20% fat cannot be homogenized at high pressure, because clusters are formed as a result of lack of membrane material. At higher homogenization temperatures the viscosity of milk is lower and therefore, separation velocity is higher. It also improves the shifting of membrane material to the fat globules. However, raw milk may develop hydrolytic rancidity due to activation of inherent milk lipase system. It is therefore, necessary to immediately pasteurize the milk in order to inactivate the lipase enzyme. Operating costs of homogenization can be reduced by the partial stream method, where smaller volumes (only the cream with a small portion of skim milk) are homogenized and then mixed with rest of the skim milk. When milk is homogenized after standardization, the full stream has to be homogenized.

Milk is pumped at 60–65°C (or products at 75–80°C) through a small orifice at high velocity at a pressure ranging from 10–25 MPa depending on the product. The high pressure created by the piston sucks the milk from the supply tank and transports it to the gap between the seat of the pressure cylinder and piston. The seat has an angle of five degrees, helping the product to accelerate in a controlled way and increasing the velocity of the liquid to ~100–400 m/s in the narrow angular gap. Homogenization takes place in a fraction of a second. The pressure energy is converted into kinetic energy, a part of which is released as heat.

### 12.6.3 EVALUATING HOMOGENIZATION EFFICIENCY

The effectiveness of the homogenization process as well as the condition of the homogenizer may be judged by assessing the homogenization efficiency. The oldest method for determination of homogenization efficiency depends on the creaming rate, where the homogenized milk sample (1000 ml) is allowed to stand for a given period of time (say 48 hours). The fat content of the top100 ml sample is determined and compared with the fat content of the remaining sample. An efficient process should not have more variation than 10% in the fat content of top and bottom layers. The NIZO method also works on the same principle, where 25 ml of the homogenized milk sample is centrifuged at 1000 rpm for 30 minutes at 40°C at a radius of 250 mm.

NIZO value = (fat content of the bottom fraction/fat content of whole sample) x 100

Higher NIZO value indicates better efficiency. For pasteurized milk, NIZO value is normally 50–80%. The size distribution method uses a specially designed microscopic slide that has a built-in micrometer. A drop of homogenized milk is taken on the slide and viewed under the microscope. The size of fat globules can be read directly for the complete field. The size distribution of fat globules in homogenized milk can also be determined with the help of a laser diffraction unit. The light passing through the sample placed in a cuvette in the path of a laser beam scatters depending on the size and number of globules in the sample. A size distribution curve is drawn between percent fat distribution and globule size. Low pressure homogenized milk has fat globules distributed over a wide size range while the distribution shifts to a narrow range with increase in homogenization pressure.

## 12.6.4 EFFECTS OF HOMOGENIZATION ON MILK CHARACTERISTICS

### 12.6.4.1 Color and Flavor

Cow milk is creamy to translucent yellow in color, while buffalo milk is white to white with a slight greenish hue. Homogenization renders the color of milk less yellow and more opaque and white, owing to the increased number of fat globules and total surface area capable of reflecting and scattering the rays of light. Homogenized milk has a richer flavor, though it is more susceptible to UV light induced off-flavor.

### 12.6.4.2 Freezing Point

Homogenization lowers the freezing point, though insignificantly. If the process is followed by sterilization, the freezing point of milk lowers to a greater extent, than any one of the treatments taken singly.

### 12.6.4.3 Curd Tension

Homogenization induces physical changes in milk proteins and increases the number of fat globules that serve as weak points in the curd. Both these phenomena result in 50–55% lower curd tension that increases digestibility of homogenized milk and products made from such milk. Curd formed from homogenized milk is much more flocculent and is highly fragile.

### 12.6.4.4   Surface Tension

Homogenization of raw milk releases sufficient quantities of surface-active substances, which results in slight decrease in the surface tension. However, if the milk is pasteurized before homogenization, the surface tension increases slightly.

### 12.6.4.5   Viscosity

Homogenization of milk increases the viscosity of milk slightly. However, this may not always perceptible visibly. The quantum of increase depends on the extent of fat globule clusters that are present, the effect being more evident in high fat products, in which when fat forms clumps more easily.

### 12.6.4.6   Acidity/pH

Homogenization of raw milk decreases the pH if lipase enzyme is present. Unless the lipase is inactivated, the lipolytic activities continue to decrease the pH. Pre-heating of milk before homogenization inhibits lipase activity and prevents changes in acidity.

## 12.7   THERMAL PROCESSING

### 12.7.1   PRINCIPLES OF HEAT PROCESSING

The main purpose of heat treatment of milk is to destroy the spoilage and dis-ease-causing organisms and so as to ensure long product life and safety for human consumption. The destruction of micro-organisms and enzymes follow well-established first order reaction kinetics. The rate of destruction depends on the concentration and is referred to as a logarithmic order of inactivation. To assess this, a homogenous suspension of a well-defined species of microbial cells is kept at a pre-set temperature. Samples are drawn at regular interval to estimate the number of surviving cells. Different numbers of cells are destroyed at varying time intervals. By plotting a graph between log of the number of organism and time, the thermal destruction rate as a function of time can be calculated. The thermal destruction rate (TDR) may be defined as the number of cells destroyed in a given unit of time and calculated as given below.

$$TDR = (C_i - C_t)/t$$

where $C_i$ = initial concentration (number) of cells, $C_t$ = concentration of surviving cells at time t and t = time in seconds.

### 12.7.1.1   D Value

Also called decimal reduction time, this is the time required to kill 90% of the microorganisms or spores in a sample at a specified temperature. In other words, it is the time required to reduce the number of microorganisms by one log cycle at a given temperature. As the D value is the time required at a certain temperature to kill 90% of the organisms being studied, after an organism is reduced by 1 D, only 10% of the original organisms remain. The population number thus reduces by one decimal place in the counting scheme. D values are expressed with the temperature at which the study is done. That is, if an organism is reduced by 90% after exposure to temperatures of 130°C for 3 minutes, then the D value would be written as $D130_C$ = 3 minutes. Absolute sterility in any food product is possible only theoretically.

### 12.7.1.2   Z Value

Z-value of an organism is the temperature required for the thermal destruction curve to move one log cycle. Expressed more simply, Z value denotes the temperature change that will bring about a 10-fold change in D value. D value represents the time needed at a certain temperature to kill an organism, the Z value indicates the resistance of an organism to different temperatures.

### 12.7.1.3   $Q_{10}$ Value

Q value indicates how much more rapidly a reaction proceeds at temperature $(t_2)$ than at a lower temperature $(t_1)$. When the value of $(t_2 - t1)$ is 10, Q reflects the change in rate for 10°C rise in temperature, and is termed $Q_{10}$. Z value and $Q_{10}$ value are interrelated and is given by the expression $Z = 18/\log Q_{10}$.

### 12.7.1.4   F Value

F value relates to the sterilization of articles by moist heat and is defined as the time in minutes at a specific reference temperature needed to kill a population of cells or spores. A relation between F and D values is expressed by the equation $F = D(\log N_0 - \log N)$, where D is the D value at 121°C and $N_0$ and N are the respective initial and final number of viable cells per unit volume.

The term thermal processing or 'heat treatment' stresses on the temperature-time combinations to which a product is exposed. The product is held at a certain preset temperature for a preset period of time in order to achieve the desired effect. A product would require longer exposure to a lower temperature than to a higher one to achieve the same effect. The processes are termed low-temperature-long-time (LTLT) or high-temperature-short-time (HTST), as per the norms followed. Relatively mild heat treatments kill most common pathogenic organisms that may appear in milk. Earlier, the most resistant organism known was *Mycobacterium tuberculosis* that causes tuberculosis (TB) in human beings and was made the index organism to achieve complete safety of milk. More recently, *Coxiellaburnetti* that causes Q fever was found to be more heat resistant and is now the index organism. Any heat treatment that which destroys this organism is intense enough to destroy all other pathogens in milk. The thermal death of pathogens such as *Mycobacterium tuberculosis, Coxiellaburnetti, Salmonella typhii* and coliforms bacteria is applied as the basis for time-temperature combinations used for thermal processing of milk by various methods. The quality aspects being of paramount importance, the retention of nutrients and other quality attributes is also considered while choosing the process.

Milk being a complex biological fluid carrying several species and varying proportions of microorganisms with different thermal resistance characteristics, in order to achieve sterility the product must be subjected to thermal conditions that are capable of destroying the most resistant organism. Although high bacterial contamination would require more severe thermal conditions, these would also impair the organoleptic properties of milk. As this requires adoption of less severe conditions, absolute sterility of milk is practically impossible to achieve. Therefore, some retail packages of milk may contain a few surviving spores, while other packages may be sterile. Based on the approach of statistical probability of percentage of sterile packages, various thermal processes have been designed with respect to time–temperature combinations to attain near sterility in the thermal processed product. The choice of time-temperature combination, therefore, depends on balancing the microbiological and quality aspects in processed milk. Table 12.2 lists the various heat treatment processes carried out in the dairy process industry, some of which are discussed in detail.

## 12.7.2 PASTEURIZATION

The term 'pasteurization' is attributed to the French inventor Louis Pasteur. It is the process of heating every particle of milk or milk product, in properly designed and operated equipment to a specified temperature and holding at that temperature for a specified period of time followed by immediate cooling and storing at low temperatures. Some pasteurization systems and equipment that have been employed are described hereunder.

### 12.7.2.1 Flash Pasteurization

These were used in early days, where, the product was forced in a thin layer over the heating surface or moved product between two heated surfaces. Milk is heated between two concentric water heated cylinders to produce pasteurized product. A revolving arm or special ridge inside the inner cylinder helps pushes he milk forward. Milk can be heated to as high as 85°C without holding. The outgoing hot product cooled in a heat exchanger.

**TABLE 12.2** Time-Temperature Combinations of Selected Dairy Processes

| Process | Temperature (°C) | Time(s) |
| --- | --- | --- |
| LTLT pasteurization (milk) | 63 | 1800 |
| HTST pasteurization (milk) | 72 | 15–20 |
| HTST pasteurization (cream) | > 80 | 1–5 |
| Thermisation | 57–68 | 15 |
| Ultra-Pasteurization | 115–130 | 2–4 |
| In-container sterilization | 115–121 | 1200–1800 |
| UHT sterilization | 135–150 | 1–6 |

## 12.7.2.2 Tubular Heater

Tubular heaters comprise of tubes, singly or in multiple units, which are used for heating, holding and cooling. They come as individually jacketed tubes and group jacketed tubes. Water is heated externally and pumped through the unit between two tubes. The circulation is in the counter current flow. More efficient heat transfer can be achieved with three concentric tubes where the product runs through the middle tube and the heating medium passes from outer and inner most tube, all in counter current flow.

## 12.7.2.3 Batch Process

The milk is heated indirectly to 63°C and held at this temperature for 30 minutes in a double jacketed vat. The process is called Low-Temperature-Long-Time (LTLT) method. Heating and cooling is done by spraying/ circulating hot water/ steam or chilled water in the vessel jacket. Uniform heating is ensured by gentle agitation by a mechanical agitator fitted on the vat. This method requires low initial cost of equipment and is suitable for small quantities ranging from 200–1000 liters.

## 12.7.2.4 Continuous Process

This process uses the plate heat exchanger, where stainless steel plates are mounted on a frame. The milk passes through different sections of the equipment to achieve heating, holding and cooling in a continuous fashion. Each section consists of varying numbers of plates depending on equipment capacity. The temperature time combination involved is high temperature and short time. High-Temperature-Short-Time (HTST) treatment for pasteurization of milk involves heating every particle of milk in a continuous flow to a minimum of 72°C for a period of at least 15 sec followed by cooling to 4°C. The process is fully automated and is ideal for handling 5000–20,000 liters per hour. The heating/cooling medium and the product flow through alternate plates. The plates are so designed as to prevent mixing of thin channels of product and heating/ cooling medium by separating the plates with rubber gaskets.

The pasteurizer design facilitates heating the incoming cold raw milk with the hot outgoing pasteurized milk, in order to save energy. To begin with, the raw cold milk (4–5°C) from the storage tank is routed to the float controlled balance tank (FCBT) of the pasteurizer assembly. It is then pumped into the regeneration section, where hot pasteurized milk (72°C) flows counter current to the raw cold milk within adjacent plates, thereby, transferring heat for pre-heating of raw milk and pre-cooling of pasteurized milk resulting in energy saving. The outgoing pasteurized pre-cooled milk enters the chilling section to be chilled to 4°C before storage. The pre-heated milk enters the heating section where it is heated to a temperature of 72°C using hot water or steam. It then passes to holding section where the temperature of milk is maintained for a specified period of time (15 sec). A flow diversion valve placed at the outlet of holding section operates on the basis of temperature sensors. All pasteurized milk is diverted forward, while under-pasteurized milk is returned to balance tank. The pasteurized milk passes to the regeneration section (as explained above) followed by cooling section where it is chilled using chilled water or glycol solution as a coolant. The cream separator and the homogenizer can be synchronized with the pasteurizer. The milk from the regeneration section 1 at a temperature of 45–50°C is sent to the separator, where skimming or standardization is done. The resultant milk is then sent to the regeneration Section 12.2, where milk is heated to 60–65°C and sent to the homogenizer. The homogenized milk enters the heating section, where it is pasteurized and subjected to further holding, precooling and chilling operations. The chilled milk is sent to the storage tanks to the filling stations.

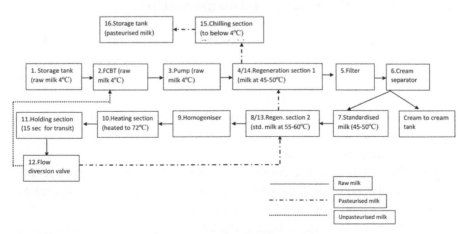

**FIGURE 12.4**  Schematic flow diagram of an HTST pasteurizer.

### 12.7.3   THERMIZATION

Milk is heated to below pasteurization temperatures to temporarily inhibit bacterial growth. The process is useful where it is not possible to immediately pasteurize all the milk and some of the milk needs to be stored for hours/days before further processing. The milk is heated to 63–65°C for 15 sec and rapidly chilled to 4°C or below. This prevents the growth of aerobic spore forming bacteria. These bacteria are destroyed upon subsequent pasteurization.

#### 12.7.3.1   Ultra-Pasteurization

Ultra-pasteurization extends the shelf life of the milk by 15–30 days by reducing re-contamination of the product during processing and packaging. Milk is heated to 125–138°C for 2–4 seconds and cooled below 4°C. As the treatment leads to extending the shelf life of the products, the process is also Extended Shelf Life (ESL) process and the products, ESL products. Very high level maintenance of hygiene and cold chain during production, storage and distribution is absolutely essential. Ultra-pasteurized products are packed in pre-sterilized containers aseptically and stored under refrigeration to extend the shelf life.

#### 12.7.4   Ultra-High Temperature Treatment (UHT)

The UHT process is applied to a product in continuous flow with the objective of destroying all micro-organisms by a very high-temperature-short-time heat treatment which also facilitates minimum chemical and physical changes in the product. The heating temperature normally ranges from 135–150°C for 1–6 seconds. UHT sterilization is achieved in a closed system to prevent airborne contamination. The product passes through heating and cooling stages in quick succession followed by aseptic filling. UHT treated products are packed aseptically in specially designed multilayer containers and can be stored at room temperature for extended period of time (2–6 months) without bacterial growth. The basis of the process is rapid heat transfer and rapid cooling, to achieve a high sterilizing effect with minimal damage to physical, chemical and nutritional qualities of the product. There are two methods of UHT treatment of products.

## 12.7.4.1   Direct Heating

Direct heating involves direct contact between heating medium and the product. The two prevalent systems to achieve this are: (a) the injection system where the steam is injected directly into the product to achieve heating at high temperatures; and (b) the infusion system, where the product is introduced or infused into the steam. The quality of the steam is important, as it comes in direct contact with the product. Use of potable water for steam generation, stainless steel piping from steam generator and heating equipment and using active carbon filter to purify the steam help to achieve high quality of superheated steam. The direct system has the following advantages.

1. Rapid heating and cooling and therefore, a low total temperature load on product.
2. Reduced deposit formation in the equipment, hence requires less frequent cleaning.
3. Low oxygen content of final product owing to removal of dissolved gases during expansion cooling.
4. More suitable for viscous products due to better resistance to pressure drops.
5. No effect on the organoleptic quality of the product.

### 12.7.4.1.1   Injection system

The product is pumped from the balance tank to a conventional tubular or plate heat exchanger, where the temperature of the product is raised to 75–80°C. It then passes to the holding cell where steam is injected into the product raising the temperature rapidly to 140–150°C with a residence time of 2–4 seconds. The rapid heat transfer occurs primarily due to condensation of steam. The heated (and diluted, owing to steam injection) product is routed to a vacuum chamber, where it is cooled through evaporation cooling. The vacuum also aids in removing steam condensate that has been mixed with the product and diluted it during heating. The sterile product is homogenized aseptically, cooled and packaged in an aseptic packaging system.

### 12.7.4.1.2  Infusion system

The product is product is infused into the steam chamber after being pre-
heated in two tubular heat exchangers. The heating chamber is kept under
constant steam pressure. The product flows from top to bottom of the pressure chamber and is brought to the sterilization temperature during this tran-
sit. It then passes through the holding cell and is routed to vacuum chamber
for removal of the steam condensate. The product is homogenized, cooled
and packaged aseptically as in the indirect system.

### 12.7.4.1.3  Indirect heating

The product is heated indirectly through the partitioning between the product
and heating medium, as is done in the pasteurizer. The heating temperatures are
slightly lower than the direct system of heating. Indirect heating during UHT has
the following advantages:
1.   No special quality of steam required.
2.   Low investment cost.
3.   Regenerative energy recovery up to 85–90%.
4.   Relatively low maintenance cost.
The dairy industry uses the plate and the tubular types of heat exchangers.

### 12.7.4.1.4  Plate heat exchanger (PHE)

The equipment is similar to the HTST pasteurizer and consists of a number of
partitioning plates fitted in a frame to demarcate various sections. Milk pumped
into the first heat exchanger is heated to ~ 75–80°C. It is then held in the holding
cell, where the temperature is maintained at 80°C for 30 seconds to 5 minutes,
depending on the design. Holding reduces deposit formation in the final heating
section. Milk then is routed to the homogenizer, which also aids to pump it into
the heating section where it is heated to 138–142°C, followed by holding in the
holding cell for 4–6 seconds. The product is cooled in the final cooler (vacuum
chamber) before leaving the plant. A thermosensor regulates the temperature
in the heating, by controlling the steam supply. If the temperature drops below
a preset value, the flow diversion valve diverts the flow of product the balance
tank through a separate cooling section. The advantages of the PHE are:

1. Better regenerative energy efficiency.
2. Easier dismantling for inspection
3. Easier to clean
4. Lower investment cost

### 12.7.4.1.5 Tubular heat exchanger

Tubular sterilizing equipments have undergone design modifications over the years. In the oldest type, the product is pushed through a high pressure piston pump to the preheating section and then to the first homogenizer head (upstream homogenization) and then to second regenerative preheater. It is heated to 100°C before entering into final heating section, from which it is back for cooling to 75°C in the regenerating section. It finally passes to the second homogenizer head (downstream homogenization) and then to final cooling and filling stations.

The second generation equipment had the final heater in the shape of a concentric double tube, where the product passes through the center tube and steam through the outer tube. After final heating, the product passes through the variable speed homogenizer, which permits a number of aseptic packaging machines to be operated by the feed from a single sterilizer. However, if the packaging machine is to work independently of the sterilizer capacity, the product from the homogenizer will have to be stored in an aseptic tank before it goes to the filling head.

The third generation sterilizer has a triple tube heater where the product passes through the middle tube and steam in outer and center tube, counter current to one another. The design allows rapid heating, energy saving and lower temperature gradient between product and heating medium.

The advantages of the tubular heaters are:
1. Better microbiological safety.
2. Low maintenance cost owing to fewer gaskets.
3. Tolerance for high pressure drops, indicating that viscous products can also be processed.

### 12.7.5 IN-CONTAINER STERILIZATION

The product is packed in clean containers and subjected to high enough temperatures (115–120°C) for long enough duration (20–30 minutes) in order to achieve microbial and enzyme destruction. Normally, glass bottles are used for milk and tin cans (200–400 g) for evaporated or sweetened condensed milk. A rotary autoclave is used for the batch process, whereas a hydrostatic tower is applied for continuous production. Sterilization increases viscosity and improves the body to provide smoother and creamier consistency to the finished product.

### 12.7.5.1   Batch Sterilizers

These consist of a large vertical or horizontal hollow drum which has openings at the top, one end or both ends. Removable single or multi-stage frame facilitate loading trays containing the packed product. A perforated steam distribution pipe having inlet at both ends extends over the entire length of the sterilizer drum. The water distribution pipe is located near the top end. There is a drain tap to remove the condensate at the bottom of the shell. The sterilizer is equipped with a pressure safety valve, water and steam gauges and a high temperature thermometer. Batch sterilizers may be of the stationary or the rotary type. In the stationary sterilizer the packed product is loaded onto the removable frame/trays and placed into the sterilizer drum. The lid is properly sealed using the kernel nuts. The sterilizer shell is filled with adequate quantity of water and steam is allowed to pass. Once the predetermined pressure is achieved, the steam inflow is controlled to maintain the pressure for the preset period of time. The time-temperature of sterilization should be such as to ensure sterility to the product without causing discoloration and cooked flavor. The heat transfer is slow in the stationary sterilizers, leading to cooked flavor in the product. The rotary sterilizer has a revolving frame which is loaded with packed product. The sterilizer rotates at a speed of 6–12 rpm during the sterilization process, which ensures rapid and uniform heat distribution throughout the product body.

### 12.7.5.2   Continuous Sterilizers

Continuous sterilizing equipment has three sections consisting of preheating, sterilization and cooling. A conveyor carries the packed product through a hot water chamber where the temperature of water is maintained to few degrees below boiling point. The product moves onto the main sterilizing section where it is heated to sterilizing temperature by high pressure steam. Finally, the product is first pre-cooled with tap water and then chilled using cold water. The ideal conditions adopted for successful sterilization are 15–20 minutes coming up time, 118–120°C heating temperature, 15–18 minutes holding time and 10–15 minutes cooling time.

## 12.8   EVAPORATION

Evaporation and drying are the two important unit operations of dairy processing primarily aimed at removal of bulk of the water from milk. The foremost objective is to increase the shelf life of milk by decreasing the water activity. Further volume of milk is also reduced so that less space is required for transportation and storage.

Evaporation is a surface phenomenon which occurs when water molecules present in a solution obtain enough energy to escape in the form of vapors. Boiling is a phenomenon which occurs when the vapor pressure of liquid equals the pressure of the surrounding air. If the pressure above the liquid is lowered, the vapor pressure of liquid at boiling is reduced and thus boiling takes place at low temperature. This principle of boiling at low temperature by creating vacuum is utilized for milk evaporation at temperature lesser than the boiling point of milk. This prevents damage to nutrients due to severe heat treatment. Equipment used for removal of water by evaporation is called evaporator.

Water is removed from milk either for milk concentration or to be used for subsequent drying. Milk is usually condensed from an initial total solid of 9–13% to a final total solid up to 40–45% before the product is fed to the inlet of the drier. Evaporators may be single effect or multiple effect. Single effect evaporators are often called as vacuum pan. Vacuum pan consist of five major parts: (i) heating surface, (ii) vapor space, (iii) entrainment separator, (iv) condenser, and (v) vacuum pump and accessories. In a vacuum pan, a heater and vapor separator are combined into a single body and attached to a condenser. The condenser continuously removes vapor by condensing and thus maintains the vacuum. A positive pump of reciprocating type is used as a vacuum pump to produce vacuum.

### 12.8.1   CLASSIFICATION OF EVAPORATORS

Evaporators can be classified on the basis of criteria such as source of heat to evaporate water, position of the heating tubes in the heater section, type of heat exchanger used to exchange heat between heating medium and the product etc. Detailed classification is given in Table 12.3.

Falling film multiple effect evaporators are most commonly used for milk evaporation. In multiple effect evaporators, a multiple set of heater

**TABLE 12.3**   Classification of Evaporators

| Criteria of classification | Evaporator types |
| --- | --- |
| Source of heat | Steam, direct fired, solar |
| Position of heating tubes | Horizontal, vertical, inclined |
| Method of product circulation | Natural, forced |
| Length of tubes | Long, short, medium |
| Direction of flow of milk film | Falling film, rising film |
| Type of heat exchanger | Plate, tubular |
| Shape of heating tube assembly | Coil, basket, straight |
| Location of steam | Inside the tube, outside the tube |
| Number of effects | Single effect, multiple effect |

and vapor separator (collectively called as calandria) are used in such a way that vapor produced in the first calandria are used as heating medium for the second calandria which is at lower temperature (Figure 12.5). Likewise, the vapors removed from the second calandria are directed to the third

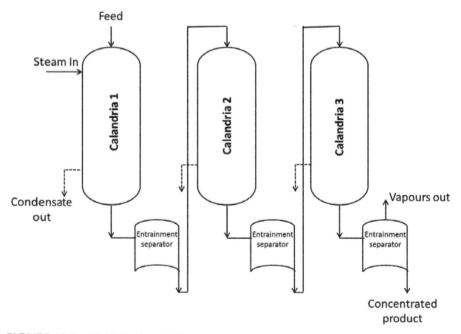

**FIGURE 12.5**   Multiple (three) effect evaporator.

calandria and so on. The temperature must decrease from the first effect towards the following effects. Thus, the corresponding decrease in boiling point is achieved by maintaining a higher vacuum. The major advantage of multiple effect evaporator is that it requires less steam to evaporate a unit of water from milk. Vapors remove from one effect are often recompressed to increase the temperature before feeding to the second effect. This is termed as vapor recompression and can be achieved either by mechanical compression or thermal steam jet which are called as mechanical vapor recompression (MVR) and thermal vapor recompression (TVR), respectively. Vapor recompressors together with multiple effect increase the energy efficiency of evaporators significantly.

Evaporators are used to manufacture sweetened condensed milk (SCM), evaporated milk and other concentrated milks. For manufacture of these products volume is reduced by evaporation and the concentrate is subjected to in-can heat sterilization. It is therefore essential to have colloidal stability during manufacture and storage. Colloidal stability of milk to heat, generally termed as heat stability, has a tremendous importance in manufacturing condensed milk. Heat stability of milk can be improved by pre-heating, homogenization or by addition of stabilizing salts.

## 12.9   DEHYDRATION

Milk or milk products are dried with the objectives of increasing the shelf life and to reduce transportation cost and storage space. The only difference in evaporation and drying is the degree of concentration. Dried milk products include whole milk powder (WMP), skim milk powder (SMP) whey powder, whey protein concentrates (WPC), whey protein isolates (WPI), casein, caseinates, coprecipitates, infant milk formula and dairy whitener.

### 12.9.1   METHODS OF DRYING

Spray drying, roller drying, fluidized bed drying (FBD), freeze drying and microwave drying are some of the methods utilized for drying of milk and milk products. Among these, spray drying is the most common and widely used method.

### 12.9.1.1   Roller Drying

Steam heated metal drums are used for roller drying. Pre-concentrated milk is applied in a thin film over the surface of these drums. The film dries rapidly due to the hot surface and the dried film is scrapped off by a stationary knife. This method of drying offers advantages such as low initial investment, less space and easy operation. However, due to severe heat treatment loss of nutrients takes place and product quality is affected. Roller drying is usually done under atmospheric pressure to avoid excessive nutrient loss due to heat. It may also be carried out under vacuum. In vacuum drum drying, the drums are enclosed and operated in a vacuum chamber.

### 12.9.1.2   Spray Drying

Spray drying is the most widely used method for milk dehydration. The principle of spray drying involves conversion of milk to a fine mist which is exposed to heated air. The vast surface area exposed to hot air accelerates the rate of evaporation and milk is dried instantaneously. For producing fog like mist of milk, atomizers are used. Type of atomizers used for spray drying are compressed air atomizers, pressure nozzle and centrifugal atomizers. Evaporation of water takes place at a lower temperature (40–50°C) and thus causes less heat damage to the product.

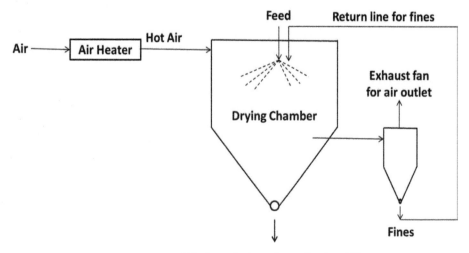

**FIGURE 12.6**   Working of a spray dryer.

The working of the spray dryer is shown in Figure 12.6. Inlet filtered air is heated in an air heater and fed into the drying chamber. This hot air mixes with the mist of milk produced by atomizers. This enables removal of moisture from milk and the dried milk particles settle at the bottom of the drying chamber. Fine particles suspended in outlet air are directed towards a fine recovery system. Most widely used systems for recovery of fines include cyclone separators, liquid dust collectors and bag filters. Fines collected from drying chamber and fluidized bed dryer (FBD) are separated from outlet air through fine recovery system and redirected from the top into the drying chamber. There exists an outlet exhaust air fan which sucks the air used for drying from the chamber. Thus, a slight vacuum is created in the drying chamber which ensures prevention of escape of hot air and product through leakages.

## 12.10   MEMBRANE SEPARATION

There exists a wide range of separation processes and finds vast applications in the area of dairy and food processing. The simplest example of separation process is the formation of cream layer when the milk is kept undisturbed for a long time. Separation of cream from milk by centrifugal separation is a mechanical separation process based on centrifugal force. Membrane separation or filtration falls under the broad category of molecular separation.

Conventional dead end filtration and membrane filtration are based on the principle of separation on the basis of molecular size. The distinguishing

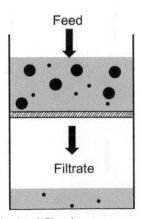

**FIGURE 12.7**   Conventional dead end filtration.

feature between the two processes is the direction of the flow of feed to that of the membrane. As depicted in Figure 12.7, the membrane is placed perpendicular to the direction of feed flow. This results in choking of the membrane due to clogging of particles on the surface of the membrane. Thus the filtration process and flux rate are affected. In contrast to the conventional dead end filtration, membrane filtration employs cross flow, that is, the direction of feed flow is parallel to the membrane surface (Figure 12.8). This prevents frequent clogging of membrane as the retained particles are continuously washed with the flow of feed stream thereby allowing longer operating times without cleaning as compared with traditional filtration.

## 12.10.1  TYPES OF MEMBRANE FILTRATION PROCESSES

Membrane filtration is a term used for processes of microfiltration (MF), ultrafiltration (UF), nanofiltration (NF) and reverse osmosis (RO). These processes are characterized by their capability of separating molecules of different sizes and characteristics. All membrane filtration processes are pressure driven and separation is based on the molecular size. Figure 12.9 depicts the order of the membrane filtration processes with respect to the size of the molecules retained by the process, pore size of the membrane and the pressure used in the process.

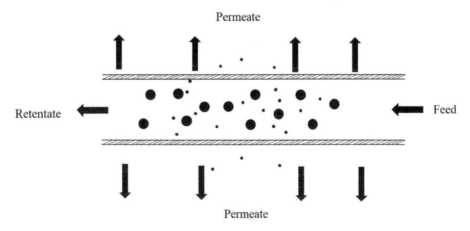

**FIGURE 12.8**   Membrane separation showing cross flow filtration.

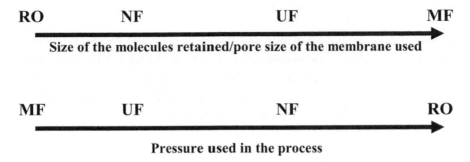

FIGURE 12.9 Order of the membrane filtration processes with respect to various characteristics.

### 12.10.1.1 Microfiltration (MF)

MF employs the membrane with most open structure typically of the pore size in the range from 0.1 to 10 μm. This allows to keep low trans-membrane pressure of 1 bar or less. The size of particles to be removed is normally between 0.1 and 10 μm. Applications for MF include fat, micro-organisms and suspended solids. MF of milk is playing an increasing role in the dairy industry, where it is used to improve the quality of milk for cheese-making and to extend the shelf life of market milk. MF has also been used to fractionate casein and to produce micellar casein.

### 12.10.1.2 Ultrafiltration (UF)

UF process utilizes a membrane with pore size lower than MF but higher than NF. The pore size vary in the range of 0.01 to 0.1 μm and the operation is carried out at a pressure of 1 to 10 bar. This membrane filtration process allows salts, organic acid, sugars and water to pass through the membrane while larger molecules including protein, fat etc. are retained fully or partly depending on the specific pore size of the membrane.

UF finds wide applications in dairy industry. Separation of whey proteins and lactose from whey is carried out using UF to manufacture whey protein concentrates from the retentate portion and lactose powder from the permeate stream. Other applications of UF include cheese-making, milk protein fractionation, and production of milk protein concentrate (MPC).

### 12.10.1.3   Nanofiltration (NF)

NF allows only monovalent ions and water to pass and retains polyvalent ions other than all the macromolecules present in the feed. NF works in the pressure range of 5–35 bars and the pore size of the membrane is less than 0.001 μm. In dairy industry, NF is usually used for demineralization of whey and also to purify and reuse the caustic cleaning solutions.

### 12.10.1.4   Reverse Osmosis (RO)

RO is a process wherein the osmotic flow, that is, the flow of solvent towards higher solute concentration is reversed by application of excess pressure. This results in reverse flow across the semi-permeable membrane and thus used to remove water from higher solute concentration to lower solute concentration. RO retains all compounds and allows only water to pass through the membrane. Thus, it can be either used for concentration of solutes (retentate stream) or for producing pure water (permeate stream).

The pore size of RO membranes are kept less than 0.01 μm and the pressure is kept in the range of 10–100 bar. RO is used to concentrate liquids and in dairy industry also, it has been used for evaporation of milk before manufacturing dairy products with lesser water contents such as evaporated milk, *khoa*, etc. It is also being widely used for purification of water.

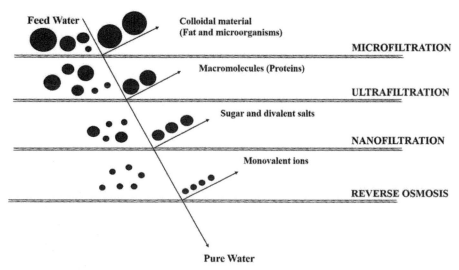

**FIGURE 12.10**   Characteristics of membrane filtration processes.

The separation ability of different membrane processes are summarized and presented in Figure 12.10 while the applications of these processes in dairy industry are summarized in Table 12.4.

## 12.11 HIGH PRESSURE PROCESSING (HPP)

High pressure processing (100–1000 MPa) is one of the most promising methods for the food processing and preservation at room temperature (Cheftel, 1995). It is also considered as "Cold processing" technique, since it involves ambient temperature for most of the food processing operations. During this treatment, food products are exposed to the pressure as high as 6000 times the atmospheric pressure which usually ranges between 300–700 MPa. Above 400 MPa, the treatment can successfully inactivate most of the vegetative bacteria. The process involves uniform application of pressure throughout the food material independent of its mass and time required for processing. This non-thermal processing technique could be applied to both liquid and solid (water-containing) foods. Operation of equipment involves filling up of pressure vessel with food product followed by application of pressure on the filled vessel for specific duration of time and finally depressurization of vessel. In most of the occasions, this processing operation uses water as pressure transmitting medium. When the food material filled vessel is pressurized for specific period of time, there is a small adiabatic rise in temperature (Myllymfiki, 1996) which results in heating up of food material.

HPP has been proved to be a successful technique for dairy applications. Its application in dairy science and technology sector involves milk homogenization (O'Reilly et al., 2000), milk fat globule size reduction, whey proteins denaturation, inactivation of some enzymes such as plasmin (Trujillo et al., 2002), inactivation of pathogens and spoilage microorganisms from milk, shelf life extension of milk (O'Reilly et al., 2000), enhancement of milk coagulation, increased yield of cheese and increased moisture retention in fresh cheese, with minimum changes in other properties important for cheese making (Lopez-Fandino et al., 1996). This treatment gives raw milk quality comparable to pasteurized milk when milk is pressurized within 400–600 MPa, since the treatment is equally effective in destroying the pathogenic and spoilage microorganisms.

Casein micelles have shown to be disintegrated into smaller particles when subjected to HPP which resulted in increased levels of casein micelle

**TABLE 12.4**   Applications of Membrane Filtration Processes in Dairy Industry

| Milk | |
|---|---|
| **RO** | • Concentration of whole milk |
| | • Concentration of milk for making *khoa* and *shrikhand* |
| **NF** | • Reducing the mineral content |
| | • Improves the heat stability of milk by reducing soluble calcium content |
| | • Salts reduction in milk used for making infant foods |
| **UF** | • Protein standardization |
| | • High protein or high calcium milk |
| | • Milk protein concentrates |
| | • Low lactose powder |
| | • Non-dairy whitener |
| | • For making insoluble and soluble components of milk |
| | • Preparation of biological peptides |
| | • Different varieties of cheese like Cheddar, Camembert, Cottage, cream, Feta, Quarg and Mozzarella |
| | • Cheese base, Cheese powder |
| | • Yoghurt |
| | • *Chhana*, *rasogolla* mix powder, long life UF-paneer, *shrikhand, dahi* |
| **MF** | • Production of pasteurized milk with extended shelf life |
| | • Fractionation of macromolecules in milk |
| | • Phosphocaseinate separation |
| Whey | |
| **RO** | • Concentration of whey |
| | • Lactose manufacturing |
| | • Concentration of UF permeate |
| **NF** | • Demineralization of whey |
| | • Fractionation of whey-protein hydrolysates for making bioactive peptides |
| | • Separation of milk oligosaccharides |
| **UF** | • Pre-treatment of whey to increase the UF permeate flux rate |
| | • Whey Protein Concentrates (WPC) |
| | • Whey protein fraction |
| | • Use of permeate for making di-calcium phosphate |
| | • Lactic acid production from whey permeate |
| | • Bio-active peptides |
| **MF** | • De-lipidization of whey protein concentrates |
| | • Phospholipids enriched fraction |

and soluble calcium phosphate content in the serum phase of milk (Law et al., 1998). This high pressure induced disruption of casein micelle resulted in increased light transmission in skimmed milk (Kromkamp et al., 1996). The re-association of casein micelles exposed to pressures above 280 MPa, with resultant micellar diameter of around 25 nm was reported by Huppertz et al. (2004). Knudsen and Skibsted (2010) suggested the formation of large number of smaller micelles on application of pressure in the range of 150–300 MPa, which get attached to the surface of larger micelles and gave appearance of perfectly spherical body with smooth and well-defined surface. HPP treatment has been reported to increase the casein micelles hydration and it is attributed to the role of denatured β-lg which forms association with casein micelles consequently favoring the micellar solvation due to an increase in the net-negative charge on the micelles (Huppertz et al., 2004).

Application of pressures >400 or >100 MPa on whey proteins denatures the most abundant, α-lactalbumin (α-la) and β-lactoglobulin (β-1g), respectively. It has been reported that β-Lg exhibits better proteolytic digestibility when it is hydrolyzed under high hydrostatic pressure. Efforts have been made to produce whey protein hydrolysates having no allergenicity by applying HPP treatment (Chicón et al., 2009). Bovine Serum Albumin (BSA) has no effect of HPP treatment possibly due to its 17 intra-molecular disulfide bonds which makes it extremely rigid molecular structure which remains largely unaffected under high pressure. Milk fat does not get affected by HPP up to 400 MPa at ambient temperatures which is attributed to the fact that phase transition of milk fat from solid to liquid is shifted to higher values under pressure (15.5°C/100 MPa). Application of high pressure (>350 MPa) also leads to reduced molecular mobility of milk fat which may results in lower degree of crystallization (Huppertz et al., 2002).

The high pressure treatments may also leads to solubilization of indigenous calcium and heat precipitated colloidal calcium phosphate which has shown to increase the levels of diffusible calcium in milk (Schrader et al., 1997). Milk enzymes are inactivated depending on their structure and processing conditions. Rademacher et al. (1999) studied the effect of high pressure-induced inactivation of the indigenous milk enzymes alkaline phosphatase (ALP), γ-glutamyltransferase (GGT) and phosphohexoseisomerase (PHI) in the pressure range 400–800 MPa at temperatures between 5–40°C and found that alkaline phosphatase (ALP) was most resistant followed by γ-glutamyltransferase (GGT) and phosphohexos-eisomerase (PHI).

Dairy products like yogurt prepared with high pressure treated milk was reported to have minor degree of syneresis, higher firmness and lower titratable acidity. It was due to the pressurization effects, which lowered the whey proteins denaturation (α-la and β-lg) and increased the dispersion of fat content into the matrix (Serra et al., 2009). Post-acidification of the yogurt during chilled storage was prevented by employing pressures over 200 MPa (Ancos et al., 2000). This study was suggested as the alternative to the use of additives for extending the shelf life of yogurt with the added advantages of improving flavor and texture of product.

For making ice-cream, HPP treatment has provided advantages by inducing early fat crystallization, shortening the time required to achieve a desirable solid fat content and thus reducing the aging time (Buchheim and Frede, 1996). It has also been reported to enhance the physical ripen-ing of cream for butter making (Buchheim and Frede, 1996).

In cheese-making, applications of HPP involves inactivation or reduc-tion of pathogenic and spoilage microorganisms, increased cheese yield and acceleration of cheese ripening. This treatment extends the shelf life of the cheese during storage, since the growth of yeasts, molds and *mesophilic* bacteria is reduced considerably in milk. High pressure treat-ment of milk prior to the manufacture of Camembert cheese resulted in higher moisture content of cheese with altered ripening pattern though having acceptable sensory quality (Sandra et al., 2004). The increase in moisture content was the result of incorporation of whey proteins into the cheese curd. The rennet coagulation process was delayed due to the interaction of the denatured whey proteins with the casein micelle, and furthermore the syneresis hindered, which results in increased moisture retention in cheese. Kolakowski et al. (1998) studied the effects of high pressure on Camembert cheese and found that pressurization at 500 MPa achieved the highest degree of proteolysis and cheese had an equivalent taste to that of 6-month-old commercial cheese. Hence it can be said that there exist a potential in HPP to develop novel and innovative products with improved shelf life and higher safety with acceptable sensory attri-butes. Also, it provides a potential to reduce energy requirements in dairy industry.

## 12.12 MICROWAVE HEATING

Microwave heating has gained attention in the area of food processing owing to its capability of volumetric heating at rapid rates thereby substantially reducing the cooking time and provides uniform heating. Higher cooking rates and uniform heating also provides the advantage of lesser damage to sensory and nutritional qualities of food. Conventional heating involves transfer of energy from the surface of the material through convection, conduction or radiation. In contrast, microwave heating employs generation of heat in the material itself through molecular interactions with the electromagnetic field (Thostenson and Chau, 1999).

Microwaves are located in a wavelength range of 1 mm to 1 m in the electromagnetic spectrum. This corresponds to the frequencies between 300 MHz to 300 GHz. As electromagnetic waves used for different other purposes like radar and cellular phones also exist in the same band, Federal Communications Commission (FCC) for industrial, scientific, and medical (ISM) purposes has reserved two frequencies, that is, 915 MHz and 2450 MHz for industrial and domestic purpose, respectively (Lauf et al., 1993).

Heat is generated inside the food in microwave heating and hence distributed volumetrically. Microwave energy penetrates the food and cause dipolar rotation of the polar solvents and conductive migration of dissolved ions. These two processes cause molecular friction and thus heat is produced inside the food. Electrical and magnetic fields are varied continuously to cause dipole rotation (Alton, 1998). The presence of a dipole, therefore becomes necessary for microwave heating. Water being dipolar in nature and present in almost all foods is the main source of heat generation inside the product. Since heat is generated throughout the material because of the presence of water as a dipole, microwave heating is faster as compared to the conventional heating (Oliveira et al., 2002).

Microwave heating finds large number of applications in food processing such as cooking, blanching, roasting, microwave assisted air, vacuum and freeze drying, etc. In milk however, applications have been limited to pasteurization, sterilization and surface treatment of products to increase the shelf life. Villameil et al. (1996) studied the microwave pasteurization of milk and found it to be an effective in pasteurizing milk with satisfactory microbial and sensory quality without excessive heat damage to milk nutrients. No advSerse effects on flavor of milk has been reported by microwave pasteurization of milk (Valero et al., 2000). Microwaves have also been used in determination of the totals solid content in sweetened condensed milk, evaporated milk, ice cream mix and yogurt. Rapid results with improved repeatability and reproducibility were obtained (Reh and Gerber, 2003). Microwaves application in surface treatment of dairy products has also been attempted successfully. Herve et al. (1998) treated cottage cheese with microwaves and reported significant increase in shelf life of the cheese.

## 12.13   PACKAGING OF MILK AND MILK PRODUCTS

Milk and milk products contain almost all essential nutrients required for the growth and development of human beings. They are highly perishable commodities and serves as a very good media for the growth of many spoilage causing microorganisms. Hence, there is an obvious need to preserve our precious food sources. Food packaging is an external means of preservation of food during storage, transportation and distribution. Hence, it forms an integral part of the product manufacturing process. Food packaging performs four major disparate functions such as containment, protection, convenience and communication. In pursuit of achieving these goals, many materials have been discovered by man for use as food packaging materials. They include wood and paper, glass, metals (such as tin, aluminum), plastics, composites, etc. The inherent properties of these packaging materials make them either highly suitable or unsuitable for a particular food product. Hence, a packaging material has to be judiciously chosen depending upon the nature of the product, availability, machinability, cost, etc. Further, with the stringent laws prevailing across the globe, in the form of food safety and standards regulations, one should be aware of what should be there on the label of a pre-packaged dairy product. The changes in the food packaging systems have been very fast in recent years, and across the world, the industry is facingchallengesin keeping pace with the fast changes in milk packaging systems. Selection of a suitable packaging depends on the product characteristics and envisaged storage conditions (Table 12.5). Packaging systems practiced for various milk and milk products are briefly discussed.

## 12.14   PACKAGING OF LIQUID MILK PRODUCTS

Milk has been packaged in different types of containers throughout the world. Although from protection point of view, milk could be packaged

**TABLE 12.5**   Characteristics of Milk and Milk Products and their Packaging Requirements

| Class of Milk and Milk Products | Examples | Characteristic attributes or contents | Packaging requirements |
|---|---|---|---|
| Fluid Milk Products | Toned milk, double toned milk, full cream milk, flavored milks, etc. | Perishable due to high moisture, ideal for microbes and presence of photosensitive components | Protection against light and oxygen |
| Dried Milk Products | WMP, SMP, dried convenience mixes, WPC, etc. | Low moisture, high fat and photosensitive | Protection against water vapor, light and oxygen |
| Fat rich products | Butter, spreads, etc. | High fat with fat soluble vitamins and pleasant aroma | Protection against oxidative and hydrolytic rancidity, aroma protection |
| Frozen dairy products | Ice Cream, frozen desserts, frozen yogurt, etc. | Frozen, high fat, pleasant aroma | Aroma protection, heat insulation, stability of the material at frozen conditions |
| Fermented dairy products | Cheese, kefir, dahi, yogurt, lassi, labneh, probiotic dairy foods, etc. | Beneficial bacteria with characteristic flavors and B-vitamins | Protection against light; oxygen, moisture and gas barrier; aroma protection |
| Traditional Indian Dairy Products | Burfi, peda, paneer, gulabjamun, rasogolla, ghee, etc. | High fat, low moisture products with characteristic flavors | Protection against moisture, light, preservation of aroma |
| Long-life products | UHT milk, Sterilized milk, retort processed products (like paneer curry), canned products (rasogolla, gulabjamun, condensed milks), etc. | Sterile, medium to high fat, characteristic flavors with minimum head space in the pack | Protection against microorganisms, light, oxygen, moisture and container surface and preservation of aroma |

under rigid packssuch as glass bottles and plastic bottles but for economical and convenience reasons, their use is discouraged. Hence, alternate flexible materials such as plastic films with more or less the same desired functional properties have been evolved. Of the total milk packed, it is estimated that flexible pouches dominate (92%), followed by bottles (5%)

and aseptic packaging cartons (3%). The unique advantages offered by the plastic packages include good barrier properties, visibility of the contents, light in weight, can be used for single-service, are easy to carry home, are more economical and attractive due to multi-color printing and ergonomic designs. Also, the use of plastic containers eliminates noise in the milk bottling plants and during delivery and also reduces water pollution caused by milk residues and detergents used in the bottle washing process.

The introduction of pasteurized milk created the need for packaging to avoid post-pasteurization recontamination. Initially, the container of choice was the recyclable glass bottle. The first glass bottle packaging for milk was used in the 1870s. Glass bottle is a multi-trip package for milk and, on an average each bottle performs as many as 90 trips to the consumer before it goes out of circulation. Bottles are cleaned in a bottle washer, filled on automatic filling machine and are capped with aluminum foil or plastic threaded closures made of either high-density polyethylene (HDPE) or polypropylene (PP). Weight of empty bottle is high and hence, requires higher transportation cost. Glass was the only container used for packaging of milk until the introduction paperboard containers in 1950s and of plastic pouches in 1960s. With the invention of polyethylene (PE) in 1930s and its wide applicability in packaging by 1960s, low-density polyethylene (LDPE) has become a major plastic used in dairy and food industry. Plastic pouches have fast replaced glass bottles for packaging of milk. Flexible pouches have proved to be a safe, quick and cost effective packaging method with a wide distribution network, providing ease of packaging and handling. A good consumer response to milk pouches paved the way for the technological changes. Presently, milk is sold in flexible packages like pouches, cartons, bags, plastic bottles and jars, etc. In India, pasteurized milk is predominantly packaged in LDPE or linear LDPE (LLDPE). In a form-fill-seal (FFS) machine, the plastic film is formed into a tube, sealed along its length, sealed at the bottom to form a pouch, filled with milk and then sealed at the top. The consumers perceive the quality deterioration of pasteurized milk through off-flavors. Among the defects, light-induced off-flavors such as burnt sunlight flavor caused due to the degradation of sulfur-containing amino acids (methionine) and metallic or cardboardy off-flavor caused due to light-induced lipid oxidation are probably the most common. Light exposure especially at wavelengths below

500 nm, also causes destruction of photosensitive vitamins mainly riboflavin and vitamin A (Kontominas, 2010). Hence, to prevent light penetration, the LDPE pouches used for liquid milk packaging are always pigmented with titanium dioxide ($TiO_2$) to give opaqueness (white) to the pouch. The Bureau of Indian Standards (BIS) has recommended pouches of LDPE or LLDPE or a mixture of both for packaging of milk. The thickness of the films recommended for 500 mL and 1000 mL pack sizes is 60 and 75 microns. Presently, pasteurized and homogenized market milks such as toned, double toned, standardized and full cream milks are all packaged in LDPE or LLDPE pouches. Liquid milk is also packaged in various capacities (500 mL to 4 L) of bottles made from $TiO_2$ pigmented HDPE or polyethylene terephthalate (PET).

The lack of a continuous cold chain, along with the prohibitive costs for having it, result in a broken cold chain for the plastic pouches of milk packaged in LDPE films in most developing countries. This limitation has become a boon for the aseptic packaging industry. The aseptic packaging market is expected to grow at a rate of 10–11% between 2015 and 2020. Aseptic packaging is a technology wherein the product and package are separately sterilized, and the product is then filled into the package and the package sealed in a sterile environment. The product is commercially sterile and shelf stable. Plastic materials used in aseptic packaging of milk products are polyethylene, polypropylene, polystyrene as tubes, bottles or plastic film laminates with paperboard or aluminum in the form of cartons. High-pressure steam is used to sterilize product lines and hydrogen peroxide with heat of UV radiation for container materials. Aseptic paperboard containers were introduced by Ruben Rausing (founder of TetraPak Company) in 1951. The triangular TetraClassic® was introduced in 1951 and the TetraBrick® in 1963; the latter is now probably the most popular milk package. Rectangular gable top carton packs (250 mL to 2 L) are the fastest-growing type within the aseptic packaging market across the globe. The popular commercial systems such as Tetra Brik® uses six layered laminate consisting of one layer of paperboard (420–500 µm), four layers of LDPE (15–20 µm each) and one layer of aluminum foil (8–12 µm). Recently, a multi-layered flexible pouch consisting of LDPE and ethylene vinyl alcohol (EVOH) had revolutionized the packaging and distribution of aseptically processed milk. Alternative to EVOH, polyvinylidene chloride (PVDC) could also function as a barrier material.

## 12.15 PACKAGING OF FERMENTED MILK PRODUCTS

### 12.15.1 DAHI AND YOGHURT

Dahi is a popular fermented dairy product in India, whereas yogurt is the best known of all cultured-milk products and the most popular worldwide. Typically it is classified as set-type, stirred-type, drinking-type, frozen-type and concentrated. Concentrated yogurt is sometimes called as strained yogurt or labneh or labaneh. Dahi and yogurt are highly perishable products and packaging protects them during handling and helps to maintain their physicochemical, nutritional and sensory characteristics. The packaging materials for yogurt and dahi must be acid-resistant, prevent the loss of volatile flavors and be impermeable to oxygen as yeasts and molds grow actively in presence of oxygen (Table 12.1). These products are soft in nature and generally require rigid or semi-rigid containers such as glass or plastic containers for set-type products while flexible packaging materials in the form of paperboard cartons and plastic pouches could be used for stirred or pourable products. Plastic materials such as high-density polyethylene (HDPE), polypropylene (PP), polystyrene (PS) and polyvinyl chloride (PVC) provide desired strength for yogurt and dahi. They are inert, low-cost and contribute no-off flavors to the products.

With a view to meet the product requirements and attract consumer attention, dahi and yogurt are packaged in materials of various shapes and designs from above mentioned materials but cups, tubs or paperboard laminates are packaging forms usually found in market. pp. containers have become the best option for yogurt packaging as they can be made with thinner walls while maintaining the same structural integrity resulting in significantly less plastic usage. The PS containers, especially the high-impact PS (HIPS) which are now extensively used by dairy industry for packaging of dahi and yogurt are very clean in appearance, gives good shining look, are light in weight and unbreakable, but the problem associated with PS and pp. cups is that of whey-off in set-type products such as dahi during storage. Preformed and cut aluminum foil and PE or PS laminates of about 100 mm diameter with a pull-tab for easy opening is usually used as lids to seal dahi and yogurt cups or tubs. In India, dahi is packaged and sold in local market by *halwais* (milk-based confectioners) in earthen pots or cups and dried leaf-based cups (Paltani and Goyal, 2007) which has a firm body. However, earthenware has the drawback of being heavy, susceptible to break and above all leads to excessive shrinkage of the product during storage due to moisture seepage through the pores of earthenware. Currently, for institutional sales, dahi is also being sold in HDPE buckets of size 1 kg, 5 kg and 10 kg.

Glass bottles are still used in some countries (France and eastern Europe) to package yogurt. The influence of pp. and PS on the sensory and physico-chemical characteristics of flavored stirred yogurts was studied by Saint-Eve et al. (2008). It was reported that PS seems to be preferable, especially for yogurts with 4% milk fat, for avoiding loss of fruity notes and for limiting the development of odor and aroma defects. Shelf stable products such as long-life drinking yogurt and stirred or set-type yogurts are manufactured by employing aseptic packaging technique using six layered laminates containing PE, aluminum foil and paperboard (PE/PE/Al foil/PE/Paperboard/PE) (Raju and Singh, 2016). Paper-based cartons made from laminates (PE/aluminum foil/PE or PE/paper/aluminum foil/PE) are usually used to package-dehydrated yogurt. Kumar and Mishra (2004) reported that the shelf life of mango-soy fortified yogurt powder packaged in aluminum foil laminated polyethylene was found to be better compared to high-density polypropylene (HDPP) pouches. Metal cans are also used for packaging of some type of dried yogurts.

Adhesion of foods and the resulting residues in packaging cause economic loss and poor product appearance and in this context, approximately 10% of all fermented milk products remain on the inside of packaging. Hansson et al. (2012) reported that adhesion of fermented milks and yogurt on packaging materials depends on the product contact time to the surface and among different packaging materials studied *viz.* PS, PE, PET, glass, etc., the adhered amount of fat was found to be lowest (4–8%) with PE contact layer package. Further, it was also reported that the product with higher fat content would adhere more to the surface after disruption or shaking of package prior to consumption.

### 12.15.2   CHEESE

Packaging of cheese is an important aspect in cheese making as it plays a vital role during curing and storage, in the final cheese shape and appearance and in its protection from the environment. Cheese packaging can be broadly grouped as (a) wrapping of cheese for storage and ripening and (b) retail packaging for consumers. Quality and water loss of finished product depend on the chemical properties and on the storage conditions in unpackaged cheese. However, in packaged cheeses, along with the storage conditions, they are dependent on the permeability and protection provided by the packaging material. Light-induced degradation of lipids, proteins and vitamins in cheeses causes both formation of off-flavors and color changes, which may swiftly damage product quality and marketability and ultimately may lead to loss in nutritional value and also formation of toxic products such as cholesterol oxides. Dalsgaard et al. (2010) reported that the contents of protein (dityrosine and dimethyl disulfide) and lipid (lipid hydroperoxides, pentanal, hexanal and heptanal) oxidation products were significantly low in vacuum packaged cheeses compared to cheeses packed in normal air. Light initiates the oxidation of fats, even at refrigerated temperatures and in unripened cheeses results in "cardboardy" or "metallic" off-flavors (Robertson, 2006). Such photooxidation changes may be prevented by either minimizing exposure to light or by controlling light barrier properties of packaging material. Depending on the reflectance, transmittance and oxygen permeability, packaging materials provide carried protection against light-induced changes in cheeses. Metals, paper and paperboard, various plastics and finally glass transmit light in increasing order (Bosset et al., 1994). Incorporation of titanium dioxide into plastic materials as an additive increases the light scattering, especially light at wavelengths of less than 400 nm, thereby reduces light transmittance. However, the reduction of light transmittance is dependent on amount of added titanium dioxide. Hence, in selecting an appropriate packaging material for a particular type of cheese, factors such as type of cheese and its textural characteristics (hard or soft); the presence of characteristic microorganisms; type of package (wholesale or retail); permeability to different fluids *viz.* water vapor, oxygen, carbon dioxide and ammonia and light; and provision for printing and labeling have to be considered.

Bulk cheeses are either paraffin wax coated or vacuum packed in flexible film. Generally, mineral waxes (paraffin) are usually applied to low-moisture fresh cheese immediately after its manufacture as they are cheap, easily applied and do not act as a nutrient for ordinary molds and other microorganisms. However, a waxed coat will not prevent mold growth if it cracks or if air gets into cheese in any way. Currently, a latex emulsion is being used for semi-hard cheese varieties in place of paraffin (Alvarez and Pascall, 2011). Also, a variety of vacuum packaging machines, gas-flushing machines, over wrapping machines and vacuum skin packaging machines are available for bulking packaging using high barrier plastic films. Co-extruded pouches having low oxygen permeability such as LLDPE-nylon-LLDPE; LLDPE-EVOH-LLDPE; LLDPE-PVDC-LLDPE or similar materials have good barrier properties can be used for vacuum packaging of cheese (Sayer, 1998). However, during storage and associated ripening changes, the properties of EVOH and nylon may change due to moisture uptake. In-bag-curing under vacuum using barrier bag technology loss of vital flavor and texture characteristics of Emmental and Parmesan cheese is prevented (Sayer, 1998). In case of surface mold-ripened cheeses such as Camembert and Brie, packing does not take place until the mold has grown to a desirable extent. In such cases, packaging material must have a limited permeability to oxygen so as to minimize the growth of anaerobic proteolytic bacteria, which can also develop if the permeability to water vapor is too low resulting in condensation inside the package. Further, the material should not adhere to the surface mold of the cheese. Oriented polypropylene (OPP) with carefully designed perforations that allow controlled quantities of water vapor permeability or paper coated with wax or laminated to perforated film are suitable for surface mold ripened cheeses (Robertson, 2006). Thermoformed HIPS coextruded or extrusion coated with PVC or PVDC and pigmented with titanium dioxide to improve the barrier properties and protection against light, thermoformed nylon-LDPE or injection molded HDPE containers with slits in the side to allow drainage of whey are usually used for fresh cheese packaging. For retail consumers, blocks of cheese are cut into small random weight size, vacuum packed in barrier bags, hot water shrunk and then weighed, labeled and reassembled as a large block and put into the carton for distribution. Also, about 8–10 pieces of rectangular or triangular chiplets of cheese or processed cheese of convenient sizes (25 g) are individually wrapped with aluminum foil and placed in an outer plastic or paperboard container. Cheese spreads may also be packed in plastic tubs (usually polypropylene) and sealed with aluminum foil laminates and closed with a press on plastic lid or plastic laminated or co-extruded squeezable tubes or bottles. Cheese is also available in slices individually wrapped in plastic films. About ten such slices are placed and sealed in an outer high barrier plastic pouch.

Although used for a wide range of products in dairy sector, modified atmosphere packaging (MAP) has been mainly applied to the packaging of cheese. Acceptable eating quality with shelf life trebled from around 4 weeks to up to 3 months can be achieved by using MAP. Hard cheeses such as cheddar are commonly packed in 100% $CO_2$ using horizontal form-fill-seal pillow pack machines. The packaging materials used include PVDC-coated PE or PET-PE. Modified atmosphere packaged cheese in pp. film has a shelf life of up to 4 weeks, compared with only 14–15 days when packaged under normal conditions. Cheeses are sold as cuts/blocks, sliced, or grated in order to accommodate consumer demands with respective to convenience. Grated cheese is packaged using flexible films similar to those used for hard cheese blocks. However, the films may be metallized. Cheese slices are packaged in similar gas mixtures but interleaved with paper for easy separation of slices. For these products it is not possible to use 100% $CO_2$ as the absorption of the gas by the product causes the packaging to collapse, crushing the product, thus interfering with the ease of separation. Therefore, $N_2$ is used as part of the gas mixture to stop the total collapse of the film around the product. The gas mixture typically used for these value-added products is 70% $N_2$ and 30% $CO_2$. Sliced or grated cheeses have a large surface area exposed to light and the surrounding atmosphere and are, thus, more susceptible to oxidatively induced color and flavor changes (Juric et al., 2003). Mortensen et al. (2004) studied the impact of MAP and storage conditions on the photo oxidation of sliced Havarti cheese and reported that when exposed to light and 0.6% residual $O_2$, the photooxidation, as quantified by the formation of 1-pentanol and 1-hexanol, significantly increased during 168 hours of storage. Mozzarella, a fresh soft unripened cheese with high moisture (50–60%), is susceptible to microbial spoilage. Alam and Goyal (2006a, 2006b) studied the effect of five different atmospheres (air, vacuum, 100% $CO_2$, 100% $N_2$ and a mixture of 50% $N_2$ and 50% $CO_2$) on the chemical quality of Mozzarella cheese made from mixed milk and stored at $7\pm1°C$ and deep freeze conditions ($-10°C$ to $-15°C$) and reported that the Mozzarella cheese packed with 100% $CO_2$, among others, could be stored for 12 weeks and 12 months, respectively with least chemical changes. The stability of sliced Mozzarella cheese packed under three different atmospheres was studied by Alves et al. (1996) who reported that the shelf life was 63 and 45 days when packed in 100% $CO_2$ and 50% $CO_2$/50% $N_2$, respectively while it was only 13 days when packed in normal air.

In a study conducted on MAP Danbo cheese (30% $CO_2$ and 70% $N_2$) blocks packaged in thermoformed, transparent polylactic acid (PLA) trays with a lid made of polylactic acid (PLA) film (500 μm) and stored in a display cabinet with exposure to fluorescent light under similar conditions to those in retail stores by Holm et al. (2006) revealed that PLA packages can be used for storage of Danbo cheese for a period of 56 days where moisture loss and lipid oxidation were limited. The oxidative stability of cream cheese stored in thermoformed trays made of amorphous PET-PE, PS-EVOH-PE and PP-PE with a transparent lid made of PET-aluminum oxide (AlOx)-PE film with different depth and color was studied by Pettersen et al. (2005). It was reported that cream cheese stored in trays made of amorphous PET-PE offered best protection against oxidation with respect to sensory flavor (acidulous and sunlight flavor notes) and color of the packaging material was an important factor regarding formation of volatile oxidation products during exposure to light.

Interest in edible with or without antimicrobial coatings for extending the shelf life of food products including fermented dairy products is increasingly growing. Cerqueira et al. (2009) reported that among three biopolymers viz. chitosan, galactomannan from *Gleditsiatriacanthos* and agar from *Glacilariabirdiae*, galactomannan presented the best properties to coat the *Saloio* cheese with decreased respiration rates and mold growth at the surface. Ricotta cheese, a fresh-type cheese, coated with chitosan-whey protein composite film and stored at 4°C under 40% $CO_2$ and 60% $N_2$ MAP conditions was found to have reduced microbial growth at the end of 30 days, suggesting potential for shelf life extension (Pierro et al., 2011). Guldas et al. (2010) investigated the shelf life of single baked mustafakemalpasa cheese-based sweets coated with different edible biopolymers viz. κ-carrageenan, chitosan, corn zein and whey protein concentrate (WPC) and reported that WPC and corn zein prolonged the shelf life of sweets from 3 to 10 days. The investigations on the antimicrobial effectiveness of sodium caseinate, chitosan and sodium caseinate-chitosan coatings on the native microflora of cheese revealed that chitosan and sodium caseinate-chitosan coatings exerted significant bactericidal effect on mesophilic, psychrotropc and yeasts and molds counts (Moreira et al., 2011).

### 12.15.3   PROBIOTIC FERMENTED DAIRY PRODUCTS

According the widely accepted definition, probiotics are defined as "live microorganisms which when administered in adequate amounts confer a health benefit on the host" (FAO/WHO, 2001). The challenge for food industry is to provide probiotic dairy products with viable probiotic organisms to consumers by adopting appropriate processing and packaging techniques including the selection of packaging material. Being anaerobic and microaerophilic in nature, probiotic organisms require lowest possible oxygen in the container. This would ensure viability of requirement number of cells in the finished product with desired functionality and avoid toxicity and death of the microorganisms. Exposure to dissolved oxygen during processing and storage is highly detrimental to *Bifidobacteriumbifidum* and *Lactobacillus acidophilus* (Cruz et al., 2007). Dave and Shah (1997) reported that the dissolved oxygen contents of yogurts containing *L. acidophilus* filled into glass remained low whereas those filled into high-density polyethylene containers significantly increased. The influence of high oxygen barrier packaging materials such as PS containers (300–350 µm) and a multilayered structure (HIPS-EVOH-PE) on the dissolved oxygen in probiotic yogurt was studied by Miller et al. (2002) and it was reported that the oxygen levels in the PS containers varied from 20–40 ppm while they decreased to 10 ppm at the end of 42 days of refrigerated storage period. The same researchers worked on incorporating oxygen absorbers into PS and reported that such oxygen scavengers integrated packaging materials created best conditions for favorable growth of anaerobic probiotic cultures (Miller et al., 2003). Talwalkar et al. (2004) investigated the effect of HIPS, HIPS-EVOH-PE and HIPS-EVOH-PE integrated with oxygen scavenging material on the survival of *L. acidophilus* and *Bifidobacterium* spp. in yogurt by monitoring the oxygen concentration during storage and reported that the dissolved oxygen depended on the type of packaging material used. At the end of 42 days of storage period, the dissolved oxygen content increased (30–38%) steadily in HIPS packaged yogurt. The oxygen levels in the HIPS-EVOH-PE declined to values lower than 4.29 ppm whereas the oxygen scavenger integrated HIPS-EVOH-PE dropped to 0.44 ppm. The effect of packaging material and storage temperature (4 and 25°C) on the viability of microencapsulated *B. longum* B6 and *B. infantis* CCRC 14633 was studied by Hsiao et al. (2004). It was reported that survival of bifidobacteriawas enhanced when they were stored at 4°C in glass bottles. The effect of glass bottles, plastic cups and clay pots on the survival of *B. longum* NCTC11818 incorporated into probiotic buffalo milk curd was investigated by Jayamanne and Adams (2004). The study concluded that bacteria survived best in the glass bottles, followed by the plastic packages and the clay pots when stored at 29±2°C. Further, it was reported that the curds packaged in glass bottles and plastic packages exhibited values of $10^6$ CFU/mL for up to 8 days. The effect of PP, PS, PE and glass containers on the progress of acid development in probiotic yogurts made from goat's and cow's milk during 21 days refrigerated storage was studied by Kudelka (2005). The study reported that the lowest acidity values of yogurt was found in yogurt stored in PS containers and glass packages favored the survival of probiotic cultures due to its extremely low oxygen permeability.

## 12.16 PACKAGING OF DRIED MILK PRODUCTS

Drying extends the shelf life of the milk, simultaneously reducing its weight and volume. The water content of milk powder ranges between 2.5% and 5.0%, and no microbial growth occurs at such low water content. However, dried milk products such as skimmed milk powder (SMP), whole milk powder (WMP), dairy whitener, infant foods, cream powder, etc. are hygroscopic and absorption of a small amount of water decreases the shelf life considerably. Several properties of milk powders containing amorphous lactose such as surface stickiness, caking, time-dependent lactose crystallization and release of encapsulated lipids, increasing rates of non-enzymatic browning and lipid oxidation can be related to their glass transition temperature. When an amorphous component is given suitable conditions of temperature and water content, powder can mobilize as a high-viscosity flow, which can make it sticky and lead to caking (Tehrany and Sonneveld, 2010). Caking is a deleterious phenomenon by which a low-moisture, free flowing powder is first transformed into lumps, then into agglomerated solid and ultimately into a sticky material. Maillard reaction or non-enzymatic browning is initiated by condensation of lactose with the free amino group of lysine in milk protein which is induced by heating during processing and long-term storage at moderate to high temperatures. During crystallization, lactose initially absorbs moisture from the surroundings due to its hygroscopic nature and subsequently release moisture as it crystallizes. Lactose crystallization modifies the microstructure and chemical composition of the surface of powder particles. Although an $a_w < 0.6$ is considered sufficient to prevent the growth of microorganisms, chemical reactions and enzymatic changes may occur at considerably lower levels. Commonly the $a_w$ of whole milk powder varies from 0.25 to 0.35 and for skimmed milk powder from 0.32 to 0.43. The moisture sorption isotherms (MSIs) for powders describe the equilibrium relationship between the moisture content of powder and the relative humidity of the surrounding environment at a specific temperature. Knowing the MSIs of powdered milk products is essential to be able to predict their stability in association with packaging characteristics. Further, among whole milk powders, lipid oxidation is a major cause of deterioration during processing and storage. The reaction of unsaturated lipids with molecular oxygen results in the formation of hydroperoxides, which then break down to off-flavor compounds. Development of off-flavors is manly caused by the formation of secondary reaction products such as alkanes, alkenes, aldehydes and ketones. The rate of lipid oxidation is greatly influenced by light, which causes development of off-flavors, a decrease in nutritional quality and the severity and speed at which these phenomenon develop. The principal light-absorbing groups of lipids are double bonds, peroxide O-O bonds and carbonyls (Schaich, 2005). Hence, packaging systems for milk powders must protect the powders from exposure to moisture, oxygen and light and anticipate the likely environmental factors, which include temperature, time, relative humidity, light and physical hazards.

For retailing to consumers, milk powder is packed into either metal cans or multilayered pouches. Metal cans have been highly popular for a long time for packaging of milk powders with varying capacities (200 g, 400 g, 500 g, 1 kg) due to their excellent physical strength, durability, absolute barrier to moisture, oxygen and light, absence of flavor or odor and rigidity (Robertson, 2006). With adequately constructed metal cans, a shelf life in excess of 5 years is realistic particularly when dried milks have been gas-flushed with nitrogen to minimize the amount of available oxygen. The metal cans are usually coated with a thin layer of tin and an additional layer of organic lacquers such as epoxy-phenolic lacquer to prevent corrosion and avoid metal-powder contact. Powdered milk including infant formulas may have hormonally active contaminants such as bisphenol A (BPA) introduced in the manufacturing process of leached from containers. Kuo and Ding (2004) detected BPA in powdered milk and infant formulas at concentrations from 45 to 113 ng $g^{-1}$. In the recent times, aluminum foil/plastic film laminates have been introduced as a replacement for the metal cans. Milk powder packed in pouches is commercially available in a capacity range of 250 g to 2.5 kg. For institutional sales such as for railways and airways, the pack size could be as small as 10 g. A multilayered construction of pouch consisting of aluminum sandwiched with two plastics layers – one on the inside such as LDPE, so that the pouch can be sealed and one on the outside such as biaxially oriented polypropylene (BOPP) or PET, to provide mechanical protection and also carry information is a common practiced to achieve a shelf life of about 2 years. Alternatively, with pouches for which shorter shelf life is acceptable, the aluminum layer may be replaced with a high-barrier plastic layer such as copolymer of EVOH or PVDC, possibly with the addition of a thin layer of metal or silica oxide deposition to enhance its oxygen barrier characteristics. Bulk packaging of milk powders is usually done in 25 kg multiwall Kraft paper sacks laminated with woven LDPE or pp. and having a liner made of HDPE. Specifications for multi-wall bags vary from country to country but generally a 4–5 ply bag with a minimum of 420 $gm^{-2}$ basis weight, with an inserted LDPE liner of 75 micron is considered to be adequate.

## 12.17 PACKAGING OF FAT-RICH DAIRY PRODUCTS

Cream and butter are important fat-rich dairy products other than ghee. Cream contains a high percentage of butterfat, so it is very susceptible to spoilage. In addition, it must be protected from water loss. The shelf life of refrigerated creams ranges between 1–2 days without proper protection. Packaging consists of units similar to those used for milk such as PE coated paperboards. Since cream is used in smaller amounts, commercial pack sizes are 500 g or smaller. Currently, cream is packaged in 250 mLasepticcartons similar to those used for aseptically processed and packaged milk. Whipped creams and synthetic formulations are sold in aerosol cans and PE tubs with snap-on lid. Imitation cream made from soybeans and vegetable oils is often marketed in wax-coated or plastic laminated paperboard cartons. Newer concepts include portion-control thermoformed packs made from LDPE, PS or PP. These may be closed with a peelable lid or snap-on cover. Tin plate containers are also being used for larger sizes. Table butter consists of milk fat (~80%), common salt (3%) and curd (1.5%). It also contains trapped moisture (15%). The natural color of butter is due to carotene and other similar fat-soluble pigments in the fat globules of milk. The flavor of butter is produced by the fermentation of cream by bacteria. Packaging must protect the butter in relation to its flavor, body and texture, appearance, moisture and color. Because of the nature of emulsion, butter is prone to rancidity caused by oxidation of the fats producing "fishy" taint. Packaging of butter is done in bulk packs of more than 5 kg and in retail packs from 10 g to 5 kg. Various types of machines are used, depending on the type of packaging. The machines are usually fully automatic and both portioning and packaging machines can often be reset for different sizes such as 250 g and 500 g or 10 g and 15 g (Singh,

2014). The wrapping or packaging material must be greaseproof and impervious to light, flavoring and aromatic substances. It should also be impermeable to moisture, otherwise the surface of the butter will dry out and the outer layers become more yellow than the rest of the butter. Butter is usually wrapped in aluminum foil for retail marketing. Parchment paper was once the most common wrapping material and still being used has now been largely replaced by aluminum foil, which is less permeable. After wrapping, the pat or bar packets continue to a cartoning machine for packing in cardboard boxes. Bulk packaging of butter is done in boxes, tubs or casks. Mostly, HDPE liners are used for wrapping the butter and corrugated fiberboard boxes of multiple plies (3, 5 or 7 ply) is used as secondary packaging material. Tubs made of HDPE or PET with press-on lids are also commonly used for packaging of butter.

## 12.18   PACKAGING OF FROZEN DAIRY PRODUCTS

Ice cream is a very popular dairy product among all age groups. It is an oil-in-water emulsion in which the disperse phase is fat. The ice cream mix is pasteurized before processing, then homogenized, frozen and air whipped in. Finally, the ice cream is further chilled and stored at –29°C. When ice cream is drawn from the freezer it is usually collected in containers which give it the desired shape and size for convenient handling during the hardening and marketing processes. Packaging of ice cream should be done as close to the freezer as possible, because longer the pipelines, greater the possible damage to the ice cream. For all type of containers, ice cream packaging equipment have common features such as an extrusion head, some with a spreader plate and a means for holding the container, moving it to the point filling and then moving it away from the filling point. Packaging filling machines designed for handling paperboard cartons may store the flat folded section in a feed cartridge, set it up for filling, filling and closing. Filling may be done on a time cycle basis or on a volume or weight basis. Closure depends up on the carton design which may be of the interlocking flat type or a type which is held closed by an adhesive applied up on closure. Earlier folding cartons were wax coated for increasing the wet strength of paperbaord and for heat-sealing purposes. However, most of the folding cartons used today are paperboard laminates containing LDPE. In India, apart from 100 mL cups, 500 mL and 1 liter home pack cartons are very popular. Other containers include cylindrical paperboard containers, plastic cups, form-seal plastic containers. General purpose polystyrene (GPPS) and high impact polystyrene

(HIPS) are widely used for ice cream cups manufacture. HIPS has additional advantage over GPPS as it improves Bulk ice cream containers of 2, 4 and 6 liters size are also available. However, they are often hand filled, but filling and lidding machines are available. Tamper evident packaging for ice cream products includes shrink banding for circular ice cream containers.

## 12.19 PACKAGING OF INDIAN TRADITIONAL MILK PRODUCTS

### 12.19.1 GHEE

It is estimated that nearly half of the total milk production in India is utilized for the manufacture of a range of traditional milk products. Approximately 25% of total milk production is converted into ghee, which is next to the liquid milk consumption; hence it has great economic significance in our country. Ghee is prepared at a temperature of around 110°C at which most of the microorganisms and enzymes such as lipases are eliminated and moisture content is left less than 0.5%; hence there is no microbial spoilage of ghee during storage. However, upon prolonged storage, ghee undergoes lipid deterioration resulting into either hydrolytic rancidity or oxidative rancidity defects. Majority of dairies in public as well as private sector are using lacquered or even unlacquered tin cans of different sizes (250 g, 500 g, 1 kg and 15 kg) for packaging of ghee. The advantages of using tin cans are manifold. The oxygen content in ghee can be reduced in case of tin cans by either hot filling or minimizing the headspace thereby preventing or delaying the oxidized flavor defect. It is very essential that tin cans be properly lacquered because rusted cans are liable to accelerate the lipid deterioration. Proper granulation in ghee is a highly desirable attribute from consumers' point of view, and ghee packaged in tin cans normally has better developed grains. The only drawback of tin cans is their high cost and involvement of foreign exchange. Glass bottles also provide excellent protection, as they do not react with the food material and can be used for high-speed operations; but are not in much use for packaging of ghee because of their fragility and high weight. Of late, semi-rigid plastic containers are replacing tin plate containers. These are mainly made from HDPE or PET. The advantages of using these containers are that they provide a moderately long shelf life (not as long as tin cans), are lightweight, economical and transport-worthy. Blow molded HDPE is available in the form of bottles (200, 400 g), jars (1 kg and 2 kg.), and jerry cans (2 kg, 5 kg, and 15 kg). PET bottles have excellent clarity are odor free and have gas barrier properties. Limited quantities of ghee are also packed in flexible pouches (less than 1 kg pack size). The most attractive feature of packaging ghee in flexible pouches is that they are most economical than any other packaging system. The selection of a laminate or a multilayer film is governed primarily by the compatibility of the contact layer, heat sealability and heat-seal strength and shelf life required apart from aroma, grease, water vapor, oxygen and light barrier properties.Laminated pouches made of PVdC/Al foil/PP are suitable for long term storage of anhydrous milk fat or ghee.

### 12.19.2 KHOA AND KHOA-BASED SWEETS

*Khoa* is a partially heat desiccated milk product obtained by continuous boiling of cow, buffalo or mixed milk or concentrated milk till desired flavor, texture and consistency is achieved. *Khoa*, normallyhas a limited shelf life of about 5 days at room temperature. The most common defects occurring in *khoa* during storage are hardening/drying of the product, fat oxidation, yeast and mold growth, staleness, etc. Selection of proper packaging system may, therefore, minimize these defects and extend the shelf life of *khoa* to a substantial longer period. According to Goyal and Rajorhia (1991) hot filling (80–90° C) of *khoa*in tin cans increases its shelf life to 14 days at 37°C, while the use of 3-ply laminate made of paper/aluminum foil/LDPE or 2-ply laminate of MST cellulose/LDPE keeps *khoa* in good condition for 10 days at 37⁰C and for 60 days under refrigerated storage. Hot filling of *khoa* at such a high temperature though helps extending the shelf life, it imparts undesirable browning in *khoa* making it unsuitable for sweets making. High barrier structures/laminates based on polyester/ethylene vinyl alcohol (EVOH)/polythene are also being used. Such laminates can be used for bulk packaging of *khoa* in cold stores for longer duration. Tin cans and rigid plastic containers of 15 kg capacity can also be advantageously used for bulk packaging of *khoa*.

Amongst the several *khoa*-based sweets, *burfi* and *peda* occupy most dominating place in terms of popularity and market demand. Currently, these sweets are largely prepared by sweetmeat makers (*halwai*) on small scale and mostly packaged in paper cartons or duplex board boxes with or without butter paper lining. The traditional packages do not provide sufficient protection to these sweets from atmospheric contamination and unhygienic handling and thus susceptible to become dry, hard and moldy and develops off flavors. Currently, *burfi* and *peda*are packaged in HDPE/polypropylene boxes and cartons of 500g and 1 kg size with an optional inserts or cavities. Palitand Pal (2000) investigated the influence of different packaging materials on the shelf life of *burfi*. *Burfi* was hot filled in pre-sterilized polyester tubs of 250 g capacity and subsequently vacuum packaged in multilayer co-extruded nylon high barrier pouches. They observed the shelf life of 52 days at 30°C in vacuum packaged samples against 16 days without vacuum packaging. Kumar et al. *(*1997) packed *peda* under modified atmospheric packaging (MAP) (80% nitrogen and 20% $CO_2$) in bags made from high barrier multilayer (EVA/EVA/PVdC/EVA) film and also at normal atmospheric conditions in barrier bags containing oxygen scavenger pouch placed in barrier bags along with indicator tablet. Samples with MAP were reported to have a shelf life of 15 days at 37°C and 30 days at 20°C. Packaging of *peda* with oxygen scavenger extended the shelf life up to 2 months at 37°C, 5 months at ambient temperature (30°C) and 6 months at 20°C.Londhe et al. (2012) reported that vacuum packaged brown *peda* could be best preserved up to 40 days at room temperature (30±1°C) without appreciable quality loss. Sharma et al. (2003) reported that *malaipeda* placed in flexible packages made from foil, paper and LDPE laminates and stored under vacuum and with $N_2$ were found to have a shelf life of 31 days of storage at 11°C. Thippeswamy et al. (2011) studied the effect of MAP ($N_2$:$CO_2$ 50:50) on the shelf stability of paneer prepared by adopting hurdle technology. Recently, MAP was used to extend the shelf life of *lalpeda* ($N_2$:$CO_2$ 70:30) (Jha et al., 2015), dietetic *rabri* (100% $N_2$) (Ghayal et al., 2015) and *Kalakand* ($N_2$:$CO_2$ 50:50) (Jain et al., 2015).

*Gulabjaman* is another*khoa* based sweet which is spherical in shape, spongy, porous and kept in sugar syrup. Its shape and porosity attributes are very critical and have to be maintained till the product reaches to the consumer. *Gulabjaman* is largely packaged without syrup in paper cartons or polyester boxes like *burfi* and *peda*. Though lacquered tin can is the most suitable packaging material for *gulabjaman*, it is expensive. Hence, Goyal and Rajorhia (1991) suggested the need to develop a plastic can extruded and laminated with a PP-Al foil material. The foil provides the necessary water vapor barrier property, smooth curved corners and good printing surface for multi-colordesigns.Such material is heat resistant and thus hot packaging of syrup and products is possible. Currently, high barrier multilayered stand-alone or gusseted pouches made from nylon and EVOH are being used for packaging of *gulabjamun*.

### 12.19.3    Paneer and Chhana-Based Sweets

Paneer, a popular traditional Indian dairy product, is obtained by acid and heat coagulation of hot milk. Paneer has a short shelf life of about 7 days at refrigeration storage and less than 24 hours at room temperature. A high moisture content (about 55%), post-manufacturing contamination and improper packaging are some of the major contributors for such a limited shelf life. The spoilage in paneer occurs due to the surface growth of microorganisms. A greenish yellow slime forms on the surface and the discoloration is accompanied by an off-flavor. Hence, it is invariably the surface that gets spoiled early while the interior remains good for a longer time at refrigeration storage. Use of an appropriate packaging material and creating air free environment, therefore, may contribute greatly in increasing the shelf life of Paneer. Sachdevaet al. (1991) vacuum packaged paneer blocks of 10 x 4 x 6 cm size in Cryovac polyethylene bags. No deterioration was reported upto 30 days at 6±1°C in vacuum packaged paneer samples. Punjrath et al. (1997) reported that vacuum packaging of paneer in specific film (EVA/EVA/PVDC/EVA) followed by heat treatment at 90°C for one minute could help extending the shelf life upto 90 days at refrigeration temperature. Also vacuum packaged paneer could be stored for extended times when nylon-EVOH based multilayered films are used. Rai et al. (2008) and Shrivastava and Goyal (2009) studied the effect of MAP on the quality of paneer during storage.Ahuja and Goyal (2013) successfully attempted to enhance the shelf life of *paneer tikka* from about one day at room temperature to 40 days at 3±1°C by adopting MAP.

*Rasogolla* is prepared from channa, which is spherical in shape, spongy (most spongy dairy product), porous and kept in sugar syrup. Its shape and porosity attributes are very critical and have to be maintained till product reaches to the consumer. Since the body and texture of *rasogolla* is very delicate and it has to be preserved in sugar syrup, it is invariably packaged in lacquered tin cans of 500 g and 1 kg capacity. The proportion of *rasogolla* and syrup is kept 40:60 and product stays in good condition for more than 6 months at ambient conditions, because hot filling (at about 90°C) technique is adopted. PP-Al foil laminate could be used for hot filling of syrup and *rasogolla*. High barrier multilayered stand-alone or gusseted pouches made from nylon and EVOH could alsobe used for packaging of *rasogolla*.

## KEYWORDS

- aseptic packaging
- bactofugation
- dehydration
- homogenization
- membrane processing
- modified atmosphere packaging
- separation
- thermal processing

## REFERENCES

Ahuja, K. K., & Goyal, G. K. (2012). Combined effect of vacuum packaging and refrigerated storage on the chemical quality of paneer tikka. *Journal of Food Science and Technology, 50*(3), 620–623.

Alam, T., & Goyal, G. K. (2006a). Influence of modified atmosphere packaging on the chemical quality of Mozzarella cheese during refrigerated storage. *Journal of Food Science and Technology, 43*(6), 662–666.

Alam, T., & Goyal, G. K. (2006b). Influence of modified atmosphere packaging (MAP) on the chemical quality of Mozzarella cheese stored in different packages at deep freeze conditions. *Indian Journal of Dairy Science, 59*(3), 139–143.

Alton, W. J. (1998). Microwave pasteurization of liquids, Society of Manufacturing Engineers Paper No. EM 98–211.

Alvarez, V. B., & Pascall, M. A. (2011). Packaging. In: J. W. Fuquay, P. F. Fox, & P. L. H. McSweeney (eds.) Encyclopedia of Dairy Sciences, 2nd edn. Academic Press, London, *4*, pp. 16–23.

Alves, R. M. V., Luca-Sarantopoulos, De, C. I. G., Dender, Van, A. G. F., & Assis-Fonseca-Faria, De. J. (1996). Stability of sliced Mozzarella cheese in modified atmosphere packaging. *Journal of Food Protection, 59*(8), 838–844.

Bosset, J. O., Gallmann, P. U., & Sieber, R. (1994). Influence on light transmittance of packaging materials on the shelf life of milk and dairy products–a review. In: M. Mathlouthi (ed.) Food packaging and preservation, Blackie Academic & Professional, *Glasgow,* 222–268.

Buchheim, W., & Frede, E. (1996). Use of high pressure treatment to influence the crystallization of emulsified fats. *DMZ LebensmindMilchwirtsch, 117,* 228–237.

Cerqueira, M. A., Lima, A. M., Souza, B. W. S., Teixeira, J. A., Moreira, R. A., & Vicente, A. A. (2009). Functional polysaccharides as edible coatings for cheese. *Journal of Agricultural and Food Chemistry, 57,* 1456–1462.

Cheftel, J. C. (1995) Review: high-pressure, microbial inactivation and food preservation/ revisión: Alta-presión, inactivaciónmicrobiológica y conservación de alimentos. *Food Science and Technology International*, *1*(2–3), 75–90.

Chicón, R., Belloque, J., Alonso, E., & López-Fandiño, R. (2009). Antibody binding and functional properties of whey protein hydrolysates obtained under high pressure. *Food Hydrocolloids*, *23*(3), 593–599.

Cruz, A. G. D., Faria, J. D. A. F., & Dender, A. G. F. V. (2007). Packaging system and probiotic dairy foods. *Food Research International*, *40*, 951–956.

Dalsgaard, T. K., Sorensen, J., Bakman, M., Vognsen, L., Nebel, C., Albrechtsen, R., & Nielsen, J. H. (2010). Light-induced protein and lipid oxidation in cheese: Dependence on fat content and packaging conditions. *Dairy Science and Technology*, *90*, 565–577.

Dave, R. I., & Shah, N. P. (1997). Viability of yogurt and probiotic bacteria in yogurt made from commercial starter cultures. *International Dairy Journal*, *7*, 31–41.

deAncos, B., Cano, M. P., & Gómez, R. (2000). Characteristics of stirred low-fat yogurt as affected by high pressure. *International dairy journal*, *10*(1), 105–111.

FAO/WHO. (2001). Joint FAO/WHO Expert Consultation on Evaluation of health and nutritional properties of probiotics in food including powder milk with live lactic acid bacteria, *5*.

Ghayal, G., Jha, A., Kumar, A., Gautam, A. K., & Rasane, P. (2015). Effect of modified atmospheric packaging on chemical and microbial changes in dietetic rabri during storage. *Journal of Food Science and Technology*, *52*(3), 1825–1829.

Goyal, G. K., & Rajorhia, G. S. (1991). Role of modern packaging in marketing of Indigenous dairy products. *Indian Food Industry*, *10*(4), 32–34.

Guldas, M., Akpinar-Bayizit, A., Ozean, T., & Yilmaz-Ersan, L. (2010). Effects of edible film coatings on shelf-life of mustafakemalpasa sweet, a cheese based dessert. *Journal of Food Science and Technology*, *47*(5), 476–481.

Hansson, K., Andersson, T., & Skepo, M. (2012). Adhesion of fermented dairy products to packaging materials. Effect of material functionality, storage time and fat content of the product. An empirical study. *Journal of Food Engineering*, *111*, 318–325.

Herve, A. G., Tang, J., Luedecke, L., & Feng, H. (1998). Dielectric properties of cottage cheese and surface treatment using microwaves. *Journal of Food Engineering*, *37*(4), 389–410.

Holm, V. K., Mortensen, G., & Risbo, J. (2006). Quality changes in semi-hard cheese packaged in a poly (lactic acid) material. *Food Chemistry*, *97*, 401–410.

Hsiao, H. C., Lian, W. C., & Chou, C. C. (2004). Effect of packaging conditions and temperature on viability of microencapsulated bifidobacteria during storage. *Journal of the Science of Food and Agriculture*, *84*(2), 134–139.

Huppertz, T., Fox, P. F., & Kelly, A. L. (2004). High pressure treatment of bovine milk: Effects on casein micelles and whey proteins. *Journal of Dairy Research*, *71*(1), 97–106.

Huppertz, T., Kelly, A. L., & Fox, P. F. (2002). Effects of high pressure on constituents and properties of milk. *International Dairy Journal*, *12*(7), 561–572.

Jain, V., Rasane, P., Jha, A., Sharma, N., & Gautam, A. (2015). Effect of modified atmospheric packaging on the shelf life of *kalakand* and its influence on microbial, textural, sensory and physic-chemical properties. *Journal of Food Science and Technology*, *52*(7), 4090–4101.

Jayamanne, V. S., & Adams, M. R. (2004). Survival of probiotic bifidobacteria in buffalo curd and their effect on sensory properties. *International Journal of Food Science and Technology*, *39*(7), 719–725.

Jha, A., Kumar, A., Jain, P., Gautam, A. K., & Rasane, P. (2015). Effect of modified atmosphere packaging on the shelf life of *lalpeda*. *Journal of Food Science and Technology, 52*(2), 1068–1074.

Juric, M., Bertelsen, G., Mortensen, G., & Petersen, M. A. (2003). Light-induced color and aroma changes in sliced, modified atmosphere packaged semi-hard cheeses. *International Dairy Journal, 13*, 239–249.

Knudsen, J. C., & Skibsted, L. H. (2010). High pressure effects on the structure of casein micelles in milk as studied by cryo-transmission electron microscopy. *Food chemistry, 119*(1), 202–208.

Kolakowski, P., Reps, A., & Babuchowski, A. (1998). Characteristics of pressurized ripened cheeses. *Polish Journal of Food and Nutrition Sciences, 7*(3), 473–483.

Kontominas, M. G. (2010). Packaging and the shelf life of milk. In: GL Robertson (ed) Packaging and Shelf Life: A Practical Guide. CRC Press, Boca Raton, USA, pp. 81–102.

Kromkamp, J., Moreira, R. M., Langeveld, L. P. M., & van Mil, P. J. J. M. (1996). Microorganisms in Milk and Yoghurt: Selective Inactivation by High Hydrostatic Pressure in Heat Treatments and Alternative Methods. *International Dairy Federation*, Brussels, 266–271.

Kudelka, W. (2005). Changes in the acidity of fermented milk products during their storage as exemplified by natural bio-yogurts. *Milchwissenschaft, 60*(3), 294–296.

Kumar, P., & Mishra, H. N. (2004). Storage stability of mango soy fortified yogurt powder in two different packaging materials: HDPP and ALP. *Journal of Food Engineering, 65*(4), 569–576.

Kumar, R., Bandyopadhya, P., & Punjrath, J. S. (1997) Shelf life of peda using different packaging techniques. *Indian Journal of Dairy Science, 50*, 40–49.

Kuo, H. W., & Ding, W. H. (2004). Trace determination of bisphenol A and phytestrogens in infant formula powders by gas chromatography-mass spectrometry. *Journal of Chromatography A, 1027*, 67–74.

Lauf, R. J., Bible, D. W., Johnson, A. C., & Everliegh, C. A. (1993). 2–18 GHz broadband microwave heating systems. *Microwave Journal, 36* (11), 24–27.

Law, A. J., Leaver, J., Felipe, X., Ferragut, V., Pla, R., & Guamis, B. (1998). Comparison of the effects of high pressure and thermal treatments on the casein micelles in goat's milk. *Journal of Agricultural and Food Chemistry, 46*(7), 2523–2530.

Londhe, G., Pal, D., & Raju, P. N. (2012). Effect of packaging techniques on shelf life of brown peda, a milk-based confection. *LWT-Food Science and Technology, 47*, 117–125.

Lopez-Fandino, R., Carrascosa, A. V., & Olano, A. (1996). The effects of high pressure on whey protein denaturation and cheese-making properties of raw milk. *Journal of Dairy Science, 79*(6), 929–936.

Miller, C. W., Nguyen, M. H., Rooney, M., & Kailasapathy, K. (2002). The influence of packaging materials on the dissolved oxygen content in probiotic yogurt. *Packaging Technology and Science, 15*, 133–138.

Miller, C. W., Nguyen, M. H., Rooney, M., & Kailasapathy, K. (2003). The control of dissolved oxygen content in probiotic yogurts by alternative packaging materials. *Packaging Technology and Science, 16*, 61–67.

Moreira, M. D. R., Pereda, M., Marcovich, N. E., & Roura, S. I. (2011). Antimicrobial effectiveness of bioactive packaging materials from edible chitosan and casein polymers: Assessment on carrot, cheese and salami. *Journal of Food Science, 76*(1), M54–M63.

Mortensen, G., Bertelsen, G., Mortensen, B. K., & Stapelfeldt, H. (2004). Light-induced changes in packaged cheeses-a review. *International Dairy Journal, 14*, 85–102.

Myllymfiki, O. (1996) High pressure food processors. In: T. Ohlsson (Ed.), High pressure processing of food and food components-a literature survey and bibliography (29–46), Goteborg: Kompendiet.

O'Reilly, C. E., O'Connor, P. M., Murphy, P. M., Kelly, A. L., & Beresford, T. P. (2000). The effect of exposure to pressure of 50 MPa on Cheddar cheese ripening. *Innovative Food Science & Emerging Technologies, 1*(2), 109–117.

Oliveira, M. E. C., & Franca, A. S. (2002). Microwave heating of foodstuff. *Journal of Food Engineering, 53*, 347–359.

Palit, C., & Pal, D. (2000). Studies on enhancement of shelf life of burfi (Abstract). Modern trends and perspective in food packaging for 21st century. ICFOST Souvenir, CFTRI Mysore.

Paltani, I., & Goyal, G. K. (2007). Packaging of dahi and yogurt–A review. *Indian Journal of Dairy Science.* 60(1), 1–11.

Pettersen, M. K., Eie, T., & Nilsson, A. (2005). Oxidative stability of cream cheese stored in thermoformed trays as affected by packaging material, drawing depth and light. *International Dairy Journal, 15*, 355–362.

Pierro, P. D., Sorrentino, A., Mariniello, L., Giosafatto, C. V. L., & Porta, R. (2011). Chitosan/whey protein film as active coating to extend Ricotta cheese shelf-life. *LWT-Food Science and Technology, 44*(10), 2324–2327.

Punjrath, J. S., Kumar, R., & Bandyopadhyay, P. (1997). Tapping the potential of traditional dairy food. In Souvenir 28th Dairy Industry Conference, Bangalore 27–29 April, 1997.

Rademacher, B., Hinrichs, J., Mayr, R., & Kessler, H. G. (1999). Reaction kinetics of ultra-high pressure treatment of milk. In: *Advances in High Pressure Bioscience and Biotechnology.* Springer Berlin Heidelberg, pp. 449–452.

Rai, S., Goyal, G. K., & Rai, G. K. (2008). Effect of modified atmosphere packaging (MAP) and storage on the chemical quality of paneer. *Journal of Dairying, Foods and Home Science.* 27(1), 33–37.

Raju, P. N., & Singh, A. K (2016). Packaging of fermented milks and dairy products. In: A. K. Puniya (ed.) *Fermented Milk and Dairy Products*, CRC Press, Boca Raton, USA. pp:637–671.

Reh, C. T., & Gerber, A. (2003). Total solids determination in dairy products by microwave oven technique. *Food Chemistry, 82*, 125–131.

Robertson, G. L. (2006). Food Packaging: Principles and Practice, 2nd edn. CRC Press, Boca Raton, Florida, USA.

Sachdeva, S., Prodopek, D., & Reuter, H. (1991). Technology of paneer from cow milk. *Jap. J. Dairy Food Sci., 40*(2), A85.

Saint-Eve, A., Levy, C., Moigne, M. L., Ducruet, V., & Souchon, I. (2008). Quality changes in yogurt during storage in different packaging materials. *Food Chemistry, 11*, 285–293.

Sandra, S., Stanford, M. A., & Goddik, L. M. (2004). The Use of High-pressure Processing in the Production of Queso Fresco Cheese. *Journal of Food Science, 69*(4), FEP153–FEP158.

Sayer, G. (1998). Packaging trends for cheese and other dairy products. In: Modern Food Packaging, Indian Institute of Packaging, Mumbai, 451–454.

Schaich, K. M. (2005). Lipid oxidation: theoretical aspects. In: F. Shahidi (ed.) Bailey's Industrial Oil and Fat Products. John Wiley, New Jersey, USA, 269–355.

Schrader, K., Buchheim, W., & Morr, C. V. (1997). High pressure effects on the colloidal calcium phosphate and the structural integrity of micellar casein in milk. Part 1. High pressure dissolution of colloidal calcium phosphate in heated milk systems. *Food/Nahrung*, *41*(3), 133–138.

Serra, M., Trujillo, A. J., Guamis, B., & Ferragut, V. (2009). Evaluation of physical properties during storage of set and stirred yogurts made from ultra-high pressure homogenization-treated milk. *Food hydrocolloids*, *23*(1), 82–91.

Sharma, H. K., Singhal, R. S., & Kulkarni, P. R. (2003). Effect of modified atmosphere packaging on the keeping quality of *malaipeda*. *Journal of Food Science and Technology*, *40*(5), 543–545.

Shrivastava, S., & Goyal, G. K. (2009). Effect of modified atmosphere packaging (MAP) on the chemical quality of paneer. *Indian J. Dairy Sci.*, *62*(4), 255–261.

Singh, S. (2014). Dairy Technology: Dairy Products and Quality Assurance, New India Publishing Agency, New Delhi, 51–54.

Talwalkar, A., Miller, C. W., Kailasapathy, K., & Nguyen, M. H. (2004). Effect of packaging materials and dissolved oxygen on the survival of probiotic bacteria in yogurt. *International Journal of Food Science and Technology*, *39*(6), 605–611.

Tehrany, E. A., & Sonneveld, K. (2010). Packaging and the shelf life of milk powders. In: GL Robertson (ed) *Packaging and Shelf Life: A Practical Guide*. CRC Press, Boca Raton, USA pp. 127–141.

Thippeswamy, L., Venkateshaiah, B. V., & Patil, S. B. (2011). Effect of modified atmospheric packaging on the shelf life stability of paneer prepared by adopting hurdle technology. *Journal of Food Science and Technology*, *48*(2), 230–235.

Thostenson, E. T., & Chou, T. W. (1999). Microwave processing: fundamentals and applications. *Composites Part A: Applied Science and Manufacturing*, *30*(9), 1055–1071.

Trujillo, A. J., Capellas, M., Saldo, J., Gervilla, R., & Guamis, B. (2002). Applications of high-hydrostatic pressure on milk and dairy products: A review. *Innovative Food Science & Emerging Technologies*, *3*(4), 295–307.

Valero, E., Villamiel, M., Sanz, J., & Martínez-Castro, I. (2000). Chemical and sensorial changes in milk pasteurized by microwave and conventional systems during cold storage. *Food Chemistry*, *70*, 77–81.

Villamiel, M., López-Fandiño, R., Corzo, N., Martínez-Castro, I., & Olano, A. (1996). Effects of continuous flow microwave treatment on chemical and microbiological characteristics of milk. *ZeitschriftfürLebensmittel-Untersuchung und Forschung*, *202*(1), 15–18.

# CHAPTER 13

# PRINCIPLES OF KINETIC MODELING OF SAFETY AND QUALITY ATTRIBUTES OF FOODS

MARIA C. GIANNAKOUROU[1] and NIKOLAOS G. STOFOROS[2]

[1]Associate Professor, Technological Educational Institute of Athens, Faculty of Food Technology and Nutrition, Department of Food Technology, Agiou Spiridonos, 12210, Egaleo, Athens, Greece, Tel.: 30-210-5385511, E-mail: mgian@teiath.gr

[2]Professor, Department of Food Science and Human Nutrition, Agricultural University of Athens, Greece, Tel.: 30-210-5294706, E-mail: stoforos@aua.gr

## CONTENTS

## 13.1  INTRODUCTION

One of the main goals of food industry is to ensure safety and maximum quality retention through proper and carefully designed processing and preservation of perishable food products. Food manufacturing and subsequent storage aims at providing products safe to consume from a microbiological viewpoint, and at the same time, delaying significantly undesirable physicochemical changes, that lead to food gradual deterioration and finally rejection by the end-user.

Optimal design of food processing is important in order to produce nutritive, safe products of acceptable sensorial quality (Valdramidis et al., 2012). During processing and storage, changes in foods due to biological, chemical and physical reactions do occur. Such changes proceed at a certain rate and with particular kinetics and might be related to the destruction of a quality attribute (e.g., vitamin content, color, etc.) or the production of an undesirable parameter (e.g., odor, microbial cells, etc.). The purpose of kinetic studies is to describe these changes and their rates quantitatively, making such an approach a valuable tool in relation to food processing, safety and quality.

The principles of microbial destruction during processing (e.g., thermal treatment etc.) were kinetically addressed a century ago and form the basis of proper designing such procedures. Kinetic modeling applied to microbial growth during food storage is referred to as predictive microbiology (Malakar et al., 2003; Poschet et al., 2003; Wijtzes et al., 1998; Zwietering et al., 1993; Zwietering and Den Besten 2011). A mathematical equation describing growth of microorganisms *vs.* time is used to predict the microbiological quality of a food as a function of several factors, such as temperature, pH, water activity. These models may then be used in order to implement risk assessment (Mataragas et al., 2010; Stella et al., 2013; Zwietering, 2015). In recent literature, there is a significant amount of published studies focusing on kinetic modeling of microbiological indices, aiming at predicting microbiological growth of the product in question and thus its safety in a variety of food matrices (McMeekin et al., 2013).

On the other, there is no doubt that food quality, in the sense of meeting consumer needs and expectations, is also an important issue (Van Boekel, 2008). The quality of processed foods cannot be expressed through a simple property; product quality depends not only on initial characteristics of the raw materials but also on the changes occurring during processing and subsequent storage (Taoukis et al., 2015) that frequently result in undesirable deterioration and potentially to decreased bioavailability and nutritional value. In this context, basic reaction mechanisms that can be important for quality evaluation and control have attracted the attention and numerous relevant works have been published recently (Van Boekel, 2008; Van Boekel and Tijskens, 2001). The aim of most studies was to optimize the conditions in order to maximize the quality of food products during processing and storage (Villota and Hawkes, 2007). Quality indicators are nor constant, since the quality of a food changes over time (Van Boekel, 2008), neither fixed for all food matrices. Some of the most unanimous quality-related changes are (Van Boekel, 2008):

- chemical reactions, mainly due to either oxidation or Maillard reactions;
- microbial reactions, involving microorganisms' growth, which may be desired (fermentation), simply undesired (spoilage microorganisms) or even dangerous (pathogens, leading to unsafe food);
- biochemical reactions catalyzed by endogenous enzymes leading to quality loss (enzymatic browning, softening, proteolysis, etc.); and
- physical reactions, where existing particles interact and phenomena such as coalescence, aggregation, and sedimentation may occur leading usually to quality loss.

Food kinetics, referring to both biological and physicochemical changes, involve the study of the rates and mechanisms that are responsible for the reactions of interest; mathematical relationships are additionally needed to adequately describe the influence of different external or internal factors, such as pH, water activity, pressure, etc. Kinetic parameters describing such changes are estimated and are potentially used to predict food microbiological and quality status at given processing and/or storage conditions. In order to build such mathematical equations, experimental data are used; the experimental design, as well as the quality of experimental data is crucial for the validity of parameters, and hence the reliability of the prediction (Van Boekel, 1996). The most widely applied procedure for food kinetic modeling (either for microbiological measurements or for chemical changes) is based on experiments conducted under different constant temperature conditions. In the traditional two-step approach, frequently adopted in literature (Huang and Vinyard, 2016; Huang 2003; Huang, 2004; Sulaiman et al., 2015a; Sulaiman et al., 2015b; Angelidis et al., 2013; Dermesonluoglu et al., 2015; Giannakourou and Taoukis, 2003b), the most representative indices are selected, and their changes are measured as a function of processing or storage time at constant temperature conditions. Then, an appropriate primary model is applied in order to describe the rate of these changes for each temperature of interest. The second step is to select a secondary model that best describes the effect of temperature on the rate of changes or any other appropriate kinetic parameter dictated by the primary model. Estimation of all parameters of primary and secondary models would allow for the prediction of quality or microbiological status of the food at any, different temperature of interest. If validated, mathematical models can be a useful tool to assess safety and quality after processing and at any stage of food storage. Instead of temperature, any other intrinsic or extrinsic parameter (such as pH, $a_w$, partial $CO_2$ pressure, etc.) can be used in a similar approach.

Alternatively, the model parameters can be determined in a single step considering the same isothermal dataset as a whole and performing a non-linear regression through appropriate mathematical equations which are developed by incorporating the secondary model equations into the primary model. The advantage of one-step analysis modeling is not only the increased number of degrees of freedom, but rather the use of more precise parameter estimates and the development of more realistic models due to the knowledge of various responses (Van Boekel, 1996). Several studies have been published in recent literature that use one-step kinetic analysis (Van Boekel, 1996, Huang and Vinyard, 2016; Huang, 2013; Huang, 2015a, 2015b; Wawire et al., 2016). However, an important issue to be addressed in this case is the selection of the fitting algorithm and the respective fit criterion that is not always well described and sustained (Van Boekel, 1996).

Another approach currently introduced involves the simultaneous determination of kinetic parameters of primary and secondary models based on experiments at dynamic conditions instead of isothermal data (Huang, 2003; Huang, 2013; Huang, 2015a; Sui and Zhou, 2014; Valdramidis et al., 2006; Valdramidis et al., 2008; Valdramidis et al., 2012). A significant advantage of this methodology is the significantly decreased number of necessary experiments. In such an approach, the selection of the appropriate temperature conditions of the non-isothermal profile is crucial for the accurate prediction of the kinetic parameters, with acceptable, low confidence intervals (Saltaouras et al., 2015).

In all above-mentioned methodologies, apart from the calculation of estimates for models' kinetic parameters, the estimation of their confidence intervals, representing the range of their uncertainty, is equally important. The conventional method of two-step procedure, besides being time-consuming, lacks reliability concerning the expression of parameters' variability (Van Boekel, 1996). Following the standard procedure of this methodology, confidence intervals of primary model parameters at each temperature are not taken into account when calculating secondary model parameters, which, consequently, can be falsely narrower than the actual ones. On the other hand, in the case of non-linear models and a one-step process, an important issue designated by Van Boekel (1996) is that the confidence intervals of the parameters involved are not symmetric and thus their estimation is not straightforward. These remarks reveal that there are weak points both in the two-step as well as in the one-step approach, mainly at the determination of the uncertainty of the secondary model parameters. Therefore, a reliable alternative is deemed necessary to account for the real uncertainty of model parameters, in order to be able to predict in a more accurate way chemical or microbiological changes, at any selected temperature and time during processing and product shelf life (at storage or distribution). Monte Carlo simulation technique is recently proposed in literature for the probabilistic assessment of stochastic variability and uncertainty associated with microbiological (Huang, 2015a; Huang and Vinyard, 2016; Aspridou and Koutsoumanis, 2015; Lianou and Koutsoumanis, 2011; Koutsoumanis and Angelidis, 2007) or chemical/sensory (Evrendilek et al., 2016; Channon et al., 2016; Wesolek and Roudot, 2016; Giannakourou et al., 2001; Giannakourou and Taoukis, 2003a; Sui and Zhou, 2014) attributes of several food matrices.

In the remaining of the chapter the basic principles that govern food kinetic modeling will be discussed. An overview of the main kinetic approaches from recent literature will be provided, and representative example calculations will be presented. Bearing in mind that the selection of the appropriate kinetic approach is of cardinal importance in order to effectively describe experimental data, some guidelines and model choice suggestions are given.

## 13.2 BASIC PRINCIPLES OF KINETICS

As already mentioned, different kind of reactions continue to occur post processing, at a rate that is influenced by several intrinsic (inherent properties of the food) and extrinsic parameters (type of packaging, conditions of storage and distribution, etc.). The first step in kinetically studying food deterioration is to determine the most important parameters that reflect the overall quality loss. Through a careful study of food composition and in relation to the process used, the reactions with the most critical effect on the deterioration rate, can be established. Assuming constant environmental conditions, a reaction scheme that expresses the effect of the concentration of the reactants, is developed. The ultimate objective is to model the change of the concentrations of constituents connected to food quality, as functions of time.

### 13.2.1 GENERAL REACTION KINETICS

Molecular, irreversible reactions are typically expressed as (Taoukis and Giannakourou, 2007):

$$v_1 A_1 + v_2 A_2 + v_3 A_3 + v_4 A_4 + ... + v_m A_m \xrightarrow{k_f} P \qquad (1)$$

where $A_i$ are the reactant species, $v_i$ the respective stoichiometric coefficients ($i = 1, 2... m$), P the products and $k_f$ the forward reaction rate constant. For such a scheme the reaction rate, $r$, is given by (Upadhyay, 2006):

$$r = -\frac{1}{v_j}\frac{d[A_j]}{dt} = k_f [A_1]^{n_1} [A_2]^{n_2} ...[A_m]^{n_m} \qquad (2)$$

where $n_j$ is the order of the reaction with respect to constituents $A_j$ and $[A_m]$ the molar concentration. For a true molecular reaction, it holds that: $n_j = v_j$.

### 13.2.2 COMPLEX REACTION KINETICS

A common case of increased complexity is when consecutive, parallel or reversible reactions occur in the food matrix (Van Boekel and Tijskens, 2001; Knol et al., 2009; Corradini and Peleg, 2006; Upadhyay, 2006). The first case (consecutive reactions) includes intermediate products, which react further. One simple example of consecutive reactions is described by the following scheme (Eq. 3):

$$A \xrightarrow{k_1} B \xrightarrow{k_2} C \qquad (3)$$

Then, the differential rate equations are (Eq. 4):

$$\frac{d[A]}{dt} = -k_1[A]$$

$$\frac{d[B]}{dt} = k_1[A] - k_2[B] \tag{4}$$

$$\frac{d[C]}{dt} = k_2[B]$$

The analytical solution when external conditions are assumed constant is (Eq. 5):

$$[A] = [A]_0 e^{-k_1 t}$$

$$[B] = [B]_0 e^{-k_2 t} + k_1[A]_0 \frac{e^{-k_1 t} - e^{-k_2 t}}{k_2 - k_1}$$

$$[C] = [C]_0 + [B]_0 \left(1 - e^{-k_2 t}\right) + [A]_0 \left(1 + \frac{k_1 e^{-k_2 t} - k_2 e^{-k_1 t}}{k_2 - k_1}\right) \tag{5}$$

where index '0' refers to the initial concentration of the different species.

On the other hand, parallel reactions happen when a reactant is simultaneously implicated in two or more different reactions, as for example (Eq. 6):

$$A + B \xrightarrow{k_1} P$$

$$A \xrightarrow{k_2} P \tag{6}$$

and the corresponding differential equation that describes the reaction is:

$$-\frac{d[A]}{dt} = k_1[A][B] + k_2[A] \tag{7}$$

which is difficult to solve analytically; instead, numerical solutions are preferred.

In the literature, one frequently assumes the so-called steady state (*i.e.*, dA/dt = 0) which simplifies significantly the above-mentioned complex equations.

In complex systems, where the degradation of crucial parameters is described by reversible, multiple-step reactions, the following equation applies (Eq. 8):

$$aA + bB \underset{k_b}{\overset{k_f}{\rightleftharpoons}} cC + dD \tag{8}$$

In this case $A$ reacts with $B$ to form products $C$ and $D$ which can back react with a rate constant of $k_b$. The reaction rate in this case would be:

$$r = \frac{-d[A]}{adt} = \frac{-d[B]}{bdt} = \frac{+d[C]}{cdt} = \frac{+d[D]}{ddt} = k_f[A]^a[B]^b - k_b[C]^c[D]^d \tag{9}$$

At a certain stage, equilibrium is reached and the equilibrium constant $K_{eq}$ is described by (Eq. 10):

$$K_{eq} = \frac{k_f}{k_b} = \frac{[C][D]}{[A][B]} \tag{10}$$

However, for most cases of food degradation systems either $k_b \ll k_f$, or for the time period of practical interest concentrations $[C]$ and $[D]$ are very small, thus allowing us to treat it as an irreversible reaction.

When studying a bimolecular reaction, Eq. (2), becomes:

$$r = -\frac{dA}{dt} = kAB \tag{11}$$

where $A$ and $B$ are the reactants.

If $B$ is assumed to be constant, then the integration of Eq. (11), gives:

$$\ln\left(\frac{[A]}{[A]_0}\right) = -k't \tag{12}$$

where $k'$ is a pseudo-order rate constant and $k'=k\times[B]$.

A common case of such reaction is the aerobic degradation of ascorbic acid, where oxygen level plays the role of reactant $B$. When studying this reaction at different levels of oxygen availability, a family of pseudo-first order plots is obtained.

## 13.3   PRIMARY MODELS

Primary models refer to mathematical equations describing the rate of change of a particular attribute in a food system as a function of time. That is:

$$\frac{dA}{dt} = f(t,[A],[B],[P],...) \tag{13}$$

where [A], [B] are some of the reactant species, [P] one of the products of the reaction.

When dealing with real, complex matrices, such as food, the most common case within the reactions previously discussed, is the one where one of the reactant (or product) species is the limiting factor, mainly affecting reaction kinetics, while all other constituents are in large excess. That allows the quality loss rate equation to be expressed in terms of the specific reactant, as:

$$r = \frac{-d[A]}{dt} = k'_f [A]^\alpha \tag{14}$$

where $\alpha$ is an apparent or pseudo order of the reaction of component $A$ and $k_f'$ is the apparent rate constant. Assuming that the aforementioned simplifications are used in complex food matrices, the change of food quality can be represented either by the loss of one or more quantifiable quality indices, symbolized by $A$ (e.g., a nutrient or characteristic flavor) or by the formation of an undesirable product $P$ (e.g., an off-flavor or discoloration). The rate of change of $A$ and $P$ can be in general represented by Eqs. (15) and (16), namely:

$$r_A = -\frac{d[A]}{dt} = k \cdot [A]^n \tag{15}$$

$$r_P = \frac{d[P]}{dt} = k' \cdot [P]^{n'} \tag{16}$$

The quality factors [A] and [P] are usually quantifiable chemical, physical, microbiological or sensory parameters, selected so as to representatively describe the quality deterioration of the particular food system. The constants k and k' are the apparent reaction rate constants and n and n' are the apparent orders of the respective reactions. The use of the term "apparent" indicates that Eqs. (15) and (16) do not necessarily describe the mechanism of the phenomenon examined. The apparent reaction orders and constants are determined by fitting the change with time of the experimentally measured values of [A] or [P] to Eqs. (15) or (16) by Differential or Integral Methods (Taoukis et al., 1997).

### 13.3.1  N^TH ORDER MODEL

One general model often used to describe reaction kinetics in food systems is the $n^{th}$ order model. Reaction order is defined as the number of molecules participating in a reaction as reactants. In other words, reaction order is the sum of the exponents of the reactant species. In food science, it is quite frequent to describe a reaction as a first order equation even though the real mechanism is not first order. The reaction order is then referred to as pseudo-first order or, equivalently, as an apparent first order. For C being the concentration of the key component, influencing reaction rate, the following general rate law is valid:

$$r = -\frac{dC}{dt} = kC^n \tag{17}$$

This differential equation (Eq. 17) is thus in the form of a power law expression, where n is the so called order of the reaction. The equation reflects the dependence of rate r on concentration for just one component; k is again the reaction rate constant. C being in mol/mL, the unit for k for a reaction and having order n is $(mol/mL)^{1-n}/min$. Depending on the particular attribute of interest, C can have different units. For example, C can be expressed in mol/min, when describing vitamin degradation, activity units/mL when dealing with enzyme inactivation, number of microorganisms per unit volume when referring to microbial destruction and so on. In the remaining of this chapter, we will often use the symbol C without particular units when not necessary (e.g., in some generalized figures and equations). Eq. (17) can be integrated with respect to time to obtain the course of the concentration as a function of time:

for $n=1$

$$C = C_0 \, e^{-kt} \tag{18}$$

for $n \neq 1$

$$C^{1-n} = C_0^{1-n} + (n-1)kt \tag{19}$$

where $C_0$ is the initial concentration at $t=0$.

A plot of Eqs. (18) and (19) illustrating the form of the concentration vs. time, for different reaction orders is presented in Figure 13.1, both in linear and logarithmic concentration scale. In this figure, the effect of reaction order, $n$, on the shape of the destruction curves on linear (top) and logarithmic scale (bottom) for a constant $k$ value is shown, revealing the importance of carefully choosing the most appropriate reaction order to the data set available. Also note that it is not appropriate to talk about logarithmic

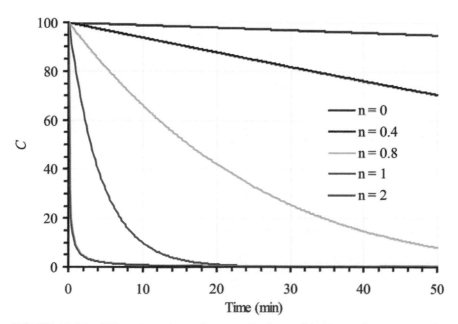

**FIGURE 13.1A** Effect of reaction order, $n$, on the shape of the destruction curves on linear (top) and logarithmic scale (bottom) for $k = 0.1$ $min^{-1} \times (mol/mL)^{1-n}$.

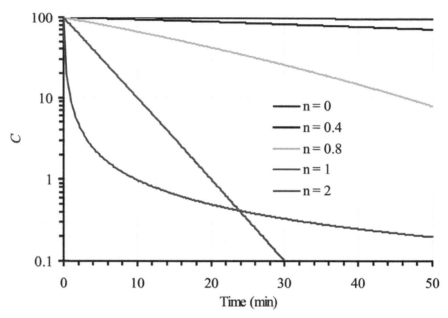

**FIGURE 13.1B**   Effect of reaction order, $n$, on the shape of the destruction curves on linear (top) and logarithmic scale (bottom) for $k = 0.1$ min$^{-1} \times$(mol/mL)$^{1-n}$.

destruction, for example "microbial population was reduced by 5 log after 10 min heating at 90°C" except for first order reactions.

More particularly, the main equations obtained for the most common reactions in foods are described in the following section:

### 13.3.1.1   First-Order Reactions (n=1)

The first order model is the most common model used and probably among the first examined in its ability to fit the experimental data during food safety and quality calculations. As already stated, the respective equations in the case of first-order reactions are:

$$r = -\frac{dC}{dt} = kC \tag{20}$$

$$\ln\left(\frac{C}{C_o}\right) = k\,t \quad \Rightarrow \quad C = C_o\,e^{-kt} \tag{21}$$

As suggested by Eqs. (20) and (21), a plot of ln$C$ (or log$C$) *vs.* time gives a straight line. The slope of this line defines the reaction rate constant, $k$.

Examples of common reactions in foods of first order include microbial death, thermal denaturation of proteins (Anema, 2001), vitamin loss in frozen, canned and dry food, oxidative color loss, fresh and dry food shelf life determining factors, microbial growth in fresh foods up to a practical shelf time, etc. In the following Figure 13.2 the effect of reaction rate constant $k$ on the rate of inactivation is illustrated.

Approximate knowledge of reaction rate constants from literature or preliminary experiments is essential in proper design of kinetic experiments. A logarithmic change of the initial concentration at the end of the experiments is rather necessary for valid conclusions. For short reaction times (in relation to the rate constant, $k$), first order cannot be distinguished from zero reactions as illustrated on the linear plot (top) of Figure 13.2. For $k = 0.001$ min$^{-1}$ and destruction data up to only 10% of the initial value, a linear relationship between concentration and time can be assumed, although data were generated assuming first order kinetics (i.e., logarithmic relationship).

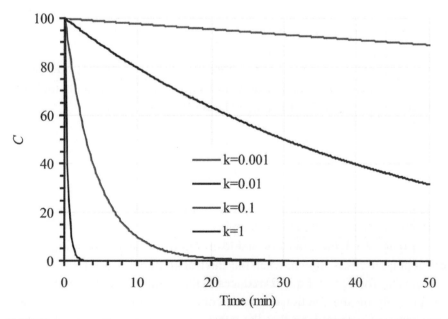

**FIGURE 13.2A**   Effect of reaction rate constant, $k$ (min$^{-1}$) on the position of the destruction curves for first order inactivation on linear (top) and logarithmic scale (bottom).

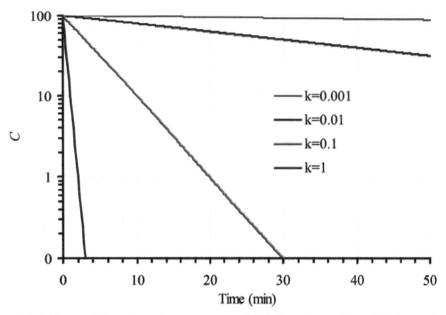

**FIGURE 13.2B**    Effect of reaction rate constant, $k$ (min$^{-1}$) on the position of the destruction curves for first order inactivation on linear (top) and logarithmic scale (bottom).

### 13.3.1.2    Fractional Model

For cases where a remaining stable fraction of the parameter under study exists, a modification of the first-order model has been proposed, namely the fractional model (Van Den Broeck et al., 2000):

$$\ln\left(\frac{C_0 - C_\infty}{C_t - C_\infty}\right) = kt \qquad (22)$$

where $C_\infty$ is a stable fraction of the parameter under consideration remaining after processing or storage for infinite time.

In Figure 13.3 the effect of stable fraction $C_\infty$ on the position of the destruction curves is shown, when the fractional model is used.

Note that for $C_\infty = 0$, Eq. (22) reduces to the regular first order model (Eq. 21). In applying the fractional model, concentration data for long enough times must be gathered, so that the constant $C_\infty$ value, if exists, should be obtained.

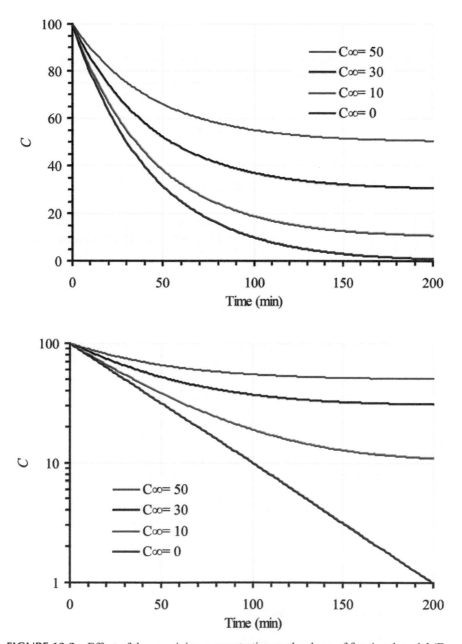

**FIGURE 13.3** Effect of the remaining concentration on the shape of fractional model (Eq. 22) curves in comparison to first order reaction predictions ($C_\infty = 0$) for $k = 0.01$ min$^{-1}$ on linear (top) and logarithmic scale (bottom).

### 13.3.1.3  Zero-Order Reactions (n=0)

When the rate of change is constant, independent of the concentration, then the equations used are (Eq. 23 and 24)

$$r = \frac{dC}{dt} = kC \tag{23}$$

$$C = C_0 + kt \tag{24}$$

Examples of common reactions in foods of zero order include frozen food shelf life, enzymatic reactions in fresh foods, non enzymatic browning, etc.

### 13.3.1.4  Second-Order Reactions (n = 2)

The respective equations in the case of second-order reactions are:

$$r = -\frac{dC}{dt} = k\,C^2 \tag{25}$$

$$\frac{1}{C} = \frac{1}{C_0} + k\,t \tag{26}$$

Second order unimolecular reactions are characterized by a hyperbolic relationship between the concentration of the reactant or product and time (Toledo, 1999). A linear plot is obtained when the term ($1/C$) is plotted *vs.* time, the slope of which gives the $k$ value.

### 13.3.1.5  $n^{th}$ Order Reactions

Apparent order reaction is not necessarily represented by an integer. The respective equations in the case of $n^{th}$-order reactions, for $n \neq 1$, are given in Eqs. (27) and (28), as already presented at the beginning of this section:

$$r = -\frac{dC}{dt} = k\,C^n \tag{27}$$

$$C^{1-n} - C_0^{1-n} = (n-1)k\,t \tag{28}$$

Note that for $n<1$, and due to the physical nature of $C$ which cannot accept negative values, calculations with the corresponding equation (Eq. 28) can be valid only up to a certain time period (that is, for $t > C^{1-n}/(n-1)k$)

### 13.3.1.6   Determining the Order of Reaction

In order to reliably select the appropriate apparent order, the experimental design should include enough measurements of concentration change of the quality index. If experiment is concluded too soon, different reaction orders might seem equivalently appropriate from a goodness of fit point of view. This is illustrated in Figure 13.4, where the effect of the extent of data collection on the shape of destruction curves is presented. As can be seen, even with data up to 90% reduction of the initial concentration, it is difficult to distinguish deviation of the linearity (valid only for the case of $n=1$) of the different curves, especially given the experimental error involved.

Practically, once the parameter that should describe a food system degradation is selected, it is necessary to proceed to a suitable experimental design, so as to produce data of the change of $C$ over time (Taoukis and Giannakourou, 2007). Analysis of this information in order to estimate the most appropriate reaction order can be accomplished by differential or integral methods (Karel and Lund, 2003).

### 13.3.1.7   Differential Methods

By taking the logarithm of Eq. (27), the equation used to describe the system change has the following form:

$$\ln(-\frac{dC}{dt}) = \ln k + n \cdot \ln C \tag{29}$$

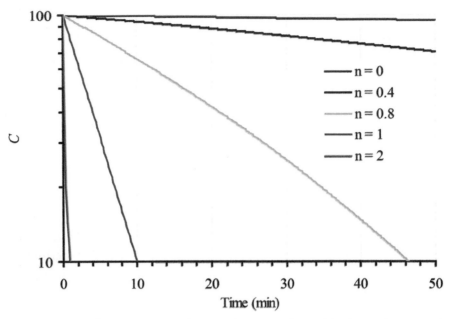

**FIGURE 13.4**  Effect of the extent of inactivation on the shape of the destruction curves for $k = 0.1 \ \mathrm{min^{-1} \times (mol/mL)^{1-n}}$.

Eq. (29) suggests that the slope of the $\ln(-dC/dt)$ vs. $\ln C$ line is equal to the order of the reaction, $n$. Thus, in determining $n$:

•    Determine $C$ vs. time

•    Calculate $-dC/dt$ between two consecutive data points as:

$$-\frac{dC}{dt} = -\frac{C_2 - C_1}{t_2 - t_1} \tag{30}$$

•    Plot $\ln(-dC/dt)$ vs. $\ln C$, draw a straight line and determine $n$ from the slope and $k$ from the intercept of the straight line.

It is suggested that the calculated $\ln(-dC/dt)$ within each time interval $(t_2 - t_1)$ to be related with the middle time $(t_1 + t_2)/2$ (that is, approximating slopes with central differences), rather than the beginning, $t_1$, or the end, $t_2$, of the particular time interval. Alternatively, $dC/dt$ can be determined from the slope of a curve (e.g., polynomial equation) fitted to $C$ vs. $t$ data. An example in using the differential method to estimate the reaction order $n$ is given below on Table 13.1 and Figure 13.5 for *E. coli* inactivation data.

**TABLE 13.1** *E. coli* Inactivation Data at 200 MPa and 25°C for Reaction Order Determination by the Differential Method

| t (min) | C (CFU/mL) | ln(C) | t (min) | ln(C) | –dC/dt | ln(–dC/dt) |
|---------|------------|-------|---------|-------|--------|------------|
| 0 | 870,000,000 | 20.58 | 10 | 16.40 | 43,490,000 | 17.59 |
| 20 | 200,000 | 12.21 | 30 | 11.50 | 7550 | 8.93 |
| 40 | 49,000 | 10.80 | 50 | 9.57 | 2240 | 7.71 |
| 60 | 4200 | 8.34 | 70 | 7.89 | 125 | 4.83 |
| 80 | 1700 | 7.44 | 90 | 7.38 | 10 | 2.30 |
| 100 | 1500 | 7.31 | 110 | 7.02 | 33 | 3.50 |
| 120 | 840 | 6.73 | 130 | 6.17 | 28.5 | 3.35 |
| 140 | 270 | 5.60 | 150 | 5.45 | 3.5 | 1.25 |
| 160 | 200 | 5.30 | 170 | 4.85 | 5.9 | 1.77 |
| 180 | 82 | 4.41 | | | | |

## 13.3.1.8 Integral Methods

This is a trial-and-error process. For $n \neq 1$, a reaction order is assumed and the quantity $(C^{1-n} - C_0^{1-n})$ is calculated and plotted *vs.* time. If the plot or the regression of this term *vs.* time yields a linear function (passing through the origin with slope equal to $-k$) then the hypothesis is correct; otherwise, a new assumption about $n$ is made and the procedure is repeated.

### *Example 1: Determination of Reaction Order*

From the following *E. coli* inactivation data at 200 MPa and 25°C, determine the order of the reaction.

From the $C$ *vs.* $t$ data we calculate $\ln(C)$ and $\ln(-dC/dt)$ values at the each mid time interval (Table 13.1). The slope of the least square regression straight line through the $\ln(-dC/dt)$ *vs.* $\ln(C)$ data gives the value of the reaction order, $n$. For the particular example $n$ is found equal to 1.4055 (Figure 13.5). Applying the integral method, that is using Eq. (19) with the data presented on Table 13.1 and non linear regression analysis we obtained $n = 1.4119 \pm 0.0805$.

### *Determining Half Life Times*

The half life time $(t_{1/2})$ is the time required for the parameter $C$ to lose half of its initial concentration. Table 13.2 summarizes the mathematical form of

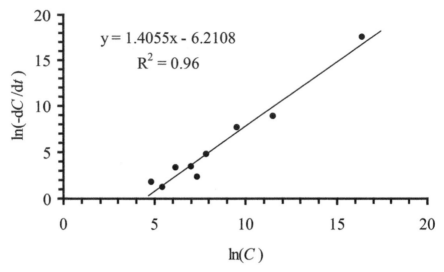

**FIGURE 13.5** Determination of the apparent inactivation reaction order for *E. coli* at 200 MPa and 25°C.

$t_{1/2}$ for the different reaction rate orders. Solving Eqs. (18) and (19) for time, processing time or shelf life can be determined for any decided fractional decrease of the critical parameter.

### 13.3.2  CLASSICAL THERMAL DEATH TIME MODEL

In the more simplistic case, inactivation of spores is generally assumed to follow log-linear kinetics (first order inactivation). According to the classical thermobacteriological approach (Bigelow et al., 1920; Ball, 1923) for thermal inactivation kinetics, inactivation rate is described by the decimal reduction time ($D_T$), as given by the following equation:

$$N = N_0 10^{-t/D_T} \Rightarrow \log N = \log N_o - \frac{t}{D_T} \tag{31}$$

where $D_T$ (decimal reduction time) is defined as the time in minutes, at constant temperature, required to reduce the initial concentration of a thermolabile substance by 90%, $N$ is a measurement of microbial load at time $t$, for example, CFU/mL, number of spores per can, etc. and $N_0$ the initial microbial load.

**TABLE 13.2**   Half Life Time of a Quality Index C for Different Reaction Orders

| Apparent reaction order | Half-life |
|---|---|
| 0 | $\dfrac{C_0}{2k}$ |
| 1 | $\dfrac{\ln 2}{k}$ |
| 2 | $\dfrac{1}{k \cdot C_o}$ |
| $n\ (n\neq1)$ | $\dfrac{1}{k \cdot [n-1]}\dfrac{2^{n-1}-1}{C_o^{\,n-1}}$ |

According to the classical thermobacteriological approach the $D$-value is defined on a semi-logarithmic plot of the logarithm of microbial population $vs.$ time, described by the survivor curve, as a straight line, as the time needed for the survivor curve to transverse a logarithmic cycle (Figure 13.6) (Stoforos, 2010). Assuming first order kinetics and for $C=0.1C_0$, $D$ can also be determined from Eq. (18):

$$D = \ln(\frac{C_0}{0.1\,C_0})/k \Rightarrow D = \frac{\ln(10)}{k} \tag{32}$$

### 13.3.3   LOG LINEAR WITH SHOULDER AND/OR TAILING

The log linear model was originally developed as two coupled differential equations, expressed as follows (Geeraerd et al., 2000):

$$\frac{dN}{dt} = -k_{\max} \times N \times \left(\frac{1}{1+C_e}\right) \times \left(1 - \frac{N_{res}}{N}\right) \tag{33}$$

$$\frac{dC_e}{dt} = -k_{\max} \times C_e \tag{34}$$

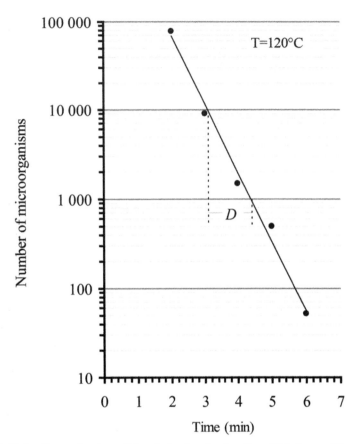

**FIGURE 13.6** Determination of the decimal reduction time ($D_T$) from a Thermal Death Time curve (or Survivor Curve) according to the classical thermobacteriological approach.

where $C_e$ is related to the physiological state of cells (dimensionless) $k_{max}$ is the specific inactivation rate (time$^{-1}$) and $N_{res}$ is the residual population density (in CFU/mL).

This model can exhibit a log-linear behavior with and without shoulder and/or tailing revealing a smooth transition between each phase (Geeraerd et al., 2005). The explicit solution of the original dynamic model when an appropriate term for shoulder description is introduced, reads as follows:

$$\log N = \log N_0 \times e^{-\mu_{max}t} \times \left( \frac{\mu_{max}S_l}{1+(e^{\mu_{max}S_l}-1)\times e^{-\mu_{max}t}} \right) \times \left(1 - \frac{N_{res}}{N}\right) \quad (35)$$

where $s_l$ is the shoulder length (time).

The model structure has been successfully applied to survival data of different microorganisms and different treatments (Greenacre et al., 2003; Marquenie et al., 2003; Ongeng et al., 2014; Valdramidis et al., 2005).

### 13.3.4 WEIBULL MODEL

A simple way to describe nonlinear curves is the Weibull equation (Barsa et al., 2012; Chakraborty et al., 2015; Smelt and Brul, 2014; Saraiva et al., 2016). This equation can properly describe inactivation curves that are either upward or downward concave. Weibull and Weibull like equations have been applied for the description of inactivation of bacterial spores as well as inactivation of vegetative cells.

$$\log\left(\frac{N}{N_0}\right) = -\left(\frac{t}{a}\right)^{\beta} \tag{36}$$

in which $N$ is the number of microorganisms at time t, $N_0$ the initial number of microorganisms, $t$ the heating time, $\alpha$ the scale parameter, and $\beta$ the shape parameter. Note that for $\beta=1$ Weibull model reduces to first order kinetics and $\alpha$ becomes the decimal reduction time. Figure 13.7 illustrates the effect of parameters $\alpha$ and $\beta$ on the shape and position of Weibull curves describing microbial inactivation.

Recently, another equation is proposed as an extension of the Weibull type curves (Albert and Mafart, 2005; Ongeng et al., 2014; Valdramidis et al., 2012), including concave, convex, and linear curves followed by a tailing effect:

$$\log_{10} N = \log_{10}\left[ (N_0 - N_{res}) \times 10^{\left(-\left(\frac{t}{a}\right)^{\beta}\right)} + N_{res} \right] \tag{37}$$

For $N_{res}=0$ and $\beta>1$, downward concave curves are obtained, while for $\beta<1$, upward concave curves are described. If $N_{res}=0$ and $\beta=1$ the model reduces to the log-linear model.

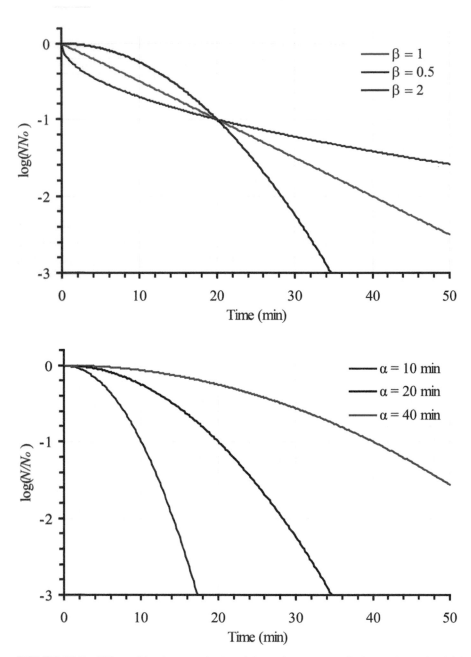

**FIGURE 13.7**    Effect of the shape, α, (top) and the scale parameter, β, (bottom) on microbial inactivation according to Weibull model. Top figure data were generated for α=20 min while the data on the bottom figure were generated for β=2

### 13.3.5 BIPHASIC INACTIVATION

One of the causes of tailing that were suggested by Cerf (1977) was the existence of two phenotypically different subpopulations. The majority is relatively heat sensitive, whereas a minority is relatively heat resistant. In that case, microbial inactivation can be described by the following equation:

$$\ln\left(\frac{N}{N_0}\right) = \ln\left((1-f)\times\exp(-k_1 t) + f\times\exp(-k_2 t)\right) \tag{38}$$

where $1-f$ represents the majority fraction of heat sensitive cells, $k_1$ the inactivation rate constant of the sensitive cells, $f$ the minority fraction of resistant cells, $k_2$ the inactivation rate constant of resistant cells. Examples of efficient use of this model or its modified form with shoulder in recent literature include studies presented by McKellar et al. (2014), Janssen et al. (2007), Ongeng et al. (2014) and Tyrovouzis et al. (2014).

### 13.3.6 LOGISTIC MODEL

In fact, the classical model implies a first-order reaction. In biological terms, it means that inactivation of a microorganism is caused by a single hit, which is a simplistic and rather untrue assumption. Moreover, when examining experimental data of microbial inactivation, there are many examples of shoulders and also sigmoidal curves (Smelt and Brul, 2014). Based on these considerations, a number of nonlinear models have been developed. In fact, these sigmoidal models have been used to describe either microbial inactivation or microbial growth. This model can be described by the following equation (McMeekin et al., 1993):

$$\frac{dN}{dt} = k \times N \times \left(1 - \frac{N}{N_{max}}\right) \tag{39}$$

where $N_{max}$ is the maximum cell density (CFU/mL).

Integrating Eq. (39) and introducing some new variables, a new form is presented, that is commonly applied in the field of predictive microbiology:

$$N = \frac{C_{GL}}{1 + \exp\{-B_L \times (t - M)\}} \tag{40}$$

where $C_{GL}$ is the asymptotic value (corresponding to the maximum density of microbial population at infinite time, log(CFU/mL)), $B_L$ the special rate of growth at time t=0 (an approximate value is assigned to this parameter) (h$^{-1}$), $M$ the time at which the absolute growth rate assumes its maximum value (h).

Since the first form of this model was presented, several modifications have been proposed so as to improve its accuracy. For example, the modified logistic model (Zwietering et al 1990; Timmermans et al., 2016; Ongeng et al., 2014) (Eq. 41) and the respective form with delay (Eq. 42) (Augustin and Carlier 2000; Guillard et al., 2016; Hereu et al., 2014) have been successfully used in many recent studies.

$$\log N = \frac{N_{max}}{1 + \exp\left[\frac{-4\mu_{max}}{N_{max}}(t - \lambda) + 2\right]} + N_0 \tag{41}$$

$$\log N = \log N_0 \qquad t \leq \lambda$$

$$\log N = \log\left(\frac{N_{max}}{1 + \left(\frac{N_{max}}{N_0} - 1\right)\exp(-\mu_{max}(t - \lambda))}\right) \qquad t > \lambda \tag{42}$$

where $N_0$ is the logarithm of the initial number of counts (CFU/mL), $\mu_{max}$ the maximum specific growth rate (log(CFU/mL)/h) and $\lambda$ the lag phase (h).

### 13.3.7 GOMPERTZ EQUATION

This model, quite popular among food microbiologists, has been extensively used in earlier (Bratchell et al., 1990; Buchanan et al., 1997; Buchanan and Bagi 1999; Gibson et al., 1988; Linton et al., 1995; Linton et al., 1996) and more recent literature (Ates et al., 2016; Simpson et al., 2015; Muñoz-Cuevas et al., 2012). Gompertz equation (Eq. 43) was found to describe adequately exponential and stationary phase of sigmoidal growth curves, but it was deemed not effective for lag phase. This weakness was overcome by the modified Gompertz model (Eq. 44) (Li et al., 2007), where a new term was added to account for the lag phase:

$$\log N = C \cdot \exp\left\{-\exp\left[\left(\frac{\mu_{max} \exp(1)}{C}\right) \cdot (\lambda - t) + 1\right]\right\} \tag{43}$$

$$\log N = A + C \cdot \exp\left\{-\exp\left[\left(\frac{\mu_{max} \exp(1)}{C}\right) \cdot (\lambda - t) + 1\right]\right\} \tag{44}$$

for $C$ being the asymptotic value (corresponding to the maximum density of microbial population at infinite time, log(CFU/mL). Representative growth curves based on Gompertz equation are presented in Figure 13.8.

### 13.3.8 BARANYI MODEL

Besides other models previously presented (e.g., logistic, Gompertz,) used to describe sigmoidal microbial growth curves, the model proposed by Baranyi and Roberts (1994), as given here by Eq. 45, can be employed, basically due to its ability for proper kinetic parameter determination even from limited number of data points at the exponential and the stationary phase of the growth curve (Tyrovouzis et al., 2014).

$$y(t) = y_0 + \mu_{max} \times A(t) - \ln\left[1 + \frac{\exp((\mu_{max} \times A(t)) - 1)}{\exp(y_{max} - y_0)}\right] \tag{45a}$$

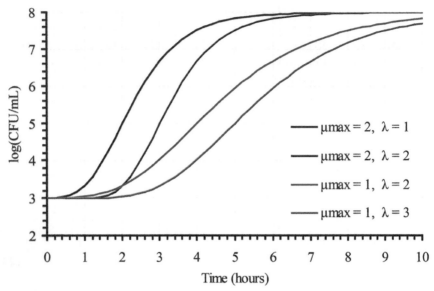

**FIGURE 13.8**   Effect of maximum specific growth rate ($\mu_{max}$ in log(CFU/mL)/h) and/or lag phase ($\lambda$ in h) on microbial growth based on modified Gompertz equation.

$$A(t) = t + \frac{1}{\mu_{max}} \ln\left(\exp\left(-\mu_{max} t\right) + \exp\left(-\mu_{max} \lambda\right) - \exp\left(-\mu_{max}(t + \lambda)\right)\right)$$

(45b)

A number of recent studies have effectively implemented these equations in order to describe microbial growth (Kim et al., 2016; Tyrovouzis et al., 2014; Dabadé et al., 2015; García et al., 2015; Smet et al., 2015) of different microorganisms, under various conditions.

### 13.3.9   OTHER MODELS

#### 13.3.9.1   Normal Distribution Model

Augustin et al. (1998) proposed a model based on log-normal distribution of the probability of microbial death as a function of time, adopting the assumption that distribution of the number of dead cells against time can be cumulative log normal. From that assumption, he developed inactivation curves that were mainly characterized by tailing (Smelt and Brul 2014).

$$N = \frac{N_0}{1 + \exp\left(\dfrac{\log(t) - m_a}{s_a^2}\right)} \tag{46}$$

where $m_a$ is the time (min) where inactivation rate is maximum and $s_a$ a parameter proportional to the standard deviation of heat resistance, in $min^{1/2}$.

### 13.3.9.2  Sapru Model

This model refers specifically to the activation of bacterial spores during the sterilization process and assumes an initial increase of an activated spore population (Sapru et al., 1992). Two types of spores are distinguished: (i) a dormant, viable population $n_d$ (CFU $mL^{-1}$) potentially able of producing colonies on an appropriate growth medium after heat activation and (ii) an active population $n_a$ (CFU $mL^{-1}$) able of forming colonies without activation on an appropriate growth medium (Smelt and Brul 2014). The model consists of two equations:

$$\frac{dN_d}{dt} = -\left(k_{d1} + k_a\right) \times N_d$$

$$\tag{47}$$

$$\frac{dN_a}{dt} = k_a N_d - k_{d2} \times N_d$$

with $N_d(t{=}0){=}N_d(0)$ and $N_a(t{=}0){=}N_a(0)$ and $k_a$ being the first-order activation constant of the dormant spores, $(min^{-1})$, $k_{d1}$ the first-order inactivation constant of dormant population, $(min^{-1})$ and $k_{d2}$ the first-order inactivation constant of $N_a$, $(min^{-1})$.

## 13.4  SECONDARY MODELS

Secondary models are used to describe primary model parameters as functions of one or more intrinsic or extrinsic conditions like temperature, pH, water activity ($a_w$), etc. In the following section, an overview of the most popular secondary models will be presented, focusing on the effect of temperature. The prevailing effect of temperature on the rate of food related reaction rates has long been the subject of research since it strongly affects post processing reaction rates.

### 13.4.1  ARRHENIUS EQUATION

The most widely used secondary model to describe the effect of temperature, $T$, on the reaction rate constant, $k$, of chemical reactions, is the Arrhenius relation, derived from thermodynamic laws as well as statistical mechanics principles (Barsa et al., 2012):

$$k = k_A \cdot \exp\left(\frac{-E_a}{RT}\right)$$

(48)

with $k_A$ representing the Arrhenius equation constant and $E_a$, in J/mol or cal/mol, is defined as the activation energy, i.e., the excess energy barrier that quality parameter $A$ needs to overcome to proceed to degradation products. $R$ is the universal gas constant (1.9872 cal/mol·K or 8.3144 J/mol·K). Temperature, $T$, in the above equation refers to absolute temperature (K). To estimate the temperature effect on the reaction rate of a specific quality deterioration mode, values of $k$ are estimated at different temperatures, in the range of interest, and ln$k$ is plotted $vs.$ the term of $1/T$ in a semilogarithmic graph. A straight line is obtained with a slope of $-E_a/R$ from which the activation energy is calculated, as shown in Figure 13.9.

It should be noted that the Arrhenius equation implies that $k_A$ is the theoretical value of the reaction rate at zero degrees K. Alternatively, the use of a reference temperature, $T_{ref}$, is recommended, corresponding to a representative value in the temperature range of the process/storage of study. By applying Eq. (48) for $T=T$ and $T=T_{ref}$ and dividing by term the resulting equation, we obtain,

$$k = k_{ref} \exp\left[\frac{-E_a}{R}\left(\frac{1}{T} - \frac{1}{T_{ref}}\right)\right]$$

(49)

where $k_{ref}$ is the rate constant at the reference temperature $T_{ref}$. The value of $E_a$ is, in that case, calculated from the linear regression of ln$k$ $vs.$ $(1/T-1/$

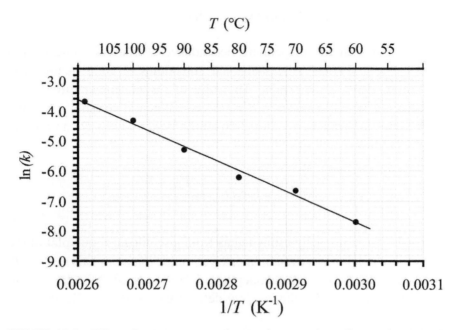

**FIGURE 13.9** Effect of temperature on the reaction rate. According to the Arrhenius equation, the slope of the straight line is equal to $-E_a/R$.

$T_{ref}$). Values of $T_{ref}$ usually employed are 255K for frozen, 277K for chilled and 295K for ambient temperature stored food products. Besides giving the $k_{ref}$ parameter a practical physical meaning, the above transformation towards Eq. (49) of the Arrhenius equation provides enhanced stability during numerical parameter estimation and integration. Equation (49) has been used as an empirical tool to describe the effect of $T$ on $k$ even when the real mechanism of the reaction is not known. Recently, the classical form of Arrhenius was reconsidered by (Peleg et al., 2012) and it was rewritten in a generic simpler form as follows:

$$\frac{k(T)}{k(T_{ref})} = \exp\left[a\left(\frac{1}{T_{ref}+b} - \frac{1}{T+b}\right)\right] \qquad (50)$$

where $T$ and $T_{ref}$ are in °C, and $a$ (in °C$^{-1}$) and $b$ (in °C) are empirical parameters to be determined from experimental data.

### 13.4.2   THE D AND Z VALUE MODEL

Another term used for temperature dependence of microbial inactivation kinetics during thermal processing and sometimes of food quality loss (Hayakawa, 1969), is the z-value. The value of z is the temperature change that causes a 10-fold change in the *D-value*. When addressing microbial and enzyme inactivation, the most frequently used model is *D* and *z-value model*, as follows:

$$\log D_T = \log D_{T_{ref}} + \frac{T_{ref} - T}{z} \qquad (51)$$

The z-value is traditionally estimated as from a semilogarithmic plot of the logD vs. T which forms a straight line, the so called *Phantom Thermal Death Time* curve (NCA 1968). The temperature difference needed for the straight line to transverse a logarithmic cycle defines the *z-value* as illustrated in Figure 13.10.

Arrhenius and *z-value* model are not compatible. However, for the limited range of temperature used to collect experimental data, either model seems adequate (Jonsson et al., 1977; David and Merson, 1990). The relationship between Arrhenius and z kinetic variables is:

$$D_{T_{ref}} = \frac{\ln 10}{k_{ref}} \qquad (52)$$

and

$$z = \frac{(\ln 10) \cdot R \cdot T^2}{E_a} \qquad (53)$$

### 13.4.3   EXPONENTIAL MODEL

Equation (49) with *b* larger than 250°C could be also described, with little or no sacrifice of the degree of fit, by the simpler exponential model (Barsa et al., 2012):

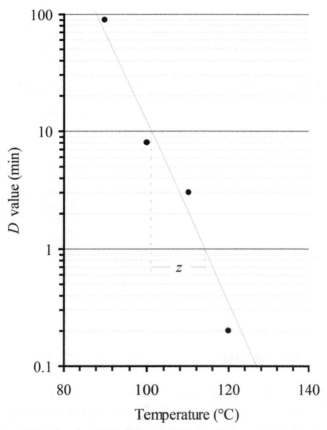

**FIGURE 13.10** Effect of temperature on the decimal reduction time according to the classical thermobacteriological approach. Determination of the $z$ value from a Phantom Thermal Death Time curve.

$$\frac{k(T)}{k(T_{ref})} = \exp\left[c\left(T - T_{ref}\right)\right]$$

(54)

where $T$ and $T_{ref}$ are both in °C and $c$ is a constant having °C units too. Examples of the interchangeability of the Arrhenius equation and the exponential model are provided in several publications. One can also examine the relationship between the two models online with the interactive Wolfram Demonstration (2015).

Equation (51) can be transformed in terms of the exponential rate constant $k(t)$:

$$\frac{k(T)}{k(T_{ref})} = 10^{\frac{T_{ref}-T}{z}} \tag{55}$$

which is similar to the exponential model, previously described (Eq. 54), for:

$$c = \frac{\ln\left[10^{\frac{T-T_{ref}}{z}}\right]}{T-T_{ref}} \tag{55}$$

or

$$z = \frac{(T-T_{ref})\times \ln 10}{c\times(T-T_{ref})} \tag{56}$$

## 13.4.4  THE WLF MODEL

In systems that are subject to glass transition, due to drastic acceleration of the diffusion controlled reactions above $T_g$, the dependence of the rate of a food reaction on temperature cannot be described by a single Arrhenius equation. In the rubbery state above $T_g$, the activation energy is not constant, but is rather a function of temperature. This behavior has been often described by an alternative equation, the Williams-Landel-Ferry (WLF) expression (Eq. 57) that empirically models the temperature dependence of mechanical and dielectric relaxations in the range of $T_g < T < T_g + 100$:

$$\log\frac{k_{ref}}{k} = \frac{C_1 \cdot (T-T_{ref})}{C_2 + (T-T_{ref})} \tag{57}$$

where $k_{ref}$ is the rate constant at the reference temperature $T_{ref}$ ($T_{ref} > T_g$) and $C_1$, $C_2$ are system dependent coefficients. Williams et al. (1955), assuming $T_{ref} = T_g$ and applying WLF equation for data available for various polymers, estimated mean values of the coefficients: $C_1 = -17.44$ and $C_2 = 51.6$. However, the uniform application of these constants is often problematic

(Barsa et al., 2012; Peleg 1992; Peleg et al., 2002; Terefe and Hendrickx 2002) and the calculation of system-specific values, whenever possible, should be preferred. Using Eq. (57) with $T_{ref}=T_g$ and rearranging the mathematical expression to the following form (Eq. 58):

$$\left[\log\frac{k_g}{k}\right]^{-1} = \frac{-C_2}{C_1\cdot(T-T_g)} - \frac{1}{C_1} \tag{58}$$

such that a plot of

$$\left[\log\frac{k_g}{k}\right]^{-1} \text{ vs } \frac{1}{(T-T_g)} \tag{59}$$

gives a straight line with a slope equal to $-C_2/C_1$ and an intercept of $-1/C_1$, if the WLF model is successfully applied. Note that WLF model introduces one more constant.

### 13.4.5 THE EYRING–POLANYI MODEL

The Eyring–Polanyi model has been recently proposed to replace the Arrhenius equation as a temperature dependence model where the plot of lnk versus $1/T$ ($T$ in K) is curvilinear (Barsa et al 2012). The original model's equation has the following form:

$$k(T) = \frac{K_B T}{h}\exp\left[-\frac{\Delta G^{\ddagger}}{RT}\right] \tag{60}$$

or equivalently:

$$\ln k = \ln\left(\frac{K_B}{h}\right) + \frac{S}{R} - \frac{H}{RT} + \ln T \tag{61}$$

where $K_B$ is the Boltzman constant ($1.381\times10^{-23}$ J K$^{-1}$), $h$ Planck's constant ($6.626\times10^{-34}$ J s), $T$ the absolute temperature in K, $DG$ the free energy of activation (kJ mol$^{-1}$), $R$ the Universal gas constant (8.314 J mol$^{-1}$ K$^{-1}$), $H$ the heat of activation (kJ mol⁻¹), and $S$ is the entropy (kJ mol⁻¹). The significance of this equation lies on that it relates the effect of temperature on the reaction rate constant to the fundamental terms of enthalpy and entropy changes and, consequently, one could estimate the enthalpy/entropy compensation in food reactions.

The Eyring–Polanyi equation can also be written in the generic form, similar to that of the Arrhenius equation, where the temperature is expressed in °C and $T+b$ and $T_{ref}+b$ replaces the absolute temperature in K, that is, (Barsa et al., 2012):

$$\frac{k(T)}{k(T_{ref})} = \frac{T+b}{T_{ref}+b} \exp\left[ a_{EP}\left(\frac{1}{T_{ref}+b} - \frac{1}{T+b}\right)\right] \tag{61}$$

where, as before, $T$ and $T_{ref}$ are in °C, $a_{EP}$ in °C⁻¹, and b (in °C) can, but no need, to be 273.16°C.

### 13.4.6  $Q_{10}$ CONCEPT

Temperature dependence has been traditionally expressed in the food industry, especially during cold storage, and the food science and biochemistry literature with the $Q_{10}$ parameter (Taoukis, 2003). $Q_{10}$ is defined as the ratio of the reaction rate constants at temperatures differing by 10°C or the change of shelf life $\theta_s$ when the food is stored at a temperature higher by 10°C:

$$Q_{10} = \frac{k_{(T+10)}}{k_{(T)}} = \frac{\theta_s(T)}{\theta_s(T+10)} \tag{62}$$

Assuming $k$ is adequately described by the exponential function, then:

$$\frac{k_{T+10}}{k_T} = \frac{k_o \cdot e^{b(T+10)}}{k_o \cdot e^{bT}} \Rightarrow Q_{10} = e^{10b} \Rightarrow \ln Q_{10} = 10b \tag{63}$$

Alternatively, shelf life $(\theta_s)$ can be plotted *vs.* temperature, following Eq. 64 and the resulting plots are often called shelf life plots:

$$\theta_s(T) = \theta_{so} \cdot e^{-bT} \Rightarrow \ln \theta_s = \ln \theta_{so} - b \cdot T \qquad (64)$$

where $b$ is the slope of the shelf life plot and $\theta_{so}$ is the intercept. The *shelf life plots* are true straight lines only for narrow temperature ranges of 10 to 20°C. Outside this interval, $Q_{10}$ and $b$ are functions of temperature, correlated to the activation energy of the food quality deterioration reaction, following Eq. (65), assuming that Arrhenius expression is the governing equation by:

$$\ln Q_{10} = 10b = \frac{E_a}{R} \frac{10}{T_K \cdot (T_K + 10)} \qquad (65)$$

### 13.4.7 THE BELEHRADEK AND SQUARE ROOT MODELS

The study of the temperature dependence of microbial growth has been an area of increased interest. As an essential part of predictive microbiology, several equations have been proposed for describing the effect of temperature on the exponential growth rate of the microorganism in question. Most prominently used in the suboptimal temperature range are the Arrhenius and the Belehradek equation (MacMeekin et al., 1993). The Belehradek equation (Belehradek, 1930) has the following generalized form (Eq. 66):

$$k = a \cdot (T - T_0)^d \qquad (66)$$

where $a$ and $d$ are parameters to be fitted and $T_0$ is regarded as the "biological zero," that is, a temperature at and below which no growth is possible. To avoid this assumption, Ratkowsky et al. (1983) changed $T_o$ to $T_{min}$ and proposed the square root model (Eq. 67):

$$\sqrt{k} = b \cdot (T - T_{min}) \qquad (67)$$

where $k$ is the growth rate, $b$ is the slope of the regression line of $k^{1/2}$ vs. $T$ and $T_{min}$ is the hypothetical growth temperature where the regression line cuts the T-axis at $k^{1/2} = 0$. $T_{min}$ is usually 2–3°C lower than the temperature at which growth is actually observed and has been described as notional or conceptual temperature. Equation (67) was later extended to the form of Eq. (68) to cover the entire growth temperature range.

$$\sqrt{k} = b \cdot (T - T_{min})\{1 - \exp[c \cdot (T - T_{max})]\} \qquad (68)$$

However, it should be stressed that, since it is very difficult to obtain accurate data at extremely low growth rates, it should be pointed out that $T_{min}$ and $T_{max}$ may differ from the true temperature limits, which should be determined through independent experiments. Despite being a merely empirical, this set of square root equations have been effectively used to many cases in the predictive microbiology literature (Barsa et al., 2012).

### 13.4.8  OTHER MODELS

#### 13.4.8.1  The Log-Logistic Model

The log-logistic model has been proposed as an alternative temperature dependence model for Weibullian microbial inactivation, which only starts at high enough temperatures and does not follow fixed order kinetics. It could also describe other food systems as well, as described by Peleg et al. (2002). The log-logistic model can be expressed as follows:

$$k(T) = \ln\{1 + \exp[c(T - T_c)]\} \qquad (69)$$

where $c$ is a constant and $T_c$ marks the temperature range where $k(t)$ starts to rise, and hence has clear physical significance (Barsa et al., 2012). It has greater flexibility than the Arrhenius, Eyring-Polanyi and exponential models, although it has two parameters instead of one.

#### 13.4.8.2  Probability Model

A probability model is a mathematical formulation which can be used to estimate a risk or possibility as a function of given intrinsic and extrinsic factors (Li et al., 2008). The establishment of a solid relationship between the growth of microbial cells and the physico-chemical properties of the environment is a prerequisite basis for probability modeling in predictive microbiology (Ratkowsky and Ross 1995).

### 13.4.8.3   Response Surface or Polynomial Model

The polynomial model of the response surface methodology was first introduced into predictive microbiology by Gibson et al. (1988) to describe growth rate and lag time as a function of storage temperature, pH, and salt concentration. It is still widely used to establish the relationship between primary model parameters and environmental factors. In its various forms is expressed as:

- Linear response surface model:

$$y = \alpha_0 + \Sigma_i b_i x_i \tag{70}$$

- Quadratic response surface model:

$$y = \alpha_0 + \Sigma_i b_i x_i + \Sigma_i \Sigma_{i<j} c_{ij} x_i x_j + \Sigma_i d_i x_i^2 \tag{71}$$

- Cubic response surface model:

$$y = \alpha_0 + \Sigma_i b_i x_i + \Sigma_i \Sigma_{i<j} c_{ij} x_i x_j + \Sigma_i d_i x_i^2$$
$$+ \Sigma_i \Sigma_{i<j} \Sigma_{j<k} e_{ijk} x_i x_j x_k + \Sigma_i f_i x_i^3 \tag{72}$$

where $y$ is the response variable, $x_i$ is the environmental factor, $c_{ij}$, $d_i$, $e_{ijk}$ and $f_i$ are regression coefficients.

Although polynomial model is flexible enough to incorporate even very strong interactive effects and provides the ability to correlate parameters of primary models with environmental factors, it is still purely empirical and has significant limitations (Li et al., 2008).

### 13.4.8.4   Neural Networks

A neural network (NN) is a computerized procedure which learns or is trained from examples through iterations and automatically derives the mathematical formulae to map the relationships between the input and output data through neurons, without any prior knowledge of their relationships (Xie et al., 1998; Li et al., 2008; Xie and Xiong, 1999). Due to the capability of

NN to operate a large number of neurons in parallel, it can handle large and complex datasets and in recent years NN has been successfully applied to correlate the primary model parameters with the environmental factors and to predict the behavior of microorganisms.

Compared with the response surface model using fixed functions, these NN models are more versatile, flexible, and less restrictive as they do not impose assumptions pertaining to the form of functions and use flexible basis functions to fit data (Geeraerd et al., 1998). In spite of their good perfor-mance, an over-fitting phenomenon always occurs during training. Geeraerd et al. (1998) indicated that overtraining and complexity of NN models led to an overfitting behavior and a worse generalization capacity. Moreover, cost and time constraints generally do not permit enormous experiments. In consequence of this scarcity of data, various problems in the application of NN models are encountered.

## 13.5   PREDICTIONS UNDER NON-ISOTHERMAL CONDITIONS

All the above discussion regarding secondary models focused on the need to use an appropriate mathematical equation to describe temperature effects on the rate of a quality or microbial change, $k(T)$. The scope of a thorough kinetic study, in a wide range of constant temperatures, of the change of dominating quality and/or microbiological indices of a food, is to obtain a validated, mathematical shelf life model. The application of this model would then allow for a reliable prediction of the quality loss or microbiologi-cal state of the product in question at time-temperature conditions that differ from the experimental ones, and need not be constant.

For variable temperature processing or storage, time, $t_p$, should be replaced by an equivalent time, $t_{eq}$, given by appropriate integrating the secondary model. This, in the case of the Arrhenius equation, would give (Stoforos, 2015):

$$t_{eq} = k_{ref} \int_0^{t_p} \exp\left[ -\frac{E_A}{R} \left( \frac{1}{T(t)} - \frac{1}{T_{ref}} \right) \right] dt \qquad (73)$$

Previous and recent surveys in the retail and the consumer stocking level confirm the detrimental temperature abuse for chilled, as well as for frozen foods (Giannakourou and Taoukis, 2003a). Food products can be exposed to a variable temperature environment that not infrequently includes stages of abusive storage or handling conditions.

For first order reactions and using the $D$ and $z$-value approach, already described, Eq. (73) can be rewritten in terms of the $F$-value, so as to correlate directly the concentration of the parameter of interest with the time-temperature conditions of the process or storage:

$$F_{T_{ref}}^{z} = \int_{t_a}^{t_b} 10^{(T-T_{ref})/z} \, dt \tag{74}$$

where $F$ (min) the equivalent processing time of a hypothetical thermal process at a constant, temperature that produces the same effect (in terms of spore destruction) as the actual thermal process, or equivalently the time at a constant temperature, $T$, required to destroy a given percentage of microorganisms whose thermal resistance is characterized by $z$.

Combining Eq. 74 with Eq. 31, one obtains the remaining microbial or other heat labile substance concentration after a thermal process at variable temperature profile T(t) as:

$$N = N_0 \, 10^{\dfrac{-\int_{t_a}^{t_b} 10^{(T-T_{ref})/z} \, dt}{D_{T_{ref}}}} \tag{75}$$

## 13.6 EFFECT OF OTHER ENVIRONMENTAL FACTORS

In the following section, the effect of the most important factors, other than temperature will be briefly discussed. Moisture content, water activity, pH, NaCl, gas composition (mostly in the case of modified atmosphere packaged products), pressure (when High Hydrostatic pressure is applied) etc. are among the most popular influencing parameters and their effect is included in secondary models, in some cases of recent literature, as shown in Table 13.3. The most simplified approach is to describe this parameter's effect on reaction rate in a way similar to the temperature effect, using the most popular secondary models (e.g., Arrhenius, Belehradek, etc.), by substituting temperature parameter with the factor in question. When the effect of more than one parameters are investigated, then more complicated equations are produced usually applying the same models in a consecutive way. For example, if the square root model is used to describe the effect of $T$ on $k$ (Eq. 67) then for different $a_w$ levels, assuming that $a_w$ effect follows again a square root model, that is assuming that:

$$\sqrt{k} = b_{a_w}\left(a_w - a_{w,min}\right) \tag{76}$$

the combined model will read:

$$\sqrt{k} = b_{T,a_w}\left(T - T_{min}\right)\left(a_w - a_{w,min}\right) \tag{77}$$

Moisture content and water activity are two amongst the most important extrinsic factors besides temperature that affect the rate of food deterioration reactions. Controlling $a_w$ is the basis for preservation of dry and intermediate moisture foods (IMF). Minimum $a_w$ values for growth can be defined for different microbial species. Mathematical models that incorporate the effect of $a_w$ as an additional parameter can be used for shelf life predictions of moisture sensitive foods (Table 13.3). Indicative models of predictive microbiology that incorporate $a_w$ effect are illustrated in (Chaix et al., 2015b).

The pH of the food system is another determining factor. The effect of pH on different microbial, enzymatic and protein reactions has been studied in model biochemical or food systems (Table 13.3). Enzymatic and microbial activity exhibit an optimum pH range and limits above and below which activity ceases, similar to the response to temperature. The functionality and solubility of proteins depend strongly on pH, with the solubility usually being at a minimum near the isoelectric point (Cheftel et al., 1985), having a direct effect on their behavior in reactions. Other examples of important acid-base catalyzed reactions include nonenzymatic browning and aspartame decomposition. Few studies consider the interaction between pH and other factors, for example, temperature (Table 13.3). Significant progress in elucidating and modeling the combined effect to microbial growth of factors such as $T$, pH, $a_w$ or salt concentration has been achieved in the field of predictive microbiology.

Gas composition also affects certain quality loss reactions. Oxygen affects both the rate and apparent order of oxidative reactions, based on its

**TABLE 13.3** Representative examples of secondary models used to describe the effect of parameters other than temperature on quality or microbial kinetics

| Parameter studied | Type of model used | | Index/System (matrix) studied | Reference |
|---|---|---|---|---|
| Water activity, $a_w$ | Modified Arrhenius | $lnk = lnk_{ref} + d(a_w - a_{w\,ref}) - \dfrac{E_a}{R}\left(\dfrac{1}{T} - \dfrac{1}{T_{ref}}\right)$ | Browning in apple concentrates | Vaikousi et al., 2008 |
| | Belehradek-based | $\sqrt{k} = b(T - T_{min})\sqrt{a_{w\,max} - a_w}$ | | |
| | Arrhenius-type | $k = k_{ref}\dfrac{a_w - a_{w,min}}{a_{w,0} - a_{w,min}}\,exp\left[\dfrac{E_a}{R}\left(\dfrac{1}{T_{ref}} - \dfrac{1}{T}\right)\right]$ | Pseudomonas spp. of osmotically pretreated gilthead seabream filets | Tsironi and Taoukis 2014 |
| | Square-root based | $\sqrt{\mu_{max}} = c\sqrt{a_w - a_{w,min}}\sqrt{1 - 10^{(pH_{min} - pH)}}\sqrt{\mu_{liquid}\dfrac{K}{K + GC} + \mu_{solid}\dfrac{GC}{K + GC}}$ | Salmonella typhimurium in broth systems | Theys et al., 2008 |

**TABLE 13.3** (Continued)

| Parameter studied | Type of model used | Index/System (matrix) studied | Reference |
|---|---|---|---|
| | Response surface type model $$k_{max}(T, a_w) = \beta_0 + \beta_1 T + \beta_2 T^2 + \beta_3 a_w + \beta_4 a_w^2 + \beta_5 T a_w$$ Arrhenius based $$\ln k_{max}(T, a_w) = C_0 + \frac{C_1}{T} + C_2 a_w^2$$ Bigelow type model $$k_{max}(T, a_w) = k_{max}(T) k_{max}(a_w)$$ | *Listeria monocytogenes* in model systems | Valdramidis et al., 2006 |
| pH | $$\Pr{}_{k_{max}}(T, a_w) = \frac{\ln 10}{AsymD_{ref}} exp\left(\frac{\ln 10}{z}(T - T_{ref})\right) exp\left(\frac{\ln 10}{z_{aw}}(a_w - 1)\right)$$ $$\ln\left(\frac{\dot{p}}{p-1}\right) = \beta_0 + \beta_1 pH + \beta_2 I' + \beta_3 LA + \beta_4 LApH + \beta_5 I'pH + \beta_6 TLA$$ | *Escherichia coli* in model systems | Haberbeck et al., 2015 |

Square root type logistic regression model

$$logit(p) = b_0 + b_1 \cdot ln(a_w - a_{w,min}) + b_2 \cdot ln\left(1 - 10^{(pH_{min} - pH)}\right) + b_3$$

$$\cdot \left(1 - \sqrt{\frac{UAc}{UAc_{min}}}\right) + b_4\sqrt{n} + b_5 n + b_6\sqrt{n}$$

$$\cdot ln(a_w - a_{w,min}) + b_7 \cdot \sqrt{n} \cdot ln\left(1 - 10^{(pH_{min} - pH)}\right) + b_8$$

$$\cdot \sqrt{n} \cdot ln\left(1 - \sqrt{\frac{UAc}{UAc_{min}}}\right)$$

*Listeria monocytogenes* in model systems

Vermeulen et al., 2009

Probability model-logistic regression

$$logit(p) = b_0 + b_1 \cdot a_w + b_2 \cdot pH + b_3 \cdot Ac + b_4 \cdot a_w^2 + b_5$$

$$\cdot pH^2 + b_6 \cdot Ac^2 + b_7 \cdot a_w \cdot pH + b_8 \cdot pH \cdot Ac$$

*Zygosaccharomyces bailii* in model systems

Dang et al., 2010

Linear regression

*Escherichia coli* in fermented meat

McQuestin et al., 2009

**TABLE 13.3**  (Continued)

| Pressure | $$ln(k) = d/\left\{1 + exp\left(a\left[\frac{t_1}{b(P_1 - P_0)}(e^{-bP_0} - e^{-bP_1})\right.\right.\right.$$ $$\left.\left.\left. + \frac{t_2 - t_1}{b(P_2 - P_1)}(e^{-bP_1} - e^{-bP_2}) + (t - t_2)e^{bP_2}\right] + c\right)\right\}$$ | Rheological properties in waxy maize starch dispersions under High Hydrostatic Pressure | Stolt et al., 1999 |
|---|---|---|---|
| | Empirical model $$k_{P,T} = k_{Pref,Tref}\, e^{A(P-P_{ref})/T}\, e^{B(T-T_{ref})/T}$$ | 5-Methyltetrahydrofolic Acid Degradation in aqueous solutions under High Hydrostatic Pressure | Verlinde et al., 2009 |
| C | Arrhenius based $$k = k_{ref_{P,T}} \cdot exp\left\{-\frac{E_{a_P}}{R}\cdot exp\left[-B\cdot(P-P_{ref})\right]\cdot\left(\frac{1}{T}-\frac{1}{T_{ref}}\right) - \frac{A\cdot(T-T_{ref})+V_{a_P}\cdot\frac{(P-P_{ref})}{T}}{R\,T}\right\}$$ | Pectin methylesterase of Greek Navel orange juice | Polydera et al., 2004 |
| Gas composition | Gamma concept (Zwietering, 1992) $$\mu_{max} = \mu_{opt}\,\gamma_T\,\gamma_{pH}\,\gamma_{aw}\,\gamma_{O_2}\,\gamma_{CO_2}\xi$$ where $\gamma_w$ (W={T, pH, $a_w$, $O_2$, $CO_2$} are dimensional weighting parameters representing the respective influences) | Listeria monocytogenes or Pseudomonas fluorescens in fresh food packed under modified atmosphere | Guillard et al., 2016; Chaix et al., 2015a |

presence in limiting or excess amounts (Table 13.3). Exclusion or limitation of $O_2$ by nitrogen flushing or vacuum packaging reduces redox potential and slows down undesirable reactions. Further, the presence and relative amount of other gases, especially carbon dioxide, and secondly ethylene and CO, strongly affects biological and microbial reactions in fresh meat, fruit and vegetables.

Another important environmental aspect, recently assuming high investigating priority, is pressure. Traditionally, processing of foods has involved thermal treatments with the specific goal of making foods microbiologically safe. With greater concerns over the nutritional benefits of foods as well as qualitative aspects such as texture and color, investigation into advanced food processing techniques has resulted in the emergence of various new technologies within the food industry. One area of recent interest is the use of ultra-high pressure (UHP) processing of foods, also referred to as high hydrostatic pressure (HHP) and high pressure processing (HPP). For HHP processing, the effect of temperature and pressure on the rate constants of food deteriorative indices can be expressed through the Arrhenius and the Eyring (Eq. 78) equations, respectively (Johnson and Eyring 1970).

$$k_P = k_{P_{ref}} \cdot \exp\left(-\frac{V_A}{R} \cdot \frac{(P - P_{ref})}{T}\right) \tag{78}$$

The effect of both pressure and temperature on the remaining concentration of a quality or safety index may be expressed by more complex equations. Using Eq. (78) for the effect of pressure and the Arrhenius secondary model for the effect of temperature, the change of a parameter $C$ under variable HHP processing conditions reads:

For $n = 1$

$$\log\left(\frac{C}{C_o}\right) = \int_0^t -k_{ref} \cdot \exp\left[-\frac{E_a(P)}{R} \cdot \left(\frac{1}{T} - \frac{1}{T_{ref}}\right) - \frac{V_A(T)}{R} \cdot \frac{(P - P_{ref})}{T}\right] dt \tag{79}$$

$$C^{1-n} - C_o^{1-n} = \int_0^t (n-1) \cdot k_{ref} \cdot \exp\left[-\frac{E_a(P)}{R} \cdot \left(\frac{1}{T} - \frac{1}{T_{ref}}\right) - \frac{V_A(T)}{R} \cdot \frac{(P - P_{ref})}{T}\right] dt \tag{80}$$

## 13.7   DETERMINATION OF KINETIC PARAMETERS

In this section, we will present alternative procedures used to determine the parameters associated with the kinetic models.

### 13.7.1   TWO-STEPS APPROACH

As already mentioned in the Introduction, the traditional approach for food kinetic modeling (either for microbiological measurements or for chemical changes) includes a specific methodology where the most representative indices are selected and their change is measured as a function or processing or storage time at constant temperatures (isothermal experiments) or constant values of the influencing intrinsic and extrinsic parameters. Afterwards, the traditional two-stage methodology is based on the application of an appropriate primary model in order to describe the rate of these changes and estimate through a regression analysis (usually linear), the relevant kinetic parameters for each temperature.

The second step is to select a secondary model that best describes the effect of temperature (and/or other influencing parameters, as described in a previous section) on the respective kinetic parameters. The parameters associated with the secondary model are determined in a second regression analysis (usually linear). Estimation of all the parameters of primary and secondary models would allow for the prediction of quality or microbiological status of the food at any, different temperature of interest. If validated, mathematical models can be a useful tool to assess safety and quality after processing and at any stage of food storage.

### 13.7.2   ONE-STEP APPROACH

#### 13.7.2.1   Isothermal Conditions

Alternatively to two-step approach, all parameters of the model can be simultaneously determined by a single, non-linear regression analysis. For example, by incorporating Eq. (49) into Eq. (21) we obtain:

$$C = C_0 \cdot \exp\left(-k_{T_{ref}} \cdot \exp\left(\frac{-E_a}{R}\left(\frac{1}{T} - \frac{1}{T_{ref}}\right)\right)\right) \tag{81}$$

From $C$ vs t and $T$ data, $k_{Tref}$, and $E_a$ can be determined in one step through a non-linear regression procedure. The advantage of one-step analysis modeling is the use of more precise parameter estimates and the development of more realistic models. As already mentioned in the Introduction, several studies have been published in recent literature that use one-step kinetic analysis based on isothermal (Van Boekel, 1996; Angelidis et al., 2013; Huang and Vinyard, 2016; Huang, 2015a, b; Wawire et al., 2016) or dynamic experiments (Huang, 2003; Huang, 2013; Huang, 2015a; Sui and Zhou, 2014; Valdramidis et al., 2006; Valdramidis et al., 2008; Valdramidis et al., 2012).

An important advantage of this methodology is the significantly decreased number of necessary experiments. An important issue designated by Van Boekel (1996) is that the confidence intervals of the parameters involved in this one step procedure are not symmetric and thus their estimation is not straightforward.

### *Example 2: Determination of kinetics of thermal degradation*

The data on the Table 13.4 refer to the reduction of the concentration of a thermo- labile agent during heat treatment at four different isothermal conditions. Describe the kinetics of the thermal degradation of the agent in question.

Using the Weibull equation as the primary model, assuming that the shape parameter, $\beta$, of the model is independent of temperature and using the Arrhenius equation as the secondary model to describe the effect of temperature on the $1/a$ parameter we end up with the following Eq. (82):

$$\log(N(t)/N_o) = -\left( \frac{t}{\alpha_{T_{ref}} e^{\frac{E_a}{R_g} \frac{(T_{ref}-T)}{T_{ref}T}}} \right)^{\beta} \tag{82}$$

Fitting Eq. (80) into the experimental data of Table 13.4, through a non-linear least square statistical routine for $T_{ref} = 348.16K$ (75°C), one should obtain the following values for the kinetic parameters appearing on Eq. (80):

**TABLE 13.4**  Thermal Degradation Data of a Heat Labile Agent at Different Constant Temperatures

|              | 60°C       | 70°C       | 80°C       | 90°C       |
| ------------ | ---------- | ---------- | ---------- | ---------- |
| Time (min)   | C (mg/mL)  | C (mg/mL)  | C (mg/mL)  | C (mg/mL)  |
| 0.0          | 100.0      | 100.0      | 100.0      | 100.0      |
| 5.0          | 76.6       | 59.4       | 39.4       | 20.1       |
| 10.0         | 67.6       | 49.0       | 25.9       | 10.3       |
| 15.0         | 62.0       | 41.0       | 19.0       | 6.0        |
| 20.0         | 58.3       | 36.8       | 15.0       | 3.6        |
| 25.0         | 53.2       | 32.1       | 13.1       | 2.6        |
| 30.0         | 51.1       | 28.9       | 10.4       | 1.8        |
| 35.0         | 48.6       | 24.7       | 8.9        | 1.3        |
| 40.0         | 45.8       | 23.1       | 7.4        | 1.0        |
| 45.0         | 43.7       | 20.4       | 6.4        | 0.7        |
| 50.0         | 41.0       | 20.5       | 5.4        | 0.6        |
| 55.0         | 39.8       | 18.4       | 4.6        | 0.4        |
| 60.0         | 38.3       | 15.9       | 4.3        | 0.4        |
| 65.0         | 36.6       | 16.1       | 3.3        | 0.3        |
| 70.0         | 35.2       | 15.4       | 3.1        | 0.2        |
| 75.0         | 34.0       | 13.0       | 2.8        | 0.1        |
| 80.0         | 34.2       | 12.1       | 2.5        | 0.2        |
| 85.0         | 32.5       | 12.3       | 2.3        | 0.1        |
| 90.0         | 32.3       | 11.8       | 1.7        | 0.1        |
| 95.0         | 30.1       | 10.2       | 1.8        | 0.1        |
| 100.0        | 28.8       | 10.2       | 1.6        | 0.1        |
| 105.0        | 28.6       | 9.2        | 1.3        | 0.1        |
| 110.0        | 26.7       | 8.9        | 1.3        | 0.1        |
| 115.0        | 26.2       | 9.0        | 1.0        | 0.0        |
| 120.0        | 26.4       | 8.0        | 0.9        | 0.0        |

$$\alpha_{T_{ref}} = 54.16 \pm 1.56 \text{ min}$$
$$E_a = 122100 \pm 14400 \text{ J/mol}$$
$$\beta = 0.480 \pm 0.015 \text{ min}$$

A comparison between experimental and fitted inactivation data is presented on Figure 13.11 (with a correlation coefficient between experimental and fitted data $R^2 = 0.9959$).

### 13.7.2.2   Dynamic Conditions

Taking as example the temperature effects, as temperature is one of the main extrinsic parameters influencing reaction rates, the response $C$ after exposure at dynamic temperature conditions will depend on both primary and secondary model parameters. For example, Eq. (81) for non-constant $T$ conditions, can be written as (in view of Eq. 73):

$$C = C_0 \cdot \exp\left(-k_{ref}\int_0^t \exp\left[-\frac{E_a}{R}\cdot\left(\frac{1}{T}-\frac{1}{T_{ref}}\right)\right]dt\right) \qquad (83)$$

Thus, knowing $C$ and $T$ as a function of time, one can use Eq. (81) in a non-linear regression routine and determine both $k_{Tref}$ and $E_a$ parameters.

Difficulties in applying this method are referring to the numerical integration of Eq. (81) for an arbitrary T(t) profile and on a second note to the selection of an appropriate T(t) profile that will produce estimates of $k_{Tref}$ and $E_a$ with narrow confidence intervals. Use of linearly increasing temperature profile (Rhim et al., 1989; Hayakawa et al., 1969) can alleviate both problems.

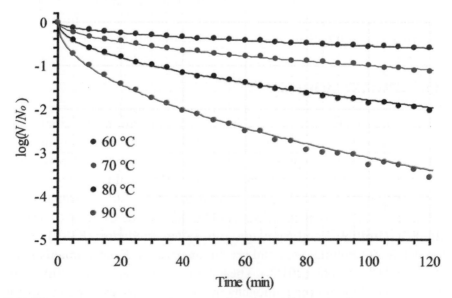

**FIGURE 13.11**   Comparison between experimental and fitted microbial inactivation data. Points refer to experimental data while lines to fitted values.

## 13.7.2.2 Recent Developments in Two-Steps Methodology – Monte Carlo Simulation

All the discussion up to this point refers to alternative approaches to model changes of a microbiological or quality attribute through appropriate mathematical equations. According to the two-step approach, the purpose is to calculate estimates for the parameters of the selected primary and secondary models. In its common application, confidence intervals of primary model parameters at each temperature are not taken into account when calculating secondary model parameters (Van Boekel, 1996), which, consequently, can be falsely narrower than the actual ones. Based on these observations, the weakness detected in confidence intervals estimation, is addressed in recent literature by using Monte Carlo technique (Duret et al., 2015; Evrendilek et al., 2016). This method can be also used to account for the variability of either an influencing factor that is not constant (e.g., product temperature in the cold chain) (Duret et al., 2015), the different properties of the initial untreated material (Channon et al., 2016; Wesolek and Roudot, 2016) and the variability of model parameters, as expressed by their 95% confidence intervals (Sui and Zhou 2014; Huang and Vinyard, 2016). The main purpose of this approach is to improve predictions of the model, regarding chemical or microbiological changes, at any selected temperature and time during processing and product shelf life (at storage or distribution).

## 13.8 TERTIARY MODELS

Tertiary models are based on the integrated application of primary and secondary models in a user–friendly software. Details about existing predictive models and comparison between different models of predictive microbiology, are provided in the work of (McDonald and Sun, 1999; Perez-Rodriguez and Valero, 2013; Swinnen et al., 2004). Some of the most popular software platforms, widely used, include ComBase, DMFit, GInaFIT, GroPIN, Food Spoilage and Safety Predictor (FSSP), Sym' Previus, NIZO Premia, etc., and are summarized in Koutsoumanis et al. (2016) and Chaix et al. (2015b). These tools offer a variety of utilities for the majority of foodborne pathogens including databases, predictions for growth, growth/no growth and inactivation, probabilistic models, and risk assessment modules (Koutsoumanis et al., 2016) and allow for decision-making in a short time.

At this point, it must be noticed that most of these software tools are designed in order to study and address safety issues while quality is scarcely the target of such integrated and fully automated systems. Therefore, it would be interesting to expand tertiary models to include data and models on quality deterioration of food matrices.

## 13.9 EPILOGUE

In this chapter, some of the most widely primary and secondary models used in recent literature were presented, along with their differences and various modifications proposed. Alternative methodologies to determine models' kinetic parameters were described and special considerations regarding reaction rate determination were presented with the aid of specific examples. A basic conclusion out of the study of reaction kinetics in food systems is that there are no particular rules or established theories in predicting food reaction kinetics. Available literature do exists and one should be aware of it when starting an application, but must be aware of that published kinetic models and parameters are based on experimental observations and are empirical in nature. As described in this chapter, numerous specifications of particular models have been used in a form deemed more suitable for the specific food matrix and the processing/storage conditions in question. Caution is needed when making assumptions for the application of particular primary or secondary models. Models with an increased number of parameters can follow closely the experimental data but they also incorporate the accompanied error. The success of a model is associated with its ability to alleviate the error always existing in the experimental data. For the selection of a particular model, its behavior at the limits of the examined variables and parameters is suggested.

## KEYWORDS

- dynamic conditions
- food safety
- isothermal

- **kinetics**
- **modeling**
- **prediction**
- **primary models**
- **quality**
- **secondary models**

## REFERENCES

Albert, I., & Mafart, P. (2005). 'A modified Weibull model for bacterial inactivation,' *International Journal of Food Microbiology, 100,* 197–211.

Anema, S. G. (2001). 'Kinetics of the irreversible thermal denaturation and disulfide aggregation of α-lactalbumin in milk samples of various concentrations,' *Journal of Food Science,* 66, no 1, 2–9.

Angelidis, A. S., Papageorgiou, D. K., Tyrovouzis, N. A., & Stoforos, N. G. (2013). 'Kinetics of *listeria monocytogenes* cell reduction in processed cheese during storage,' *Food Control,* 29, 18–21.

Aspridou, Z., & Koutsoumanis, K. P. (2015). 'Individual cell heterogeneity as variability source in population dynamics of microbial inactivation,' *Food Microbiology, 45,* Part B, 216–221.

Ates, M. B., Skipnes, D., Rode, T. M., & Lekang, O. I. (2016). 'Comparison of spore inactivation with novel agitating retort, static retort and combined high pressure-temperature treatments,' *Food Control, 60,* 484–492.

Augustin, J. C., & Carlier, V., 2000, 'Mathematical modeling of the growth rate and lag time for *Listeria monocytogenes*,' *International Journal of Food Microbiology, 56,* 29–51.

Augustin, J. C., Carlier, V., & Rozier, J., 1998, 'Mathematical modeling of the heat resistance of *Listeria monocytogenes*,' *Journal of Applied Microbiology, 84,* 185–191.

Ball, C. O. (1923). Bulletin of the national research council No. 37, 7, Part 1, Natl. Res. Council, Washington, DC.

Baranyi, J., & Roberts, T. A. (1994). 'A dynamic approach to predicting bacterial growth in food,' *International Journal of Food Microbiology. 23,* 277–294.

Barsa, C. S., Normand, M. D., & Peleg, M. (2012). 'On Models of the temperature effect on the rate of chemical reactions and biological processes in foods,' *Food Engineering Reviews, 4,* 191–202.

Belehradek, J. (1930). 'Temperature coefficients in biology.' *Biological Reviews, 5,* no 1., 30–58.

Bigelow, W. D., Bohart, G. S., Richardson, A. C., & Ball, C. O. (1920). Natl. Canners Assoc. Res. Lab., Bull. 16–L, Washington, DC.

Bratchell, N., McClure, P. J., Kelly, T. M., & Roberts, T. A. (1990). 'Predicting microbial growth: graphical methods for comparing models,' *International Journal of Food Microbiology, 11,* 279–287.

Buchanan, R. L., & Bagi, L. K. (1999). 'Microbial competition: Effect of fluorescents on the growth of *Listeria monocytogenes*,' *Food Microbiology, 16,* 523–529.

Buchanan, R. L., Whiting, R. C., & Damert, W. C. (1997). 'When is simple good enough: A comparison of the Gompertz, Baranyi, and three-phase linear models for fitting bacterial growth curves,' *Food Microbiology, 14*, 313–326.

Cerfk, O. (1977). 'Tailing of survival curves of bacterial spores: A review,' *Journal of Applied Microbiology, 42*, 1–9.

Chaix, E., Broyart, B., Couvert, O., Guillaume, C., Gontard, N., & Guillard, V. (2015a). 'Mechanistic model coupling gas exchange dynamics and *Listeria monocytogenes* growth in modified atmosphere packaging of non respiring food,' *Food Microbiology, 51*, 192–205.

Chaix, E., Couvert, O., Guillaume, C., Gontard, N., & Guillard, V. (2015b). 'Predictive microbiology coupled with gas ($O_2/CO_2$) Transfer in Food/Packaging Systems: How to Develop an Efficient Decision Support Tool for Food Packaging Dimensioning,' *Comprehensive Reviews in Food Science and Food Safety, 14*, 1–21.

Chakraborty, S., Rao, P. S., & Mishra, H. N. (2015). 'Empirical model based on Weibull distribution describing the destruction kinetics of natural microbiota in pineapple (Ananas comosus L.) puree during high-pressure processing,' *International Journal of Food Microbiology, 211*, 117–127.

Channon, H. A., Hamilton, A. J., D'Souza, D. N., & Dunshea, F. R. (2016). 'Estimating the impact of various pathway parameters on tenderness, flavor and juiciness of pork using Monte Carlo simulation methods,' *Meat Science, 116*, 58–66.

Cheftel, J. C., Cuq, J. L., & Lorient, D. (1985). 'Protéines alimentaires,' Biochimie–propriétés fonctionnelles–valeur nutritionnelle–modifications chimiques. Paris: Technique et documentation–Lavoisier, 1–295.

Corradini, M. G., & Peleg, M. (2006). 'Linear and non-linear kinetics in the synthesis and degradation of acrylamide in foods and model systems,' *Critical Reviews in Food Science and Nutrition, 46*, 489–517.

Dabadé, D. S., Azokpota, P., Nout, M. J. R., Hounhouigan, D. J., Zwietering, M. H., & den Besten, H. M. W. (2015). 'Prediction of spoilage of tropical shrimp (Penaeus notialis) under dynamic temperature regimes,' *International Journal of Food Microbiology, 210*, 121–130.

Dang, T. D. T., Mertens, L., Vermeulen, A., Geeraerd, A. H., Van Impe, J. F., Debevere, J et al. (2010). 'Modelling the growth/no growth boundary of Zygosaccharomyces bailii in acidic conditions: A contribution to the alternative method to preserve foods without using chemical preservatives,' *International Journal of Food Microbiology, 137*, 1–12.

David, J. R., & Merson, R. L. (1990). 'Kinetic parameters for inactivation of Bacillus stearothermophilus at high temperatures,' *Journal of Food Science, 55*(2), 488–493, 515.

Dermesonluoglu, E., Katsaros, G., Tsevdou, M., Giannakourou, M., & Taoukis, P. (2015). 'Kinetic study of quality indices and shelf life modeling of frozen spinach under dynamic conditions of the cold chain,' *Journal of Food Engineering, 148*, pp. 13–23.

Duret, S., Gwanpua, S. G., Hoang, H-M, Guillier, L., Flick, D., Laguerre, O., Verlinden, B. E., De Roeck, A., Nicolai, B. M., & Geeraerd, A 2015, 'Identification of the significant factors in food quality using global sensitivity analysis and the accept-and-reject algorithm. Part III: Application to the apple cold chain,' *Journal of Food Engineering, 148*, 66–73.

Evrendilek, G. A., Avsar, Y. K., & Evrendilek, F., 2016, 'Modelling stochastic variability and uncertainty in aroma active compounds of PEF-treated peach nectar as a function of physical and sensory properties, and treatment time,' *Food Chemistry, 190*, 634–642.

García, M. R., Vilas, C., Herrera, J. R., Bernárdez, M., Balsa-Canto, E., & Alonso, A. A. (2015). 'Quality and shelf-life prediction for retail fresh hake (*Merluccius merluccius*),' *International Journal of Food Microbiology, 208*, 65–74.

Geeraerd, A. H., Herremans, C. H., & Van Impe, J. F. (2000). 'Structural model requirements to describe microbial inactivation during a mild heat treatment,' *International Journal of Food Microbiology, 59*, 185–209.

Geeraerd, A. H., Herremans, C. H., Cenens, C., & Van Impe, J. F. (1998). 'Application of artificial neural networks as a non-linear modular modeling technique to describe bacterial growth in chilled food products,' *International Journal of Food Microbiology, 44*, 49–68.

Geeraerd, A. H., Valdramidis, V. P., & Van Impe, J. F. (2005). 'GInaFiT, a freeware tool to assess non-log-linear microbial survivor curves,' *International Journal of Food Microbiology, 102*, 95–105.

Giannakourou, M. C., & Taoukis, P. S. (2003a). 'Application of a TTI-based distribution management system for quality optimization of frozen vegetables at the consumer end,' *Journal of Food Science, 68*(1), 201–209.

Giannakourou, M. C., & Taoukis, P. S. (2003b). 'Kinetic modeling of vitamin C loss in frozen green vegetables under variable storage conditions,' *Food Chemistry, 83*, 33–41.

Giannakourou, M. C., Koutsoumanis, K., Dermesonlouoglou, E., & Taoukis, P. S. (2001). 'Applicability of the shelf life decision system (slds) for control of nutritional quality of frozen vegetables,' *Acta Horticulturae.*

Gibson, A. M., Bratchell, N., & Roberts, T. A. (1988). 'Predicting microbial growth: growth responses of *salmonella*e in a laboratory medium as affected by pH, sodium chloride and storage temperature,' *International Journal of Food Microbiology, 6*, 155–178.

Greenacre, E. J., Brocklehurst, T. F., Waspe, C. R., Wilson, D. R., & Wilson, P. D. G. (2003). '*Salmonella* enterica serovar Typhimurium and *Listeria monocytogenes* acid tolerance response induced by organic acids at 20°C: Optimization and modeling,' *Applied and Environmental Microbiology, 69*, 3945–3951.

Guillard, V., Couvert, O., Stahl, V., Hanin, A., Denis, C., Huchet, V., et al., 2016, 'Validation of a predictive model coupling gas transfer and microbial growth in fresh food packed under modified atmosphere,' *Food Microbiology, 58*, 43–55.

Haberbeck, L. U., Oliveira, R. C., Vivijs, B., Wenseleers, T., Aertsen, A., Michiels, C., & Geeraerd, A. H. (2015). 'Variability in growth/no growth boundaries of 188 different *Escherichia coli* strains reveals that approximately 75% have a higher growth probability under low pH conditions than *E. coli* O157:H7 strain ATCC 43888,' *Food Microbiology, 45*, Part B, 222–230.

Hayakawa, K., Schnell, P. G., & Kleyn, D. H. (1969). 'Estimating Thermal Death Time Characteristics of Thermally Vulnerable Factors by Programd Heating of Sample Solution or Suspension,' *Food Technology II, 23*, 1090–1094.

Hereu, A., Dalgaard, P., Garriga, M., Aymerich, T., & Bover-Cid, S., 2014, 'Analysing and modeling the growth behavior of *Listeria monocytogenes* on RTE cooked meat products after a high pressure treatment at 400 MPa,' *International Journal of Food Microbiology, 186*, 84–94.

Huang, L. (2003). 'Dynamic computer simulation of clostridium perfringens growth in cooked ground beef,' *International Journal of Food Microbiology, 87*, 217–227.

Huang, L. (2004). 'Numerical analysis of the growth of Clostridium perfringens in cooked beef under isothermal and dynamic conditions,' *Journal of Food Safety, 24*, 53–70.

Huang, L. (2015a). 'Dynamic determination of kinetic parameters, computer simulation, and probabilistic analysis of growth of Clostridium perfringens in cooked beef during cooling,' *International Journal of Food Microbiology, 195*, 20–29.

Huang, L. (2015b). 'Growth of *Staphylococcus aureus* in cooked potato and potato salad-a one-step kinetic analysis,' *Journal of Food Science, 80*, no 12, M2837–2844.

Huang, L., & Vinyard, B. T. (2016). 'Direct dynamic kinetic analysis and computer simulation of growth of clostridium perfringens in cooked turkey during cooling,' *Journal of Food Science, 81*(3), M692–701.

Huang, L., 2013, 'Determination of thermal inactivation kinetics of *Listeria monocytogenes* in chicken meats by isothermal and dynamic methods,' *Food Control, 33*, 484–488.

Janssen, M., Geeraerd, A. H., Cappuyns, A., Garcia-Gonzalez, L., Schockaert, G., Van Houteghem, N., et al. (2007). 'Individual and combined effects of pH and lactic acid concentration on *Listeria innocua* inactivation: Development of a predictive model and assessment of experimental variability,' *Applied and Environmental Microbiology, 73*, 1601–1611.

Johnson, F. H., & Eyring, H. (1970). 'The kinetic basis of pressure effects in biology and chemistry,' in AM Zimmerman (ed.), High Pressure Effects on Cellular Processes. New York: Academic Press, Chapter 1, pp. 1–44.

Jonsson, U., Snygg, B., Härnulv, B. G., & Zachrisson, T. (1977). 'Testing two models for the temperature dependence of the heat inactivation rate of Bacillus stearothermophilus spores,' *Journal of Food Science, 42*(5), 1251–1252, 1263.

Karel, M., & Lund, D. B. (2003). 'Reaction kinetics,' in: *Physical Principles of Food Preservation*, New York: Marcel Dekker, Inc.

Kim, B. S., Lee, M., Kim, J. Y., Jung, J. Y., & Koo, J. (2016). 'Development of a freshness-assessment model for a real-time online monitoring system of packaged commercial milk in distribution,' *LWT – Food Science and Technology, 68*, 532–540.

Knol, J. J., Viklund, G. A. I., Linssen, J. P. H., Sjöholm, I. M., Skog, K. I., & van Boekel, M. A. J. S. (2009). 'Kinetic modeling: A tool to predict the formation of acrylamide in potato crisps,' *Food Chemistry, 113*, 103–109.

Koutsoumanis, K. P., Lianou, A., & Gougouli, M. (2016). 'Last developments in foodborne pathogens modeling,' *Current Opinion in Food Science, 8*, 89–98.

Koutsoumanis, K., & Angelidis, A. S. (2007). 'Probabilistic modeling approach for evaluating the compliance of ready-to-eat foods with new European union safety criteria for *Listeria monocytogenes*,' *Applied and Environmental Microbiology, 73*, 4996–5004.

Li, H., Xie, G., & Edmondson, A. S. (2008). 'Review of secondary mathematical models of predictive microbiology,' *Journal of Food Products Marketing, 14*, 57–74.

Li, H., Xie, G., & Edmondson, A., (2007). 'Evolution and limitations of primary mathematical models in predictive microbiology,' *British Food Journal, 109*, 608–626.

Lianou, A., & Koutsoumanis, K. P. (2011). 'A stochastic approach for integrating strain variability in modeling *Salmonella* enterica growth as a function of pH and water activity,' *International Journal of Food Microbiology, 149*, 254–261.

Linton, R. H., Carter, W. H., Pierson, M. D., & Hackney, C. R. (1995). 'Use of a modified gompertz equation to model nonlinear survival curves for *listeria monocytogenes* scott a,' *Journal of Food Protection, 58*, 946–954.

Linton, R. H., Carter, W. H., Pierson, M. D., Hackney, C. R., & Eifert, J. D. (1996). 'Use of a modified Gompertz equation to predict the effects of temperature, pH, and NaCI on the inactivation of *Listeria monocytogenes* Scott A heated in infant formula,' *Journal of Food Protection, 59*, 16–23.

Malakar, P. K., Barker, G. C., Zwietering, M. H., & Van't Riet, K. (2003). 'Relevance of microbial interactions to predictive microbiology,' *International Journal of Food Microbiology, 84*, 263–272.

Marquenie, D., Geeraerd, A. H., Lammertyn, J., Soontjens, C., Van Impe, J. F., Michiels, C. W., et al. (2003). 'Combinations of pulsed white light and UV-C or mild heat treatment to inactivate conidia of Botrytis cinerea and Monilia fructigena,' *International Journal of Food Microbiology, 85*, 185–196.

Mataragas, M., Zwietering, M. H., Skandamis, P. N., & Drosinos, E. H. (2010). 'Quantitative microbiological risk assessment as a tool to obtain useful information for risk managers-Specific application to *Listeria monocytogenes* and ready-to-eat meat products,' *International Journal of Food Microbiology, 141*, S170–S79.

McDonald, K., & Sun, D. W. (1999). 'Predictive food microbiology for the meat industry: a review,' *International Journal of Food Microbiology, 52*, 1–27.

McKellar, R. C., Peréz-Rodríguez, F., Harris, L. J., Moyne, A-l, Blais, B., Topp, E., Bezanson, G., Bach, S & Delaquis, P (2014). 'Evaluation of different approaches for modeling *Escherichia coli* O157:H7 survival on field lettuce,' *International Journal of Food Microbiology, 184*, 74–85.

McMeekin, T. A., Olley, J. N., Ross, T., & Ratkowsky, D. A. (1993). *Predictive Microbiology: Theory and Application.* 1st ed. England: Research Studies Press Ltd. p 340.

McMeekin, T., Olley, J., Ratkowsky, D., Corkrey, R., & Ross, T. (2013). 'Predictive microbiology theory and application: Is it all about rates?,' *Food Control, 29*, 290–299.

McQuestin, O. J., Shadbolt, C. T., & Ross, T. (2009). 'Quantification of the relative effects of temperature, pH, and water activity on inactivation of *Escherichia coli* in fermented meat by meta-analysis,' *Applied and Environmental Microbiology, 75*, 6963–6972.

Muñoz-Cuevas, M., Metris, A., & Baranyi, J. (2012). 'Predictive modeling of *Salmonella*: From cell cycle measurements to e-models,' *Food Research International, 45*, 852–862.

NCA. (1968). Laboratory Manual for Food Canners and Processors, 1, Microbiology and Processing. National Canners Association Research Laboratory, The Avi Publishing Company, Inc., Westport, CT.

Ongeng, D., Haberbeck, L. U., Mauriello, G., Ryckeboer, J., Springael, D., & Geeraerd, A. H. (2014). 'Modeling the Fate of *Escherichia coli* O157:H7 and *Salmonella* enterica in the Agricultural Environment: Current Perspective,' *Journal of Food Science, 79*, no 4, R421–R27.

Peleg, M. (1992). 'On The Use of The WLF Model in Polymers and Foods,' *Critical Reviews in Food Science and Nutrition, 32*, 59–66.

Peleg, M., Engel, R., Gonzalez-Martinez, C., & Corradini, M. G. (2002). 'Non-Arrhenius and non-WLF kinetics in food systems,' *Journal of the Science of Food and Agriculture, 82*, 1346–1355.

Peleg, M., Normand, M. D., & Corradini, M. G. (2012). 'The Arrhenius equation revisited,' *Critical Reviews in Food Science and Nutrition, 52*, 830–851.

Perez-Rodriguez, F., & Valero, A. (2013). 'Predictive microbiology in foods,' Springer. New York.

Polydera, A. C., Galanou, E., Stoforos, N. G., & Taoukis, P. S. (2004). 'Inactivation kinetics of pectin methylesterase of greek Navel orange juice as a function of high hydrostatic pressure and temperature conditions,' *Journal of Food Engineering,.62*, no 3, 291–98, doi: http://dx.doi.org/10.1016/S0260–8774(03)00242–0245.

Poschet, F., Geeraerd, A. H., Scheerlinck, N., Nicolaï, B. M., & Van Impe, J. F. (2003). 'Monte Carlo analysis as a tool to incorporate variation on experimental data in predictive microbiology,' *Food Microbiology, 20*, 285–295.

Ratkowsky, D. A., & Ross, T. (1995). 'Modelling the bacterial growth/no growth interface,' *Letters in Applied Microbiology, 20*, 29–33.

Ratkowsky, D. A., Lowry, R. K., McMeekin, T. A., Stokes, A. N., & Chandler, R. E. (1983). 'Model for bacterial culture growth rate throughout the entire biokinetic temperature range,' *Journal of Bacteriology, 154*, 1222–1226.

Rhim, J. W., Nunes, R. V., Jones, V. A., & Swartzel, K. R. (1989). 'Determination of Kinetic Parameters using Linearly Increasing Temperature,' *Journal of Food Science, 54*(2), 446–450.

Saltaouras, K. F., Yanniotis, S., & Stoforos, N. G. (2015). 'Optimum dynamic temperature profiles for thermal destruction kinetics determination,' 12th International Congress on Engineering and Food, Québec City, Canada, Paper No. P1.134.

Sapru, V., Smerage, G. H., Texeira, A. A., & Lindsay, J. A. (1993). 'Comparisosn of predictive models for bacterial spore population resources to sterilization temperatures,' *Journal of Food Science, 58*(1), 223–228.

Saraiva, C., Fontes, M. C., Patarata, L., Martins, C., Cadavez, V., & Gonzales-Barron, U. (2016). 'Modelling the kinetics of *listeria monocytogenes* in refrigerated fresh beef under different packaging atmospheres,' *LWT-Food Science and Technology, 66*, 664–671.

Simpson, R., Nuñez, H., & Almonacid, S. (2015). 3-Modeling thermal processing and reactions: sterilization and pasteurization A2-Bakalis, Serafim. In: K Knoerzer & PJ Fryer, (eds.) Modeling Food Processing Operations. Woodhead Publishing.

Smelt, J. P. P. M., & Brul, S. (2014). 'Thermal inactivation of microorganisms,' *Critical Reviews in Food Science and Nutrition, 54*, 1371–1385.

Smet, C., Van Derlinden, E., Mertens, L., Noriega, E., & Van Impe, J. F. (2015). 'Effect of cell immobilization on the growth dynamics of *Salmonella typhimurium* and *Escherichia coli* at suboptimal temperatures,' *International Journal of Food Microbiology, 208*, 75–83.

Stella, P., Cerf, O., Hugas, M., Koutsoumanis, K. P., Nguyen-The, C., Sofos, J. N., et al. (2013). 'Ranking the microbiological safety of foods: A new tool and its application to composite products,' *Trends in Food Science and Technology, 33*, 124–138.

Stoforos, N. G. (2010). 'Thermal process calculations through ball's original formula method: A critical presentation of the method and simplification of its use through regression equations,' *Food Engineering Reviews, 2*, 1–16.

Stoforos, N. G. (2015). Thermal Processing.' In: T. Varzakas and C. Tzia, (eds.), *Handbook of Food Processing: Food Preservation*, CRC Press, Taylor & Francis Group, Boca Raton, FL, USA, Chapter 2, pp. 27–56. http://dx.doi.org/10.1201/b19397–9393.

Stolt, M., Stoforos, N. G., Taoukis, P. S., & Autio, K. (1999). 'Evaluation and modeling of rheological properties of high pressure treated waxy maize starch dispersions,' *Journal of Food Engineering, 40*, 293–298.

Sui, X., & Zhou, W. (2014). 'Monte Carlo modeling of non-isothermal degradation of two cyanidin-based anthocyanins in aqueous system at high temperatures and its impact on antioxidant capacities,' *Food Chemistry, 148*, 342–350.

Sulaiman, A., Soo, M. J., Farid, M., & Silva, F. V. M. (2015a). 'Thermosonication for polyphenoloxidase inactivation in fruits: Modeling the ultrasound and thermal kinetics in pear, apple and strawberry purees at different temperatures,' *Journal of Food Engineering, 165*, 133–140.

Sulaiman, A., Soo, M. J., Yoon, M. M. L., Farid, M., & Silva, F. V. M. (2015b). 'Modeling the polyphenoloxidase inactivation kinetics in pear, apple and strawberry purees after High Pressure Processing,' *Journal of Food Engineering, 147*, 89–94.

Swinnen, I. A. M., Bernaerts, K., Dens, E. J. J., Geeraerd, A. H., & Van Impe, J. F. (2004). 'Predictive modeling of the microbial lag phase: a review,' *International Journal of Food Microbiology, 94*, 137–159.

Taoukis, P. S. (2003). 'Q10,' in D Heldman (ed.), Encyclopedia of Agricultural, Food and Biological *Engineering,* Marcel Dekker, Inc., New York.

Taoukis, P. S., & Giannakourou, M. C. (2007). 'Reaction kinetics,' in: S. S. Sablani, M. Shafur Rahman, A. K. Datta, & A. S. Mujumdar (eds.), Handbook of Food and Bioprocess modeling techniques, CRC Press, Taylor and Francis Group, pp. 235–263.

Taoukis, P. S., Labuza, T. P., Saguy, S. (1997). 'Kinetics of food deterioration and shelf-life prediction,' in: K. J. Valentas, E. Rotstein, R. P., Singh (eds.), Handbook of food engineering practice, CRC Press, New York, pp. 361–403.

Taoukis, P. S., Tsironi, T. S., & Giannakourou, M. C. (2015). 'Reaction Kinetics,' in: C. Tzia & T. Varzakas (eds.), Handbook of Food Processing and Engineering, Vol. I: Food Engineering Fundamentals, CRC Press, Boca Raton, Florida, USA, pp. 529–570.

Terefe, N. S., & Hendrickx, M. (2002). 'Kinetics of the pectin methylesterase catalyzed de-esterification of pectin in frozen food model systems,' *Biotechnology Progress, 18,* 221–228.

Theys, T. E., Geeraerd, A. H., Verhulst, A., Poot, K., Van Bree, I., Devlieghere, F., et al. (2008). 'Effect of pH, water activity and gel micro-structure, including oxygen profiles and rheological characterization, on the growth kinetics of *Salmonella typhimurium,'* *International Journal of Food Microbiology, 128,* 67–77.

Timmermans, R. A. H., Nederhoff, A. L., Nierop Groot, M. N., van Boekel, M. A. J. S., & Mastwijk, H. C. (2016). 'Effect of electrical field strength applied by PEF processing and storage temperature on the outgrowth of yeasts and molds naturally present in a fresh fruit smoothie,' *International Journal of Food Microbiology, 230,* 21–30.

Toledo, R. T., (1999). 'Kinetics of Chemical Reactions in Foods,' in: Fundamentals of Food Process Engineering, Kluwer Academic/Plenum Publishers, New York, pp. 302–314.

Tsironi, T. N., & Taoukis, P. S. (2014). 'Effect of processing parameters on water activity and shelf life of osmotically dehydrated fish filets,' *Journal of Food Engineering, 123,* 188–192.

Tyrovouzis, N. A., Angelidis, A. S., & Stoforos, N. G. (2014). 'Bi-phasic growth of *listeria monocytogenes* in chemically defined medium at low temperatures,' *International Journal of Food Microbiology, 186,* 110–119.

Upadhyay, S. K. (2006). Chemical Kinetics and Reaction Dynamics. Springer, Anamaya Publishers, India.

Vaikousi, H., Koutsoumanis, K., & Biliaderis, C. G. (2008). 'Kinetic modeling of non-enzymatic browning of apple juice concentrates differing in water activity under isothermal and dynamic heating conditions,' *Food Chemistry, 107,* 785–796.

Valdramidis, V. P., Belaubre, N., Zuniga, R., Foster, A. M., Havet, M., Geeraerd, A. H., et al. (2005). 'Development of predictive modeling approaches for surface temperature and associated microbiological inactivation during hot dry air decontamination,' *International Journal of Food Microbiology, 100,* 261–274.

Valdramidis, V. P., Geeraerd, A. H., Bernaerts, K., & Van Impe, J. F. M. (2008). 'Identification of non-linear microbial inactivation kinetics under dynamic conditions,' *International Journal of Food Microbiology, 128,* 146–152.

Valdramidis, V. P., Geeraerd, A. H., Gaze, J. E., Kondjoyan, A., Boyd, A. R., Shaw, H. L., et al. (2006). 'Quantitative description of *listeria monocytogenes* inactivation kinetics with temperature and water activity as the influencing factors. Model prediction and methodological validation on dynamic data,' *Journal of Food Engineering, 76,* 79–88.

Valdramidis, V. P., Taoukis, P. S., Stoforos, N. G., & Van Impe, J. F. M. (2012). 'Modeling the Kinetics of Microbial and Quality Attributes of Fluid Food During Novel Thermal and Non-Thermal Processes,' in P. J. Cullen, B. K., Tiwari, & V. P., Valdramidis (eds.), Novel Thermal and Non-Thermal Technologies for Fluid Foods, Elsevier Inc. USA.

Van Boekel M. A. J. S., & Tijskens, L. M. M. (2001). 'Kinetic modeling.' In: L. M. M. Tijskens, T. M. Hertog, & B. M., Nicolai (Eds.), *Food Process Modeling*. CRC Press, USA.

Van Boekel, M. A. J. S. (1996). 'Statistical aspects of kinetic modeling for food science problems,' *Journal of Food Science, 61*(3), 477–486.

Van Boekel, M. A. J. S. (2008). 'Kinetic modeling of food quality: A critical review,' *Comprehensive Reviews in Food Science and Food Safety, 7*, 144–158.

Van Den Broeck, I., Ludikhuyze, L. R., van Loey, A. M., & Hendrickx, M. E. (2000). 'Inactivation of orange pectinesterase by combined high-pressure and-temperature treatments: A kinetic study,' *Journal of Agricultural and Food Chemistry, 48*, 1960–1970.

Verlinde, P. H. C. J., Oey, I., Deborggraeve, W. M., Hendrickx, M. E., & Van Loey, A. M. (2009). 'Mechanism and related kinetics of 5-wlethyltetrahydrofolic acid degradation during combined high hydrostatic pressure-thermal treatments,' *Journal of Agricultural and Food Chemistry, 57*, 6803–6814.

Vermeulen, A., Gysemans, K. P. M., Bernaerts, K., Geeraerd, A. H., Debevere, J., Devlieghere, F., et al. (2009). 'Modelling the influence of the inoculation level on the growth/no growth interface of *Listeria monocytogenes* as a function of pH, aw and acetic acid,' *International Journal of Food Microbiology, 135*, 83–89.

Villota, R., & Hawkes, J.G. (2007). 'Reaction kinetics in food systems,' in D. R. Heldman, D. B., Lund (Eds.) *Handbook of Food Engineering,* 2nd edition, CRC Press. New York, USA, pp. 125–287.

Wawire, M., Oey, I., Mathooko, F. M., Njoroge, C. K., Shitanda, D., & Hendrickx, M. (2016). 'Kinetics of thermal inactivation of peroxidase and color degradation of african cowpea (vigna unguiculata) leaves,' *Journal of Food Science, 81*(1), E56–E64.

Wesolek, N., & Roudot, A. C. (2016). 'Assessing aflatoxin B1 distribution and variability in pistachios: Validation of a Monte Carlo modeling method and comparison to the Codex method,' *Food Control, 59*, 553–560.

Wijtzes, T., Van'T Riet, K., Huis In'T Veld, JHJ & Zwietering, M. H. (1998). 'A decision support system for the prediction of microbial food safety and food quality,' *International Journal of Food Microbiology, 42*, 79–90.

Williams, M. L., Landel, R. F., & Ferry, J. D. (1955). 'The temperature dependence of relaxation mechanisms in amorphous polymers and other glass forming liquids,' *Journal of Chemical Engineering, 77*, 3701–3707.

Wolfram Demonstration, (2015). http://demonstrations.wolfram.com, accessed on July 11th, 2016.

Xie, G., & Xiong, R. (1999). 'Use of hyperbolic and neural network models in modeling quality changes of dry peas in long time cooking,' *Journal of Food Engineering, 41*, 151–162.

Xie, G., Xiong, R., & Church, I., 1998, 'Comparison of kinetics, neural network and fuzzy logic in modeling texture changes of dry peas in long time cooking,' *LWT-Food Science and Technology, 31*, 639–647.

Zwietering, M. H. (2015). 'Risk assessment and risk management for safe foods: Assessment needs inclusion of variability and uncertainty, management needs discrete decisions,' *International Journal of Food Microbiology, 213*, 118–123.

Zwietering, M. H., & den Besten, H. M. W. (2011). 'Modelling: One word for many activities and uses,' *Food Microbiology, 28*, 818–822.

Zwietering, M. H., Jongenburger, I., Rombouts, F. M., & Van't Riet, K. (1990). 'Modeling of the bacterial growth curve,' *Appl Environ Microbiol, 56*, 1875–1881.

Zwietering, M. H., Rombouts, F. M., & Van't Riet, K. (1993). 'Some aspects of modeling microbial quality of food,' *Food Control, 4*, 89–96.

# CHAPTER 14

# VALUE ADDITION AND PRESERVATION OF FISHERY PRODUCTS

C. O. MOHAN,[1] G. NINAN,[1] J. BINDU,[1] A. A. ZYNUDHEEN,[1] and C. N. RAVISHANKAR[2]

[1]Fish Processing Division, Central Institute of Fisheries Technology, Matsyapuri, Willingdon Island, Cochin – 682029, India

[2]ICAR-Central Institute of Fisheries Technology (Indian Council of Agricultural Research), Matsyapuri, Cochin – 682029, India

## CONTENTS

## 14.1   INTRODUCTION

Fish is considered as 'poor peoples rich food' as it provides the essential nourishment, especially quality proteins and fats (macronutrients), vitamins and minerals (micronutrients). Proteins are important for growth and development of the body, maintenance and repairing of damaged tissues and for production of enzymes and hormones required for many body processes. The importance of fish in providing easily digested protein of high biological value is well documented. On a fresh-weight basis, fish contains a good quantity of protein, about 16–22%, and contains all the eight essential amino acids including the sulfur-containing lysine, methionine and cysteine. Fish contributes up to 180 kcal per capita per day, but reaches such high levels only in a few countries where there is a lack of alternative protein foods grown locally or where there is a strong preference for fish (examples are Iceland, Japan and some small island states). More typically, fish provides about 20–30 kcal per capita per day. Fish proteins are essential in the diet of some densely populated countries where the total protein intake level is low, and are very important in the diets of many other countries. Worldwide, about a billion people rely on fish as their main source of animal proteins. Dependence on fish is usually higher in coastal than in inland areas. About 20% of the world's population derives at least one-fifth of its animal protein intake from fish, and some small island states depend almost exclusively on fish. Among the animal's proteins, fish protein is one of the cheapest animal protein sources and it accounts for about 40% of the total animal protein intake of an average person in the tropics. However, the protein intake from fish is not uniform throughout the world. It is highest in most of the islands, Japan, Canada and some regions of Africa (Figure 14.1). Apart from proteins, fish also contains a considerable amount of fat content. The fat content of fish varies depending on the species as well as the season but, in general, fish have less fat than red meats. The fat content ranges from 0.2% to 25%. However, fats from fatty fish species contain the polyunsaturated fatty acids (PUFAs) namely EPA (eicosapentaenoic acid) and DHA (docosahexaenoic acid) (omega 3 fatty acids) which are essential for proper growth of children and are associated with the prevention of cardiovascular diseases such a coronary heart disease. In pregnant women, the presence of PUFAs in their diets helps in proper brain development among unborn babies. The fat also contributes to energy supplies and assists in the proper absorption of fat-soluble vitamins namely A, D, E, and K. Fish is also a rich source of vitamins, particularly vitamins A and D from fatty species, as well as thiamin, riboflavin and niacin (vitamins $B_1$, $B_2$ and $B_3$). As compared to plant sources, Vitamin A from fish is more readily available to the body and the fatty fish contains more vitamin A than lean species. Vitamin A is required for normal vision and for bone growth and it reduces the mortality for children under 5. Vitamin D present in fish liver and oils is important for bone growth since it is essential for the absorption and metabolism of calcium. Thiamin, niacin and riboflavin helps in energy metabolism. Various minerals present in fish include iron, calcium, zinc, iodine (from marine fish), phosphorus, selenium and fluorine. These minerals are highly 'bioavailable.' Iron is important in the synthesis of hemoglobin in red blood cells which is important for transporting oxygen to all parts of the body. Iron deficiency is associated with anemia, impaired brain function and in infants it is associated with poor learning ability. Due to its role in the immune system, its deficiency may also be associated with increased risk of infection. Zinc is required for most body processes as it occurs together with proteins in essential enzymes. Zinc plays an important role in growth and development as well in the proper functioning of the immune system and for a health skin. Zinc deficiency is associated with poor growth, skin problems and loss of hair among other problems. Iodine, which is present in seafood, is important for hormones that regulate body metabolism and in children it is required for growth and normal mental development. A deficiency of iodine may lead to goiter and mental retardation in children. Due to these nutritional benefits and increased purchasing power of the consumers, the worlds per capita fish consumption has increased considerably. In 2004, the world average per capita fish consumption reached 16.1 kg from 9 kg in 1960s and at present it is around 17.4 kg (http://www.globefish. org/dynamisk.php4?id=4787). However, this per capita consumption is not uniform everywhere and it shows a marked continental, regional and national differences (Figure 14.2). Among the vast diversity of fish and shellfish population, only very few are commercially important and they fetch higher price compared to others. This could be mainly due to the non-familiarity of the fish species to the consumers. At present the world trade of fish and fishery products is approaching US$ 100 billion (http://www.globefish.org/dynamisk.php4?id=4787). Both per capita consumption and fish trade can be increased significantly by adopting various value addition techniques.

## 14.2  VALUE ADDITION

Value addition in general, is the process of changing or transforming a product from its original state to a more valuable state. A broad definition of value addition is to economically add value to a product by changing its current

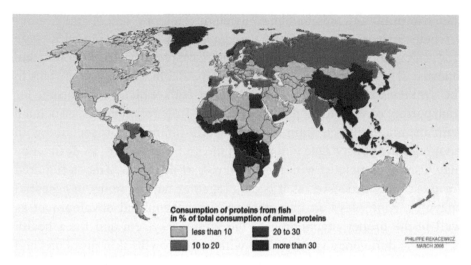

**FIGURE 14.1** World fish protein consumption (Source: Earthtrend database, World Resources Institute (WRI), Washington: Faostat, Food and Agriculture Organization of the United Nations (FAO) http://www.grida.no/publications/vg/water2/page/3283.aspx).

place, time, and form characteristics to the market place preferences. Value addition can be accomplished in a number of different ways, but for fishery products the most important one is the innovative approach. Innovation focuses on improving existing processes, procedures, products, and services

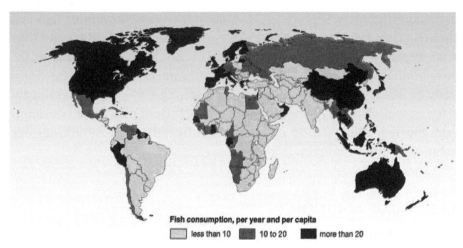

**FIGURE 14.2** World per capita fish consumption (Adapted from State of the world's fisheries and aquaculture 2006, Food and Agriculture Organization of the United Nations (FAO) http://maps.grida.no/go/graphic/world-fish-consumption-per-capita-and-per-year).

or creating new ones. Often, successful value-added ideas focus on very narrow, highly technical, geographically large markets where competition is sparse. Innovative value-added activities developed on fish processing factories or at research stations are sources of national growth through changes either in the kind of product in the technology of production. By encouraging innovative ideas, adding value becomes a reality. Value-addition should start with intelligent market information on customers and competitors to make sure an opportunity exists. A recipe for success is to begin with a basic commodity and add a healthy dose of ingenuity to create a product desired by consumers that also has a valuable edge on the competition.

The contribution of fisheries to the economy of a country will be greater than the value of the catch at the place of landing if the value addition is practiced. The value addition also results in the added costs and these costs need to be recovered if the operation is to be economically feasible. This requires consumers to have enough purchasing power to recover the costs. World over the general wealth and purchasing power of consumers is showing an increasing trend, which can be exploited for marketing diversified value added fish products. The Value addition for fishery products is different from the other food items. While for all other food items, the value addition takes place in expanded processing, for fishery products, the most important value addition is in the freshness, or even more importantly in supplying the fish alive for the high value live fish market. Many raw commodities have intrinsic value in their original state. For example, freshly harvested or cultured fish and shellfish have a value of its own. While harvesting, if the quality is maintained by proper handling and chilled storage, it increases its value. In culture varieties, feeding fish with carotenoid pigments improves the color of flesh, is also considered as adding value. A wide range of diversified fishery products in live, chilled, frozen, heat sterilized, dried, salted, smoked, fermented and combinations of these processes have entered the market to meet the consumers demand.

## 14.2.1 LIVE FISHERY PRODUCTS

There is a great demand for live fish and shell-fishes world over. These products fetches maximum price compared to all the other forms of value added products as it maintains the freshness. The candidate species for live transportation include high value species, cultured grouper, red snapper, seabreams, seabass, red tilapia, reef fish, and air breathing fishes, shrimp, crabs, lobster, clams, oyster and mussels. These are normally transported in air cargo maintained at low temperature to lessen the metabolic activities of the animals. Although its a better way of supplying in its fresh form, it's highly cost intensive and the problem of mortality has to be tackled effectively.

## 14.2.2  CHILLED FISHERY PRODUCTS

Chilling is an effective method of maintaining the freshness of fishery products. This normally involves keeping fishes in melting ice or slurry ice to maintain the fish temperature around 1–4°C, which delays the enzymatic action and microbial activity there by extending the shelf life of products. Traditionally chilling is carried out using melting ice either flake ice or crushed block ice. Of late, slurry ice has been introduced for chilling. Chilling fishery products using slurry Ice has several advantages as compared to traditional melting ice: (i) the sub-zero storage temperature of slurry ice, (ii) its faster chilling rates due to the higher heat exchange power, (iii) the limited physical damage caused to the fish surface due to its microscopic spherical crystals, and (iv) the prevention of dehydration events due to the full coverage of the fish surface. A wide range of fish products varying from whole, gutted, headless gutted fish, filet, steaks, loins, cubes can be used for chilling. An important chilled fish product of delicacy in many countries is the Sashimi. Sashimi is a Japanese term used for raw fresh fish. Although sashimi can be prepared from most of the fishes, Sashimi from tuna particularly Blue fin and Big eye tuna are in great demand. Like wise for shrimp, whole, headless shell on, peeled and de-veined products are having good market demand. Apart from these chilled crab, crab legs, pasteurized crabs, lobster whole and tail, cleaned and pasteurized mussels are also in good demand in many countries. Chilled products are superior to frozen products in the quality however they have limited shelf life which varies from few days to only few weeks depending on the fish species. Chilling in ordinary air packs will have limited shelf life, which can be increased by using advanced vacuum packaging, modified atmosphere and active packaging techniques.

In vacuum packaging, the air inside the package is removed by applying vacuum using machineries. The entrapped air in the ordinary air pack will accelerate the oxidation of fishery products and encourages the growth of aerobic spoilage organisms. This can be minimized by adopting vacuum packaging technique. However, it creates anaerobic environment which encourages the growth of anaerobic spoilers and the most dangerous Clostridium botulinum growth and toxin production. This can be minimized by using good quality raw material and maintaining the temperature of storage less than 3.3°C throughout, which prevents the growth and toxin production of this organism. In modified atmosphere packaging (MAP), the air inside the package is replaced with the favorable gasses like carbon dioxide ($CO_2$), nitrogen ($N_2$) and oxygen ($O_2$) depending on the fat content of the fishes. Based on fat content, fishes are grouped as fatty (fat content > 5%), medium fatty (2.5–5%) and lean (< 2.5%). Fatty fishes are normally packed with 60: 40 ($CO_2$:$N_2$) and lean fishes with 40:30:30 ($CO_2$:$N_2$:$O_2$) in the ration of 1:2 or 1:3 (fish to gas volume). The $CO_2$ is known to have bacteriostatic effect which helps in preserving fish freshness. As $N_2$ is an inert gas, its mainly used as filler to prevent the package collapse. In MAP products, $O_2$ is introduced in the package to prevent the anaerobic atmosphere as a preventive measure to reduce the C. botulinum risk. $O_2$ is also introduced to maintain the reddish color of red meat fishes, whereas its use is avoided normally for fatty fishes to prevent the fat oxidation. Both vacuum and modified atmosphere are effective in extending the shelf life of fishes compared to ordinary air packs (Manju et al., 2008; Rajesh et al., 2006; Ravishankar et al., 2008). However, they are cost intensive as they require costly equipments and pure gasses, which can be overcome by adopting active packaging technology. It is a type of packaging that changes the condition of the packaging and maintains these conditions throughout the storage period to extend shelf-life or to improve safety or sensory properties while maintaining the quality of packaged food (Vermeiren et al., 1999). It uses either chemicals or natural compounds extracted from plants or animals to excert its action. Active packaging technique is either gas-flushing or more recently gas-scavenging or emitting systems added to emit (e.g., $N_2$, $CO_2$, ethanol) and/or to remove (e.g., $O_2$, $CO_2$, odor) gases during packaging or distribution. In case of a gas-scavenging or emitting system, reactive compounds are either contained in individual sachets or stickers associated to the packaging material or, more recently, directly incorporated into the packaging material. Major active packaging techniques are concerned with substances that absorb oxygen, ethylene, moisture, carbon dioxide, flavors/odors and those which release carbon dioxide, antimicrobial agents, antioxidants and flavors. Active packaging techniques, mainly $O_2$ scavenger (Mohan, 2008; Mohan et al., 2009a,b, 2010) and $CO_2$ emitters (Mohan, 2008) has reported to maintain the freshness, safety and microbial quality for longer period for Seer fish (Scomberomorus commerson) steaks during chilled conditions. Edible coating with natural antimicrobial and antioxidant compound in combination with active packaging systems is also useful in extending the shelf life of fish products (Mohan et al., 2012; Remya et al., 2016, 2017; Renuka et al., 2016). By adopting these advanced techniques storage life of fish in chilled conditions can be improved considerably.

### 14.2.3  FROZEN FISHERY PRODUCTS

Freezing is an age-old practice to retain the quality and freshness of fishery products for a long time. This involves the conversion of water present in fishery products to ice, that is, a phase change from liquid to solid phase takes place in freezing. This retards the microbial and enzymatic action by reducing the water available for their action. Freezing can be either slow freezing or quick freezing. Slow freezing affects the quality whereas quick freezing preserves the quality. Quick freezing is normally accomplished using any of the following four methods: air blast freezing, indirect contact freezing (plate freezing), immersion freezing and cryogenic freezing. Normally products are frozen till it attains a core temperature of $-18°C$ or low and are stored in cold storage maintained at this temperature. The freezing and frozen storage of fish have been largely used to retain their sensory and nutritional properties before consumption. Frozen products forms one of the largest portions of fishery products traded all over the world. A wide variety of fish, shrimp and cephalopods in various forms are preserved in frozen form. Among fish products, whole, whole gutted, headed and gutted, fileted, loins, steaks, breaded and battered and minced fish are frozen stored. Shrimp based products range from head on individual quick frozen (IQF) to headless, peeled and de-veined, peeled and un de-veined, easy peel shrimp, butterfly style, stretched shrimp, skewered shrimp, breaded or tempura shrimps and cooked shrimps. Various IQF packs of whole squid, squid tubes, squid filets, rings, stuffed squid, cuttle fish, octopus, sushi and sashimi, squid filets, pine cut blanched and cooked products, seafood mix and analog products from surimi are frozen and marketed. The recent trend in the marketing of frozen products is the consumer retail packs. Mainly fish steaks, filets and fish fingers, cleaned and blanched shrimps and cephalopods in small individual packs are frozen and marketed in supermarkets in many countries (Lakshmisha et al., 2008). Apart from these various value added ready to fry products like balls, cutlets, specialty products and curry products are also frozen and marketed.

## 14.2.4 DRIED AND SALTED FISHERY PRODUCTS

Drying is probably one of the oldest methods of food preservation. It consists of removal of water to a final desired concentration, which in turn reduces the water activity of the product thereby assures microbial stability and guarantees extended shelf-life of the product. In some cases common table salt (sodium chloride) is also used to prolong the shelf life of fish. Salt absorbs much of the water in the food and makes it difficult for micro-organisms to survive. Lean fishes are mainly preserved by dry or kench salting in which 3–4% of coarse salt is normally used. For fatty fishes such as herring, sardine, anchovy, mackerel, wet salting or pickle curing (3–3.5%) is normally followed. In some cases brining is also practiced as a preparation for smoking or drying in which fishes are immersed in saturated salt solution for 30 – 90 min depending on the size of fishes. Sun drying is the most economic way of fish drying. With the increased developments various advanced equipments have evolved for effective drying to monitor the quality of fish products. The drying form will vary with the fish species and size (Table 14.1). Normally small varieties of fish and shell fishes like Anchovies, Bombay duck, lizard fish shark, ribbon fish, thread fin breams, lesser sardines, small varieties of freshwater fishes, medium and small sized shrimps are dried whole after proper washing whereas some large varieties of fishes like sharks, catfishes etc. are washed, degutted and spread open to dry for human consumption. Some fish species are also dried for using in poultry feed. These are dried without any pre-treatment. Traditionally, dry fish is considered as an item usually consumed by people from lower income strata. Therefore, the prices paid for this product was rather very low in recent past. The drying activity is normally woman centric in many under developed and developing nations. In majority of the countries, sand drying was the method predominantly used for drying. As the consumers are aware of the quality of dry products, these products fetch low prices in local as well as

**TABLE 14.1**  Different Species Used for Drying and the Product Forms

| Species Dried | Product Form |
| --- | --- |
| Anchovy | Whole |
| Abalone | Whole half shell or shelled, sliced |
| Bombay duck | Whole |
| Clams | Whole, shelled |
| Cuttlefish | Whole, sliced |
| Fish bladders (maws) | Whole |
| Fish skins | Bits and pieces |
| Hake | Butterfly, filets |
| Horse mackerel | Whole, sliced |
| Jellyfish | Shredded |
| Jewfish (croaker) | Salted, whole, filets, sliced |
| Mackerel | Whole, sliced |
| Mussels | Whole, half shell and shelled |
| Needlefish | Whole |
| Octopus (small-size) | Whole |
| Oysters | Whole |
| Pomfret (small size) | Whole |
| Ribbonfish | Whole and sliced |
| Salmon | Whole, filet |
| Sardine | Whole |
| Scallops | Whole, broken |
| Sea cucumbers | Whole |
| Seaweed | Pressed sheets, shredded |
| Shark | Sliced, chunks, filets, steaks |
| Shark fins | Whole, skinned, cartilage removed |
| Shrimp | Whole, head-on, broken, shell-on and peeled |
| Small variety of fishes and shrimps | Whole |
| Squid | Whole, shredded |

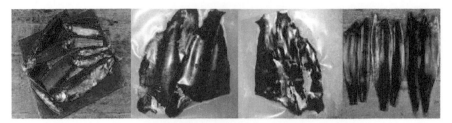

Smoked fish products (L to R whole cleaned salmon, Ventral split
opened rainbow trout dorsal and ventral views, salmon fillet)

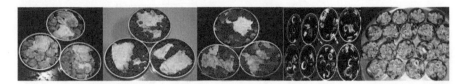

Canned yellow fin tuna along with babycorn, green peas and broccoli
and Canned seafood mix in Tomato sauce and brine medium in Tin-free-
steel (TFS) cans (L to R).

Value added shrimp and fish products in different styles (batterfried fish
fillet, shrimp in butterfly style and fish skewers; L to R).

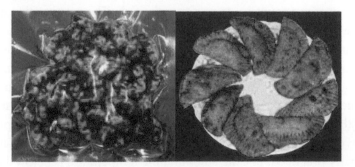

Indian white shrimp (*Fenneropeneaus indicus*) in Sous-vide pack and
fried stuffed fish product (*Momo*).

Fish mince and coated fish fingers

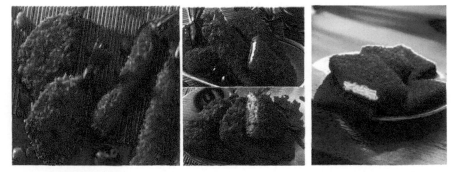

Coated shrimp in butterfly and nobashi style and coated fish fillet.

international markets. However, with the introduction of modern automatic hygienic drying techniques and attractive packaging it is possible to fetch better price.

## 14.2.5  SMOKED FISHERY PRODUCTS

Smoking is one of the most widely used traditional fish processing methods employed in many countries to preserve the fish. The preservation effect of the smoke is a result of drying (withdrawal of moisture) of the product during the smoking as well as due to smoke particles absorption in to the flesh. The smoke particles mainly phenolic compounds, carbonyl and organic acids, being absorbed by the product, inhibit bacterial growth on the surface of the product. The smoke particles also have a positive effect on the taste and color of the product and in many instances, smoking is normally practiced to improve these sensory characteristics. Three different smoking methods are normally practiced. In Cold smoking, the temperature is below 30°C and the product is not cooked in this method. In hot smoking, the temperature varies between 65–100°C, which leads to cooking of the product but not dried. In smoke drying method, the product is first hot smoked, so that it gets cooked, and then, with continued smoking the product is dried (temperatures vary between 45–85°C). Salmon, trout, mahseer, mackerel, small varieties of marine and freshwater fishes, oyster, clams and many other groups are normally preferred for smoking. '*Masmin*' is a cooked, smoked and dried product of commercial importance in Lakshadweep islands of India. Very fresh Skipjack tuna are normally preferred for this product.

## 14.2.6  THERMAL PROCESSED FISHERY PRODUCTS

Thermal sterilization is one of the most efficient methods of food preservation widely practiced worldwide. The main objective of the thermal processing is to achieve long-term shelf stability. Thermal processing generally involves heating the food products packaged in hermatically sealed containers for a predetermined time at a pre-selected temperature to eliminate the pathogens of public health significance as well a those microorganisms and enzymes that deteriorate the food during storage. Fishery products being low acid food (pH > 4.5), the microorganisms of significance is *Clostridium botulinum*, which is a rod shaped, anaerobic mesophilic pathogen capable of producing highly heat resistant spores and produces '*botulinum*' toxin. These products have to be processed such that all the points in the container should achieve a minimum lethality of 2.52 minutes when processed at 121.1°C (250°F), which corresponds to 12 decimal reduction of *C. botulinum* which is known as '*botulinum cook*.' In practice, fish products are processed beyond this lethality for safety reasons. Historically heat processing started in glass containers. Over the years different containers like metal, rigid plastic containers and flexible retortable pouches are developed for thermal processing. The selection of prime quality fish and shell fishes is important for heat processing. If the small fishes like sardine, mackerel, trout are used for thermal processing, they should be gutted, descaled and washed properly before cutting into suitable size. They can be given cold blanching treatment (immersing in salt solution (4–10%) for a desired period) and then filled in pouches for pre-cooking and further processing. If the large fishes like tuna are used, they are gutted, washed and cut into small pieces and pre-cooked for definite time interval, skin and dark meat is removed, cut into desired size and filled in to containers for further processing. For shrimps/prawns, headless, peeled and de-veined prawns are used for hot blanching and then filled in to the containers. For cephalopods, like squid and cuttlefish, rings prepared and packed either in brine or masala medium. For oyster, clams and mussels, they are fried and packed with masala or in brine medium. The different styles of packs based on the filling medium are natural or dry pack (processed in its own juice or without added filling medium), brine, oil, tomato sauce, or curry packs. The desired quantity of solid to liquid ratio is maintained as per the written procedures or as per buyer's specifications. Thermal processing of ready to eat fish products like traditional kerala style mackerel curry (Gopal et al., 2001), rohu curry (Sonaji et al., 2002), seer fish curry (Gopal et al., 2002; Ravishankar et al., 2002), seer fish moilee (Manju et al., 2004), mussel (Bindu et al., 2004), sardine in oil (Ali et al 2005), prawan kuruma (Mohan et al., 2006 and 2008), tuna in different filling medium, filling ingredients (Mohan et al., 2014 & 2015), shrimp in curry, brine and natural pack (Mallick, 2008) and calcium and iron fortified shrimp soup (Shashidhar, 2007) have been reported.

### 14.2.7 FERMENTED FISHERY PRODUCTS

Fermentation is the most important way of preserving fish. Fermented fish pastes and sauces have a much more important place in the daily diet of many North East Indian states and South East Asian countries than salted or dried fish. During the fermentation of fish, protein is broken down by endogenous enzymes in the presence of a high salt concentration (20–30%). The high salt concentration also inhibits the growth of microorganisms. However, salt tolerant microorganisms will grow in the fermented products which are responsible for the characteristic taste and flavor. In some cases, boiled or roasted rice is added as a source of carbohydrate for fermentation. Although fermented fish products are a good source of protein, they can be consumed only in limited quantities because of the high salt content of these products. Fermentation of fish is especially used in situations where drying of fish is not possible because the climate is too wet and where cooling and sterilization of the product is too expensive. The species used for fermentation includes anchovies, herring, deep-bodied herring, fimbriated herring, mackerel, round scad, slipmouth, carps, catfish, climbing perch, gourami, mudfish, shrimp, mussels, oysters and octopus.

Fermented fish can be in the form of sauce (Nuoc-mam, *Nampla, Patis* and *Shottsuru*), paste (*Bagoong, Balao-balao, Belachan, Ngapi, Prahoc* and *Trassi*) or whole or pieces of fish which retains its structure as much as possible (*Colombo cure, Pedah-siam, Sushi, Anchoa* and *Momone*).

## 14.2.8  SPECIALTY FISHERY PRODUCTS

Due to increased fishing pressure, many common fish stocks are becoming exhausted. As a result, the search for new fishing grounds for new stocks is continuously increasing which lands some strange fishery resources. As these resources are not common, consumers will be reluctant to purchase. These can be used for the preparation of some specialty products. A wide variety of specialty products produced and consumed in different countries, which include, fish based products like mince, surimi, balls, cutlets, fingers, patties, burger, coated products and many imitation products, specialty shrimp based products like whole shrimp, peeled shrimp, peeled and de-veined (PD), peeled and un-de veined (PUD), cooked shrimp, Stretched shrimp (Nobashi), Barbecue, Sushi (Cooked butterfly shrimp), Skewered shrimp, Shrimp head-on (center peeled), Shrimp head-on cooked (center peeled) and coated shrimps in different forms. The range of products also includes lobster tail, crab claws, coated oyster and mussels and pickle, wafers, fish with mixed vegetable products, etc. The range of product also varies with the region and eating habit of the people.

### 14.2.8.1  Fish Mince

Fish mince is prepared by concentrating only edible muscle part removing all the other parts like scale, skin, gut and bones. Fish mince acts as base material for majority of the value added products. It is prepared by split opening the fish ventrally and passing through the meat bone separator. Larger bones are normally removed manually and fine pin bones are then separated by passing the mince through a pin bone remover, so that the bone content in the final mince should not exceed 2% as it affects the quality of mince. Due to very fine nature of the mince, the surface area will be more compared to muscle of filet as such and hence it is susceptible to spoilage. As fish mince is used as base material for the preparation of various other products, monitoring the quality of mince is very important. So the temperature should be maintained below 10°C during the preparation of mince. Normally, fish mince is stored under frozen condition. Fish mince is used for the preparation of surimi, fish balls, fish finger and other specialty products.

### 14.2.8.2　Fish Finger

Fish fingers are regular rectangular sized fish portions made from either frozen fish filet or fish mince. Skinless and bone less filet are partially frozen to get the correct shape of the finger. For ease of cutting operation, frozen slabs of 1.5 cm thickness are used. The frozen slabs are passed through a motor operated band saw to cut into suitable size. A typical British fish finger weights around one ounce (28 g) and in Asian countries it varies from 20–25 g each. They are battered and breaded before freezing or cooking. They are normally frozen stored and consumed as snack food after frying in vegetable oil.

### 14.2.8.3　Fish Balls

Fish balls are restructured convenient product which is believed to have originally come from China. These are similar to the products like Kofter of India, Polpette of Italy, Koningsberger Klopse of Germany, Swedish meat balls, Koefte of Turkey, Nunh Hoa of Vietnam. It is prepared from a mixture of fish, fat particles, water, carbohydrate, ginger, garlic, pepper and salt. During the processing, meat is mixed with the ingredients and carbohydrate source, which will bind the particles directly or indirectly. The mixture is then formed to the desired shape and this shape is retained after freezing or cooking. Balls are also prepared from crab, shrimp and clams. They are normally battered and breaded before freezing or cooking like other specialty products and are frozen stored.

### 14.2.8.4　Fish Cutlet

Cutlet is a spicy snack food popular in many Asian, European and South American countries. However, the spiciness will be more in Asian countries than in other countries. Fish cutlets are prepared by mixing the cooked fish meat with the cooked potato and fried onion, chilly, ginger, garlic, pepper, turmeric, coriander, oil, salt and other ingredients based on the consumers taste. After proper mixing, they are shaped desirably and battered and breaded. Cutlet can also be prepared from cooked meat from skeletal frame remaining after fileting the fish. Frozen storage is normal practice and is consumed after frying in oil.

### 14.2.8.5    Fish Burgers (Fish Patties)

Burgers are similar to cutlets except the spiciness, which is less in burgers. White meat from lean variety of fish is known to give better product. In the preparation, fish meat or mice is cooked and mixed with cooked potato, fried onion, carbohydrate source, spices, herbs and salt and formed into desired shape. It is normally frozen stored and whenever required it is thawed, heated and eaten as sandwiches with fresh vegetables, leaves and plain buns.

### 14.2.8.6    Surimi

Surimi is a myofibrillar protein concentrate obtained from mechanical deboning of fish flesh which is washed with chilled water and with added cryoprotectants (Park, 2005). In the preparation, the fish is split into two and washed thoroughly and the mince is separated mechanically using meat bone separating machine. The mince thus obtained is washed number of times to remove water soluble proteins, pigments, enzymes and lipid and to concentrate myofibrillar protein which gives characteristic gelling and elastic properties. This surimi can be used as base material for the preparation of textured imitation products like lobster tail, crab legs, shrimp, scallop, crab stick etc. The utmost care has to be taken while preparing surimi. Not all the fishes give same textural properties and lean fishes are preferred over their fatty counterparts. Alaskan Pollack is the most preferred species for surimi preparation. In Asian countries thread fin breams is the preferred species. The temperature should be always less than 10°C throughout the process of surimi preparation, as a slight abuse in temperature causes great changes in the textural properties. It is frozen stored till it is used for the product preparation. Although there is considerable loss of many important nutrients in the process of surimi preparation, the high demand and attractive prices are magnetizing many fish processing establishments to adopt surimi processing.

Apart from the few products discussed above, there are very large numbers of value-added products produced and consumed all over the world. As it is highly impossible to cover every product in detail only the commercially important and most commonly used fishery products are dealt in this chapter.

## KEYWORDS

- fish
- nutritional
- significance
- value addition
- chilled fish
- thermal processed fish
- coated fish

## REFERENCES

Ali, A., Sudhir, B., & Gopal, T. K. S. (2005). Effect of heat processing on the texture profile of canned and retort pouch packed oil sardine (*Sardinella longiceps*) in oil medium, *J. Food Sci., 70*(5), S350–S354.

Bindu, J., Gopal, T. K. S., & Nair, T. S. U. (2004). Ready-to-eat mussel meat processed in retort pouches for the retail and export market. *Packaging Technol. Sci., 17* **(3), 113–117.**

Gopal, T. K. S., Ravishankar, C. N., Vijayan, P. K., Madhavan, P., & Balachandran, K. K. (2002). Heat processing of seer fish curry in retort pouch. In: Riverine and Reservoir Fisheries of India, by Boopendranath, M. R., Meenakumari, B., Joseph, J., Sankar, T. V., Pravin, P. & Edwin, L. (eds.). Society of Fisheries Technologists (India), Cochin, India, 211–216.

Gopal, T. K. S., Vijayan, P. K., Balachandran, K. K., & Madhavan, P. (1998). Heat penetration of fish curry in retort pouch. In: Advances and Priorities in Fisheries Technology, by Balachandran, K. K., Iyer, T. S. G., Joseph, J., Perigreen, P. A., Raghunath, M. R. & Varghese, M. D. (eds.). Society of Fisheries Technologists (India), Cochin, India, pp. 236–241.

Gopal, T. K. S., Vijayan, P. K., Balachandran, K. K., Madhavan, P., & Iyer, T. G. S. (2001). Traditional Kerala style fish curry in indigenous retort pouch. *Food Cont.*, *12*, 523–527.
http://maps.grida.no/go/graphic/world-fish-consumption-per-capita-and-per-year.
http://www.globefish.org/dynamisk.php4?id=4787.
http://www.grida.no/publications/vg/water2/page/3283.aspx.
Lakshmisha, I. P., Ravi Shankar, C. N., Ninan, G., Mohan, C. O., & Srinivasa Gopal, T. K. (2008). Effect of freezing time on the quality of Indian mackerel (*Rastrelliger kanagurta*) during frozen storage. *Journal of Food Science*, *73*(7), S345–353.
Mallick, A. K. (2008). Retort pouch processing of Indian white shrimp (Fenneropenaeus indicus) using flexible packaging materials. PhD Thesis, Central Institute of Fisheries Education, Mumbai, India, p. 244.
Manju, S., Mohan, C. O., Mallick, A. K., Ravi Shankar, C. N., & Srinivasa Gopal, T. K. (2008). Influence of vacuum packaging and organic acid treatment on the chilled shelf life of pearlspot (*Etroplus suratensis*, Bloch 1790). *Journal of Food Quality*, *31*, 347–365.
Manju, S., Sonaji, E. R., Leema, J., Gopal, T. K. S., Ravishankar, C. N., & Vijayan, P. K. (2004). Heat penetration characteristics and shelf life studies of seer fish moilee packed in retort pouch. *Fishery Technol*, *41*(1), 37–44.
Mohan, C. O., Ravi Shankar, C. N., Bindu, J., Geethalakshmi, V., & Srinivasa Gopal, T. K. (2006). Effect of thermal processing on texture and subjective sensory characteristics of Prawn kuruma in retortable pouches and aluminum cans. *J. Food Sci.*, *71*(6), S496–S500.
Mohan, C. O., Ravi Shankar, C. N., Srinivasa Gopal, T. K., & Bindu, J. (2008). Thermal processing of Prawn kuruma in retortable pouches and aluminum cans. *Int. J. Food Sci. Technol*, *43*, 200–207.
Mohan, C. O., Ravishankar, C. N., Srinivasa Gopal, T. K, Ashok Kumar, K., & Lalitha, K. V. (2009a). Biogenic amines formation in Seer fish (*Scomberomorus commerson*) steaks packed with $O_2$-scavenger during chilled storage. *Food Res. Int*, *42*, 411–416.
Mohan, C. O., Ravishankar, C. N., Srinivasa Gopal, T. K., & Ashok Kumar, K. (2009b). Nucleotide breakdown products of seer fish (*Scomberomorus commerson*) steaks stored in $O_2$ scavenger packs during chilled storage. *Innov. Food Sci. Emerg. Technol*, *10*, 272–278.
Mohan, C. O., Ravishankar, C. N., Srinivasa Gopal, T. K., Lalitha, K. V. (2010). Effect of reduced oxygen atmosphere and sodium acetate treatment on the microbial quality changes of Seer fish (*Scomberomorus commerson*) steaks stored in ice. *Food Mic, 27*, 526–534.
Mohan, C. O., Ravishankar, C. N., Srinivasa Gopal, T. K., Lalitha, K. V. (2012). Effect of chitosan edible coating on the quality of double fileted Indian oil sardine (Sardinella longiceps) during chilled storage. *Food Hydrocolloids*, *26*(1), 167–174.
Mohan, C. O., Remya, S., Murthy, L. N., Ravishankar, C. N., & Ashok Kumar, K. (2015). Effect of filling medium on histamine content and quality of canned yellowfin tuna (Thunnus albacares). *Food Control, 50*, 320–327.
Mohan, C. O., Remya, S., Ravishankar, C. N., Vijayan, P. K., & Srinivasa Gopal, T. K. (2014). Effect of filling ingredient on the quality of canned yellowfin tuna (Thunnus albacares). International *Journal of Food Science and Technology*, *49*(6), 1557–1564.
Ravishankar, C. N., Gopal, T. K. S., & Vijayan, P. K. (2002). Studies on heat processing and storage of seer fish curry in retort pouches. *Packag. Technol. Sci.*, *15*, 3–7.
Remya, S., Mohan, C. O., Bindu, J., Sivaraman, G. K., Venkateswarlu, G., & Ravishankar, C. N. (2016). Effect of chitosan based active packaging film on the keeping quality of chilled stored barracuda fish. *Journal of Food Science & Technology, 53*(1), 685–693.

Remya, S., Mohan, C. O., Venkateswarlu, G., Sivaraman, G. K., & Ravishankar, C. N. (2017). Combined effect of $O_2$ scavenger and antimicrobial film on shelf life of fresh cobia (*Rachycentron canadum*) fish steaks stored at 2°C. Food Control, *71*, 71–78.

Renuka, V., Mohan, C. O., Kriplani, Y., Sivaraman, G. K., & Ravishankar, C. N. (2016). Effect of Chitosan edible coating on the microbial quality of Ribbonfish, *Lepturacanthus savala* (Cuvier, 1929) steaks. *Fishery Technology, 53*, 146–150.

Shashidhar, K. (2007). Studies on Ready-to-serve calcium and iron fortified shrimp soup in retortable pouches. MFSc Thesis, Central Institute of Fisheries Education, Mumbai, India, p. 128.

Sonaji, E. R., Manju, S., Rashmy, S., Gopal, T. K. S., Ravishankar, C. N., Vijayan, P. K., et al. (2002). Heat penetration characteristics of rohu curry. In: Riverine and Reservoir Fisheries of India, by Boopendranath, M. R., Meenakumari, B., Joseph, J., Sankar, T. V., Pravin, P. & Edwin, L., (eds.). Society of Fisheries Technologists (India), Cochin, pp. 320–324.

Vermeiren, L., Devlieghere, F., Van Beest, M., De Kruijf, N., & Debevere, J. (1999). Developments in the active packaging of foods. *Trends Food Sci. Technol, 10*(3), 77–86.

# PART III

# CANDELILLA

# CHAPTER 15

# CANDELILLA PLANT (*EUPHORBIA ANTISIPHYLITICA* ZUCC.) APPROACH AND MARKET: AN OVERVIEW

JORGE A. AGUIRRE-JOYA,[1] ROMEO ROJAS,[2]
JANETH M. VENTURA-SOBREVILLA,[1]
MIGUEL A. AGUILAR-GONZALEZ,[3] HELIODORO DE LA GARZA,[4]
RUTH E. BELMARES,[1] RAUL RODRÍGUEZ-HERRERA,[1] and
CRISTOBAL N. AGUILAR[1]*

[1]*Department of Food Science and Technology, School of Chemistry, Autonomous University of Coahuila, 25280 Saltillo, Coahuila, Mexico*

[2]*Research Center and Development for Food Industries, School of Agronomy, Autonomous University of Nuevo Leon, Escobedo, Nuevo Leyn, 66054, Mexico*

[3]*CINVESTAV, Center for Research and Advanced Studies, IPN Unit Ramos Arizpe, Coahuila, Mexico*

[4]*Basic Science Department, Antonio Narro Agrarian Autonomous University, Saltillo, 25315, Coahuila, Mexico,*
*E-mail: cristobal.aguilar@uadec.edu.mx*

## CONTENTS

## ABSTRACT

Candelilla (*Euphorbia antisyphilitica* Zucc.) plant is a natural source which wax is obtained; their commercial exploitation began in 1900 and for lest 100 years has been an important economic source for 3,500 producers, called "candelilleros" distributed in 230 commons of 33 municipalities in the north part of Mexico. However, the traditional extraction process involves the use of sulfuric acid, which represents a high risk for the health of candelilleros and is dangerous for the environment. Throughout these 114 years of exploitation the scheme of candelilla approach has passed to be controlled by the Mexican government to the private industries in actuality. Whatsoever keep equal thought this time are the life conditions of poverty of the candelilleros due principally at the intervention of intermediates and the low prices that the exportation companies buy the candelilla wax to the producers. Nowadays there is an effort by the Universidad Autónoma de Coahuila and Comision Nacional Forestal to substitute the sulfuric acid for and organic one to protect the producers health and environment quality in a sustainable and eco-friendly process that permit obtain a completely organic and high quality wax.

## 15.1   INTRODUCTION

Candelilla (*Euphorbia antisyphilitica* Zucc.) is one of the few plants that produce an abundant quantity of hydrocarbons (Saucedo-Pompa et al., 2009) and because its physiology and photosynthetic mechanisms, with an adequate science and technology applications can be one of the most photo production capacity plants (CIQA-CONAZA, 1981). Fact that is corroborated since the candelilla wax extraction has been an important economic source to a considerable number of residents of the Northeast of Mexico, specifically 3,500 producers, called "candelilleros" distributed in 230 commons of 33 municipalities have been exploited candelilla since 100 years ago, who obtain the wax by a traditional method of extraction. The discovery and exploitation of candelilla plant is attributed to the prehispanic indigenous of the Northeast of Mexico, whom used the branches to extract the wax with boiling water in clay pots. They used the wax mixed with natural colorants for decorative purposes (Romahn, 1992) and for cover the thread of their arches to protect them of climate changes (CIQA-CONAZA, 1981). The natives Apaches uses too the plant in medicinal preparations to treat toothache and as laxative (Stanley, 1961).

The origin of the name "candelilla" has two possible explanations, one of them refers that the shape of the bush, which are long, straight, erect and covered by wax looking like little candles (Saucedo-Pompa et al., 2007), and the other indices that the prehispanic indigenous used to burn the branches to illuminate in night, working like a candle (Ochoa-Reyes, 2010). However both explanations are related to the candle either for shape or function. Candelilla is endemic of the Chiuahuan desert zone that includes estates like: Chihuahua, Coahuila, Durango, Hidalgo, Nuevo Leon, Tamaulipas and Zacatecas, and in minor proportion in Oaxaca, Puebla, Queretaro and San Luis Potosi, in Mexico, and in the states of New Mexico and Texas; in the Unites States of America. Nevertheless, in Coahuila (Mexico) represents and invaluable natural source due to this polity is the main producer of candelilla wax (CONAFOR, 2013)

After Spanish invasion the remaining water of wax extraction was used to treat sexual transmitted infections like syphilis. Scora et al. (1995) fact that influence to Zuccarini who in 1829 classifies the candelilla like Euphorbia antisyphilitica (CIQA-CONAZA, 1981). In 1907 the professor McConnel Sanders reports a series of physical constants for candelilla wax; thereafter in 1909 Gerardo Alcocer modified the classification of Zuccarini and classifies the candelilla like Euphorbia cerifera highlighting the capacity of wax production by the plant (CIQA-CONAZA, 1981; Ochoa-Reyes, 2010).That is the reason for what candelilla is appreciated due to the candelilla wax is applied in diverse industries like food, cosmetic, pharmaceutics and materials ones. The 90% of total wax production is exported to countries like: United States of America, Japan, Germany, Spain, France, Holland, England, Ireland, Italy, Colombia and Argentina (CONAFOR, 2013).

## 15.2   CANDELILLA WAX EXTRACTION

The commercial exploitation of candelilla wax begins in 1900, after The Journal of the Royal Society of Arts declares that "the wild plant known as candelilla contains wax of excellent quality, and enough quantity, which makes it in an invaluable species" (Aguilar et al., 2008). It leads the settling of thousands of people in the desert part of the north of Mexico, specifically in the area of Chihuahuan desert, such groups were responsible of introduce the candelilla wax in an incipient market of commodities. Later at the beginning of the word ward, the demand of candelilla wax was increased, principally because the United States of America's army used the wax to cover the fabrics of tents to protect them from climate and prevent mosquitos prick (CIQA-CONAZA, 1981).

The first pilot plant build to extract candelilla wax was installed in New Mexico in 1910; these extraction was done using hot water. In 1912 the candelilla wax becomes popular in the US market and the idea came of design an equipment to obtain the wax by solvents and refine it with diatomaceous earth and other equipment to extract the wax of the plant through steam stripping. Later in 1913 the distribution of the wax in the United States was done by the use of intermediates. In the same year two industrial plants of wax extraction by steam stripping were installed in Mexico, one in Torreón and the other in Monterrey, both resulted unaffordable (CIQA-CONAZA, 1981).

In 1914, Borrego and Flores designs and use the extraction with sulfuric acid. This method becomes popular quickly because the simple of the technique and the high production volumes of wax, of 2.4 percent of total plant (Ochoa-Reyes, 2010). The traditional process of extraction consisted in placing the plant of candelilla in iron cauldrons called "pailas" buried in the ground to which water is added (300 L) and then the plant is immersed (192–256 Kg) and sulfuric acid is added to obtain an approximated concentration of 0.3% in the boiling water, the wax is then recovery with a performed spoon called "espumadera" and then they put the crude wax in an iron circular container, this part of the process is called "charquear" and then they transfer the wax to another circular metallic container known as "cortador" where the wax is boiled again with water to separate the wax from ground, then they let the wax cooling all night. For the next day the wax is ready to be sealed. The extraction with sulfuric acids turns in the only way to extract candelilla wax. The first candelilla workers or "candelilleros" were recruiting from ranches or mining centers.

## 15.3   COMMERCIAL ANTECEDENTS

During the first word war period (1914–1918) the demand of candelilla wax was increased by the industries of Unites States. Since 1914 up to 1937 the properties with candelilla were rented to dealers, whom controlled the candelilleros. The people start to take this form of organization around this new economic activity. In this time the market of candelilla wax was consolidated by the work of candelilleros, the organization of wax production by the dealers and distribution by intermediates (CIQA-CONAZA, 1981)

Although the candelilla was commercially exploited since 1900, it is up to 1918 that the Mexican federal government permits the free exploitation of the plant. In order to control the wax production, in 1920 was created the "Banco HipotecarioEjidal" (Mortgage Ranch Bank). One interesting fact is that in 1928 to 1931 the price of crude wax, (cerote or tejo) was one third higher than the price of cotton. In 1935 the forest tax for one ton of candelilla was of 5 pesos or 250 Mexican pesos by ton of crude wax (CIQA-CONAZA, 1981)

For the year 1936 the principal collectors centers are: (i) Ojinaga in Chihuahua, (ii) Torreon, (iii) Ocampo, and (iv) Saltillo in Coahuila. In 1937 is integrated the Union of Credits for Candelilla Wax Producers (UCPCC, for its name in Spanish) the UCPCC was integrated only by the dealers in order to eliminate the intermediates. Bancomex supports in 1938 the decision of the UCPCC to distribute the wax and eliminate the dealers and the Mexican government exempts the union to pay the exportation taxes so the exportation rate of wax was 98% of the total production (CIQA-CONAZA, 1981).

## 15.4   CONSOLIDATION OF THE CANDELILLA WAX MARKET

During the second war (1939–1945) the demand of candelilla wax was estimated in six or seven thousands ton per year, the army of the US used the wax to cover and protect some parts of the air planes, also to develop explosives (Aguilar et al., 2008) the UCPCC decides that the 61% of the total production will be performed in Coahuila and the other 39% between

Chihuahua, Durango and Nuevo Leon. Thereafter due to a politic movement in 1940 the UCPCC loses their autonomy and controls only the 80% of the exportations. In 1943 the taxes for candelilla were incremented to 5.85 Mexican pesos for ton of candelilla plant and 292.50 Mexican pesos per ton of crude wax (CIQA-CONAZA, 1981).

One of the first attempts to modify the contaminating and dangerous wax extraction method with sulfuric acid was made in 1953, when was proposed the use of solvents for the wax extraction. In the same year Bancomex takes the control of wax commercialization and the illegal sale of candelilla wax was incremented. Thereafter in 1954 were integrated, with help of the government, the union of ranches producers of candelilla wax (UEPCC in Spanish) and they decree that the national found for ranch development concentrates the production of wax and deliver it to Bancomext for its distribution. It creates diverse committees pro-development of arid zones of Mexico; the committee was in charge to invest in candeliller zone, with the gains of wax sale.

In 1955 the South West Research Institute (San Antonio, Texas) designs a portrait wax extractor; they reports good wax yields. In this year the national bank for rural credit will take the control of the wax production but Bancomex stills controlling the blanching and distribution of wax and destinies the 2.5% of the gain to elaborate a grant to develop the candeliller region but is in 1956 that the grant is created. In 1957 the federal taxes for ton of crude wax was of 428.00 Mexican pesos. Due to the price that they pay for candelilla wax the candelilleros start a strike calling for an increment in the price of the wax, finally the increment is approved (CENAMEX, 2007).

Afterthe Second World War ending, and during the decades of 50's and 60's, the demand for candelilla wax decreases and the exploitation was wobbling. But a cause of the scares or null economic opportunities of the people of the arid zones of Mexico they were taken in the candelilla wax exploitation as the only economic activity to solvent their basic human needs like food, home and clothing.

In 1960 the Mexican government declines to keep buying candelilla wax so the candelilleros start a political movement to reincorporate the wax in the national market, the government try to stop the movement with false promises of solution. In 1961 the movement starts again and the government and candelilleros found a solution; increase the pay per kilogram of crude wax and establish a limit of production of 3,000 tons per year.

## 15.5 CANDELILLA WAX RESURGENCE

At the middle of 1970 resurge the interest for natural and renewable commodities, principally because the crisis of petroleum and their sub products market, and surge the need and interest for innovate their production processes. During these time candelilla wax was obtained by the same and archaic method of extraction intact since beginning of the century (CIQA-CONAZA, 1981)

In 1979, Bancomex loses the control of distribution of the wax; Banrural takes the control of the production and distribution of the wax, it is permitted the candelilla exploitation to smallholders. Table 15.1 summarizes the most relevant scientific and/or technical researches about candelilla, either plant or wax that have been reported during the candelilla exploitation period.

During the next years the candelilla exploitation experienced a series of changes until the actual system of exploitation, where the candelilleros approach the plant located in the immediate hills or neighbor properties, were they travel in donkeys or trucks to gather the wild plant. They can stay 15 days or more collecting the plant, the collecting process consist in take manually the braches and pull it from the ground. After they had collected enough plant they back to the places were the pailas are, to start the extraction process, in this point all family members (if is necessary) including wife, sons or child participates. When the wax is cool it is ready to be sold. Actually the market of candelilla wax is controlled by a small number of companies which recollect the wax in collection centers in the cities near to the rural places that produces the wax or collect the wax directly in the production areas, for later purify it by diverse processes and offer different and specific products for a wide market of waxes either in Mexico or other countries like Japan, U.S.A., France and Germany, etc.

## 15.6 CONCLUSIONS

Candelilla wax extraction is a hard work that represents for 3,500 candelilleros of 33 localities of the Chiuauan desert, the main economic source. Most of them are located in medium or high marginalization zones, with scarcest of water and other sources and some of them do not have access to education and the only way to solve economic needs is by candelilla exploitation, however, national market in Mexico pays candelilla wax very low

**TABLE 15.1**    Scientific and/or Technical Researches About Candelilla

| Research | Institute | Country | Reference |
|---|---|---|---|
| Analyses of the wax (Properties and chemical structure) | NHI | Mexico | CIQA-CONAZA, 1981, 1907 |
| Analyses of the wax (Properties and chemical structure) | TAES | U.S.A. | TAES, 1910 |
| Analyses of the wax (Properties and chemical structure) | LMP | France | CIQA-CONAZA, 1981, 1925 |
| Blanching and refining of the wax | ICT | Germany | CIQA-CONAZA, 1981, 1936 |
| Analyses of the wax (Properties and chemical structure) | U of W | U.S.A. | Baldinus, 1949 |
| Wax determination of n-acids and n-alcohols | U of W | U.S.A. | Schuette and Baldinus, 1948 |
| Studies of Wax Composition | DPC-OSU | U.S.A. | Wagner and Brown, 1953 |
| Use of the wax to cover fruits | | Mexico | Paredes-López et al., 1974 |
| Determination of water vapor and oxygen permeability | U of W | U.S.A | Donhowe and Fennema., 1993 |
| Identification of hydrocarbons from candelilla growth in different localities | U of C | U.S.A. | Scora et al., 1995 |
| Detection of paraffin and wax coatings in apples | IL/UM | Germany | Ritter et al., 1999 |
| Physicochemical characterization of candelilla wax | IF/UAP | Mexico | Dosseti-Romero et al., 2002 |
| Wax extraction by solvents | CIQA | Mexico | CIQA-CONAZA, 1981 (1978) |
| Characterization of physical properties of an candelilla wax/mesquite gum emulsions | UAM-I | Mexico | Bosquez-Molina et al., 2003 |
| Elaboration of a healing cream | DIA-UAdeC | Mexico | Ventura-Sobrevilla et al., 2006 |

**TABLE 15.1** (Continued)

| Research | Institute | Country | Reference |
|---|---|---|---|
| Production of candelilla seedlings by cuttings | INIFAP-UACh | Mexico | Villa-Castorena et al., 2010 |
| Elaboration of products like margarine, vegetable creams and hedges with candelilla wax | UASLP | Mexico | Toro-Vazquez et al., 2013 |
| Application of candelilla wax to prolong shelf life of fresh-cut fruits | DIA-UAdeC | Mexico | Saucedo-Pompa et al., 2007 |
| Extraction of ellagic acid from candelilla branches | UAM-I | Mexico | Aguilera-Carbó et al., 2008 |
| Elaboration of edible coatings to prolong shelf life of avocados with candelilla wax | DIA-UAdeC | Mexico | Saucedo-Pompa et al., 2009 |
| Extraction of ellagic acid from residual branches of candelilla branches | DIA-UAdeC | Mexico | Ascacio-Vadez et al., 2010 |
| Extraction of candelilla wax with organic acids in pilot plant scale | DIA-UAdeC | Mexico | UAdeC-CONAFOR, 2012 |
| Design and installation of an equipment in candeliller regions to extract wax with organic acids | DIA-UAdeC | Mexico | UAdeC-CONAFOR., 2013 |
| Isolation, identification and characterization of a new ellagitanin | DIA-UAdeC | Mexico | Ascacio-Valdez et al., 2013 |
| Chemical and structural characterization of candelilla plant | DIA-UAdeC | Mexico | Rojas et al., 2013 |
| Social, economic and environmental opportunities offcandelilla wax approach | GUB | Belgium | Arato et al., 2014 |

NHI – National Health Institute; TAES – Texas Agricultural Experimental Station; LMP – Laboratoire Municipal de París; ICT – Institute fur Chemische, Technologie II der Deutschen, Technischen Hochschle; U of W – University of Wisconsin; DPC-OSU – Departament of Physiologica Chemestry-Ohio State University; U of C – University of California; IL/UM – Institut für Lebensmittelchemie/Universität Münster; IF/UAP – Instituto de Fisica/Universidad Autónoma de Puebla; CIQA – Centre for Research in Applied Chemistry; UAM-I – Universidad Autónoma Metropolitana-Iztapalapa; DIA-UAdeC – Food Research Department, Universidad Autónoma de Coahuila; INIFAP – Instituto Nacional de Investigaciones Forestales, Agricolas y Pecuarias; UACh – Universidad Autónoma Chapingo; UASLP – Universidad Autónoma de San Luis Potosi; GUB – Ghent University of Belgium.

compared with the price of exportation, resulting in poverty conditions for candelilleros, to say nothing of the actual process is very dangerous for the health of the producers due to the use of sulfuric acid and it is dangerous to environment too.

At the present time there is an effort by the Mexican forestry commission, Comisión Nacional Forestal (CONAFOR) and the Food Research Department of the Autonomous University of Coahuila (DIA-UAdeC) to change the dangerous and contaminating process with sulfuric acid by a extraction process with organic acids that permits count with a secure process for the candelillero, innoxiously for the environment, and a completely organic and traceable wax in an ecofriendly and sustainable process.

## KEYWORDS

- **candelilla wax**
- **exploitation**
- **extraction**
- **history**
- **market**

## REFERENCES

Aguilar, C. N., Rodríguez-Herrera, R., Saucedo-Pompa, S., & Jasso-Cantú, D. (2008). Outstanding phytochemicals from Mexican semidesert: from plant to natural chemicals and the biotechnology (In spanish). Autonomous University of Coahuila, Mexico.

Aguilera-Carbo, A. F., Augur, C., Prado-Barragan, L. A., Aguilar, C. N., & Favela-Torres, E. (2008) Extraction and analysis of ellagic acid from novel complex sources. *Chemical Papers*, *62* (4), 440–444. DOI: 10.2478/s11696-008-0042-y.

Arato, M., Speelman, S., & Van Huylenbroeck, G. (2014). The contribution of non-timber forest products towards sustainable rural development: The case of candelilla wax from the chihuahuan desert in mexico. *Natural Resources Forum*.

Ascacio-Valdés, J., Burboa, E., Aguilera-Carbo, A. F., Aparicio, M., Pérez-Schmidt, R., Rodríguez, R., et al. (2013). Antifungal ellagitanin isolated from *Euphorbia antisyphilitica* Zucc. *Asian Pacific Journal of Tropical Biomedicine*, *3*(1), 41–46.

Ascacio-Valdés, J. A., Aguilera-Carbó, A., Martínez-Hernández, J. L., Rodríguez-Herrera, R., Aguilar, C. N. (2010) *Euphorbia antisyphilitica* residues as a new source of ellagic acid. *Chemical Papers*. 64(4), 528–532. DOI: 10.2478/s11696-010-0034-0036.

Baldinus J. G. (1949). Candelilla wax: An x-ray and a chromatographic study. Thesis (Ph. D.), *University of Wisconsin, Madison*. Pp. 109.

Bosquez-Molina, E., Guerrero-Legarreta, I., & Vernon-Carter, E. J. (2003). Moisture barrier properties and morphology of mesquite gum-candelilla wax based edible emulsion coatings. *Food Research International, 36*(9), 885–893.

CENAMEX. (2007). Mexican Natural Waxes Corp. informative bulletin, printed in Mexico.

CIQA. (2007) Research Center for Applied Chemistry. http://www.ciqa.mx/

CIQA-CONAZA. (1981). Candelilla. Serie Del Desierto. Vol. 5. Mexico

CONAFOR. (2013). National Forestry Commission of Mexico. www.conafor.gob.mx

Dosseti-Romero, V., Méndez-Bermúdez, J. A., & López-Cruz, E. (2002). Thermal diffusivity, thermal conductivity and resistivity of candelilla wax. *Journal of Physics Condensed Matter, 14*(41), 9725–9732.

Ochoa, E., Saucedo-Pompa, S., Rojas-Molina, R., De la Garza, H., Charles-Rodríguez, A. V., & Aguilar, C. N. (2011). Evaluation of a candelilla wax-based edible coating to prolong the shelf-life quality and safety of apples. *American Journal of Agricultural and Biological Sciences, 6*(1), 92–98.

Ochoa-Reyes, E., Saucedo-Pmpa, S., De la Garza, H., Martínez, D. G., Rodríguez, R., & Aguilar-Gonzalez, C. N. (2010) Traditional extraction of *Euphorbia antysiphilitica* wax. *Acta Química Mexicana., 2*(3), 1-9.

Paredes-López, O., Camargo-Rubio, E., & Gallardo-Navarro, Y. (1974). Use of coatings of candelilla wax for preservation of limes. *Journal of the Science of Food and Agriculture, 25*(10), 1207–1210.

Ritter, B., Schulte, E., & Their, H. P. (1999). Thin layer chromatography with FID-Analysis or the detection of paraffin and wax coating in apples (In german). German Food Journal *(Deutsche Lebensmittel-Rundschau), 95*(3), 90–94.

Rojas-Molina, R., De León-Zapata, M. A., Saucedo-Pompa, S., Aguilar-Gonzalez, M. A., & Aguilar, C. N. (2013). Chemical and structural characterization of Candelilla (*Euphorbia antisyphilitica* Zucc.). *Journal of Medicinal Plants Research. 7*(12), 702–702. doi: 10.5897/JMPR11.321.

Romahn, C. F. (1992). Top non-timber forest products from Mexico (In spanish). Autonomous University of Chapingo, México, 376 pp.

Saucedo-Pompa, S., Jasso-Cantu, D., Ventura-Sobrevilla, J., Sáenz-Galindo, A., Rodríguez-Herrera, R., & Aguilar, A. C. (2007). Effect of Candelilla Wax With Natural Antioxidants on the Shelf Life Quality of Fresh-cut Fruits. *Journal of Food Quality, 30*, 823–836.

Saucedo-Pompa, S., Rojas-Molina, R., Aguilera-Carbó, A. F., Saenz-Galindo, A., De la Garza, H., Jasso-Cantú D., et al. (2009) Edible film based on candelilla wax to improve the shelf life and quality of avocado. *Food Research International, 42*(4), 511–515. doi: 10.1016/j.foodres.2009.02.017.

Schuette, H. A., & Baldinus, J. G. (1948). Studies on candelilla wax. Its n-acids and n-alcohols. *Journal of the American Oil Chemists Society, 26*(10), 530–532.

Scora, G. A., Ahmed, M., & Scora, R. W. (1995). Epicuticular hydrocarbons of candelilla (*Euphorbia antisiphylitica*) from three different geographical areas. *Industrial Crops and Products, 4*(3), 179–184.

TAES. (1911). Texas agricultural experimental station, annual report of the office of experiment stations. *U.S. Agricultural Department.* pp. 242.

Toro-Vazquez, J. F., Mauricio-Perez, R., González-Chávez, M. M., Sánchez-Becerril, M., Ornelas-Paz, J. J., & Pérez-Martínez, J. D. (2013). Physical properties of organogels and wather in oil emulsions structured by mixtures of candelilla wax monoglycerides. *Food Research International, 54,* 1360–1368.

Ventura-Sobrevilla, J. M., Aguilar-Gonzales, C. N., Belmares-Cerda, R. E., Aguilera-Carbo, A. F., Saucedo-Pompa S. (2005). Effect of Mexican Desert Plants Extracts Creams in the Cicatrization of Skin Wounds in Rabbits. March, La Habana, Cuba.

Villa-Castorena, A., Catalán-Valencia, E. A., Inzunza-Ibarra, M. A., González-López, M. de L., & Arreola-Ávila, J. G. (2010). Production of Candelilla Seedlings (*Euphorbia anti-siphyllitica* Zucc.) By Cuttings. *Chapingo Journal Series Forestry and Environmental Sciences, 16* (1), 37–47.

Wagner, F. T., & Brown, J. B. (1953). The compositions of some unhydrolyzed naturally occurring waxes, calculated using functional group analysis and fractionation by molecular distillation, with a note on the saponification of carnauba wax and the composition of the resulting fractions. *Journal of the American Oil Chemists Society, 30*(7), 291–298.

# CHAPTER 16

# CANDELILLA (*EUPHORBIA ANTISIPHYLITICA* ZUCC): GEOGRAPHICAL DISTRIBUTION CLIMATE AND EDAPHOLOGY

FRANCISCO HERNANDEZ-CENTENO[1]*,
HAYDEE YAJAIRA LÓPEZ-DE LA PEÑA[1] and ROMEO ROJAS[2]

[1]*Antonio Narro Agrarian Autonomous University, Buenavista, Saltillo, Coahuila, 25315, Mexico, Tel. +(844) 411 0200, Ext 2009, E-mail: francisco.hdezc@gmail.com*

[2]*Research Center and Development for Food Industries, School of Agronomy, Autonomous University of Nuevo Leon, Escobedo, Nuevo Leyn, 66054, Mexico*

## CONTENTS

## ABSTRACT

Within the means of production in arid and semiarid areas of Mexico, specifically in the chihuahuan desert, the activities of exploitation of natural resources are, which account for many communities the basis of its economy. One of those resources is the candelilla, which is exclusive to this ecosystem, distributed from southwestern Texas, in the US, throughout the Mexican highlands, to the San Luis Potosi state, highlighting the Coahuila de Zaragoza state as the largest producer of this species exploited to extract the wax, which has very special properties and increasingly diverse applications in the world. This plant grows in areas of semi-desert climate, with highly adaptable to drought conditions. The candelilla area is among the most unfavorable climates to develop agricultural activities, but is suitable for the production of waxes. The climate is steppe or desert type, with erratic rainfall regime with rainfall of 150–500 mm annually and extreme temperatures of 44°C to –2°C. The geology for candelilla are sort litosol floors, regosol and xerosol mainly with medium texture (90%), the type of relief would plains, hills and mountains preferably with exposure to the east and south and altitudes below 2,000 m. It is noteworthy that in some regions there have been limiting factors caused or not by man, where this species is considered endangered, so they should take action in areas where abundant ensure that harvesting is sustainable and not put and at risk species in its entire range to, in this way, continue to have this noble raw material for the benefit of present and future humanity.

## 16.1  INTRODUCTION

Among the aspects to be taken into account in the study of a species are the three essential described in this chapter: geographical distribution, soil and climatic characteristics of growth as for the management of a species within an ecosystem specific it is essential to know the relationships between them, so that their study, management and use is provided, generating a sustainable benefit for those who intend used it (Challenger, 1998), mainly because biotic resources in desert and semi-desert are limited because, or it has not been discovered, or developed any technology package for sustainable use existing species in this geographical environment, or it has not made the proper transfer of the knowledge to producers, as mentioned by Barsch (2004).

The traditional historical exploitation of plant species in the Mexican desert and semi-desert has been a result of having to meet the survival needs of the human population living in this ecosystem (Challenger, 1998; Rodríguez-Hernández, 2012), finding a livelihood and development, in most cases precarious, given the low income they receive for their products because of the middlemen and not fair trade (Canales et al., 2006; Hernández-García, 2013).

Much of the land area occupied by our country belongs to this type of ecosystem (Rzedowski, 1998; Gómez-Lorence, 2011), distributed between the two of North America's largest deserts, the Chihuahuan and Sonoran (Flores-Del Angel, 2013), and smaller, as the Poblano-Oaxaca. In this environment factors are limited to the cultivation and production of commercial plant species.

Species such as lechuguilla agave, oregano, damiana, sotol, jojoba, mesquite and candelilla, to mention some of the most significant have been the basis of the economy of many rural communities of the medium (Cervantes, 2005; CONABIO, 2008), as grows in conditions of climate and soil extremely stunted, often unsuitable for other crops that are more demanding in terms of soil and water concerns (Villa-Salas, 1981).

In the case of the candelilla species in this review paper focuses, have been studied and cataloged the soil and climatic characteristics in which grows and develops to expand the knowledge base for better use of this renewable biotic resource, focusing on sustainable exploitation, but allows to exploit the potential of plant compounds usable, in this case waxes present in the stems (De la Garza and Berlanga, 1993); attention to the harmful effects of over-exploitation to cause thereby an imbalance in the delicate relationship in the ecosystem it belongs to the noble species (Cervantes-Ramirez, 2006) from which, indeed, is very recent is also provided the discovery of its modern potential (Maldonado, 1983), which even today continue to diversify.

## 16.2 GEOGRAPHICAL DISTRIBUTION

### 16.2.1 GENERAL DISTRIBUTION

Scora et al. (1995) mention that the candelilla is native to the American desert, stretching from Arizona, New Mexico and Texas in the United States of America to Zacatecas, along the highlands in Mexico, which coincides with that reported by Flores-Del Angel (2013) and Cabello et al. (2013). In Figure 16.1, the distribution of this ecosystem is illustrated.

Cervantes-Ramírez (2006) warns that this is a plant with a wide range of distribution corresponding to the chihuahuan desert area between parallels 22° 00' and 30° 40' N and meridians 99° 25' W and 105° 50' W; rural units located collection of this plant in 22 of the 32 states. Martínez et al. (2002) mention that the highest concentrations of candelilla are among the meridians 102° and 104° W, from the Bravo River to the parallel 24° N, but particularly marked in the meridians 102° and 103° W (Flores-Del Angel, 2013), placing the Coahuila state and the one with the best concentrations of this species. CITES (2009) reports the potential populations in the chihuahuan desert candelilla shown in Figure 16.2 distribution.

Meanwhile, Villa-Castorena et al. (2010) reported to Mexico a distribution that covers the states of Durango, Zacatecas, Chihuahua, Nuevo Leon, San Luis Potosi, Tamaulipas and Coahuila, the latter being the most important area and production as it participates with 80% of national production. This information is consistent with that reported in previous form by Canales et al. (2006).Roblero-Pérez (2012) located the distribution of this species in the states of Coahuila, Chihuahua, Durango, Zacatecas, Nuevo Leon, Guanajuato, Hidalgo, Aguascalientes, Queretaro, San Luis Potosi,

**FIGURE 16.1**   Chihuahuan desert region and its three subregions (Granados-Sánchez et al., 2011).

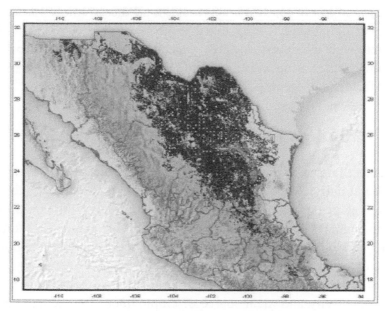

**FIGURE 16.2** Candelilla potential distribution in the Chihuahuan desert (CITES (2009), with information of National Workshop on candelilla (2008)).

Tamaulipas, south of Texas (Martínez et al., 2002), Puebla and Oaxaca, with commercial operation only in the first 5 states.

For his part, Pérez-López (2014) agrees to refer to the geographical distribution of the Candelilla is located in the chihuahuan desert of Mexico and the southeastern United States; in Mexico in the states of Coahuila, Chihuahua, Durango, Zacatecas, Nuevo Leon, Tamaulipas and San Luis Potosi. Gomez-Lorence (2011) also reports that is found throughout the chihuahuan desert, stretching south to Hidalgo state, Chihuahua west and north to the region of Big Bend National Park in Texas, USA; In the east of Nuevo Leon and Tamaulipas and in the south in Puebla and Oaxaca states. Flores-Del Angel (2013) reports that in New Mexico, USA, this plant is considered endangered, a situation that usually indicates an extreme situation.

CONAFOR (2009) and CONABIO (2009) indicated that the potential area ranges from 30'129,896 and 32'420,133 Ha, according to CITES (2009), although the results of potential stock assessment conducted by Perez et al (2013) reported an approximate potential for use in those states until 1 '348, 297.15 Ha by 2015, Martínez-Salvador et al. (2015) estimate a potential area of 8'526,336 Ha between the aforementioned states. Although the candelilla is present in all these regions, it does not develop in all habitats, it is notable that the range of the plant is not fully consistent with the desert-like area, especially in the north. Probably in the United States is on the edge of the distribution caused by environmental factors such severe frosts (Flores-Del Angel, 2013).

## 16.2.2 CANDELILLA AREA

The candelilla area is among the most unfavorable climates to develop agricultural activities, but is suitable for the production of waxes. The climate is steppe or desert type, with erratic rainfall regime rainfall 150–500 mm annually and temperature extremes of 44° to –2° C (Alonso & Álvarez, 2011). Villa-Salas (1981) reported that the candelilla region has an area of 14 million hectares, which decreases it by intensive exploitation of this species over decades, and is called so because so far the economic activity and main source of employment for the habitants of these geographic areas is precisely in collecting candelilla (Flores-Del Angel, 2013). It is subdivided, according to Peña-Contreras (1998), in four areas:

1. Las Coloradas.
2. Ocampo.
3. Coyame.
4. Tlahualillo.

The Candelilla zone comprises producing states of Coahuila, Zacatecas, Durango, Chihuahua and Nuevo Leon, which cover 28.5% of a country's desert strata territory (Padilla-Fuentes, 1959) (Figure 16.3).

This region is bounded on the east by the Sierra Madre Oriental, to the west by the Sierra Madre Occidental, and south by the mountains of Zacatecas and San Luis Potosi (Flores-Lopez, 1995). In all states covered by the "candelilla area," the largest producer is Coahuila de Zaragoza (Barsch, 2004), state that the communities are most productive in the country (Hernández-García, 2013); because, according to Villarreal-Barrera (1995), this area is located by the partial surfaces of the municipalities of Castaños, Cuatrociénegas, Francisco I. Madero, General Cepeda, Ocampo, Parras, Ramos Arizpe, Saltillo, San Pedro, Sierra Mojada and Viesca. Ávila-Rebollar (2007) identified the municipalities concerned candelilla production in the region (Table 16.1).

**FIGURE 16.3** Mexican candelilla area. From: Instituto de la candelilla (Candelilla Institute, 2015).

**TABLE 16.1** Candelilla Municipalities in the Chihuahuan Desert

| Cuahuila | Chihuahua | Durango | Zacatecas |
|---|---|---|---|
| Castaños | Aldama | Cuencamé | Concepción del Oro |
| Cuatrociénegas | Coyame | Lerdo | Mazapil |
| Francisco I. Madero | Manuel Benavides | Mapimí | Melchor Ocampo |
| General Cepeda | Ojinaga | Nazas | |
| Juárez | | Peñón Blanco | |
| Matamoros | | Rodeo | |
| Ocampo | | San Juan Guadalupe | |
| Parras | | San Luis del Cordero | |
| Ramos Arizpe | | San Pedro del Gallo | |
| Sabinas | | General Simón Bolívar | |
| San Buenaventura | | Tlahualillo | |
| San Pedro | | | |
| Sierra Mojada | | | |
| Torreón | | | |
| Viesca | | | |

*Source*: Avila-Rebollar (2007), with information of CERAMEX (2006).

## 16.3 CLIMATOLOGICAL CHARACTERISTICS

As reported by Ávila-Rebollar (2007), the chihuahuan desert is a desert areas with greater wealth and biological diversity in the world. Their particular conditions of humidity, soil composition and temperature allow the growth of almost a quarter of the 1, 500 cacti known to science as well as the various floral species that can only be grow in this region of the world, such is the candelilla plant case. Candelilla plant is part of the community called crasilosulifolio thorny scrub, and is associated with lechuguilla (*Agave lechuguilla*), guapilla (*Hechtia glomerata*), Sotol (*Dasylirion berlandieri*) and espadín (*Agave striata*), mainly; sometimes it is part of scrubland helpless parvifolius associated with gobernadora (*Larrea tridentata*) and rosulifolius oligocilyndercaulescentis forest associated with *Yucca carnerosana* (Villa-Salas, 1981).This plant grows in areas of semi-desert climate, with highly adaptable to drought conditions. He has developed physiological mechanisms to tolerate extreme temperatures, covering the epidermis of stems with wax, allowing it to retain moisture that captures soil, preserve the chlorophyll and protected from extreme heat energy (Flores-Del Angel, 2013). The climatic characteristics of the ecosystem where it develops are described in the following subsections.

### 16.3.1 CHIHUAHUAN DESERT'S CLIMATOLOGICAL CHARACTERISTICS

The area of the Chihuahuan desert climate, or desert of northern Mexico, includes in its variety of climates those very dry type, semi-warm, with predominantly summer rains, high temperatures and cool winters. The warmest month is June and the coldest January. The scarce average annual rainfall between 100–400 mm, which occur mostly in summer; the percentage of rain is between 5% and 10.2%, the average annual temperature is between 18°C and 22°C (INEGI, 1983; CONABIO, 1998). Importantly, in addition to the annual average rainfall also it influences the distribution of it throughout the year, so plant species have adapted both its morphology and physiology and reproduction conditions available water. Also the extreme maximum and minimum temperatures determine the distribution of many species (Gómez-Lorence, 2011).

## 16.3.2  CANDELILLA AREA'S CLIMATE CHARACTERISTICS

### 16.3.2.1  Type of Climate

Candelilla is a native plant resources in arid and semi-arid areas of northern Mexico where the average annual temperature is above 20°C and annual rainfall ranges from 50–350 mm (Villa-Castorena et al., 2010). Scora et al. (1995) refer to the candelilla thrives in desert areas with only 100–500 mm of annual rainfall at elevations of around 800 meters, although Pérez-López (2014) mentions altitudinal ranges from 240 to 2.400 m, where the best production areas have medium altitudes above sea level of 700 to 1,500 m, and Gómez-Lorence (2011) places the height of growth between 1000 and 2000 m. Pérez-López (2014) refers to favorable climatic specification for the growth of the candelilla plant is BW (very arid with rains in summer), in which an average rainfall ranging from 120–200 mm per year occurs, an average temperature annual of 18–22°C, maximum resisting minimum temperatures of −14-C and 47°C, and altitude levels where it is situated between the heights of 250 to 1,400 meters. The BW climates are located in the northern part of the Mexican plateau at altitudes less than 1,500 m (Gómez-Lorence, 2011).

The amount of water that falls annually in the specific areas where the candelilla grows, according to De la Cruz (1958), varies from 50 to 200 mm per year. It is reported that the waxy secretion is more abundant in the drier months of winter and the plants that grow in humid regions produce less wax than those living in drier sites (Gómez-Lorence, 2011). They have significant populations of candelilla located in Coahuila in areas where stormwater runoff are 10–20 mm per year, corresponding to the basin of the Salado River in the chihuahuan desert (Roblero-Pérez, 2012). The climate plays an important role in soil formation, which was recognized by Dokuchaev in 1883, who said all the planet's soils were formed by the harmonious interaction of natural factors: bedrock or parent material, topography, climate, living organisms and age or time of formation thereof (Reyna, 2008).

## 16.4  SOIL CHARACTERISTICS

### 16.4.1  SOIL DEFINITION

Soil is the layer that lies between the crust and atmosphere, and its importance is that it is a dynamic and living natural element, which is a fundamental part in biogeochemical cycles, which develops several essential functions in nature, which are of ecological, economic, social and cultural. In addition, the soil provides nutrients, water and the hardware necessary for plant growth and biomass production in general, play a key role as a power source for living beings (Saucedo, 2010).

Currently the ground is defined as a natural body made up of solids (mineral matter and organic matter), liquids and gases, which contribute to the surface of the earth, they occupy a place in space and having one or both: horizons or layers, which differ from the initial material because of additions, losses, transfers and transformations of energy and matter or the ability to support plants in a natural environment (Soil Survey Staff, 1998). Bautista et al. (2004) report that the soil is the basic substrate for plants; captures, retains and water issues; and is an effective environmental filter. Consequently, this concept reflects the soil's ability to operate within the limits of the ecosystem of which it forms part and with which it interacts.

## 16.4.2  SOILS CHARACTERISTICS

The soil is formed from fracture minerals by weathering action, resulting from these primary components, which are combined with the organic matter. This process of constant and depends on the interaction of five factors: climate, organic matter, bedrock, relief and time (Cepeda, 2007). The soil is considered a non-renewable resource, because the rate of formation and regeneration is slower than the speed of the processes that contribute to degradation, deterioration and destruction (Saucedo, 2010).

### 16.4.2.1  Soil Composition

The soil matrix is a system of three phases: solid, liquid and gas. In an ideal soil, the solid matter comprises 50% (45 to 48% inorganic and 2 to 5% material organic matter) and the pore space corresponding to 50% (25% air and 25% water). The proportion of water and air varies according to the conditions, when it rains or is irrigated, the pores are filled with water displacing air and emptied when water infiltrates or evaporates allowing the air back to his place (Plaster, 1999). The exchange of air and water in the soil depends on factors such as texture, porosity and thickness of layers or horizons (Foster, 1991). For a soil has a favorable ratio of air and water, you should be able to absorb water quickly during periods of rain or irrigation and retain sufficient to meet the needs of plants moisture (Saucedo, 2010).

## 16.4.2.2 Physical Properties of Soil

The mechanical behavior or physical properties of soil, determine the use that can be given to it. The main physical properties are described in the following subsections.

### 16.4.2.2.1 Texture

It is an indicator of the relative proportions of sand, silt and clay constituting a floor, the texture is considered the most important physical property that influences other properties such as structure, color, consistency and moisture retention. The texture determines the particle size of a ground, which affects retention capacity and aeration thereof. The particle size affects the inner surface and the number and size of pores. The inner surface is important because the reactions occur in ground soil particles; well, you have to soils with a majority of smaller particles like silt and clay have a higher internal surface and retain more water and nutrients but also pollutants (Volke & Velasco, 2002).

### 16.4.2.2.2 Permeability

It refers to the ease with which air, water and roots of the plants in the soil move. In a very permeable soil water it infiltrates rapidly and aeration maintains a good level of oxygen. The permeability depends on the number of pores in the soil, but is affected more by the size and distribution thereof (Palmer and Troeh, 1995).

### 16.4.2.2.3 Slope

Defined as the angle of the soil surface with the horizontal. This property affects the rate of water flow and the degree of soil erosion (Camp and Daugherty, 2000).

### 16.4.2.2.4   Consistency

It is the resistance of a soil to deformation or break; It is the degree of cohesion or adhesion of the soil mass and depends on the moisture content and the carburizing of solid particles. This property is used to determine the appropriate time for farm work and estimate the moisture content (Saucedo, 2010).

### 16.4.2.2.5   Apparent Density

The apparent density of the soilis the mass ratio of solids and the total volume they occupy, including the pore space between the solid particles. The value of the apparent density may range from 1.0 $g/cm^3$ in clay soils, up to 1.8 $g/cm^3$ in compacted sandy soils (Saucedo, 2010). There is a close relationship between the value of the apparent density of soil and other properties there of, such as texture, organic matter content, porosity, thermal conductivity and soil resistance to penetration, which affect step water on the floor and hisavailabilitytoplants (Preciado et al., 2004). Density depends on soil moisture and perform calculations necessary for the material transport (Van Deuren et al., 1997).

### 16.4.2.2.6   Real Density

It refers to the relationship between the mass of a unit volume of soil solids, it allows to know the minerals in soil and calculate % pore space occupied by a particular soil. Because the organic matter is much lighter than an equal volume of mineral solids, the amount of it present in a soil affects the density of the particle, surface soils which have a lower density than the subsoil (Preciado et al., 2014).

### 16.4.2.2.7   Porosity

It is the total percentage of voids between the solid particles of the soil, allowing the circulation of air and water. Soil pores are irregular in size, shape and direction. Clay soils have a higher porosity than sandy soils. For plant growth, the pore size is more important than the total volume, because a clays oil with numerous small pores, retains a lot of water, but aeration is difficult, while a sandy soil with few large pores, it has a good air circulation but retains very little water (Preciado et al., 2004).

### 16.4.2.2.8   Moisture

Water is essential for all living things that participates in various cellular metabolic reactions, serving as carrier solvent and nutrients from the soil to plants and within them. Water also ionizes the macro and micronutrients plant and allows the organic matter to be degraded more easily. Excess water promotes soil leaching of some compounds, so that water is an important regulator of the physical, chemical and biological soil activities (Topp, 1993).

### 16.4.2.2.9   Depth

It refers to the thickness of the topsoil and subsoil favorable for the growth and penetration of the plant roots. Soil depth may be limited by physical or chemical barriers as well as by high groundwater levels, layers of gravel or toxic levels of certain substances (Plaster, 1997).

### 16.4.2.3   Chemical Properties of Soil

Most plants required for growth of 16 or 17 essential elements, of which carbon, oxygen and hydrogen are taken from air and water. The remaining elements are absorbed by the roots of the plants from the soil (Saucedo, 2010). The most important soil chemical properties are listed in the following subsections.

### 16.4.2.3.1   Cation Exchange

It is the ability of soil to retain nutrients and is related directly to the number of cations that may attract soil colloids. The cation exchange capacity (CEC) expressed in mg equivalents per 100 g soil. Different cations are retained in the colloids of different strength, so that the $H^+$, $Ca^{+2}$, $Mg^{+2}$, $K^{+1}$ $Na^{+1}$ are a series of high to low relative to the degree of retention (Cepeda, 2007).

### 16.4.2.3.2   Hydrogen Potential (pH)

The hydrogen potential, or pH, soil depends on a balance between hydrogen and hydroxyl ions in the soil solution (Saucedo, 2010). The alkaline soil having a pH greater than 7 and is produced by the reaction of sodium and calcium in the water. It is an acid soil if having a pH less than 7 and is the result of leaching hydroxyls. Acid soils affect plant growth because the availability of phosphorus and other nutrients is reduced. Soil pH is related not only to the nature of the cations absorbed, but also by the relative amounts of each. Sodium is considered as the most active in enhancing cation in soil pH, followed calcium; while magnesium and potassium have lower influence. Thus, they saturated Na, soils have a higher pH than when saturated Ca or Mg (Navarro and Navarro, 2003).

### 16.4.3   THE SOILS OF THE ARID AND SEMI-ARID AREAS

Regarding the climate, are temperature and precipitation items mostly related to soil formation. His influence is very important to the existence of arid or semi-arid land occupied regions where rainfall is scarce (Gómez-Lorence, 2011; Reyna, 2008). In general, soils in arid regions correspond to Aridisols, calcisols and gypsosoles; There are different kinds of them, depending on weather conditions and the substrate that gives rise to them, as reported by Gómez-Lorence (2011). Thus, in the chihuahuan desert soils are accumulating layers of calcium carbonate or caliche ranging from a few millimeters to 40 cm, and often have a thickness of several tens of centimeters to 2.30 m; in this case, they called calcimorphous (Cervantes, 2002).

In some places of Zacatecas, San Luis Potosi, Coahuila, Durango, Chihuahua, Tamaulipas and Aguascalientes, there are soils known as mokisoles, phaeosem, dark color and high fertility. Another type of deep, heavy, clay soils of high productivity and unwieldy, are vertisols. In the Chihuahuan desert soils are often found high in soluble salts, these soils belong to the solonchak (Gómez-Lorence, 2011).

## 16.4.4 CHARACTERISTICS OF SOILS FAVORABLE FOR CANDELILLA

Generally, the soils where the candelilla grows are highly variable in texture, from silty loam, clay loam till, clay, low areas of poor desert and sandy plains with content of 0.96 to 2.64% organic matter (WAFLA, 2006). Romahn (1992), for its part, argues that grows on slopes with chalky soil, alluvial, shallow (less than 25 cm) of sandy loam, shallow, stony, well drained, rich in calcium carbonate, with pH ranging from 7.4 to 8.4 and poor in nitrogen. Rojas et al. (2009) reports that a bush abundant in places with altitudes of 700–1200 m, with slopes of 1–3%, where it contributes to the prevention of soil erosion. Cardenas (1988) says that a limiting factor nutrient uptake by the plant candelilla throughout the seasons is the pH, the same that is related to the production of wax. Furthermore, carbonates mentioned influence the hydrogen concentration in the soil and consequently increases or decreases the pH, which affects the assimilation of phosphorus.

Gomez-Lorence (2011) specifies pH ranges from 6.3 to 8, the color from dark to gray, soils are shallow to deep, growing the plant in the past, that is, in the mountains where it is more organic matter due to the decomposition of agave lechuguilla leaves or other plants. The slope is very variable too. This plant prefers the slopes where drainage is good, and is rarely found in the lower parts of the valleys or on the clay soils of low areas of desert plains. In a study by Roblero-Perez (2012), in Cuatro Cienegas, Coahuila., Soil types in land abounding candelilla plant, corresponding to lithosols (shallow soils, residual source and medium texture) that are described in all ranges, these soils are presented associated with other a little deeper and darker (rendzinas); or, to clear medium textured soils (calcaric regosols). These soils are limited by a physical phase, either burdensome, stony or soft petrocalcic, cover most of the ground of the plots studied and downs abound; yellowish brown soils are also found in the plains or deeper soils (xerosols haplic gypsicos). Flores-Del Angel (2013) agrees with this when describing the candelilla grows on limestone slopes and rocky ground.

Martinez-Salvador et al. (2015) argue that the soil characteristics for the candelilla are sort litosol soils, regosol and xerosol mainly with medium (90%) texture, the kind of relief would plains, hills and mountains with exposure preferably east and south and altitudes below 2,000 m. Roblero-Perez (2012) describes a high-production candelilla belonging to the province of the Sierra Madre Oriental, formed by layers of ancient marine sedimentary rocks (cretaceous and upper jurassic), including prominently dominated limestones, plaster and sedimentary rocks cluster, remaining in the background argillaceous rocks (shales), reports that these properties are located within the boundaries of the sub-province of the Sierra de Paila and saws subprovince and Coahuila plains, where the limestone dominate in the mountains, while in the valleys emerge basaltic volcanic rocks, according to CTENAL (1975).

## 16.5  CONCLUSIONS

Among the activities that the man realizes, exploitation of natural resources necessarily leads to the processing industry that, in one way or another, it makes our life easier, often regardless of where or how to reach our hands all those amenities that enhance our quality of life. The exploitation of Candelilla wax for his is not the exception, for it reaches its destination must go through a whole process of logistics and transformation does not always have as a foundation a sustainable exploitation of this noble species.

Many communities in the chihuahuan desert, concentrated in four regions of the Candelilla area, its economy based on the exploitation of this species because the climate of the ecosystem where they live does not allow the development of agricultural and livestock activities to increase sources of income for their daily lives, so, in many cases, the exploitation of natural resources in the medium becomes the only viable productive activities for survival.

Full knowledge of the characteristics of climate and soil conditions for the development and production of this kind could allow them to develop, as has been done, efforts to sustainably exploit this plant, keeping it to maximize biomass production of this species, thus mitigating the impact caused by the careless exploitation of natural populations, especially in those regions where it is scarce or becoming scarce, and remembering that each species has a specific and irreplaceable role in the perfect machinery of nature. Although not only the Candelilla is the only usable species in the desert and semi-desert ecosystems, sustainable management of natural resources in them, many limited by the conditions, sometimes display a sense of respect for nature, and of course, to the same humanity as part of that, preserving a balance of all lights need to perpetuate the coexistence and survival of all species.

## KEYWORDS

- **candelilla**
- **climate**
- **distribution**
- **edaphoology**

## REFERENCES

Alonso-Narro, J. L., & Álvarez-Rodríguez, G. (2011). Diagnosis of the production chain candelilla wax in the state of Coahuila. Rural Finance, *State Government of Coahuila. Saltillo, Coahuila, Mexico,* 8.

Ávila-Rebollar, F. (2007). Candelilla (*Euphorbia antisyphilitica* Zucc) in northern Zacatecas. Thesis. *Antonio Narro Agrarian Autonomous University. Saltillo, Coahuila, México,* 67.

Barsch, F. (2004). Candelilla (*Euphorbia antisyphilitica*): Utilization in Mexico and international trade. Medicinal Plant Conservation, *9*(10), 46–50.

Bautista, C. A., Etchevers, B. J., Del, Castillo, R. F., & Gutiérrez, C. (2004). Soil quality and indicators. *Ecosystems: Scientific and Technical Journal of Ecology and Environment. 13* (2), 90–97.

Cabello, A. C. J., Sáenz, G. A., Barajas, B. L., Pérez, B. C., Ávila, O. C., & Valdez, G. J. A. (2013). Candelilla wax and applications. *Advances in Chemistry, 8* (2), 105–110.

Camp, W. G., & Daugherty, T. B. (2000). Managing our natural resources. 4th Edition. Cengage Learning (Ed.), USA 448 p.

Canales, G., E., Canales Martínez, V., & Zamarrón R. E. M. (2006). Candelilla, from the Mexican desert to the world. CONABIO. *Biodiversitas, 69,* 1–5.

Cárdenas, R. F. (1988). Agro-climatic influence of some characteristics in the production of wax Candelilla (*Euphorbia antisyphilitica* Zucc). Thesis. *Antonio Narro Agrarian Autonomous University. Saltillo, Coahuila,* México.

Cepeda, D. J. M. (2007). Soil chemistry. Second edition. Editorial Trillas, México, 167 p.

Cervantes-Ramírez, M. C. (2005). Economically important plants in arid and semiarid areas of Mexico. *Annals of the X meeting of geographers of Latin America,* University of Sao Paulo, Brazil. pp. 3388–3407.

Cervantes-Ramírez, M. C. (2006). IV. Waxes producing plants. Economically important plants in arid and semiarid areas of Mexico. *National Autonomous University of Mexico,* 125–143

Cervantes, R. M. C. (2002). Economically important plants in arid areas of Mexico. Institute of Geography, *National Autonomous University of Mexico,* 1–21, 125–137.

CETENAL (Study Committee Homeland). (1975). Geological Map G14A61 y G13B59, scale 1:50,000. INEGI, Coahuila, México.

Challenger, A. (1998). Use and Conservation of terrestrial ecosystems of Mexico. Past, present and future. National Commission for the Knowledge and Use of Biodiversity. Institute of Biology. *National Autonomous University of Mexico,* México, pp. 689–713.

CITES. (2009). Status assessment *Euphorbia antisyphilitica* in Mexico within the CITES Appendices. Convention on International Trade in Endangered Species of Wild Fauna and Flora. Eighteenth Meeting of the Plants Committee. Buenos Aires, Argentina, March 17–21, (2009). PC18 Inf. 10.

CONABIO (National Commission for the Knowledge and Use of Biodiversity). (2008). Conservation, use and trade of the candelilla. National Workshop. Press Release No. 11. December 17, (2008). México, D. F.

CONAFOR (National Forestry Commission). (2009). Technical report. Management of Commercial Forest Plantations. Zapopan, Jalisco, México.

De la Cruz, C. J. A. (1958). Morphological study of the candelilla. Thesis. Antonio Narro Agrarian Autonomous University. *Saltillo, Coahuila, México.* pp. 5–17.

De La Garza, De La, P. F. E., & Berlanga, R. C. A. (1993). Methodology for the evaluation and management of candelilla under natural conditions. Technical Leaflet No. 5. INIFAP Experimental field "La Sauceda." Saltillo, Coahuila, México. 46 p.

Flores Del Ángel, M. L. (2013). Current status of stocks candelilla (*Euphorbia antisiphilytica* Zucc): inventory, sexual and asexual spread in the state of Coahuila, Mexico. PhD thesis. *Autonomous University of Nuevo León. Nuevo León, México,* 134.

Flores-López. C. (1995). Seed viability, seedling emergence and plantations Candelilla (*Euphorbia antisyphilitica* Zucc.) In Ramos Arizpe, Coahuila. Antonio Narro Agrarian Autonomous University. *Saltillo, Coahuila, México,* .

Foster, A. B. (1991). Approved methods in soil conservation. Editorial Trillas. México. 411 p.

García, E., & CONABIO (National Commission for Biodiversity). (1998). Climate (Koppen, modified by García), scale 11,000,000, México.

Gómez-Lorence, F. (2011). Exploitation, management and evaluation of economically important plants in arid and semiarid areas of Mexico. Doctoral Dissertation. *National Autonomous University of Mexico,*DF, p. 208.

Granados-Sánchez, D., Sánchez González, A., Granados Victorino, R. L., & Borja de la Rosa, A. (2011). Vegetation ecology of the chihuahuan desert. Chapingo Journal. *Series Forestry and Environmental Sciences,*17, 11–130

Hernández-García, G. (2013). The possibilities of sustainable industrialization of candelilla in the Chihuahuan Desert. Graduate thesis. National Polytechnique Institute, CIDIIR-Michoacán. Jiquilpan, Michoacán, México. 136.

INEGI (National Institute of Statistics, Geography and Informatics). (2004). *Guide to Interpreting Maps. Edaphology. Aguascalientes, Ags.* México. p. 8.

INEGI (National Institute of Statistics, Geography and Informatics). (1983). *Geographical Synthesis of the State of Coahuila.* México, p. 163.

Instituto de la Candelilla (Institute of candelilla). (2015). Candelilla. Chihuahuan desert. Disponible en: http://www.candelilla.org/?page_id=530 & lang=es.

M. Villa-Castorena, E. A., Catalán-Valencia, M. A., Inzunza-Ibarra, M. de L. González-López; J. G. Arreola-Ávila. (2010). Production of candelilla seedlings (*Euphorbia antisiphyllitica* Zucc.) by cuttings. Chapingo Journal. *Series Forestry and Environmental Sciences,* 16(1), 37–47.

Maldonado, A. L. J. (1983). "Candelilla research," Second national meeting on ecology, management and domestication of crop plants desert. Forestry Under Secretariat, INIF, México, pp. 11–1

Martínez Salvador, M., Hermosillo Rojas, D. E., Mojica Guerrero, A. S., & Prieto Amparán, J. A. (2015). Production potential and zoning for the use and management of dryland forest species. First edition. National Institute of Forestry, Agricultural and Livestoch Research. P.18.

Martínez, G. M., Jiménez, R. J., Cruz, D. R., Juárez, A. E., García, R., Cervantes, A., et al. (2002). The genera of Euphorbiaceae family in Mexico. Institute of Biology. *UNAM Botanical Series, 73*(2), 155–281.

Navarro, B. S., & Navarro, G. G. (2003). Agricultural Chemistry, soil and chemical elements essential for plant life. 2nd Edition. Mundiprensa Editions. Madrid, España.

Padilla-Fuentes, G. (1959). Determination of plants candelilla wax (Euphorbia spp.) of the six producing areas of northern Mexico. Thesis. Antonio Narro Agrarian Autonomous University. *Saltillo, Coahuila, México.*

Palmer, R. G., & Troeh, F. R. (1995). Introductory Soil Science Laboratory Manual. 3rd Edition. Oxford University Press, NY.

Peña-Contreras, A. R. (1998). Some methodologies utility room genotype analysis in measuring the recovery rate in the growth of interaction Candelilla (*Euphorbia antisyphillitica* Zucc.) Under different ecological conditions. Graduate thesis. Antonio Narro Agrarian Autonomous University. *Saltillo, Coahuila, México.*

Pérez-López, J. G. (2014). Use of Candelilla wax (*Euphorbia antisiphilitica* Zucc) in northern Mexico. Thesis. Antonio Narro Agrarian Autonomous University. *Saltillo, Coahuila, México, 97.*

Pérez, M. R., Zamora, M. M. C., Cortés, B. N. B., & Torres, R. G. (2013). Current geographical distribution and potential Candelilla (*Euphorbia antisyphilitica* Zucc) in Mexico. Memories Foorestal VII National Meeting of Innovation: Innovation for food sustainability. September 10–13, (2013). Veracruz, Veracruz, México. p. 24 ISBN: 978–607–37–0063–4.

Plaster, E. J. (1999). SOIL Science and management. Paraninfo Editions. Madrid, España. 419 p.

Preciado, R. P., Zermeño, G. H., González, A. M. A. & Segura, C. M. A. (2004). Edaphology (anthology). COSNET-DGETA. México, D. F.

Reyna, T. T. (2008). Climates and soil formation in Mexico. Nicolas Aguilera International Diploma Course of Soil Science. CISC, España. pp. 1–17

Roblero-Pérez, E. F. (2012). Biomass prediction models and performance Candelilla wax (Euphorbia antisiphylitica Zucc) in natural populations of Cuatro Cienegas, Coahuila. Thesis. Antonio Narro Agrarian Autonomous University. *Saltillo, Coahuila, México,* pp. 55

Rodríguez-Hernández, M. G. (2012). Induction of plant growth *Euphorbia antisyphilitica* through *Bacillus* sp. Graduate thesis. National Polythechnique Institute, CIIDIR Michoacán, *México,* pp. 75.

Rojas, M. R., Saucedo, P. S., De León, Z. M. A., Jasso, C. D., & Aguilar, C. N. (2009). Essay: Past, present and future of the candelilla. *Mexican Journal of Forestry Sciences, 2*(6), 7–18.

Rzedowski, J. (1988). "Analysis of the geographical distribution of Prosopis Complex in North America," *Mexican Botanical Acta.* 3, México, pp. 7–9.

Saucedo, V., M. E. (2010). Remediation of soil contaminated with oily wastewater from the cheese industry in the Mennonite community of Nuevo Ideal, Durango. Graduate thesis. National Polytechnique Institute, CIIDIR, Durango, México, pp. 3–8.

Scora, G. A., Ahmed, M., & Scora, R. W. (1995). Epicuticular hydrocarbons of candelilla (*Euphorbia antosophylitica*) from three different geographical areas. Industrial crops and products, 4(1995) 179–184.

Soil Survey Staff. (1998). Keys to soil taxonomy. Second edition. USDA. Natural Resources Conservation Service. U. S. Government Printing Office.

Topp, G. C. (1993). Soil water content. In: Soil Sampling and Methods of Analysis. Martin R. Carter Editor. Canadian Society of Soil Science. Lewis Publisher. pp. 541–557.

Van Deuren, J., Wang, Z., & Ledbetter, J. (1997). Remediation Technologies Screening Matrix and Reference Guide. 3rd Edn. Technology Innovation Office, EPA. http://www.epa.gov/tio/remed.htm

Villa-Castorena, M., Catalán Valencia, E. A., Inzunza Ibarra, M. A., González López, M. L., Arreola Ávila, J. G. (2010). Seedling production Candelilla (*Euphorbia antisyphilitica* Zucc) by cuttings. Chapingo Journal. Series Forestry and Environmental Sciences ciencias forestales y del ambiente, *16* (1), 37–47.

Villa-Salas, A. B. (1981). Research Candelilla (*Euphorbia antisyphilitica*) in the area of CIFNE. Ciencia Forestal, No. *32*, 6, 3–18.

Villarreal-Barrera, A. R. (1995). Poverty and marginalization of Candelilla wax producers. Thesis. Antonio Narro Agrarian Autonomous University. *Saltillo, Coahuila, México.*

Volke-Sepúlveda, T., & Velasco-Trejo, J. A. (2002). Remediation technologies for contaminated soils. National Institute of Ecology. México, D. F. p. 64.

WAFLA. (2006). Integrated water resources management and agroforestry in arid, semi-arid and sub-humid Latin America. http://www.wafla.com.

# CHAPTER 17

# MORPHOANATOMIC, TAXONOMIC, AND PHYSIOLOGICAL ASPECTS OF *EUPHORBIA ANTISYPHILITICA* ZUCC.

JORGE ARIEL TORRES-CASTILLO,[1] ADRIANA GUTIÉRREZ-DÍEZ,[2] EDILIA DE LA ROSA-MANZANO,[1] and ARTURO MORA-OLIVO[1]

[1]*Universidad Autonoma de Tamaulipas-Institute of Applied Ecology. Av. Division del Golfo 356, Col. Libertad, 87019, Tel. +52 (834)3162721, 3181800, Ext. 1606, Ciudad Victoria, Tamaulipas, Mexico, E-mail: jorgearieltorres@hotmail.com*

[2]*Universidad Autonoma de Nuevo Leon – Faculty of Agronomy, Francisco Villa. Ex Hacienda el Canada, 66050, General Escobedo, Nuevo Leon, Mexico*

## CONTENTS

## ABSTRACT

The Candelilla plant (*Euphorbia antisyphilitica*) has been used as a tradi-tional source of hydrocarbons to elaborate candles but recently the potential applications of several compounds has expanded the perspectives for this crop. Phylogenetic analyzes of the *Euphorbia* genus placed to *E. antisyphilitica* in the *Chamaesyce* subgenus, where most of American *Euphorbia* species are located, including to plants adapted to dry land conditions. In this work a brief anatomical description of aerial stem, rhizome and root is showed and also a comparison of the variable wax compositions is presented to illustrate that exploitation of this resource should consider variations at environmental, genetic and agronomical levels. An integrative knowledge of physiology, biochemistry, genetic improvement, pest and disease studies of this crop would lead to increase yielding for future massive production of Candelilla.

## 17.1  INTRODUCTION

The Candelilla *(Euphorbia antisyphilitica* Zucc.), native plant to the desert area of Mexico and part of the Southern United States, has received special attention in recent decades due to the many potential applications of this species as a source of hydrocarbon compounds, antioxidant compounds and production of biodegradable films or covers (Saucedo-Pompa et al., 2007; Padmaja et al., 2009; Ascacio-Valdes et al., 2010; Burboa et al., 2014). Traditionally, this plant has been used in obtaining wax for candle manufacturing, as an alternative to using paraffin; however, the extraction of these hydrocarbon compounds is inefficient and a polluting waste is generated (Rojas-Molina et al., 2012). The study trends Candelilla have focused on optimizing processes for extracting wax, besides the increase in applications in areas such as food, cosmetic, pharmaceutical and its use for biofuels (Rojas-Molina et al., 2012; Johari and Kumar, 2013). This has motivated the development of different extraction processes that maximize the various components of this plant (De León-Zapata, 2008; Rojas-Molina et al., 2012).

## 17.2  TAXONOMY

The scientific name of Candelilla is *Euphorbia antisyphilitica* Zucc. and belongs to the *Euphorbiaceae*, one of the most widely distributed families on the planet and with representatives of economic importance such as: *Jatropha curcas* L., *Hevea brasiliensis* (Willd ex A. Juss.) Müll. Arg., *Manihot esculenta* Crantz, *Ricinus communis* L., *Euphorbia pulcherrima*

Willd. ex Klotzsch and *E. antisyphilitica* Zucc. *Euphorbia* is the type genus of the family, and is regarded as one of the giants in the group of angiosperms because of its great diversity with about 2000 known species. The main characteristic of plants of the genus *Euphorbia* is its specialized inflorescence called cyathium, whose structure resembles a flower, because of the dense clustering of stamens and reduced pistils within an cup-shaped involucre, structured of bracts and male flowers groups (Steinmann, 2002).

Within the phylogeny of the genus *Euphorbia* four subgeneric clades based on molecular, morphoanatomic, physiological and ecological studies are distinguished (Horn et al., 2012). Subgenus *Rhizanthium* generally includes species adapted to arid climates, with many morphological variants and it is distributed in the African region. The subgenus *Esula* is distributed in temperate climates throughout the Mediterranean through Central Asia, encompassing predominantly herbaceous species, but may also occur succulent and tree species. The *Euphorbia* subgenus, which includes the greater diversity of the entire genus, is distributed in Africa, Madagascar and Neotropical regions. Finally, the subgenus *Chamaesyce,* with about 600 species encompasses most of the species of *Euphorbia* of America, including growth forms ranging from small herbs to shrubs, through plants with large leaves, even those with small leaves, this group includes few succulent or tree species, presenting erect or decumbent, and rarely prostrate stems. Most species of this group have petaloid appendages subtending the inflorescence, which have been lost several times along the evolutionary history of the group. Physiologically, noted that this subgeneric group includes all known variants of photosynthesis, the $C_3$, $C_4$ and CAM pathways and $C_2$ pathway. This subgenus is widely distributed in the Americas, from Canada to Argentina and has a center of diversity in Mexico and Central America, ranking from tropical forests to deserts and areas of disturbance with altitudes ranging from sea level to 3000 m (Sage et al., 2011; Yang and Berry, 2011; Horn et al., 2012; Yang et al., 2012).

## 17.3 MORPHOLGICAL AND ANATOMIC DESCRIPTION

Within the *Chamaesyce* subgenus the Candelilla plant is located, which has generally straight and devoid of branches plants that grow grouped to form clumps, with a height of 20 to 110 cm, cylindrical stems usually have a greenish color when young stems and gray-ash appearance when mature;

aerial stems emerge from a fleshy rhizome which arise fibrous and fleshy roots. The phyllotaxis is alternating, with sessile, linear and foliar append-ages 2 to 3 mm long and 1 mm wide, grooved, with some pubescence and from green to brown color; leaves are deciduous and occur most frequently in young parts. The flowers are included in reddish to pink cyathium, the inflorescence is in the form of head spike. The fruit is a trilocular capsule that opens when mature seed, these are light brown, rough and four-angled (Villas-Salas, 1981).

The most conspicuous part is the Candelilla stem, which specializes in photosynthesize and is evolutionarily adapted to decrease the loss of water by the presence of a wax coating. The stem has a monolayered epidermis with a relatively thick, waxy cuticle (Figures 17.1A) (Rojas-Molina, 2013); the epidermis is composed of tabular roughly isodiametric cells without intercellular spaces and present of ciclocytic stomata, which are exposed and regularly distributed over the entire surface, not shown in subsidence or folds (Figures 17.1B–17.1D). Immediately below the epidermis, is located the clorenchyma with large cell lumens, numerous intercellular spaces and thin cell walls of cellulosic nature according to the staining with Fast Green ®. This strategic organization allows the plant to expose their photosyn-thetic machinery throughout the stem length, thereby increasing the surface exposed to light. Chlorenchyma is most evident in the first 5–6 cell lay-ers of the subepidermal and cortical region (Figure 17.1E), that gradually results in cortical parenchyma, which may or may not have few chloroplasts. Laticifers are immersed in the matrix of the cortical parenchyma region, in cross section they have similar cellular diameters to parenchyma cells, but difference is on the thickened cell wall and dark tones content, the distribu-tion seems be somewhat irregular and in longitudinal section appreciate as continuous tubular ducts branching or simple, with curves that pass between parenchymal cells (Figure 17.1F). At the boundary between the cortical region and the vascular region there are bundles of fibers that precede on the phloem, vascular region has numerous bundles arranged in a circular pattern, which can gather as the stem grows and thickens; phloem is located at the outer of vascular bundle or rounding to xylem in older stems (Figure 17.1G), and parenchymatous cells followed by xylem to the inner, which is dominated by sclerenchyma surrounding xylem elements, characterized by lignin thickenings including annular, helical and scalariform deposition pat-terns (Figure 17.1I). The pith is of parenchymal nature with wide and clear cellular lumens, rarely with chloroplasts or inclusions and thin cell walls

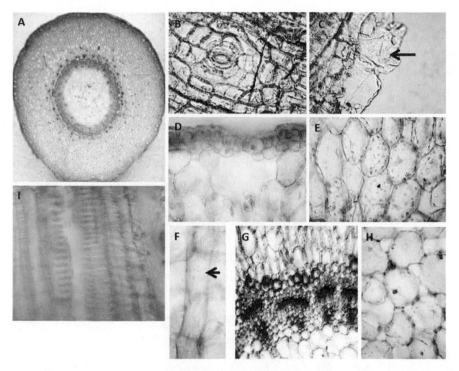

**FIGURE 17.1** Anatomy of *Euphorbia antisyphilitica* stem. A. Cross section of aerial stem of Candelilla, circle arrange of vascular region is observed; B. Frontal view of the stem epidermis, a typical stoma and epidermal cells is appreciated; C. Rising of the stem epidermis, the arrow indicates the waxy coating; D. Cross of stem, where epidermal and subepidermal region is observed; E. Stem cortical parenchyma with numerous chloroplasts; F. Stem laticifer in longitudinal section; G. Close view of vascular region in cross section of a young shoot, note the presence of laticifers on top of the vascular bundles; H. Pith parenchyma with no colored content; I. Vascular xylem elements in longitudinal section where deposition patterns lignin are presented.

(Figure 1H); within the pith, laticifers also occur similarly to the cortical region.

Another characteristic organs of this species is the rhizome (Figure 17.2A), which is a fleshy axis, brown to reddish brown colored with a periderm where the phellem is the outer cell layer of dark brown tone, probably due to the accumulation of metabolites and waxes with a protective function; under this, the phellogen is located, with tabular, thin-walled cells and deeper to this, the phelloderm appears with a rounded and small oval cells, which gradually blend with the cortical parenchyma cells (Figure 17.2B), composed of cells of thin walls and large lumens, lacking of specialized structures or obvious colorations, which is indicative of a simple fundamental parenchyma (Figure 17.2C), laticifers are distributed near the vascular region and not so much to the cortex zone. The vascular system is composed of vascular bundles arranged in a circle, with a bundle of fibers oriented toward the outside and covering the phloem, which is oriented towards the inside, then a series of 2–3 tabular cells layers organized in a continuous loop is observed between phloem and xylem positions, which seems to correspond to an area of vascular cambium (Figure 17.2D). After this, the xylem vessel elements represented by elements and sclerenchyma are observed (Figure 17.2F). Finally, a parenchymatous pith with larger lumen cells than cortical parenchyma are observed, unlike the aerial stem, the laticifers not seem common to the pith (Figure 2E).

Root in cross section (Figure 17.3A) has a periderm, where the presence of phellum, phellogen and phelloderm is appreciated; under this, it

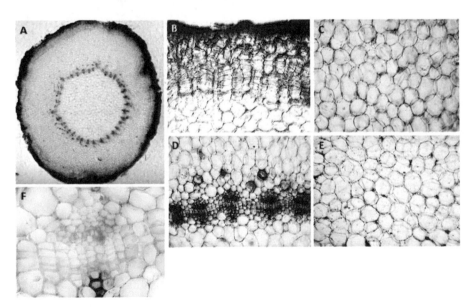

**FIGURE 17.2**  Anatomy of the rhizome of *Euphorbia antisyphilitica*. A. Cross section of young rhizome, note that the periderm is in the periphery, followed by abundant cortical parenchyma, vascular bundles arranged in a circle and a parenchymatous pith in the center; B. Detail of the periderm, towards the part outer phellem, the phellogenas tabular cells and the phelloderm; C. Cortical parenchyma without color content or other structures; D. Approach the vascular region, laticifers appreciate very close to the vascular area; E. Medullary parenchyma, similar to cortical; F. Approaching a vascular bundle, on top parenchyma, phloem, cambium and xylem vascular towards the bottom.

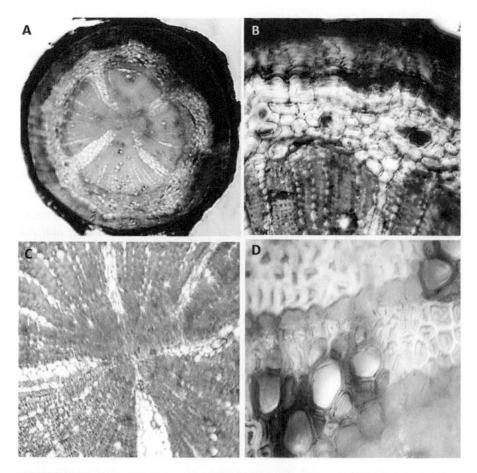

**FIGURE 17.3** Anatomy of the root of *Euphorbia antisyphilitica*. **A.** Root cross section, which is seen tetrarch root and xylem presents secondary growth; **B.** Peridermal region approach of the cortical region parenchymal and vascular region **C.** Vascular region approach, note the presence of rays through the xylem tissue parenchyma; **D.** Close view of the xylem vascular elements in the vicinity of a ray parenchyma and sclerenchyma elements.

is a cortical region with groups of fibers underlying the phloem, which forms a ring surrounding the xylem, which features prominent vascular elements embedded between sclerenchyma (Figure 17.3B). Root is tetrarch type, with four prominent parenchyma rays traversing between the xylem (Figure 17.3C), occasionally minor parenchyma rays between the xylem (Figure 17.3D) were also observed. Laticifers were detected in the periphery of the phloem.

## 17.4  PHYSIOLOGICAL ASPECTS

The economically relevance of *E. antisyphilitica* is in the stem waxy coating, which varies in thickness depending on the age and irrigation. However, despite the economic, industrial and ecological importance of the waxy coating of Candelilla, little is known about its biosynthetic enzymes and chemical composition, similarly to what happens with other species of the genus. The presence of abundant waxy covering is a common feature of several species of the genus *Euphorbia* and as a secondary component influenced by the environment it presents variation between species, which has been noted by Hemmers and Gülz (1986), who compared the yielding and approximate composition of wax from five species of *Euphorbia*. They found that the total production of wax ranged from 0.6–2.1%; while hydrocarbon fraction ranged from 4–17%, 8–18% wax esters, aldehydes (2–6%), free fatty acids (1–8%), free primary alcohols (13–39%) and triterpenols and other unidentified compounds are presented in high concentrations; being, for example, *E. characias* the species with the highest oil content and lower content of free fatty acids; *E. peplus* more content waxy esters and less aldehydes; *E. cyparissias* had low oil content and high concentration of triterpenols and unidentified compounds. This shows that the quantity and composition of the wax varies among different species. Such variation can be quite specific and as an example the case of *Euphorbia esula* and *E. aphylla* could be mentioned, where the chemical composition highlighted significant differences as: 1-Hexacosanol ($C_{26}$) the aldehyde octacosanol ($C_{28}$), five triterpenes (α-amyrin, β-amyrin, δ-amyrenone, 24-methylenecycloartenol and lupeyl acetate), methyl esters of $C_{26}$–$C_{34}$ are the predominant compounds in the epicuticular wax *E. esula* (Manners and Davis, 1984). For *E. aphylla,* the alkanes with $C_{23}$ to $C_{35}$ were the main components, along with $C_{36}$ to $C_{53}$ esters and alcohol derivatives of $C_{26}$–$C_{28}$ and fatty acids of $C_{16}$, while aldehydes and acetates were detected in small quantities between 0.8 and 0.9%. The predominant triterpenoids were the α-amyrin, β-amyrin, lupeol, both esters and their acetates, together with simiarenol, Oleanen-3-one, Ursen-3-one and Lupen-3-one (Gülz et al., 1988). In the particular case of *E. antisyphilitica,* the wax composition are summarized as 45–52% of hydrocarbon components and the rest as wax esters, acid esters, alcohols, resinous material and pentacyclic triterpenoles; the hydrocarbon fraction was composed of $C_{14}$ to $C_{38}$ molecules, which showed variation dependent on the position in the plant and the region where plant was collected, indicating that not only chemical differences occur between species but also variations in composition and concentrations can occur within the same species, such as *E. antisyphilitica* (Scora et al., 1995; Nass et al., 1998). Therefore, it must considered that production and characteristics of the wax may be influenced by plant physiological responses to environmental factors (light, temperature, vapor pressure deficit of water), and then to assurance quality wax Candelilla it should consider the complex genetic and physiological aspects of the plant to optimize resource use, so a holistic understanding of the physiological responses and wax biosynthetic pathways integrated with crop management will lead to improve, increase yields.

On the Candelilla, knowledge of wax production is quite limited, but there are large gaps in knowledge of plant physiology, which undoubtedly are some areas that should be explored, and as priority we could include plant tolerance to water stress, changes in carbon sequestration, more efficiently water use, responses to growth regulators, resistance to pests and diseases, which support the bases for an integrated crop management. Thus, the holistic understanding of the physiological and morphological features of the plant not only must be glimpsed as processes to increase production of the plant or its economically important components, but in part also need to be associated with potential in crop improvement in order to protect the most important varieties, which could be susceptible to pests and diseases that may appear in future, mainly due to the increase in the mass distribution of Candelilla for extensive agriculture, industry or for the production of bioenergy purposes.

## KEYWORDS

- **anatomical description**
- **anatomy**
- **candelilla**
- **laticifers**
- **morphology**
- **physiology**

## REFERENCES

Ascacio-Valdés, J. A., Aguilera-Carbó, A., Martínez-Hernández, J. L., Rodríguez-Herrera, R., & Aguilar, C. N. (2010). *Euphorbia antisyphilitica* residues as a new source of ellagic acid. *Chemical Papers, 64,* 528–532.

Burboa, E. A., Ascacio-Valdés, J. A., Zugasti-Cruz, A., Rodríguez-Herrera, R., Aguilar, C. N., & Coahuila, S. (2014). Capacidad antioxidante y antibacteriana de extractos de residuos de candelilla. *Revista Mexicana de Ciencias Farmacéuticas* (Antioxidant and antibacterial capacity of extracts from candelilla residues. *Mexican magazine of Farmaceutical Sciences*), *45,* 51–56.

De León-Zapata, M. A. (2008). Mejoras tecnológicas al proceso de extracción de cera de candelilla. Tesis profesional. (Technological improvements to the candelilla wax extraction process. Professional dissertation) Faculty of Chemical Sciences, *Universidad Autonoma of Coahuila. Saltillo, Coahuila. Mexico, 58.*

Gülz, P. G., Bodden, J., Muller, E., & Marner, F. J. (1988). Epicuticular wax of *Euphorbia aphylla* Brouss.ex Willd., Euphorbiaceae. Z. Naturforsch.C, *43,* 19–23.

Hemmers, H., & Gülz, P. G. (1986). Epicuticular waxes from leaves of five *Euphorbia* species. Phytochemistry, *25,* 2103–2107.

Horn, J. W., van Ee, B. W., Morawetz, J. J., Riina, R., Steinmann, V. W., Berry, P. E., et al. (2012). Phylogenetics and the evolution of major structural characters in the giant genus *Euphorbia* L. (*Euphorbiaceae*). *Molecular Phylogenetics and Evolution, 63,* 305–326.

Johari, S., & Kumar, A. (2013). *Euphorbia antisyphilitica*: Effect of growth regulators in improving growth and productivity of hydrocarbon yielding plant. *International Journal of Life Science & Pharma Research, 3,* 25–28.

Manners, G. D., & Davis, D. G. (1984). Epicuticular wax constituents of North American and European *Euphorbia esula* biotypes. *Phytochemistry, 23,* 1059–1062.

Nass, R., Markstädter, C., Hauke, V., & Riederer, M. (1998). Quantitative gas chromatographic analysis of plant cuticular waxes containing long-chain aldehydes. *Phytochemical Analysis, 9,* 112–118.

Padmaja, K. V., Atheya, N., & Bhatnagar, A. K. (2009). Upgrading of Candelilla biocrude to hydrocarbon fuels by fluid catalytic cracking. *Biomass and Bioenergy, 33,* 1664–1669.

Rojas Molina, R., Saucedo Pompa, S., De León Zapata, M. A., Jasso Cantú, D., & Aguilar, C. N. (2012). Ensayo pasado, presente y futuro de la candelilla. *Revista Mexicana de Ciencias Forestales* (Essay past, present and future of Candelilla. *Mexican magazine of Forest Sciences*), *2,* 7–18.

Rojas-Molina, R., De León-Zapata, M. A., Saucedo-Pompa, S., Aguilar-Gonzalez, M. A., & Aguilar, C. N. (2013). Chemical and structural characterization of Candelilla *(Euphorbia antisyphilitica Zucc.)*. *Journal of Medicinal Plants Research, 7,* 702–705.

Sage, T. L., Sage, R. F., Vogan, P. J., Rahman, B., Johnson, D. C., Oakley, J. C., et al. (2011). The occurrence of C2 photosynthesis in Euphorbia subgenus Chamaesyce (Euphorbiaceae). *Journal of Experimental Botany, 62,* 3183–3195.

Saucedo-Pompa, S., Jasso-Cantu, D., Ventura-Sobrevilla, J., Sáenz-Galindo, A., Rodríguez-Herrera, R., & Aguilar, C. N. (2007). Effect of Candelilla wax with natural antioxidants on the shelf life quality of fresh-cut fruits. *Journal of Food Quality, 30,* 823–836.

Scora, G. A., Ahmed, M., & Scora, R. W. (1995). Epicuticular hydrocarbons of candelilla (*Euphorbia antisiphylitica*) from three different geographical areas. *Industrial crops and products, 4,* 179–184.

Steinmann, V. W. (2002). Diversidad y endemismo de la familia Euphorbiaceae en México. *Acta Botanica Mexicana* (Diversity and endemism of Euphorbiaceae in Mexico. *Mexican Botánical Acta*), *61*, 61–93.

Villa-Salas, A. B. (2012). Investigación sobre Candelilla (Euphorbia antisyphilitica) en el Área del CIFNE. Revista Mexicana de Ciencias Forestales (Research on Candelilla (Euphorbia antisyphilitica) in CIFNE área. *Mexican magazine of Forest Sciences*), p. 6.

Yang, Y., & Berry, P. E. (2011). Phylogenetics of the *Chamaesyce* clade (*Euphorbia*, Euphorbiaceae): reticulate evolution and long-distance dispersal in a prominent C4 lineage. *American Journal of Botany, 98*, 1486–1503.

Yang, Y., Riina, R., Morawetz, J. J., Haevermans, T., Aubriot, X., & Berry, P. E. (2012). Molecular phylogenetics and classification of *Euphorbia subgenus Chamaesyce* (Euphorbiaceae). Taxon, 61, 764–789.

# CHAPTER 18

# EXTRACTION METHODS AND COMMON USES OF CANDELILLA WAX

MIGUEL A. DE LEÓN-ZAPATA and CRISTOBAL N. AGUILAR

*Department of Food Science and Technology, School of Chemistry, Autonomous University of Coahuila, 25280 Saltillo, Coahuila, Mexico, E-mail: cristobal.aguilar@uadec.edu.mx*

## CONTENTS

## ABSTRACT

Candelilla wax is one of the most used natural products in industry, from cosmetics to electronics, besides being a substance generally recognized as safe (GRAS) by the Food and Drug Administration (FDA). At the present time, the candelilla wax extraction process is carry out by a traditional method with high human and environmental risks associated. This process is almost a century old and consists of immersions plant parts in high concentration sulfuric acid solutions. This has resulted in serious health problems to "foresters or candelilleros" and pollution caused by the deposition of sulfuric acid. Because the traditional extraction process has not been changed for almost a century and the negative effects that this represents, it has been tried to implement other processes. So in this chapter is shown new extraction process of candelilla wax as well as the outlook for actual applications.

## 18.1  INTRODUCTION

Candelilla wax is one of the natural products most widely used in industry due to their unique properties such as transparent yellow color, hardness, brightness and is easy to digest without being toxic, because it is a substance generally recognized as safe (GRAS) by the Food and Drugs Administration (Rojas-Molina et al., 2011).

Their physicochemical properties such as melting point, impermeability, its low rate of contraction and dielectric properties allow it to function effectively in the process of precision casting or lost wax (Canales et al., 2006), also used as curing agent of other waxes and synthetic products (Torres & Román-Domínguez, 1980), manufacture of polishers and brighteners (Pérez, 1985), for the transport and storage of products (Cervantes-Ramirez, 2002) and in various industries such as food, cosmetic, electrical, mechanical and plastic production (Canales-Gutiérrez et al., 2005). Methodologies have also been generated for use in other industries, including the manufacture of edible films (Rojas-Molina et al., 2011). The harvest of candelilla plant for producing natural wax has been one of the most important forest economic activities in certain regions of northern Mexico, being Coahuila the main state where this activity takes place in the municipalities of Ocampo, Cuatro Cienegas and Sierra Mojada (Ochoa-Reyes et al., 2010).

The traditional candelilla wax extraction process dates back to 1914, when Mr. Borrego and Mr. Flores invented a simple method for the extraction (Maldonado, 1979). The technique of obtaining candelilla in the field basically is manual by removing the plant, which is accomplished by pulling it, and then the root is shaken in order to remove dirt, dry garbage, and impurities, forming lumps of approximately 30 kg, which are transported by trucks or animals to the place where the wax is obtained (Romhan, 1984). Subsequently candelilla wax extraction is accomplished by using sulfuric acid as extractant, which is very corrosive and polluting at the same time, representing a serious threat to both the personnel responsible for this activity and the environment. Besides, iron containers rust in contact with sulfuric acid; therefore, candelilla wax obtained by foresters present a red color, characteristic of the presence of iron (Ochoa-Reyes et al., 2010).

The FAO states that it is necessary to find new technologies that reduce the emission of pollutants for this process (Saucedo-Pompa, 2007a). Therefore, have been several studies to find and implement new methods of extraction, which are mentioned in this chapter, and the various uses and applications of wax in various areas of the industry to date.

## 18.2 PROCESSES OF EXTRACTION OF CANDELILLA WAX

The traditional extraction and refining process of candelilla wax proposed by Mr. Borrego and Mr. Flores (Maldonado, 1979) is inefficient and insecure (Figure 18.1). The volumes of production and extraction yields of wax are extremely low, which is directly reflected in low productivity and high cost of manufacture which leads to search different extraction methods more effectives. There have been several studies than have led to innovation and developed of different extraction processes of candelilla wax (Figure 18.1).

### 18.2.1 TRADITIONAL PROCESS OF EXTRACTION OF CANDELILLA WAX

The candelilla wax extraction begins with the harvesting of candelilla plant, which is rooted out without respecting the provisions of NOM-018-

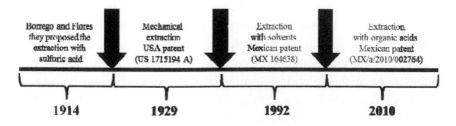

**FIGURE 18.1**  Extraction processes of candelilla wax (IMPI, 2015).

SEMARNAT-1999. Therefore, the intensive collection has caused, despite the large capacity of plant regeneration, that many natural populations have been eliminated as a result of the traditional practice of resource exploitation, which involves the total destruction of the plant, including roots (Maldonado, 1979; Del Campo, 1986; Canales et al., 2006). For extracting iron containers called "pots" are used, with an approximate capacity of 500 liters of water, to which were added with water to approximately 85% of the capacity of the kettle, then proceed to heating by fire direct to the boiling point using the previously blanched plants that were allowed to dry in the sun, which serve as fuel for the pots (Ochoa-Reyes et al., 2010; Rojas-Molina et al., 2011).

In the pots two charges of candelilla (eight thirds) are horizontally placed, once the water was heated to its boiling point, the plant is compressed in the pots and is pressurized with two racks until the plant is completely submerged in water (Ochoa-Reyes et al., 2010). A "third" is the unit of measure equivalent to a "brazada" (all that can be taken between open arms), which is equivalent from 24 to 32 kg of the plant; this variation is explained by the humidity content of the plants (Rojas-Molina et al., 2011).

When the water in the pots reaches the boiling point the sulfuric acid is added, so a violent reaction occurs because the sulfuric acid has a boiling point lower than water therefore drastically increase the temperature and the boiling dissolution, allowing the separation of the wax in the form of foam that floats at the surface, to which foresters call "wax Cut" (Ochoa-Reyes et al., 2010). The foaming wax is removed from the pots using aluminum or iron utensils with holes, this tool is the brainchild of foresters, which is called "skimmer." The holes allows remove the water-sulfuric acid solution and collect as much wax as a foam and then pass is to an iron container with a hole on a lower side portion thereof, covered with a timber, said opening allows the subsequently drained (Ochoa-Reyes et al., 2010).

In the iron vessel, the hot foam (cerote) is separated by decanting of the solution, which is drained by removing a piece of wood that is embedded in the bottom of the container. The solution has a dark color due to the large amount of impurities contained, and these precipitates to the bottom of the container. Then the solution is returned to the pots to reuse it in another extraction (Saucedo-Pompa, 2007a; De Leon-Zapata, 2008; Ochoa-Reyes et al., 2010). The cerote is allowed cooling to solidify at room temperature about 12 hours, after the hardened wax is fragmented by blows in big chunks of easily handled and part of the impurities adhered to the solid fragment of cerote are removed manually (Saucedo-Pompa, 2007a; De Leon-Zapata, 2008; Ochoa-Reyes et al., 2010).

Subsequently, in an iron vessel pour water which was used for extraction of the wax in the pots and heated to boiling point using plant candelilla as fuel after having extracted the wax and let dry. Once the water is boiling, cerote chunks are added to re-melt by adding sulfuric acid in order to remove impurities that are between the solid wax (Ochoa-Reyes et al., 2010). The decanted wax free of solid impurities, known as "crude candelilla wax," is cooled to solidify again for 24 hours (Ochoa-Reyes et al., 2010).

## 18.2.2 MECHANICAL EXTRACTION

U.S. Patent (1929) (U.S. 1715194 A1) (Figure 18.1), describes a candelilla wax extraction process by mechanical processes using roll crushers through which are passed candelilla stems, thus releasing the crystals cuticular wax of the plant in the form of a fine powder then melted at a temperature of 60–80°C. Similarly, the residual plant is also heated to the same temperature to melt the wax that was not removed by the mechanical process. In this process, candelilla plant do not is subjected to several methods of separating wax crystals to melt later (IMPI, 2015).

## 18.2.3 EXTRACTION WITH SOLVENTS

In 1992 was described a method and equipment for candelilla wax extraction with selective solvents in the Mexican patent MX164638 (Figure 18.1), that consisting of an extractor-evaporator equipment that consists of several stages: (a) load the batch plant in the reservoir between the jet spray and the bottom of the equipment, (b) washing plants with a spray of water or steam with the jets, (c) wax extraction with solvent and steam or thermal oil of heating, (d) filtering the solution and sending to a tank step, (e) recovering the solvent by heating (40 to 100°C) and a capacitor, (f) removing bagasse, (g) evaporating the solvent by heating (40 to 100°C) and condensing solvent. Selective solvents are: n-heptane, carbon tetrachloride, n-hexane and benzene. The whole process takes from 180–290 min. All this takes place in a steam heated cylindrical equipment, thermal oil or electric heaters (IMPI, 2015).

## 18.2.4　EXTRACTION WITH ORGANIC ACIDS

In the Department of Food Science and Technology of the School of Chemistry of the Autonomous University of Coahuila in charge of Dr. Cristóbal Noé Aguilar, Saucedo-Pompa et al. (2007a) and De León-Zapata et al. (2008) proposed an alternative technology, which replaces the sulfuric acid by organic acids, these agents have the characteristic of not to emit toxic gases. Based on these studies it was possible protect this technology which is described in the Mexican patent MX/a/2010/002764 (2010) (Figure 18.1). Is an alternative process that is adapted to the needs of foresters and is characterized as an organic process that consisting in water-organic acid solution subjecting the candelilla plant by immersion in the solution and heating it to its boiling point.

It is noteworthy that the organic acids are intermediate natural products of animal metabolism, of the fermentation of carbohydrates by microorganisms (Rodriguez-Willows, 2011) and is commonly found in fruits. Mainly stand out citric acid, malic acid, acetic acid, lactic acid and tartaric acid (Master-Duran and Borja-Padilla, 1993). Act as preservatives, antimicrobials and neutralizing by modifying the pH in the food and agricultural industry mainly (Rodríguez-Sauceda, 2011). With this process, a new alternative for obtaining a good wax yield with higher purity is proposed without affecting health and the environment. Therefore, the new extraction process is a viable option, as it would improve the quality and the conditions under which the activity takes place.

Subsequently Rojas-Molina (2010), elaborated a design for the construction of an equipment for candelilla wax extraction which is in process of patent by the IMPI, based on the above mentioned patent that consisting on a "Separation machinery and refining model of candelilla wax." It is noteworthy that this equipment is being implemented in the candelillera zone of Cuatro Ciénegas, Coahuila and in others northern states of Mexico, with the support of CONAFOR, which aims to improve working conditions, health of foresters and the environment, which leads to a positive effect on income, since due to the purity of the wax obtained by an organic process the wax is very much demand and valued in the market, mainly in cosmetics and food.

Besides in the department of environmental engineering of the Iberoamerican University in charge of PhD. Alicia Rodriguez, is trying to create a cooperative with the women of the local communities of San Francisco del Barrial and La Constancia to produce a candelilla emulsion that serves as polisher furniture wooden, also they are looking for a method of candelilla wax extraction based on the use of acetic acid as a substitute for sulfuric acid (Hernández-García, 2013).

## 18.3  MAIN USES AND APPLICATIONS OF CANDELILLA WAX

Candelilla is one of the most important natural resources of northern Mexico, for both habitants of the region, as various national and international industries (Canales et al., 2006). In 1941, Nature magazine published an article by John Whitaker, in which it was mentioned that candelilla wax may be is the raw material with the largest number of commercial applications of all the substances extracted from wild plants growing on the American continent. Today it is used in more than 20 industries around the world (Rojas-Molina et al., 2011), mainly in the cosmetic and food industry.

### 18.3.1  COSMETIC INDUSTRY

Candelilla is listed in the International Nomenclature of Cosmetic Ingredients (INCI) of the European Commission, as cosmetic ingredient and therefore is to be named as candelilla wax in the label declaration of constituents (Löhnert, 1997). In cosmetic industry, candelilla wax is a product of high quality that can not be replaced by other chemicals and vegetable waxes, due to their special and unique properties, since it has a special structure, partially crystalline to mask other waxes, oils and pigments, besides improves the thermal stability, increases the shelf life of the product without changing its structure (Mazgareanu, 2004) and also functions as an emollient agent and film formation (Löhnert, 1997).

The main cosmetic products made from candelilla wax, are the mask followed by lipstick and cosmetic pencil (Seo et al., 1999). Candelilla wax, also has the property to show less tendency to excretion of oil on the surface of the lipsticks after large temperature changes (Seo et al., 1999), making this property an advantage attributed to candelilla wax respect to other vegetable and synthetic waxes, for the production of cosmetic products.

### 18.3.2  FOOD INDUSTRY

Candelilla wax is a substance Generally Recognized as Safe (GRAS) by the Food and Drug Administration (FDA) (Rojas-Molina et al., 2011) .Therefore, candelilla wax is used extensively in the food industry in the manufacture of various products, as shown in Table 18.1.

## 18.3.2.1 Edible Coatings and Films with Candelilla Wax

Edible coatings (Table 18.2) are a particular form of films directly applied to the surface of materials (Kester, and Fennema, 1986; Han, and Gennadios, 2005). However the edible films (Table 18.3) must be composed of a film-forming material, typically a water-soluble hydrocolloid or polymer, further, these films must be manufactured on equipment and in a facility suitable for processing food products (Rossman, 2009). Edible films and coatings can be formulated from biopolymers (such as polysaccharides and proteins) and lipids, which present desirable characteristics as biodegradability, edibility, esthetic appearance, good oxygen and vapor barriers; often they are obtained from wastes, which improves their profitability (Miranda et al., 2003). The

**TABLE 18.1**  Main Uses Permitted of the Candelilla Wax in the Food Industry

| | Uses | Reference |
|---|---|---|
| European Union (EU) | • Coating agent for bakery. | EU, 1995 |
| | • Coating agent of chocolate for fine bakery. | |
| | • Coating agent for dried fruits and coffee. | |
| | • Snacks. | |
| | • Coating agent for fresh fruits. | |
| | • Component of dietary supplements. | |
| México | • Coating agent for citrus. | Lakshminarayana et al., 1974; Bósquez-Molina, 2005 |
| | • Chewing gum. | |
| | • Candy. | |
| | • Matchstick. | |
| USA | • Coating agent and finishing surfaces (hard candy and chewing gum). | SCOGS, 1981; USA, 2004; CODEX, 2004; IOFI, 2005; Cabello-Alvarado et al., 2013 |
| | • Carrier agent of food additives (flavorings (25–50 mg/kg of flavoring) and dyes). | |
| | • Release agent and loading. | |
| | • Emulsifier. | |

**TABLE 18.2** Investigated Applications and Functions of Edible Coatings with Candelilla Wax

| Fruits | Coating material | Primary functions | References |
|---|---|---|---|
| Grapefruit "Ruby Red" | Candelilla wax, gelatin, soy isolate, hydroxypropyl methylcellulose, zein, antifoam, fatty acids and hystrene | Decreased weight loss ($H_2O$ barrier) and good gas barrier ($O_2$) | Hagenmaier and Baker, 1996 |
| Limes persian | Mesquite gum, candelilla wax, mineral oil, sorbitol and calcium | Decreased weight loss ($H_2O$ barrier) | Bósquez-Molina and Vernon-Carter, 2005 |
| Mandarin "Clemenules" | Natural waxes (bees, candelilla and carnauba) and shellac gum | Decreased weight loss ($H_2O$ barrier) and improved appearance and firmness | Cáceres et al., 2003 |
| Oranges "Valencia" | Polihetylene and candelilla wax | Decreased weight loss ($H_2O$ barrier) and improved flavor and appearance | Hagenmaier, 2000 |
| Apples "Delicious" "Braeburn" and "Fuji" | Candelilla wax-shellac | Decreased weight loss ($H_2O$ barrier) and good gas barrier ($O_2$, $CO_2$ and ethanol) | Alleyne and Hagenmaier, 2000; Bai et al., 2003 |
| Apples "Delicius," "Fuji," "Braeburn" and "Granny smith" | Candelilla wax, oleic acid, morpholine and polydimethylsiloxane antifoam | Decreased weight loss ($H_2O$), good gas barrier (ethanol, $CO_2$ and $O_2$) and improved appearance and firmness | Bai et al., 2003 |
| Oranges "Valencia" and tangerines "Marisol" | Oxidized polyethylenes homopolymers, candelilla wax, oleic acid, myristic acid, ammonium hydroxide and silicone | Decreased weight loss ($H_2O$ barrier) and improved appearance, firmness and flavor | Navarro-Tarazaga et al., 2007 |
| Green guavas cv. "Media china" | Propylene glycol, xanthan gum, candelilla wax, carnauba wax and sodium hydroxide | Decreased weight loss ($H_2O$ barrier) and improved appearance and firmness | Zambrano-Zaragoza et al., 2013 |

waxes as candelilla wax in the edible coatings (Table 18.2) are more effective in blocking the migration of moisture (Bósquez-Molina, and Vernon-Carter, 2005).

**TABLE 18.3**   Effect of Candelilla Wax on the Properties of Edible Films

| Composition | WVP (g mm m$^{-2}$ d$^{-1}$ KPa$^{-1}$) | TS (MPa), EM (MPa), E (%) | References |
|---|---|---|---|
| Whey protein isolate, sorbitol or glycerol and candelilla wax (0.8%) | – | – | Kim and Ustunol, 2001 |
| Mesquite gum, chitosan, mineral oil, candelilla wax (66%), glacial acetic acid, | 3.7 | – | Ruíz-Ramos et al., 2006 |
| High amylose starch, glycerol and candelilla wax (10%) | 0.34 | 29.8, 982.9, 6.4 | Muscat et al., 2013 |
| Sodium carboxymethyl cellulose, oxidized potato starch, soy protein isolate, pork gelatin, candelilla wax (0.5%), sorbitol and tween 80. | 22 | 4.2, 22.5, 71.2 | Kowalczyk, and Baraniak, 2014 |
| Whey protein isolate, vegetable glycerol, candelilla wax (5%) and magnesium nitrate | 40 | 2.3, 24, 65 | Janjarasskul et al., 2014 |

Lipids as the candelilla wax are compounds of low volatility and function as plasticizers, which are added to the film (Kester, and Fennema, 1986), considering two forces, one between the molecules film forming, called cohesion and other in the coating and the substrate, called adhesion (Guilbert, 1986), affecting the mechanical and physical properties of the film (Table 18.3) (elasticity, flexibility, wettability, permeability, strength and shear strength) (Park, and Chinnan, 1995; Koelsch, 1994; Lazaridou, and Biliaderis, 2002) because decrease intramolecular forces of the polymer chains, thereby producing a decrease in the cohesive strength, tension and the glass transition temperature (Koelsch, 1994; Lazaridou and Biliaderis, 2002).

TS = tensile strength; EM = elastic modulus; E = elongation.

## 18.3.3. Other Uses and Applications

Candelilla wax has other uses and applications in various areas of industry, however the number of applications for Candelilla wax has been decreasing since the 1940s (Table 18.4) principally because of substitution by other wax products in less specialty applications (Arato et al., 2014).

## 18.4  NEW USES AND APPLICATIONS OF CANDELILLA WAX

The commercial extraction of wax from candelilla plant *Euphorbia antisyphilitica* is one of the main economic activities of the Mexican semi-desert, so they continue emerging new applications and products from candelilla wax.

### 18.4.1  EDIBLE COATINGS WITH ANTIOXIDANT ADDITIVES

The latest in the field of edible coatings relates to the implementation of bioactive elements as antioxidants and antimicrobials. The current trend for obtaining these compounds is from industrial and vegetable waste of various plants (Sánchez-González et al., 2011) that they have proved useful in protecting the fruit against pathogens (Campos et al., 2011). The antioxidants are most effective on the food surface, since oxidation is a surface-air phenomenon (Min, and Krochta, 2007), therefore the addition of antioxidant additives on edible coatings based on candelilla wax improved the shelf life of fruits (Table 18.5) in order to avoid oxidative process and the attack of microorganisms.

This researches indicate that whole fruit and fresh cut fruits the reduction in weight loss is very important and the use of edible films carriers natural antioxidants are an excellent tool for monitoring the weight loss (Baez et al., 2001; Ghaouth et al., 1991; González-Aguilar et al., 2005) and prevent the growth of microorganisms. The antioxidant concentration influences in the rate of water loss, because is directly proportional to the increase of the concentration of the antioxidant (Saucedo-Pompa et al., 2009). Because the natural antioxidants as phenolic compounds of tarbush interact by hydrogen bonds with each of the chemical structures of the components of edible coating (De León et al., 2013), mainly polysaccharide chains and proteins contained in gum arabic (Pasquel, 2001) because to the ability of the hydroxyl groups of the phenolic compounds of tarbush to form complexes with polysaccharides and proteins (Aguilar et al., 2007) and thereby create a more closed structure (De León-Zapata et al., 2015).

### 18.4.2  ORGANOGELS

An innovative application is the development of organogels, which are bi-continuous colloidal systems coexisting in the form of micro solid heterogeneous (gelling) and organic liquid phases, which have shown good thermomechanical

**TABLE 18.4**  Other Uses and Applications of Candelilla Wax

| | Uses | Reference |
|---|---|---|
| 1910–1940 | Sealing wax | University of Texas at Austin, 2004; Arato et al., 2014 |
| | Phonograph records | |
| | Candles | |
| | Waterproofing World War I | |
| | Lubricants (Auto, floor, leather, etc.) | |
| | Ointments | |
| | Housing | |
| | Mattress stuffing | |
| | Leather polishes | |
| | Folk remedies (venereal, kidney, wart removal) | |
| | Dental molding compounds | |
| | Anatomical models | |
| | Hardener | |
| | Lamp fuel | |
| | Sealing wax | |
| | Paint remover thickener | |
| | Electrical insulations | |
| | Coating to strengthen and waterproof stitching thread tannin leather | |
| | Solvent to treat leather shoe soles to prevent squeaking | |
| | Lacquers for metals | |
| | Acid-proofing agent for metal etching | |
| | Plastics as plasticizer, filler or base rubber softener and preservative | |
| | Varnishes | |
| | Agent in manufacturing celluloid | |
| | Crayons and lead pencils | |
| | Carbon paper, stencil & tracing papers | |
| | Parchment, paper sizing | |
| | Adhesives, cements | |
| | Paints, coatings and shines | |
| | Acid-proofing agent for metals | |
| | Rubber softener and preservative | |
| | Matches | |

**TABLE 18.4** (Continued)

|  | Uses | Reference |
|---|---|---|
|  | Thermal printing |  |
|  | Casting for precision instruments |  |
|  | Pharmaceutics |  |
| 1940–1960 | Waterproofing munitions | University of Texas at Austin, 2004; Arato et al., 2014 |
|  | Thermal printing |  |
|  | Casting for precision instruments |  |
|  | Pharmaceutics |  |
| 1960–2010 | Thermal printing | University of Texas at Austin, 2004; Arato et al., 2014 |
|  | Casting for precision instruments |  |
|  | Pharmaceutics |  |

properties using candelilla wax (Cabello-Alvarado et al., 2013). Compounds that are used to physically modify oils are known as organic compounds of low molecular mass gelling, such as fatty acids and n-alkanes. The main components of candelilla wax are hydrocarbons, leading to the possibility for the development of edible organic gels by dispersions of candelilla wax in vegetable oil (Hagenmaier, 2000), as margarines, creams and hedges (Rojas-Molina et al., 2011). Dispersions of Candelilla wax in fatty oil (e.g., safflower oil) results in formation of anorganogel with textures of potential use by the food industry (Toro-Vazquez et al., 2007;Alvarez-Mitre et al., 2013).

## 18.4.3 PRODUCTS FROM WASTE CANDELILLA

There have been few studies of waste plant material generated after the extraction process of candelilla wax (*Euphorbia antisyphilitica* Zucc.), to give added value in the development and production of new products. Therefore, is noted the work realized by Ascacio-Valdes et al. (2010) who reported that residues of fibrous plant tissue of *Euphorbia antisyphilitica* Zucc., obtained after the process of candelilla wax extraction, are a new source of ellagic acid, which is a powerful natural antioxidant, obtaining about 20 mg of ellagic acid per gram of fibrous tissue of candelilla plant.

Ellagic acid is a molecule of phenolic nature present in free form in some plant species as products of secondary metabolism (Kaponen and Happonen,

**TABLE 18.5**   Examples of Edible Coatings with Candelilla Wax and Antioxidant Additives on Fruits that Have Been Investigated

| Antioxidant additive | Fruits | Primary functions | References |
|---|---|---|---|
| Ellagic acid | Fresh cut fruits (apples, avocados and bananas) | Prevents oxidation of fruit | Saucedo-Pompa et al., 2007b |
| Aloe vera | Fresh cut fruits (apples, avocados and bananas) | Decreased weight loss, prevents oxidation of fruit and improved firmness | Saucedo-Pompa et al., 2007b |
| Ellagic acid | Whole apples | Decreased weight loss, prevents oxidation of fruit and improved appearance | Ochoa-Reyes, 2009 |
| Ellagic acid | Whole avocados | Decreased changes in the appearance and weight loss, and reduced growth of *C. gloeosporioides* | Saucedo-Pompa *et al.,*2009 |
| 1-methylcyclo-propene | Whole guanabanas | Decreased weight loss and improved firmness | Tovar-Gomez et al., 2011 |
| Aqueous extract from *Flourensia cernua* | Whole peppers | Decreased weight loss and improved appearance and firmness | Ochoa-Reyes et al., 2013 |
| Aqueous extract from *Flourensia cernua* | Whole tomatoes | Reduced weight loss and improved firmness and reduced growth of *Botrytis cinerea* and *C. gloeosporioides* | Ruíz-Martínez, 2013 |
| Polifenols of candelilla plant | Papaya | | |
| Fresa | Decreased weight loss, prevents oxidation of fruit and improved appearance | Télles-Pichardo et al., 2012; Télles-Pichardo et al., 2013 | |
| Aqueos extract fromtarbush fermented | Whole apples | Decreased weight loss, prevents oxidation of fruit and improved appearance, firmness and color | De León-Zapata et al., 2015 |
| Polyphenoles of pomegranate | Pears | Decreased weight loss and improved appearance | Cruz et al., 2015 |

2007) and reacts to form complexes with other molecules such as proteins, polysaccharides and alkaloids (Cruz-Atonio et al., 2010). Ellagic acid is present in grapes, strawberries, blackberries, blueberries, nuts and other foods is a type of phenolic compound that acts as an antioxidant; however, they are not easily available and can vary in effectiveness depending on if they are used in their natural form or purified (Saucedo-Pompa, 2009). Ellagic acid has been reported to have antimutagenic, anticarcinogenic, antioxidant, antiviral and anti-inflamatory activities (Meyer et al., 1998: Privadarsini et al., 2002).

The properties of ellagic acid are also exploited in the food industry, developing nutraceutical beverages and dietary supplements. However, the application for food preservation is of a great impact for the perishable food industry using it as an inhibitor of microorganisms (Saucedo-Pompa et al., 2009; Cruz-Atonio et al., 2010). Is also stand out the work done at the Institute of Candelilla (2011), where carried out studies that determined that the bagasse of candelilla is mostly composed of lignin and hemicellulose molecules, thus is propose to use the new carbon fiber, with high strength characteristics, in sustainable community projects such as the building of ecological housing and elaboration of furniture.

## 18.5  CONCLUSIONS

Since nearly a century ago the production of candelilla wax has been directed in accordance with the dynamics of urban markets and external, that little have responded to the needs of the inhabitants, who continue to live in conditions of extreme poverty.Today there are new methods for the extraction of candelilla wax, and is now entering a new phase of growth in the candelillera region, due to the implementation of a new process of candelilla wax extraction with organic acids and with help of CONAFOR, and this will open a new market of organic candelilla wax, which will have an impact on the growth and development of the communities of the Mexican semi-desert.While the candelilla wax continues represent an indispensable material for the industries of various branches, will continue innovating new products and applications to satisfy their demands, so that should be taken into account with strategies that are sustainable projects with the environment and the integration of regional development, for which this activity will continue to carry out in the Mexican semi-desert.

## ACKNOWLEDGEMENTS

This study is part of the Project "Design of a high yield process for extraction of high quality candelilla wax and formulation of new products from the wax extracted" CONAFOR-CONACYT-S0002-2008-C01-91633. Financial resources are provided by CONAFOR and CONACYT Mexico.

## KEYWORDS

- applications
- *Euphorbia antisyphilitica*
- extraction
- organic process

## REFERENCES

Aguilar, C. N., Rodríguez, R., Gutiérrez-Sánchez, G., Augur, C., Favela-Torres, E., & Prado-Barragan, L. A. (2007). Microbial tannases: Advances and perspectives. *Applied Microbiology and Biotechnology, 76*, 47–59.

Alleyne, V., & Hagenmaier, R. D. (2000). Candelilla-Shellac: An alternative formulation of coating apples. *HortScience*, 35, 691–963.

Alvarez-Mitre, F. M., Toro-Vásquez, J. F., & Moscosa-Santillán, M. (2013). Shear rate and cooling modeling for the study of candelilla wax organogels rheological properties. *Journal of Food Engineering, 119*, 611–618.

Arato, M., Speelman, S., & Van-Huylenbroeck, G. (2014). The contribution of non-timber forest products towards sustainable rural development: The case of Candelilla wax from the Chihuahuan Desert in Mexico. *Natural Resources Forum, 38*(2), 141–153.

Ascacio-Valdes, J. A., Aguilera-Carbo, A., Martinez-Hernandez, J. L., Rodriguez-Herrera, R., Aguilar, C. N. (2010). *Euphorbia antisyphilitica* residues as a new source of ellagic acid. *Chemical Papers, 64*(4), pp. 528–532.

Báez, R., Bringas, E., González, G., Mendoza, T., Ojeda, J., & Mercado, J. (2001). Postharvest behavior of the Mango "Tommy Atkins" treated with hot wáter and waxes. *Proc. Interamer. Soc. Trop. Hort., 44*, 39–43.

Bai, J., Hagenmaier, R. D., & Baldwin, E. A. (2003). Coating selection "Delicious" and others apples. *Postharvest Biology and Technology, 28*, 381–390.

Bósquez-Molina, E., & Vernon-Carter, E. J. (2005). Effect of plasticizers and calcium in wáter vapour permeability of films base on Mesquite rubber and Candelilla wax. *Mexican Journal of Chemical Engineering, 4*, 157–162.

Cabello-Alvarado, Ch. J., Sáenz-Galindo, A., Barajas-Bermúdez, L., Pérez-Berumen, C., Ávila-Orta, C., & Valdés-Garza, J. A. (2013). Cera de candelilla y sus aplicaciones. *Avances en química, 8*(2), 105–110.

Cáceres, I., Martínez, J., Cuquerella, J., Del Rio, M., Navarro, P. (2003). Influencia del encerado en la calidad de la mandarina "Clemenules" procedente de sistemas de producción integrada. *Revista Iberamericana de Tecnología Postcosecha, 5,* 113–116.

Campos, C. A., Gerschenson, L. N., & Flores, S. K. (2011). Development of edible films and coatings with antimicrobial activity. *Food Bioprocess Technology, 4,* 849–875.

Canales-Gutiérrez, E., Canales-Martínez, V., & Zamarrón-Rodríguez, E. M. (2005). Candelilla: drops of wax from the mexican deserta round the world, In: Lopez C and Segura G, The richness of the Mexican forests: beyond the wood, experiences of rural communities, Center for International Forestry Research, México, pp. 101–107.

Canales, E., Canales, V., Zamarrón, E. M. (2006). Candelilla, from the mexican desert to the world. *Biodiversitas, 69,* 1–5.

Cervantes-Ramírez, M. C. (2002). Plants with economical relevance in semiarid and arid zones of Mexico. Geography Institute, UNAM, México, pp. 3388–3407.

Codex. (2004). Report of the 36th Session of the Codex Committee on Food Additives and Contaminants, Alinorm 04/27/12, Appendix IX. FAO, Rome.

Cruz-Antonio, F. V., Saucedo-Pompa, S., Martínez-Vázquez, G., Aguilera, A., Rodríguez, R., & Aguilar, C. N. (2010). Chemical and industrial properties of ellagic acid. *Mexican Chemical Acta, 2*(3), 1–12.

Cruz, V., Rojas, R., Saucedo-Pompa, S., Martinez, D. G., Aguilera-Carbó, A. F., Alvarez, O. B., Rodríguez, R., et al. (2015). Improvement of shelf life and sensory quality of pears using a specialized edible coating. *Journal of Chemistry,* 1–7.

De León-Zapata, M. A. (2008). Technological improvements to the candelilla wax extraction process. Thesis, *Autonomous University of Coahuila, México, 58.*

De León-Zapata, M. A., Sáenz-Galindo, A., Rojas-Molina, R., Rodríguez-Herrera, R., Jasso-Cantú, D., & Aguilar, C. N. (2015). Edible candelilla wax coating with fermented extract of tarbush improves the shelf life and quality of apples. *Food Packaging and Shelf Life, 3,* 70–75.

De León, M. A., Sáenz, A., Jasso-Cantu, D., Rodríguez, R., Pandey, A., & Aguilar, C. N. (2013). Fermented Flourensia cernua extracts and their *In Vitro* assay against Penicillium expansum and Fusarium oxysporum. *Food Technol. Biotechnol, 51*(2), 233–239.

Del, Campo. (1986). Utilización y comercialización de la cera de candelilla, en Patton D, González C, Medina A, Segura L. y Hamre R. (eds.), Management and utilization of arid land plants. Symposium proceedings, Feb. 1822, 1985, Saltillo, Mexico, USDA, Forest Service General Technical Report RM135, *Fort Collins,* 113.

EU. (1995). European Commission Directive 95/2/EC on food additives other than colors and sweeteners. Brussels.

Ghaouth, E. L., Arul, J., & Ponnampalam, R. (1991). Use of chitosan coating to reduce water loss and mantain quality of cucumber and bell pepper fruits. *Journal of Food Processing and Preservation, 15,* 359–368.

Gonzales-Aguilar, G. A., Monroy-Garcinia, I. N., Goycoolea-Valencia, F., Diaz-Cinco, M. E., Ayala-Zavala, J. F. (2005). Chitosan edible coatings. An alternative to prevent microbial damage and to prolong the quality of fresh-cut papaya. Proceedings of the Simposium "New trends in preservation and packaging of fruits and vegetables. La Habana, Cuba, pp. 121–133.

Guilbert, S. (1986). Technology and application of edible protective films. En: Mathlouthi, M. (ed.). Food Packaging and Preservation. Theory and Practice. Ed. Elsevier Applied Science Publishers (New York, USA), pp. 371–394.

Hagenmaier, R. D. (2000). Evaluation of a polyethylene-candelilla coating for 'Valencia' oranges. *Postharvest Biology and Technology, 19*(2), 147–154.

Hagenmaier, R. D., & Baker, R. A. (1996). Edible coatings from candelilla wax microemulsions. *Journal of Food Science, 61*(3), 562–565.

Han, J. H., & Gennadios, A. (2005). Edible films and coatings: a review. In: J. Han, Innovations in Food Packaging (pp. 239–259): Elsevier Science & Technology Books.

Hernández-García, G. (2013). The possibilities of sustainable industrialization of the candelilla in the Chihuahuan desert. Thesis. National Polytechnique Institute, CIIDIR Michoacán, *Michoacán, Mexico,* 1–124.

IMPI, 2015, http://lp.espacenet.com/searchResults?page=1 & ST=singleline & compact=false & query=candelilla & locale=es_LP & DB=lp.espacenet.com

Institute of Candelilla, 2011, http://www.candelilla.org.

IOFI. (2005). Communication (email, 2005-3-14) from T. Cachet, *International Organization of the Flavour Industry* (Brussels), to FAO-JECFA Secretary (Rome).

Janjarasskul, T., Rauch, D. J., McCarthy, K. L., & Krochta, J. M. (2014). Barrier and tensile properties of whey protein–candelilla wax film/sheet, *LWT Food Science and Technology,* 377–382.

Kaponen, J. M., & Happonen, A. M. (2007). Contents of anthocyanins and ellagitannins in selected foods consumed in Finland. *Journal of Agricultural and Food Chemistry, 55,* 1612–1619.

Kester, J., & Fennema, O. (1986). Edible films and coatings: a review. *Food Technology, 40,* 47–59.

Kim, S. J., & Ustunol, Z. (2001). Sensory attributes of whey protein isolate and candelilla wax emulsion edible films. *Journal of Food Science, 66*(6), 909–911.

Koelsch, C. (1994). Edible water vapor barriers properties and promise. *Trends in Food Science and Technology, 51,* 76–81.

Kowalczyk, D., & Baraniak, B. (2014). Effect of candelilla wax on functional properties of biopolymer emulsion films-A comparative study. *Food Hydrocolloids, 41,* 195–209.

Lakshminarayana, S., Sarmiento, L., Ortiz, J. I., & Siade, G. (1974). Extension of storage life of citrus fruits by application of candelilla wax emulsion and comparison of its efficiency with Tag and Flavorseal. *Proc. Fla. State Hortic. Soc., 87,* 325–330, as cited In Evaluation of the Health Aspects of Candelilla Wax as a Food Ingredient: Report of the Select Committee on GRAS Substances, Life Sciences Research Office of the Federation of American Societies for Experimental Biology, Bethesda, Maryland (1981).

Lazaridou, A., & Biliaderis, C. G. (2002). Thermophysical properties of chitosan, chitosan starch and chitosan pullulan films near the glass transition, *Carbohydrate Polymers, 48,* 179–190.

Löhnert, G. (1997). Candelilla wax as raw material in traditional and new cosmetic applications.*SÖFW-Seifen, Öle, Fette, Wachse, 123*(15), 1009–1014.

Maestro-Durán, R., & Borja-Padilla, R. (1993). Actividad antioxidante de esteroles y acidos orgánicos naturales.*Grasas y aceites, 44*(3), 208–212.

Maldonado, L. (1979). La investigación desarrollada sobre la candelilla (Euphorbia antisyphilitica Zucc.).*Ciencia Forestal, 4*(18), 3–10.

Mazgareanu, N. (2004). Cosmetic pencil-simple and complicated. *Kosmetik International Magazine, 6,* 2004,http://cosmalux.ru/eng/press2.html

Meyer, A. S., Heinonen, M., Frankel, E. N. (1998). Antioxidant interactions of catechin, cyanidin, caffeic acid, quercetin and ellagic acid on human LDL oxidation. *Food Chem*, *61*, 71–75.

Min, S., & Krochta, J. M. (2007). Ascorbic acid-containing whey protein film coatings for control of oxidation. *J. Agric. Food. Chem*, *55*(8), 2964–2969.

Miranda, S. P., Cárdenas, G., López, D., Lara, S. V. L. A. (2003). Chitosan film behavior composed of an avocado storage model. *Journal of the Mexican Society of Chemistry*, *47*, 331–336.

Muscat, D., Ahikari, R., McKnight, S., Guo, Q., & Adhikari, B. (2013). The physicochemical characteristics and hydrophobicity of high amylose starch–glycerol films in the presence of three natural waxes. *Journal of Food Engineering*, *119*, 205–219.

Navarro-Tarazaga, M. L., Pérez-Gago, M. B., Goodner, K., & Plotto, A. (2007). A new composite coating containing HPMC, beeswax, and shellac for "Valencia" oranges and "Marisol" tangerines. *Proc. Fla. State Hort. Soc*, *120*, 228–234.

Ochoa-Reyes, E. (2009). Evaluation of edible active coatings from candelilla wax on the shelf life quality of apples. (Golden delicious). *Thesis*. Antonio Narro Agrarian Autonomous University, *Saltillo, Coahuila*, 1–98.

Ochoa-Reyes, E., Martínez-Vazquez, G., Saucedo-Pompa, S., Montañez, J., Rojas-Molina, R., De León-Zapata, M. A., et al. (2013). Improvement of shelf life quality of green bell peppers using edible coating formulations. *Journal of Microbiology, Biotechnology and Food Sciences*, *2*, 2448–2451.

Ochoa-Reyes, E., Saucedo-Pompa, S., De La Grarza, H., Martínez, D. G., Rodríguez, R. Y., Aguilar-Gonzalez C. N. (2010). Traditional extraction of candelilla wax. *Mexican Chemical Acta*, *2*(3), 1–13.

Park, H. J., & Chinnan, M. S. (1995). Gas and water vapor barrier properties of edible films from protein and cellulosic materials. *J. Food Engr.*, *25*, 497–507.

Pasquel, A. (2001). Gumms: and approximation to the food industry. *Amazonic Journal of Food Research*, *1*(1), 1–8.

Priyadarsini, K. I., Khopde, S. M., Kumar, S. S., & Mohan, H. (2002). Free radical studies of ellagic acid, a natural phenolic antioxidant. *J. Agric. Food Chem.*, *50*, 2200–2206.

Rodríguez-Sauceda, E. N. (2011). Use of antimicrobial natural agents for preservation of fruits and vegetables. *Ra Ximhai*, *7*(1), 153–170.

Rojas-Molina, R. (2010). Design and prototype of an extraction plant of candelilla wax. *Thesis. Autonomous University of Coahuila. Coahuila*, Mexico.1–79.

Rojas-Molina, R., Saucedo-Pompa, S., De León-Zapata, M. A., Jasso-Cantú, D., & Aguilar, C. N. (2011). Past, present and future of candelilla. *Rev. Mex. Cien. For, 2*(6), 7–18.

Romhan, DLVCF. (1984). Top non-timber forestry products from Mexico. *Chapingo Journal*, México, *561*.

Rossman, J. M. (2009). Commercial Manufacture of Edible Films Huber (eds.), Edible Films and Coatings for Food Applications. Springer Science (USA), 367–390.

Ruíz-Martínez, J. (2013). Edible oleo-proteic coatings to prolong the shelf life of tomato. Cubiertas comestibles óleo proteicas para prolongar la vida de anaquel del tomate. *Thesis. Autonomous University of Coahuila. Coahuila*, Mexico. 1–42.

Ruíz-Ramos, J. O., Pérez-Orozco, J. P., Báez-González, J. G., Bósquez-Molina, E., Pérez-Alonso, C., & Vernon-Carter, E. J. (2006). Interrelationship between the viscoelastic properties and effective moisture diffusivity of emulsions with the water vapor permeability of edible films stabilized by mesquite gum–chitosan complexes. *Carbohydrate Polymers*, *64*, 355–363.

Sánchez-González, L., Vargas, M., González-Martínez, C., Chiralt, A., & Cháfer, M. (2011). Use of essential oils in bioactive edible coatings: a review. *Food Eng. Rev.*, *3*, 1–1610.

Saucedo-Pompa, S. (2007a). Development of edible films from candelilla wax and antioxidant additives. *Thesis. Autonomous University of Coahuila. Coahuila,* Mexico. 1–99.

Saucedo-Pompa, S., Jasso-Cantu, D., Ventura-Sobrevilla, J., Sáenz-Galindo, A., & Aguilar-Gonzales, C. N. (2007b). Effect of candelilla wax with natural antioxidants on the shelf life quality of cut fresh fruits. *Journal of Food Quality*, *30*, 823–836.

Saucedo-Pompa, S., Rojas-Molina, R., Aguilera-Carbo, A., Saenz-Galindo, A., De La Garza, H., Jasso-Cantú, D., et al. (2009). Edible film based on candelilla wax to improve the shelf life and quality of avocado. *Food Research International*, *42*, 511–515.

SCOGS. (1981). Evaluation of the Health Aspects of Candelilla Wax as a Food Ingredient: Report of the Select Committee on GRAS Substances, Life Sciences Research Office of the Federation of American Societies for Experimental Biology, Bethesda, Maryland.

Seo, S. Y., Lee, I. S., Shin, H. Y., Choi, K. Y., Kang, S. H., & Ahn, H. J. (1999). Observation of the Sweating in Lipstick by Scanning Electron Microscopy. *International Journal of Cosmetic Science*, *21*(3), 207–216.

Télles-Pichardo, R., Cruz-Aldaco, K., Ochoa-Reyes, E., Aguilar, C. N., & Rojas, R. (2013). Edible Coatings of Wax and Polyphenols from Candelilla: An Alternative of Conservation of Papaya (*Carica papaya* L.). *Acta Química Mexicana*, *5*(10), 1–7.

Télles-Pichardo, R., Rojas-Molina, R., Buenrostro-Figueroa, J. J., Cruz-Aldaco, K., F. J., & Hernández-Campos, Aguilar, C. N. (2012). Formulation and functionality of an edible Candelilla wax cover applied in strawberry. XXXIII National Meeting and II International Congress AMIDIQ, 1 to 4 May 2012, San José del Cabo, BCS, Mexico.

Toro-Vázquez, J. F., Morales-Rueda, J. A., Dibildox-Alvarado, E., Charó-Alonso, M., Alonso-Macías, M., & González-Chávez. (2007). Thermal and Textural Properties of Organogels Developed by Candelilla Wax in Safflower Oil. *J. Am. Oil. Chem. Soc.*, *84*, 989–1000.

Torres, N., & Román-Domínguez, M. (1980). The Mexican candelilla wax an its comercial perspectives.. *Thesis*, National Autonomous University of Mexico, 1–84.

Tovar-Gómez, B., Mata-Montes de Oca, M., García-Galindo, H. S., & Montalvo-González, E. (2011). Effect of wax emulsions and 1 methylciclopropane in the post-harvest conservation of soursop. *Chapingo Journal. Series Horticulture*, *17*, 53–61.

University of Texas at Austin, 2004. http://www.texasbeyondhistory.net/waxcamps/index.html USA. (2004). Candelilla wax. Title 21, United States *Code of Federal Regulations*, Section 184.1976. U. S. Government Printing Office, Washington, DC.

Zambrano-Zaragoza, M. L., Mercado-Silva, E., Ramirez-Zamorano, P., Cornejo-Villegas, M. A., Gutiérrez-Cortez, E., & Quintanar-Guerrero, D. (2013). Use of solid lipid nanoparticles (SLNs) in edible coatings to increase guava (*Psidium guajava* L.) shelf-life, *Food Research International*, *51*, 946–953.

# CHAPTER 19

# CHARACTERISTICS AND NEW APPLICATIONS OF THE CANDELILLA WAX

OLGA BERENICE ALVAREZ-PÉREZ,[1] CRISTÓBAL N. AGUILAR,[1] JORGE A. AGUIRRE-JOYA[1] and ROMEO ROJAS[2*]

[1]Department of Food Science and Technology, Autonomous University of Coahuila, Blvd. Venustiano Carranza S/N Col. República Oriente, 25280 Saltillo, Coahuila, Mexico, Tel.: +528444161238, Fax: +528444159534

[2]Autonomous University of Nuevo Leon, School of Agronomy, Research Center and Development for Food Industries, Francisco Villa s/n, Ex-Hacienda el Canada, Escobedo, Nuevo Leon, 66054, Mexico, E-mail: romeo.rojasmln@uanl.edu.mx

## CONTENTS

## ABSTRACT

The common name of the Candelilla plant appears to be derived from the particular shape of the plant stalks: long, straight, erect and wax covered, given the appearance of little candles. Candelilla wax may be obtained from several species of *Euphorbiacea*, the primary source is *Euphorbia antisyphilitica*. The plant grows as a bush or shrub in dense stands, principally in the Chihuahuan desert in northeastern Mexico. To yield economical amounts of waxes, a desert like climate is a required condition. Most of the components of Candelilla wax are naturally found in vegetables and fruits. Its chemical composition is characterized by a high hydrocarbon content (around 50%) and a relatively low amount of volatile esters. Its resin content can reach 40% by weight, a major factor contributing to its sticky consistency. The principal applications of candelilla wax include its uses as a glazing and surface finishing agent and as a carrier for food additives. The objective of this review is to replicate the information about the characteristics and applications of candelilla wax to elucidate that is a natural product with a big opportunity into global market.

## 19.1 INTRODUCTION

The candelilla plant belongs to the *Euphorbia* species, this family species are differentiated by the latex contained by the *Euphorbia plants*. This plant is originally from arid zones (Molina et al., 2003). Normally grows on lime soil slopes or hillsides associated with formations of rocky material. The plant root is relatively small, although a plant moderate size can develop more than 100 stalks of a grayish-green color, with typical dimensions of 30–60 cm long and 0.1–1.0 cm in diameter, forming bushes of approximately 90 cm in diameter. It could be founded at elevations of around 800 m and are commonly associated with growths of such plants as lechuguilla, sotol palm, etc. In 1829, the botanist J. G. Zuccarini described the Candelilla plant for the first time with the scientific name *Euphorbia antisyphilitica* (Candelilla Institute, 2015).

The *Euphorbia* species that produce candelilla wax are indigenous from Mexico, specially in the northern states of Coahuila, Chihuahua, Nuevo Leon and Zacatecas. They also occur in the southern states of USA, as Texas, California and Arizona. Mexico has the potential to be the main producer of candelilla wax, however the knowledge about its chemical composition and plant anatomy its a serious limiting factor for more applications (Rojas Molina et al., 2013). The commercial exploitation of this plant for produce wax, dates back to the beginning of the 20th century, and has since become one of the main economic activities of the Mexican region. There are more than 3,500 small producers who are known locally and nationally as "Candelilleros." Some of them complement their income by gathering other plant species or by agriculture (Candelilla Institute, 2015).

Waxes are, chemically, fatty acid esters with alcohols of high molecular weight. This molecules are obtained by a chemical reaction between an acid and an alcohol. Physically, are highly insoluble particles in aqueous media and at room temperature are shown in solid state and with intermediate hardness (Molina, 2008b). In animals this material is in the skin and feathers, making them have a hydrophobic character. In vegetables, waxes lining the epidermis fruits and/or stems which prevents water loss by evaporation (Kuznesof, 2005; Resolucao, 2005).

Candelilla wax is a complex substance of plant origin, insoluble in water, but highly soluble in acetone, chloroform, benzene and other organic solvents. It is characterized by a high content of hydrocarbons (around 50%) and a relatively low quantity of volatile esters and is recognized by the Food and Drug Administration (FDA) as a substance Generally Recognized as Safe (GRAS) under regulations 21CFR, 175.105, 175.320, 176.180, and in accordance with regulation 184.1976, might be used in food with no limitation other than current good manufacturing practice (Howe and Williams, 1990; Toro Vazquez et al., 2007; Ochoa Reyes et al., 2010).

In its raw form, candelilla wax is a yellowish-brown hard, brittle, lustrous solid with an aromatic odor when heated. It consists primarily of odd-numbered saturated straight-chain hydrocarbons ($C_{29}$ to $C_{33}$), together with esters of acids and alcohols with even numbered carbon chains ($C_{28}$ to $C_{34}$). The most abundant n-alkane, $C_{31}$, comprises more than 80% of total n-alkanes. Free acids, free alcohols, sterols, neutral resins, and mineral matter (<1%) are also present (Kuznesof, 2005). The proportion of the components determines the hardness, water repellency, brightness and other characteristics of the wax as its melting temperature.

Candelilla wax (*Euphorbia antisyphilitica*) stalks are of great economic importance in Northern Mexico. Traditionally is extracted from this plant which is harvested in wild nature. Into the plant, wax is a protection system to reduce excessive water loss. During wet season, wax deposits on the stem surface are minimal because there is no need for plant desiccation. Crude and candelilla wax is obtained by first boiling the dried stalks of the candelilla in water acidified with sulfuric acid to release the wax. The molten wax, known as cerote, is then skimmed off and allowed to solidify. It is transferred to lead lined tanks for refining by treatment with acid and subsequent passage through filter presses. Its uses in confectionery and food has no limitations beyond good manufacturing practices, and also is used in the manufacture of edible coatings (FAO, Kuznesof, 2005).

## 19.2 CHARACTERISTICS

Candelilla wax is insoluble in water, but soluble in chloroform and toluene. It has a specific gravity of about 0.98, and there re some physical properties characteristics for a food grade quality candelilla wax (Table 19.1).

Wolfmeier et al. (1996) have reported the "average composition" of candelilla wax base on a large number of values found in literature (Table 19.2). Candelilla wax, which is often compared with carnauba wax, differs dramatically from the latter in its high hydrocarbon content (max. 45%) and resin content (20%). Specifications of identify and purity for candelilla wax were established by the Joint FAO/WHO Expert Committee on Food Additives (JECFA) in 1992 (FAO, 1992). Recent monographs are published in Japan (Japan, 2000) and in the USA (FCC, 2003).

In a study of thermal and textural organogels developed by candelilla wax and safflower, an analysis of candelilla wax showed the main characteristic of it, a high proportion of n-alkanes with 29–33 carbons (Table 19.3). The main component of the candelilla wax was hentriacontane (78.9%), a n-alkane with the molecular formula $CH_3(CH_2)_{29}CH_3$, molecular weight of 436.85, melting point (99.5% purity) of 67.05°C, and $\Delta H_M$ of 73.3 kJ/mol (Candelilla Institute, 2015; Grant, 2005; Lipid Bank, 2006; INFOTHERM, 2006; Toro Vazquez et al., 2007). The melting point founded was at 64.42°C, that is closet to the

**TABLE 19.1** Physical Properties of Candelilla Wax

| Physical properties | Value range |
|---|---|
| Melting range (°C) | 68.5–72.5 |
| Acid value | 12 – 22 |
| Saponification value | 43–65 |
| Ester value | 31–43 |
| Iodo value | 14–27 |

**TABLE 19.2** Average Composition of Candelilla Wax

| Components | % (w/w) |
| --- | --- |
| Hydrocarbons (98% paraffins and 2% alkenes) | 42.0 |
| Wax, resin and sitosteroyl esters | 39.0 |
| Lactones | 6.0 |
| Free wax and resin acids | 8.0 |
| Free wax and resin alcohols (terpene alcohols) | 5.0 |
| Saponifiable components | 23.0–29.0 |
| Unsaponifiable components | 71.0–77.0 |

melting temperature reported for 99.5% pure hentriacontane (67.05°C) (18). Toro Vasquez et al., 2007 mention that it is important to point out that besides triterpenoids, other n-alkanes such as nonacosane and tritriacontane were present as minor components in candelilla wax. These compounds might develop a mixed molecular packing with hentriacontane during cooling.

When the candelilla wax is incorporated in a emulsion system its behavior is associated with the gelation and melting of candelilla wax components, mainly hentriacontane, also is the similar behavior as the one observed by several n-alkanes when is dissolved in different organic solvents (Abdallah et al., 1999; Abdallah and Weiss, 2000; Vazquez et al., 2007). Organogelation is a promising alternative to provide structure and texture to vegetable oils without the use of saturated fatty acids (Bot et al., 2009; Hughes et al., 2009; Vazquez et al., 2013). The candelilla wax and FDA approved gelator, forms organogels with vegetable oils. These organogels have shown physical and textural properties with potential use in food products (Toro Vazquez et al., 2007). One example is the interaction between tripalmitin and the n-alkanes present in the candelilla wax that seemed to develop mixed self assembled structures, resulting in organogels with thermal and mechanical properties different from the ones observed by the organogels obtained with the separated systems (Toro Vazquez et al., 2009). The experiments with candelilla

**TABLE 19.3** Characteristics of Candelilla Wax (Toro Vasquez et al., 2007)

| Name | Condensed formula | % |
| --- | --- | --- |
| Nonacosane | $C_{29}H_{60}$ | 4.16 |
| Hentriacontane | $C_{31}H_{64}$ | 78.86 |
| Tritriacontane | $C_{33}H_{68}$ | 8.01 |
| Alcohols of penta-cyclic triterpenoids | $C_{30}H_{49}OH$ | 7.38 |
| Unidentified | – | 1.60 |

wax using safflower oil as the liquid phase, showed that the application of a continuous shear during organogelation of the n-alkanes of candelilla wax, resulted in the formation of sol-type system with weaker gel structure than the organogels developed under quiescent conditions (Chopin Doroteo et al., 2011). In the particular case of candelilla wax, they investigated the effect of shearing using different percentages of candelilla wax solutions in safflower oil. Pre-sheared candelilla wax solutions showed enhanced nucleation and crystal growth suggesting the development of flow induced liquid structures. Additionally, the corresponding gels had higher rheological properties compared with candelilla wax organogels developed under quiescent conditions or with the application of continuous shear. This indicated that the application of shearing just until achieving metastable conditions still enables the formation of permanent junction zones during the development of the three-dimensional crystal network (Chopin Doroteo et al., 2011).

### 19.2.1  CHEMICAL COMPOSITION AND MICROSTRUCTURAL CHARACTERISTICS

Some different analysis has been made to candelilla plant to determinate its characeristics. In a proximal chemical composition it was found a reduced content of moisture, about 0.4%, total solids (99.6%). lipids (15.9%) of which 30% are esters of candelilla wax, crude fiber (9%), proteins (2.3%), ashes (10.9%), total sugars (0.27%), reducing sugars (0.16%). Also it was reported that candelilla wax posses bioactive compounds such as antioxidants in different proportions, hydrolysable tannins (0.56%), condensed tannins (0.16%), ellagic acid (2.2%), gallic acid (0.6%) and chatequin (0.2%), the last three components, the values are reported as the milligrams of antioxidant per gram of plant in dry base (Martínez, 2002; López Guerra, 2006; Rojas Molina et al., 2013).

On the other hand, the microstructural analysis, demonstrate that in a surface of a young candelilla stem where wax was exposed after manual wringing of stems, the wax is easily recovered as thin layers. After removing these wax layers a green cuticular surface of epidermis was observed. In the stem external surface the wax is deposited as small granules, which grow protecting the steam as thick layers. This explains the candelilla tolerance to cold and hot temperatures in the desert, its disease and insect resistance.

## 19.3 APPLICATIONS

The main uses of the candelilla is due to the pre-hispanic indigenous in the North of Mexico, the wax is extracted raw wax boiling the stems in containers of mud and was then mixed with colorants for ornamental purposes. Candelilla wax is one of the natural products more considerate in different industries and has a great demand in the market, being currently used in over 20 different industries all over the world, mainly in the United States, the European Union and Japan for its unique characteristics of high quality. by their distinctive properties candelilla wax gives the category of raw materials essential for the manufacture of various products (Maldonado, 1979; Hernández, 2006; Carranza González, 2007; CONAFOR, 2008; Semarnat, 2008a; Ochoa Reyes et al., 2010).

Candelilla wax has a wide range of applications in the industry food is used for being a non-toxic natural product, is considered as GRAS substances and is used in decoration of cakes and biscuits, bakery products, polish sweets, popsicles, coffee, substitutes for coffee, tea, infusions of herbs, hot cereals, drinks containing high content of cocoa drinks sugary artificial, in product imitation of chocolate or chocolate substitutes, as well as for the coating of chocolate and fruit, in beverages aromatized water such as serum and drinks electrolyte, on the surface of fresh fruits, vegetables, legumes, pulses, marine algae, nuts and seeds (Hagenmaier and Baldwin, 2003; Molina, 2008a; Castorena et al., 2010; Alvarado et al., 2013).

This wax is an ideal product to cover, protect and prolong the shelf life of the cheeses, which are exposed to decay and losses if are not protected properly, wax meets three functions in the protection of the cheeses, a better presentation of the product, protection against attack by fungi and moisture retention (Multiceras, 2014). The permeability of the wax prevents the loss of moisture, also for lack of nutrients the wax prevents the growth of any type of microorganisms on the surface of the cheese (Pompa, 2007).

On the other hand the growing demand by consumers for fruit and vegetables has generated the need to preserve the quality of these products, in order to extend its shelf life. Candelilla wax emulsions comply with the function of slow maturation and aging, migration of oils and the transport of solutes, allow ensuring impregnation and a uniform deposit of the wax on the surface of the product, slowing efficiently its dehydration, maintaining a controlled atmosphere on the outer surface, which allows the protection of product environmental conditions of transportation and storage (Kowalczyk

& Baraniak, 2014). In addition, these actions allow to improve the retention of essential volatile components, control the transfer of additives, as well as maintain a glossy finish on the surface of the product (Molina, 2008a).

Is used in the production of edible coatings for fruits and vegetables such as limes, apples, soursop, avocado among other fruits, one of the main advantages of the candelilla wax is its high permeability to $O_2$ and $CO_2$ which allows that the covered products have adequate internal levels of $O_2$ and $CO_2$, which reduces the development of flavors due to the growth of micro-organisms and also help prolong the shelf life of these fruits (Pompa, 2007). In the formulations of gum bases, it is used for the manufacture of chewing gums or gums moscar because contain high percentages of polymers, rubbers and waxes (Cosmogaceta, 2011). These components confer greater stability rubber base, which allows the formulation of the typical bubbles. The amount of wax that is used in the manufacture of rubber base confers plasticity, hardness consistency or smoothness and appearance shine (Ramírez, 2005; Candelilla Institute, 2015).

Also it has an application on the cartons as wax coating in order to protect its integrity as packaging, primarily for use in the food industry, whose products must reach the final consumer without deterioration in their essential properties. Wax meets the function provide rigidity, flexibility and protection against moisture (Multiceras, 2014).

Candelilla wax is essential in the cosmetic industry, such as in eyeliner, mascara for lashes, eye shadows, formulation of make-up, body creams and products for the care of hair, lotions, cosmetic pencil, lipstick, among others (Tapia and Chilpa, 2008; Semarnat, 2008b); thanks to its unique properties of protection, brightness and consistency. The latter may represent up to 20% of the total number of wax in the formulation, properties that gives is softness, hardness, gloss, softening point, they show less tendency toward sweating because a phenomenon of excretion of oil on the surface of the lipstick. The skin creams properties are improved with the addition of wax are consistency, emollient, protection against water loss (Alvarado, 2013).

The modeling is to add or develop ways through the application of different materials, soft and flexible, which can be without difficulty and allow a rapid uptake of prints taken by the artist is the manual development usually in clay or wax (Candelilla Institute, 2015). Modeling in wax in a very complex process, are essential the visual and manual skill of the artist, as well as the precise knowledge of the materials are essential. In the industry of molding of precision and foundry, where, due to its extraordinary properties of molding the candelilla wax allows you to accurately reproduce every detail, as required by these strict manufacturing processes. The use of wax in this process is by its capacity and contraction ratio, as well as its malleability and ductility, which allow to operate very efficiently (Multiceras, 2014).

Chemical stability, high melting point, water resistance and low electrical resistance, possessing candelilla wax makes it an extremely useful material in the manufacture of products for the electrical and electronics industry (Torres and Domínguez, 1980; Gutíerrez et al., 2005). The property of insulation that account this wax is very useful in the manufacture of electrical and electronic devices in this industrial sector wax is used for the coating of cables, in the manufacture of capacitors, as binder of other materials such as paper, wood, asbestos that can improve the properties of insulation. Candelilla wax works as an excellent barrier against moisture, so they are very effective in systems of protection against corrosion (Semarnat, 2008b).

Candelilla wax plays an important role in the production of adhesives because of their properties, reduces the viscosity of the adhesive, flexibility and elongation (Canales et al., 2006). Also it's applications are very useful in the manufacture of chipboard of wood confer the agglomerate, water repellency, reduction in the absorption of water and resistance to moisture-induced swelling (Multiceras, 2014).

Candelilla wax it is a raw material essential to the manufacture of matches, applies on paper match through a wet system, the paper waxed with wax, fulfills the function protect the moisture match, prolongs the duration of the flame, facilitates the ignition of the match (Cervantes Ramírez, 2005).For being a component fundamental candelilla wax is used in the manufacture of crayons, for its properties distinctive malleability and flexibility, but above all by its non-toxic nature. The wax may contribute up to 60% of the total of the formulation. The characteristics in which participates the wax are ability sliding, consistency and plasticity softness or hardness and adhesion to the role in normal application (Villa Castorena et al., 2010; Cabello Alvarado, 2013).

Since antiquity, the waxes have been used in fireworks as the favorite oxidizing agent that provides the resource of carbon required at the stage of ignition, allowing to generate a wide range of colors that cannot be achieved with other carbon sources (Valera, 2004). Candelilla wax emulsions are used to impermeability preserve in wicks and detonation, ensuring the absence of humidity during the ignition phase. On the other hand, some varieties of rich hydrocarbon waxes, as it is the case of the candelilla wax, found an important application in the military sector, can be used as an active ingredient in the processes of detonation or as a coating or film waterproof for the helmet protection, explosives, and impregnation of gas masks and in airplanes fabric coverings (Candelilla Institute, 2015; Multiceras, 2014).

Candelilla wax presents several applications in the industry of construction as for example release for metal molds and wood, regulators in the concrete box, as wrapper and waterproofing of various types of containers (Barsch, 2004). In the dolomitic nature refractory brick industry, problems may occur due to the presence of humidity, this can cause fractures in the material, once installed in the oven, which translates into high rates of waste, pores of plants and costly loss of production. To prevent these problems, wax is used during the manufacturing process of the refractory, which seals the cracks of the brick, protecting it against the presence of moisture (Multiceras 2014; Candelilla Institute, 2015).

In the pharmaceutical industry candelilla wax is used to give brightness and consistency to medicaments, mainly in its presentation in the form of tablets, due to the distinctive properties of the wax and its small particle size, can be applied both in the vests carried by specialists of this industry, in walls and doors that make up the protective cabin (Canales et al., 2006). Artificial logs, in addition to decorate fireplaces for homes, meet the primary function as combustible material, providing heat by radiation, usually are made of sawdust and wax. Wax is integrated into the body of the artificial logs by immersion and it helps that the combustion process is conducted safely and evenly (Barsch, 2004; Valera, 2004).

Due to its inherent compatibility and high solubility, candelilla wax is widely used as additives in the formulation of fats and oils, in order to improve some features and specific properties, including viscosity, sealed, point of freezing, heat dissipation, corrosion resistance, reduction of friction and wear, among others. It is used in very sophisticated propulsion devices, including spacecraft (Ramírez, 2005).

By their physical and chemical properties wax is a raw material essential in the manufacture of polishes for shoes, performs the function keep the skin of the footwear in good condition and shine. The application of wax to the preservation of floors dates back to several centuries ago, it retards the penetration of air and moisture thus increasing the life of materials and preventing the deterioration of the surface by abrasion and scratches (Villa Castorena et al., 2010).

The paper products are coated with candelilla wax to preserve the integrity of the goods that are packed or wrapped with this material, wax improves the properties and characteristics of the paper in its function of packaging, giving structure, seal and protection. Wax modified the structure of the paper, allowing to improve adhesive characteristics in cold and hot, the ability to glide, brightness, capacity as sealer or insulation, resistance to moisture and fat (Canales et al., 2006). On the other hand, the inks that are used in the manufacture of carbon paper is formulated with waxes, oils and pigments. In what refers to waxes, two different types are normally used for this application: amorphous waxes, which give body to the inks and allow the retention of oil; high hardness waxes, like candelilla, which enable keep ink in solid state. One of the main functions that meet the wax is the act as a vehicle for transporting the color and avoid the complete absorption of the ink on the inner surface of the paper. Typical formulations of carbon paper inks, waxes can represent up to 40% of the total formulation (Gupta and Mehrotra, 1997).

The use of the candelilla wax has experienced a growth in its application as an additive for the industry of inks, for example inks for printer, improving the ability to slip, scratch resistance, resistance to the rubbed, gloss, resistance to abrasion, anti-blocking properties (Semarnat, 2008b). Waxes are widely used in the formulations of cleaning products, both for personal use and for industrial applications. The candelilla wax, with a high content of saturated hydrocarbons and esters, presents a great compatibility with the ingredients used in the formulation of shampoo, soaps and polishes for floors and cars. Waxes are also produced for polishing cars and tires providing better protection (Torres and Domínguez, 1980). In the plumbing industry, for example, candelilla wax collars are used as a seal between the toilet and the floor, for its properties of malleability, these products enable to ensure an excellent seal (Candelilla Institute, 2015).

The excellence of a candle depends on the nature of the wick, but above all the quality of its combustible material. The base of the candle wax significantly improves the following consistency properties hardness or softness, melting point and appearance, there is wide variety of candles such as candles against insects, tea candles, floating candles, decorative candles among others (Castorena and García, 2013).

The natural fibers contain waxy substances that protect them from inclement weather and biological attacks (Gupta and Mehrotra, 1997). The removal of these substances during the process of manufacture of textiles alters the characteristics of friction and absorbency of the fibers, resulting in a loss of softness, flexibility and elasticity. Because of this situation, the external application of waxes which allow to obtain a finish suitable textile products is required. Waxes give textile materials a lubricating effect, allowing them to achieve a uniform and pleasant texture to the touch, in addition to reducing friction and the generation of static electricity (Canales et al., 2006).

On the other hand, the majority of the plastiline are basically composed of dyes, non-stick materials and waxes being the latter which give an excellent finish in texture, malleability and color retention (Candelilla Institute, 2015). Candelilla wax is also used as hardener agent of other softer waxes and with lower melting point as well as diluent for the beeswax and carnauba (Ramírez, 2005).

Finally, it can be mentioned among its multiple uses of the candelilla wax in developing (Canales et al., 2005; Zapata, 2008; Cosmogaceta, 2011; Bellasté and Mandujano, 2013): rinse aid, agents of microencapsulation, colors varnishes, wax plugs, tapes for machines, compounds to protect metals, production of cement, gummed paper, disposable containers, manufacture of phonograph records, rubber, lacquers, paintings, plastics, skin protectants, polishes, paint removers, and ointments.

## 19.4 CONCLUSIONS

Candelilla wax is a complex substance that is extracted from the plan *Euphorbia antysiphilitica* that is founded mainly in the north of Mexico. Due to their widely applications for its chemical composition and properties has a big opportunity in the global market.

## KEYWORDS

- applications
- candelilla wax
- characteristics
- chemical composition

## REFERENCES

Abdallah, D. J., Lu, L., & Weiss, R. G. (1999) Thermoreversible organogels from alkane gelators with one heteroatom. *Chem Mater, 11*, 2907–2911.

BARSCH, F. (2004). Candelilla (*Euphorbia antisyphillitica*): utilization in Mexico and international trade. *Medicinal Plant Conservation, 9,* 46–50.

Bosques Molina, E,. (2008a). Development of edible coatings formed with mesquite gum and candelilla wax for fruit preservation. *Metropolitan Autonomous University – Iztapalapa, Biotech,* 28–31.

Bosques-Molina, E. (2008b). Development of edible coatings formulated with mesquite gum and candelilla wax for fruit preservation. Food world, *1.* 28–31.

Bosques-Molina, E., Guerrero Legarreta, I., & Vernon Carter, E. J. (2003).Moisture barrier properties and morphology of mesquite gum–candelilla wax based edible emulsion coatings.*Food Research International, 36,* 885–893.

Bot, A., Veldhuizen, S. J. Y., den Adel, R., & Roijers, C. E. (2009). Non-TAG structuring of edible oils and emulsions. *Food Hydrocolloids, 23*(4), 1184–1189.

Brasil Resolucao, R. D. C. (2005). No. 43, de 1 de Marzo.Diario Oficial: 2 de marzo.

Cabello Alvarado, C. J., Sáenz Galindo, A., Barajas Bermúdez, L., Pérez Berumen, C., Ávila Orta, C., & Valdés Garza, J. (2013). Candelilla wax and its applications Advances in Chemistry, 8, num. 2, may-august, 105–110. Universidad de los Andes Mérida, Venezuela.

Canales E. V., Canales Martínez., & Zamarrón, E. M. (2006). Candelilla of the Mexican desert towards the world. CONABIO. Biodiversitas num, *69*:1–5.

Canales Gutiérrez, E., Canales Martínez, V., & Zamarrón Rodríguez, E. M. (2005). Candelilla: Wax drops of the Mexican desert around the world, en López C, Chanfón S. y Segura G. The wealth of the Mexican forests: beyond the wood. Experiences of rural communities. Center for Education and Training for Sustainable Development, Semarnat, Procymaf, Conafor, *Center for International Forestry Research, México,* 101–107.

Candelilla Institute (2015) http://www.candelilla.org

Carranza González, E. (2007). Review of "Life in the Mexican deserts" from Héctor M. Hernández. Acta Botanica Mexicana. Institute of Ecology, A. C. México. Núm, *80,* 89–90.

Cervantes Ramírez, M. C. (2005). Plants of economic importance in arid and semi-arid zones of Mexico. Anais do X Encontro de Geografos da América Latina, *Universidade de São Paulo,* 3388–3407.

National Commission for the Knowledge and Use of Biodiversity (CONABIO). (2008). Conservation, use and trade of the candelilla. Press release CONABIO Núm. 11 Mexico, D. F. 1–3, DC.

COSMOGACETA. (2015). Online available at: http://cosblog.com/test/wp-content/uploads/2009/05/COSMOGACETA-ENERO-11.pdf. Año 2, No.24, Enero (2011).

De león-Zapata M. (2008). Technological improvements to the process of extraction of candelilla wax (Euphorbia antysyphilitica Zucc.). Professional Thesis. School of Chemistry, Autonomous University of Coahuila. Saltillo, Coahuila, Mexico.

FAO. (1992). Candelilla wax. In *Compendium of Food Additive Specifications, Food and Nutrition.*

FCC. (2003). Candelilla wax. In *Food Chemicals Codex*, 5th edn., National Academies Press, Washington.

Grant, D. L. (2005) Candelilla wax, first draft. WHO food additive series 30, http://www.inchem.org/documents/jecfa/jecmono/v30je12.htm

Gupta, N., & Mehrotra, N. K. (1997). Potassium nutrition related biomasa and wax productivity of *Euphorbia antisyphillitica* Zucc in sand culture. *Journal of the Indian Botanical Society 76,* 99–101.

Hagenmaier, R. D., & Baldwin E. (2003). Coating selection for Delicious and other apples. *Postharvest biology and technology, 28,* 381–390.

Hernandez García, G. (2013). The possibilities of sustainable industrialization of the candelilla in the desert of Chihuahua. Master's thesis, National Polytechnic Institute.

Hernández, M. H. (2006). Life in the Mexican deserts. Economic Culture Fund, *Mexico.*

Howe, G. F., & Williams, E. L. (2001). Euphorbia antisyphilitica (The candelilla plant). Demonstrates Providence, *Design and Typology in Creation. CRSQ, 27,* 3. 86–91.

Hughes, E. N., Marangoni, A. G., Wright, A. J., Rogers, A. M., & Rhus, J. W. E. (2009). Potential food applications of edible oil organogels. *Trends in Food Science and Technology, 20*(10), 470–480.

INFOTHERM (2006). http://www.fiz-chemie.de/infotherm/servlet/infothermSerch

Japan. (2000). Candelilla wax.*In Japan's Specifications and Standards for Food Additives,* 7th ed.

Kowalczyk, D., & Baraniak, B. (2014). Effect of candelilla wax on functional properties of biopolymer emulsion films-A comparative study. *Food Hydrocolloids, 41,* 195–209.

Kuznesof, P. M. (2005). Candelilla Wax. Chemical and Technical Assessment 65th JECF.

LipidBank (2006). http://www.lipidbank.jp/cgi-bin/detail.cgi?id=WWA1101.

Lopez-Guerra, N. L. (2006).Physical, chemical and microbiological evaluation of polyphenols-rich plant sources (In Spanish). Professional thesis dissertation, Autonomous University of Coahuila, *Saltillo, Mexico.*

Maldonado, A. L. (1979). Research developed on candelilla wax. Forestry Science num. 18, National Institute of Forestry Research, Mexico.

Martínez Ballesté, A., &. Mandujano, M. (2013). The Consequences of Harvesting on Regeneration of a Non-timber Wax Producing Species (Euphorbia antisyphilitica Zucc.) of the Chihuahuan Desert. *Economic Botany, 67* (2), 121–136.

Martínez, J. C. (2002). Production of Aspergillus niger tannase using Tarbush as sole carbon source and energy (In Spanish). Professional thesis dissertation, *Universidad Autónoma de Coahuila, Saltillo, México.*

Multiceras. (2014). Instituto de la candelilla. http://www.multiceras.com/es/.

Ochoa Reyes, E., Saucedo Pompa, S., De la Garza, H., Martínez, D., Rodríguez, R., & Aguilar C. N.,(2010). Extracción tradicional de cera de Euphorbia antysiphilitica. Acta química Mexicana. *Revista Científica de la Universidad Autónoma de Coahuila, 2,* No. 3, 1–13.

Rojas Molina, R., De Leon Zapata, M. A., Saucedo Pompa, S., Aguilar González, M. A., & Aguilar C. N. (2013). Chemical and structural characterization of Candelilla (Euphorbia antisyphilitica Zucc.). *Journal of Medicinal Plants Research, 7*(12), 702–705.

Rojas-Molina, R., Saucedo-Pompa, S., De León-Zapata, M.A., Jasso-Cantú, D., & Aguilar, C. N. (2011). Past, Present and Future of Candelilla. *Rev. Mex. Cien. For, 2*(6), 7–18.

Saucedo-Pompa S. (2007). Development of edible films made from candelilla wax and antioxidant actives. Professional Thesis. School of Chemistry, Autonomous University of Coahuila. Saltillo, Coahuila, p. 99.

Semarnat. Ministry of the Environment and Natural Resources SEMARNAT. (2008b) Catalog of timber and non-timber forest resources. http://www.conafor.gob.mx/biblioteca/ Catalogo_de_recursos_forestales_M_y_N.pdf 9–11 pp.

Semarnat. Ministry of the Environment and Natural Resources. SEMARNAT (2008a). Forest Products Trade Information Sheet. http://www.conafor.gob.mx/biblioteca/fichas-de-informacion-comercial-productos-forestales.pdf 1–2 pp.

Tapia Tapia E. C., & Reyes Chilpa, R. (2008). Non-timber forest products in Mexico: Economic aspects for sustainable development timber and forests. Institute of Ecology, A. C. Mexico. *14*, Núm. 3, 95–11.

Toro-Vazquez, J. F., Alonzo-Macias, M., Dibildox-Alvarado, E., Charó-Alonso, M. A. (2009). The effect of tripalmitin crystallization on the thermo-mechanical properties of candelilla wax organogels. *Food Biophys 4*, 199–212. doi:10.1007/s11483–009–9118–7.

Toro Vazquez, J. F., Morales Rueda, J. A., Diblidox Alvarado, E., Charó Alonso, M., Alonso Macías, M., & González Chávez, M. M. (2007). Thermal and Textural Properties of Organogels Developed by Candelilla Wax in Safflower Oil. *Journal of American Oil Chemist's Society, 84*(11), 989–1000. DOI 10.1007/s11746–007–1139–0.

Torres, N., & Domínguez, M. D. (1980). Candelilla wax in Mexico and its marketing prospects. Bachelor Thesis, *UNAM, Mexico.*

United States *Code of Federal Regulations.* (2004). Candelilla wax. Title 21, United States *Code of Federal Regulations*, Section 184.1976. U.S. Government Printing Office, Washington, DC.

Valera, M. R. (2004). Exploitation of the Candelilla in the State of Coahuila, Journal Hipatía from Alejandría Num. 4, Saltillo, Mexico.

Villa-Castorena; M., Catalán-Valencia. E. A., Inzunza-Ibarra, M. A., González-López, M. de L., & Arreola-Ávila, J. G. (2010). Production of candelilla seedlings (Euphorbia antisyphilitica Zucc.) by cuttings. *Journal of Chapingo Forestry and Environmental Sciences Series, 16*(1), 37–47.

Wolfmeier, U., Schmidt, H., Heinrichs, F.-L., Michalczyk, G., Payer, W., Dietsche, W., et al. (1996). Waxes. In *Ullmann's Encyclopedia of Industrial Chemistry, 28A*, 103–122.

# CHAPTER 20

# EXTRACTION OF BIOACTIVE COMPOUNDS, CHARACTERIZATION, AND ITS USE

GUILLERMO CRISTIAN G. MARTÍNEZ-ÁVILA,[1]
CRISTOBAL N. AGUILAR,[2] ERNESTO J. SANCHEZ-ALEJO,[1]
J. A. ASCACIO-VALDES,[2] A. F. AGUILERA-CARBÓ,[3] and
ROMEO ROJAS[1*]

[1]Autonomous University of Nuevo Leon, School of Agronomy,
Research Center and Development for Food Industries, Francisco Villa
s/n, Ex-Hacienda el Canada, Escobedo, Nuevo Leon, 66054, Mexico,
*E-mail: romeo.rojasmln@uanl.edu.mx

[2]Department of Food Science and Technology, Autonomous
University of Coahuila, Blvd. Venustiano Carranza S/N Col.
RepъblicaOriente, 25280 Saltillo, Coahuila, Mexico

[3]Department of Food Science and Nutrition, Agrarian Autonomous
University Antonio Narro, Calzada Antonio Narro, 1923 Buenavista,
Saltillo, Coahuila, Mexico

## CONTENTS

## ABSTRACT

The extraction process candelilla wax generates a lot of stems without wax. These stems are used as fuel for the process of extracting wax and livestock feed. Still, a large number of stems stored outdoors generating environmental pollution. In recent years, some research has focused on finding an application. Recently, methods have been reported extraction and purification using eco friendly technologies by using water and ethanol, likewise, they have been achieved quantify bioactive compounds such as ellagic acid (EA), gallic acid (GA) and catechin at concentrations of $2.2 \pm 0.15$ mg g$^{-1}$, $0.6 \pm 0.03$ mg g$^{-1}$ and $0.2 \pm 0.02$ mg g$^{-1}$, respectively. In addition, it has been possible to identify a new molecule named candelitannin. The importance of these bioactive compounds is that they exhibit antitumor, antiviral, antioxidant activities, etc. Thus, stems candelilla an alternative for preparing compounds of high biological-industrial interest.

## 20.1 INTRODUCTION

Tannins are high molecular weight plant polyphenols divided into two chemically and biologically distinct groups: condensed tannins or proanthocyanidins and hydrolyzable tannins or ellagitannins. (Seeram et al., 2005). Naturally, the plants produce a wide variety of bioactive compounds like flavonoids, phenolic acids, lignans salicylates, stanols, sterols, glucosinolates, among others (Hooper and Cassidy, 2006). However, the qualitative and quantitative studies of bioactive compounds from plant materials mostly rely on the selection of proper extraction method (Azmir et al., 2013).

To the recovery of bioactive compounds from plants and agroindustrial wastes are used several techniques like Solid State Fermentation (León et al., 2013; Martins et al., 2011), infusion (De León-Zapata et al., 2016), microwave (Rojas et al., 2015), ultrasound (Muñiz-Márquez et al., 2013; Tomšik et al., 2016; Wong Paz, Muñiz Márquez, Martínez Ávila, Belmares Cerda, & Aguilar, 2015), heat-reflux system (Wong-Paz et al., 2015), supercritical fluids (da Silva et al., 2016), hydrodistillation, pressurized liquids (Machado et al., 2015), subcritical fluids, enzyme assisted extraction (Boulila et al., 2015; Sánchez-Camargo et al., 2016), sequential extraction (Bitencourt et al., 2014), liquid-liquid extraction (Burin et al., 2014), pulsed electric fields (Medina-Meza & Barbosa-Cánovas, 2015), pulsed ohmic heating (Darra et al., 2012), etc. Also, several types of solvents are used like ethanol (Ruiz-Martínez et al., 2011), $CO_2$, methanol (Al-Saikhan et al., 2015; Karabegović et al., 2011), water (Ascacio-Valdés et al., 2010; Ascacio-Valdés et al., 2013; Saifudin et al., 2016), acid solutions (Rojas et al., 2015), acetone (Netshiluvhi and Eloff, 2016), etc.

Candelilla plant has a great economic importance in the northwest of México due to their higher content of wax which is used for various purposes. Recently, candelilla stems has been reported as the potential source of phenolic compounds like EA, GA, catechin and one new compound called candelitannin (Ascacio-Valdés et al., 2011, 2013). This has been possible through of a clean and eco-friendly methodology reported by Ascacio-Valdés et al. (2010, 2013) and Rojas-Molina et al. (2013). This has opened an area of opportunity for the use of waste from the extraction process candelilla wax (stems) for obtaining bioactive compounds that have attracted much recent attention due to their numerous biological activities and implications in potential benefits to human health.

## 20.2 EXTRACTION METHODS

Nowadays, there are methodologies to recover bioactive compounds from candelilla waste. These methodologies are specific and with only one purpose, make this residue a source of bioactive compounds. The first report of method for extraction of bioactive compounds (EA) from *Euphorbia antisyphilitica* branches (Candelilla) was the reported by Aguilera-Carbo et al. (2008). The method consists in dried the plant were transferred to tubes with screw caps containing 4.95 mL of methanol and 50 µL of 6 M HCl. The tubes were closed and the content was incubated for 168 h at 90°C. After cooling this material was filtered through a Whatman No. 41 in a vacuum system. An aliquot of 1.5 mL of the filtered fraction was centrifuged at 36000 min$^{-1}$ for 20 min. The supernatant was decanted and the precipitate was re-suspended in ethanol (1.5 mL).

The second method reported for extractionof bioactive compounds (EA) from this specie was reported by Ascacio-Valdés et al. (2010). The method consists in clean the stalks according to the traditional method (DeLeón-Zapata, 2008; Rojas-Molina et al., 2011) using sulfuric acid. The residual vegetal tissue was dehydrated at 60°C for 48 h. After that, the tissue was ground. For extract the total polyphenols of candelilla (TPC) was used the best conditions for EA extraction (20 mg mL$^{-1}$ mass/volume ratio). Dehydrated and pulverized candelilla stalks were placed in a baker with distilled water at 60°C for 30 min. After that, the obtained extract was filtered using filter paper (Whatman No. 41) to eliminate the biggest residue particles. For purification, a column chromatography was performed used an Amberlite XAD-16. They used water to discard undesirable compounds, and then, ethanol as the eluent to obtain a TPC fraction according to Seeram et al. (2005). Finally, the solvent was evaporated from the fractionated extract and TPC was recovered as a fine powder.

The next reports made by Ascacio-Valdés et al. (2013), Rojas-Molina et al. (2013) and Télles-Pichardo et al. (2013) was carried out according to the reported by Ascacio-Valdés et al. (2010). Today, they are the only methods of extraction of bioactive compounds from candelilla stems.

## 20.3 CHARACTERIZATION AND ITS USE

Characterization of bioactive compounds candelilla was carried out according to what reported by Aguilera-Carbo et al. (2008) using the following methodology: the sample was sonicated for 30 min, the material was filtered through a 0.45 μm nylon membrane and injected into HPLC system. The detection is carried out using a Varian Pro-Star 330 with photodiode array (PDA) system with detection at 254 nm using a modification of the Amakura et al. (2000) method. Fractionation of the injected material is carried out on an Optisil ODS column (5 μm, 250 × 4.6 mm) at 25°C and an isocratic mobile phase consisting of acetonitrile (solvent A) and water:phosphoric acid (pH 2.5, solvent B), in the ratio of 30:70 is applied at the flow rate of 0.7 mL min$^{-1}$. Sample injection volume was 10 μL, retention time was 5.1 min. After that, the method is validated with a linearity of 0.9947 ± 0.0025. The concentration of EA in the candelilla branches was 2.18 ± 0.39 mg g$^{-1}$ (dry basis).

Ascacio-Valdés et al. (2010) hydrolyzed sample with sulfuric acid (96%) and methanol. After that, are placed the samples (10 mg) in closed assay tubes with a screw top with a plastic ring for hermetic closing. An aliquot (1.5 mL) of the hydrolysis mixture is added to the tubes which were closed with a screw top; hydrolysis was carried out in a stove (80°C for 30 h). After the hydrolysis, is removed the solvent by evaporation to remove volatile components and is centrifuged at 3300 x g for 30 min, are discarded the supernatant and is add ethanol to dissolve EA. EA quantification was carried out using

HPLC equipment according to the usage conditions reported by Aguilera-Carbo et al. (2008).

In 2013, Ascacio-Valdés et al., reported the identification of a new molecule from candelilla called candelitannin (Figure 20.1) according to the methodology proposed by Ascacio-Valdés et al. (2010) and Aguilera-Carbo et al. (2008). After that, an anionic interchange column (Hi-Trap Q XL) using a FPLC equipment and a Varian mass spectrum model 500-MS was used to mass analysis with 1 mL/min, mass rate 400–2000 by 10 min. Finally, the antifungal activity of candelitannin was measured against four important phytopathogenic fungus.

The last study indicated that the candelilla stalks that is a potential source of EA ($2.2\pm0.15$ mg g$^{-1}$) catechin ($0.2\pm0.02$ mg g$^{-1}$) GA ($0.6\pm0.03$ mg g$^{-1}$) and lipids ($15.9\pm1.119\%$) obviously (Rojas-Molina et al., 2013). The methodology used was the reported by Aguilera-Carbo et al. (2008), Ascacio-Valdés et al. (2010, 2013). Below the compounds, characterization and applications of bioactive compounds extracted from the stems of candelilla are listed (Table 20.1).

## 20.4 CONCLUSIONS

The recovery process of bioactive compounds from candelilla stems is an area little explored. However, recent research suggests that a potential bioactive compounds that can be used in various fields because of their biological properties source. The search for alternative recovery and studies to identify

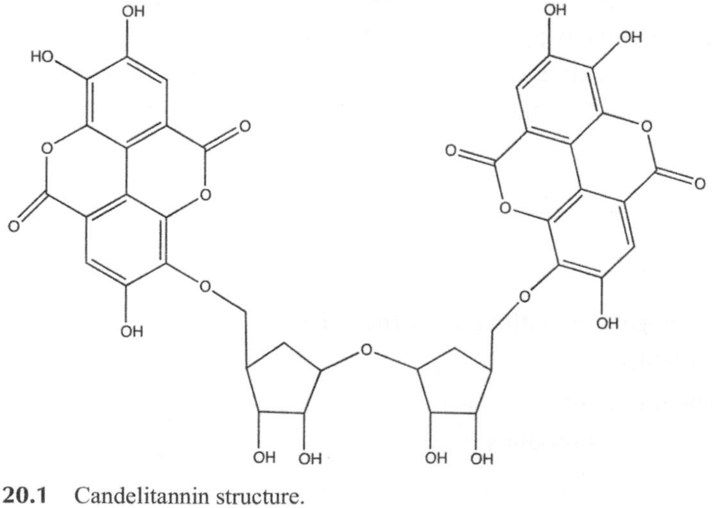

FIGURE 20.1   Candelitannin structure.

**TABLE 20.1**   Compounds, Characterization and Use of Bioactive Compounds from Candelilla Stems

| Compound(s) | Characterization or Use | Reference |
|---|---|---|
| Ellagic acid | Extraction, quantification and improved the detection method | Aguilera-Carbo et al., 2008 |
| Ellagic acid | Optimization of the extraction process | Ascacio-Valdés et al., 2010, 2013 |
| Total purified polyphenols | Extraction to make edible films to prolong the self life of whole Papaya | Télles-Pichardo et al., 2013 |
| Candelitannin | Identification, quantification and antifungal activity against *Alternariaalternata, Fusariumoxysporum, Colletotrichum gloeosporioides*, and *Rhizoctoniasolani* | Ascacio-Valdés et al., 2013 |
| Proximal content, sugars, hydrolysable and condensed tannins, ellagic acid, gallic acid and catechin | Chemical and micro-structural characterization of candelilla stems | Rojas-Molina et al., 2013 |
| Ellagic acid | Candelilla stalks was probed as support to produce ellagitannase enzyme to continuous production of ellagic acid in a packed-bed reactor | Buenrostro-Figueroa, Ascacio-Valdés et al., 2014; Buenrostro-Figueroa, Huerta-Ochoa et al., 2014 |

more molecules present within the total polyphenols extracted is necessary. Thus, a comprehensive utilization of the plant will be achieved by reducing air pollution and will add value to a residue which is currently used as livestock feed and fuel for the extraction process candelilla wax.

## KEYWORDS

- **antifungal and antioxidant properties**
- **candelilla**
- **candelitannin**
- **phenolic compounds**

# REFERENCES

Aguilera-Carbo, A., Augur, C., Prado-Barragan, L., Aguilar, C., & Favela-Torres, E. (2008). Extraction and analysis of ellagic acid from novel complex sources. *Chemical Papers, 62*(4), 440–444. doi: 10.2478/s11696-008-0042-y.

Al-Saikhan, F. I., & Ansari, M. N. (2016). Evaluation of The Diuretic and Urinary Electrolyte Effects of Methanolic Extract of *Peganum harmala* L. in Wistar albino rats. *Saudi Journal of Biological Sciences.* doi: http://dx.doi.org/10.1016/j.sjbs. 01.025.

Ascacio-Valdés, J. A., Aguilera-Carbó, A., Martínez-Hernández, J. L., Rodríguez-Herrera, R., & Aguilar, C. N. (2010). *Euphorbia antisyphilitica* residues as a new source of ellagic acid. *Chemical Papers, 64*(4), 528–532.

Ascacio-Valdés, J. A., Buenrostro-Figueroa, J. J., Aguilera-Carbo, A., Prado-Barragán, A., Rodríguez-Herrera, R., & Aguilar, C. N. (2011). Ellagitannins: Biosynthesis, biodegradation and biological properties. *Journal of Medicinal Plants Research, 15*(19), 4696–4703.

Ascacio-Valdés, J. A., Burboa, E., Aguilera-Carbo, A. F., Aparicio, M., Pérez-Schmidt, R., Rodríguez, R., et al. (2013). Antifungal ellagitannin isolated from *Euphorbia antisyphilitica* Zucc. *Asian Pacific. Journal of Tropical Biomedicine, 3*(1), 41–46. doi: http://dx.doi.org/10.1016/S2221–1691(13)60021-0.

Azmir, J., Zaidul, I. S. M., Rahman, M. M., Sharif, K. M., Mohamed, A., Sahena, F., et al. (2013). Techniques for extraction of bioactive compounds from plant materials: A review. *Journal of Food Engineering, 117*(4), 426–436. doi: http://dx.doi.org/10.1016/j.jfoodeng.2013.01.014.

Bitencourt, R. G., Queiroga, C. L., Duarte, G. H. B., Eberlin, M. N., Kohn, L. K., Arns, C. W., et al. (2014). Sequential extraction of bioactive compounds from Melia azedarach L. in fixed bed extractor using $CO_2$, ethanol and water. *The Journal of Supercritical Fluids, 95*, 355–363. doi: http://dx.doi.org/10.1016/j.supflu.2014.09.027.

Boulila, A., Hassen, I., Haouari, L., Mejri, F., Amor, I. B., Casabianca, H., et al. (2015). Enzyme-assisted extraction of bioactive compounds from bay leaves (*Laurus nobilis* L.). *Industrial Crops and Products, 74*, 485–493. doi: http://dx.doi.org/10.1016/j.indcrop.2015.05.050.

Buenrostro-Figueroa, J., Ascacio-Valdés, A., Sepúlveda, L., De la Cruz, R., Prado-Barragán, A., Aguilar-González, M. A., et al. (2014). Potential use of different agroindustrial byproducts as supports for fungal ellagitannase production under solid-state fermentation. *Food and Bioproducts Processing, 92*(4), 376–382. doi: http://dx.doi.org/10.1016/j.fbp.2013.08.010.

Buenrostro-Figueroa, J., Huerta-Ochoa, S., Prado-Barragán, A., Ascacio-Valdés, J., Sepúlveda, L., Rodríguez, R., et al. (2014). Continuous production of ellagic acid in a packed-bed reactor. *Process Biochemistry, 49*(10), 1595–1600. doi: http://dx.doi.org/10.1016/j.procbio.2014.06.005.

Burin, V. M., Ferreira-Lima, N. E., Panceri, C. P., & Bordignon-Luiz, M. T. (2014). Bioactive compounds and antioxidant activity of Vitis vinifera and Vitis labrusca grapes: Evaluation of different extraction methods. *Microchemical Journal, 114*, 155–163. doi: http://dx.doi.org/10.1016/j.microc.2013.12.014.

da Silva, R. P. F. F., Rocha-Santos, T. A. P., & Duarte, A. C. (2016). Supercritical fluid extraction of bioactive compounds. *TrAC Trends in Analytical Chemistry, 76*, 40–51. doi: http://dx.doi.org/10.1016/j.trac.2015.11.013.

Darra, N., Grimi, N., Vorobiev, E., Louka, N., & Maroun, R. (2012). Extraction of Poly-phenols from Red Grape Pomace Assisted by Pulsed Ohmic Heating. [journal article]. *Food and Bioprocess Technology, 6*(5), 1281–1289. doi: 10.1007/s11947-012-0869-7.

De León-Zapata, M. A., Pastrana-Castro, L., Rua-Rodríguez, M. L., Alvarez-Pérez, O. B., Rodríguez-Herrera, R., & Aguilar, C. N. (2016). Experimental protocol for the recovery and evaluation of bioactive compounds of tarbush against postharvest fruit fungi. *Food Chemistry, 198*, 62–67. doi: http://dx.doi.org/10.1016/j.foodchem.2015.11.034.

DeLeón-Zapata, M. A. (2008). Technological improvements to the process of extraction of candelilla wax. Chemist Pharmaceutical Biologist, Autonomous University of Coa-huila, Saltillo, Coahuila, México.

Eid, H. H., Labib, R. M., Hamid, N. S. A., Hamed, M. A., & Ross, S. A. (2015). Hepato-protective and antioxidant polyphenols from a standardized methanolic extract of the leaves of *Liquidambar styraciflua* L. *Bulletin of Faculty of Pharmacy, Cairo University, 53*(2), 117–127. doi: http://dx.doi.org/10.1016/j.bfopcu.2015.05.002.

Hooper, L., & Cassidy, A. (2006). A review of the health care potential of bioactive com-pounds. *Journal of the Science of Food and Agriculture, 86*(12), 1805–1813. doi: 10.1002/jsfa.2599.

Jyoti, M. A., Nam, K. W., Jang, W. S., Kim, Y. H., Kim, S. K., Lee, B. E., et al. (2015). Anti-mycobacterial activity of methanolic plant extract of *Artemisia capillaris* containing ursolic acid and hydroquinone against Mycobacterium tuberculosis. *Journal of Infec-tion and Chemotherapy*. doi: http://dx.doi.org/10.1016/j.jiac. 11.014.

Karabegović, I., Nikolova, M., Veličković, D., Stojičević, S., Veljković, V., & Lazić, M. (2011). Comparison of Antioxidant and Antimicrobial Activities of Methanolic Extracts of the Artemisia sp. Recovered by Different Extraction Techniques. *Chinese Journal of Chemical Engineering, 19*(3), 504–511. doi: http://dx.doi.org/10.1016/S1004-9541(11)60013-X.

León, M. A. D., Sáenz, A., Jasso-Cantu, D., Rodríguez, R., Pandey, A., & Aguilar, C. N. (2013). Fermented Flourensia cernua Extracts and Their in vitro Assay Against Penicil-lium expansum and Fusarium oxysporum. *Food Technology and Biotechnology, 51*(2), 233–239.

Machado, A. P. D. F., Pasquel-Reátegui, J. L., Barbero, G. F., & Martínez, J. (2015). Pressur-ized liquid extraction of bioactive compounds from blackberry (*Rubus fruticosus* L.) residues: a comparison with conventional methods. *Food Research International, 77, Part 3*, 675–683. doi: http://dx.doi.org/10.1016/j.foodres.2014.12.042.

Martins, S., Mussatto, S. I., Martínez-Avila, G., Montañez-Saenz, J., Aguilar, C. N., & Teixeira, J. A. (2011). Bioactive phenolic compounds: Production and extraction by solid-state fermentation. A review. *Biotechnology Advances, 29*(3), 365–373. doi: http://dx.doi.org/10.1016/j.biotechadv.2011.01.008.

Medina-Meza, I. G., & Barbosa-Cánovas, G. V. (2015). Assisted extraction of bioac-tive compounds from plum and grape peels by ultrasonics and pulsed electric fields. *Journal of Food Engineering, 166*, 268–275. doi: http://dx.doi.org/10.1016/j.jfoodeng.2015.06.012.

Msaada, K., Jemia, M. B., Salem, N., Bachrouch, O., Sriti, J., Tammar, S., et al. (2013). Anti-oxidant activity of methanolic extracts from three coriander (*Coriandrum sativum* L.) fruit varieties. *Arabian Journal of Chemistry*. doi: http://dx.doi.org/10.1016/j.arabjc. 12.011.

Muñiz-Márquez, D. B., Martínez-Ávila, G. C., Wong-Paz, J. E., Belmares-Cerda, R., Rodrí-guez-Herrera, R., & Aguilar, C. N. (2013). Ultrasound-assisted extraction of phenolic compounds from *Laurus nobilis* L. and their antioxidant activity. *Ultrasonics Sonochemistry, 20*(5), 1149–1154. doi: http://dx.doi.org/10.1016/j.ultsonch.2013.02.008.

Nayak, G., Honguntikar, S. D., Kalthur, S. G., D'Souza, A. S., Mutalik, S., Setty, M. M., et al. (2016). Ethanolic extract of Moringa oleifera Lam. leaves protect the pre-pubertal spermatogonial cells from cyclophosphamide-induced damage. *Journal of Ethnopharmacology*. doi: http://dx.doi.org/10.1016/j.jep.2016. 02.003.

Netshiluvhi, T. R., & Eloff, J. N. (2016). Influence of annual rainfall on antibacterial activity of acetone leaf extracts of selected medicinal trees. *South African Journal of Botany, 102*, 197–201. doi: http://dx.doi.org/10.1016/j.sajb.2015.04.008.

Pereira, P., Cebola, M. J., Oliveira, M. C., & Bernardo-Gil, M. G. (2015). Supercritical Fluid Extraction vs Conventional Extraction of Myrtle Leaves and Berries: Comparison of Antioxidant Activity and Identification of Bioactive Compounds. *The Journal of Supercritical Fluids*. doi: http://dx.doi.org/10.1016/j.supflu.2015.09.006.

Rojas-Molina, R., León-Zapata, M. Á. d., Saucedo-Pompa, S., Aguilar-González, M. Á., & Aguilar-González, C. N. (2013). Chemical and structural characterization of Candelilla (*Euphorbia antisyphilitica* Zucc.). *Journal of Medicinal Plants Research, 7*(12), 702–705.

Rojas-Molina, R., Saucedo-Pompa, S., León-Zapata, M. Á. d., Jasso-Cantú, D., & Aguilar-González, C. N. (2011). Essay: Past, present and future of the candelilla. *Mexican Journal of Forestry Sciences, 2*(6), 7–18.

Rojas, R., Contreras-Esquivel, J., Orozco-Esquivel, M., Muñoz, C., Aguirre-Joya, J., & Aguilar, C. (2015). Mango Peel as Source of Antioxidants and Pectin: Microwave Assisted Extraction. *Waste and Biomass Valorization*, 1–8. doi: 10.1007/s12649-015-9401-4.

Ruiz-Martínez, J., Ascacio, J. A., Rodríguez, R., Morales, D., & Aguilar, C. N. (2011). Phytochemical screening of extracts from some Mexican plants used in traditional medicine. *Journal of Medicinal Plants Research, 5*(13), 2791–2797.

Saifudin, A., Usia, T., AbLallo, S., Morita, H., Tanaka, K., & Tezuka, Y. (2016). Potent water extracts of Indonesian medicinal plants against PTP1B. *Asian Pacific. Journal of Tropical Biomedicine, 6*(1), 38–43. doi: http://dx.doi.org/10.1016/j.apjtb.2015.09.021.

Sánchez-Camargo, A. D. P., Montero, L., Stiger-Pouvreau, V., Tanniou, A., Cifuentes, A., Herrero, M., et al. (2016). Considerations on the use of enzyme-assisted extraction in combination with pressurized liquids to recover bioactive compounds from algae. *Food Chemistry, 192*, 67–74. doi: http://dx.doi.org/10.1016/j.foodchem.2015.06.098.

Seeram, N., Lee, R., Hardy, M., & Heber, D. (2005). Rapid large scale purification of ellagitannins from pomegranate husk, a by-product of the commercial juice industry. *Separation and Purification Technology, 41*(1), 49–55. doi: http://dx.doi.org/10.1016/j.seppur.2004.04.003.

Silva, C. M. D., Zanqui, A. B., Souza, A. H. P., Gohara, A. K., Gomes, S. T. M., Silva, E. A. D., et al. (2016). Extraction of oil and bioactive compounds from Araucaria angustifolia (Bertol.) Kuntze using subcritical n-propane and organic solvents. *The Journal of Supercritical Fluids*. doi: http://dx.doi.org/10.1016/j.supflu..02.003.

Télles-Pichardo, R., Cruz-Aldaco, K., Ochoa-Reyes, E., Aguilar, C. N., & Rojas, R. (2013). Edible Coatings of Wax and Polyphenols from Candelilla: An Alternative of Conservation of Papaya (*Carica papaya* L.). *Acta Química Mexicana, 5*(10), 1–7.

Tomšik, A., Pavlić, B., Vladić, J., Ramić, M., Brindza, J., & Vidović, S. (2016). Optimization of ultrasound-assisted extraction of bioactive compounds from wild garlic (*Allium ursinum* L.). *Ultrasonics Sonochemistry, 29*, 502–511. doi: http://dx.doi.org/10.1016/j.ultsonch.2015.11.005.

Wong Paz, J. E., Muñiz Márquez, D. B., Martínez Ávila, G. C. G., Belmares Cerda, R. E.,
& Aguilar, C. N. (2015). Ultrasound-assisted extraction of polyphenols from native
plants in the Mexican desert. *Ultrasonics Sonochemistry, 22*, 474–481. doi: http://dx.
doi.org/10.1016/j.ultsonch.2014.06.001.

Wong-Paz, J. E., Contreras-Esquivel, J. C., Rodríguez-Herrera, R., Carrillo-Inungaray, M. L.,
López, L. I., Nevárez-Moorillón, G. V., et al. (2015). Total phenolic content, in vitro
antioxidant activity and chemical composition of plant extracts from semiarid Mexican
region. *Asian Pacific Journal of Tropical Medicine, 8*(2), 104–111. doi: http://dx.doi.
org/10.1016/S1995–7645(14)60299-6.

# CHAPTER 21

# REGULATION AND SOCIOECONOMIC IMPACT OF EXPLOITATION OF CANDELILLA PLANT

ROMEO ROJAS, GUILLERMO CRISTIAN G. MARTÍNEZ-ÁVILA, and ERNESTO J. SANCHEZ-ALEJO

*Autonomous University of Nuevo Leon, School of Agronomy, Research Center and Development for Food Industries, Francisco Villa s/n, Ex-Hacienda el Canada, Escobedo, Nuevo Leon, 66054, Mexico, E-mail: romeo.rojasmln@uanl.edu.mx*

## CONTENTS

## ABSTRACT

Candelilla (*Euphorbia antisyphilitica* Zucc.) is one of the non-timber specie of the Chihuahuan desert. For two centuries, large quantities of non-timber candelilla have been harvested from semiarid lands of Northeast of Mexico, mainly in Coahuila state. The traditional method for wax extraction consists in submerged the plant into boiling water with sulfuric acid in an iron cauldron known as "paila." However, this archaic method was developed in 1914 without any modification until now that exist a new eco-friendly method for wax extraction. Due to its application, it represents a reliable source of income to communities and contributes to economic development, but should be considered the regulation that exists for exploitation. Thus, decisions may be taken comprehensive support for producers to help improve the extraction conditions, environmental impact and profitability of the product (candelilla wax). And intervention programs which will regulate their operation and opening new markets for innovative products.

## 21.1    INTRODUCTION

Around the world, the commercialization of non-timber forest products (NTFP) is regarded as an important contributor to rural development, providing income sources and generating better living conditions for poor farmers (Arato et al., 2014). However, it is necessary to consider socio-economic benefits generated for communities and the environmental impact generated by its exploitation. For them, harvesting NTFP is their main job since there are limited opportunities for agriculture and other economic activities (Marshall et al., 2006). In México, the commercialization of NTFP represents an important income contributor for rural families, principally for those that live in arid regions. Considering its importance, the Candelilla plant is an interesting example of the exploitation of NTFP as it is an important source of income for the Candelilla suburbs of north-eastern region of the country.

Candelilla plant (*Euphorbia antisyphilitica* Zucc.) is a North American desert shrub native to south-west Texas to Zacatecas regions on the Mexican Altiplano (Rojas-Molina et al., 2013; Rojas-Molina et al., 2011; Scora et al., 1995). In 1932 was described by Zuccarini and in 1909 by Alcocer referred as *Euphorbia cerifera*, both mentioned that produces a natural wax. In 1914, Mr. Borrego and Mr. Flores invented and reported a simple method for the extraction of wax from that raw material (candelilla plant) (Rojas-Molina et al., 2011). In 2013, the plant was chemical characterized by Rojas-Molina et al. (2013)

Nowadays, this wax is widely used around the world in several industries like electrical (Scora et al., 1995), cosmetic, pharmaceutical industry

(Rojas-Molina et al., 2011) and because is generally recognized as safe (GRAS) (Zhang et al., 2016) to use in the food industry (Jang et al., 2015; Toro-Vazquez et al., 2011) is widely used to preserve the shelf life of several food products like apples (De León-Zapata et al., 2015; Ochoa et al., 2011), avocado (Saucedo-Pompa et al., 2009), pears (Cruz et al., 2015), oranges (Hagenmaier, 2000), etc., to make organogels (Alvarez-Mitre et al., 2012; Alvarez-Mitre et al., 2013; López-Martínez et al., 2014; Rocha et al., 2013; Toro-Vazquez et al., 2013). Although this activity is recognized with a potential rural development, this review aims to provide an overview from regulation to the socioeconomic impact because of its many applications, the product increases its value.

## 21.2 REGULATION

### 21.2.1 NATIONAL REGULATION

During the last decade, public policies on biodiversity conservation and utilization based on ecological criteria, have allowed for the improvement of resource conservation in México (Arato et al., 2014). In 2003, the Government issued the General Law for Sustainable Development of Forest (GLSDF) that include the approach of candelilla, the law defines the roles and responsibilities of government environmental agencies according to their particular expertise and area of action; the Regulation of the General Law for Sustainable Development of Forest (RGLSDF) that includes the development and the application of the Mexican Official Norms (NOM), like NOM-018-RECNAT (1999) that contain the specifications, requirements, procedures and parameters of preservation and utilization targets, which need to be respected when developing commercial activities with natural resources. In the section 3 of GLSDF requirements for the use of non-timber forest resources are specified. It must show the "Notice of Use" to inform and justify how Candelilla plant will extract from their natural environment and the methods used according to the article 53 of RGLSDF. To prepare a "Notice of Use" is required from an accompanying technical study prepared by a technical manager registered with the national forest registry, where you must ensure that the use of candelilla be sustainable and not detrimental to the survival of the species in the wild. This 'Notice of Use' is presented to SEMARNAT (that is the government agency responsible for the regulation, promotion and control of the commercialization and preservation of natural resources), that assigned an ID (art. 54 of RGLDFS) (I.C., 2013).

When an ejido has a Notice of Use, shall prepare an annual report based on Article 27 of RLGDFS indicating how the cerote was produced. When candelilla wax is transported or sold, the community must emit a remission note or a sales invoice. The article 100 of LGDFS indicated that no authorization is granted if the use could jeopardize the populations and environmental ecosystem functions (I.C., 2013). The criteria and the technical and administrative measures to realize sustainable utilization of grass candelilla cerote transport and storage specifications are described in the NOM-018-RECNAT (1999). All this in the different agencies that focus on particular objectives such as the National Forest Commision (CONAFOR) which promotes sustainable forestry development and enhances the conservation and restoration of Mexico's natural resources and is also responsible for the administration of utilization permits for the harvesting and commercialization of wild forest resources, and it provides different economic funds and support services to the rural communities (Arato et al., 2014; CONAFOR, 2016).

### 21.2.2 INTERNATIONAL REGULATION

The Convention on International Trade in Endangered Species of Wild Fauna and Flora (CITES) is an international agreement between governments signed in Washington D.C. on 1973 and coming into effect July 1, 1975. Its aim is to ensure that international trade in species of wild animals and plants enlisted in its Appendices does not threaten their survival. Now, CITES has 181 memberships. Mexico joins the 02/07/1991 and takes effect from 30/09/1991 (CITES, 2016). CITES regulates the exportation, re-exportation, importation, and introduction from the sea of plants and animals dead or alive included in its Appendices. This regulation is based on a permits and certifications system that can only be issued if certain conditions are met, and must be presented to get out or get in a country. For this it is essential that each signatory country designates one or more Administrative Authorities, that regulate and issue the permits and certifications, and one or more Scientific Authorities that advise on the effects of the commercial trade on the species.

In Mexico, there are three CITES authorities: first, the General Direction of Wild Life of the Secretary of Environment and Natural Resources (DGVS-SEMARNAT) that is the administrative authority; second, the National Commission for the Knowledge and Use of the Biodiversity (CONABIO) which is the scientific authority; and third, the Federal Attorney for Environmental Protection (PROFEPA) which is the law enforcement authority and is responsible for ensuring environmental justice using law enforcement. It controls the proper utilization of natural resources in general (Arato et al., 2014; CONABIO, 2016; PROFEPA, 2016; SEMARNAT, 2016).

Since 1975, Candelilla (*Euphorbia antisyphilitica*) has been listed in Appendix II of the Convention on International Trade in Endangered Species of wild Fauna and Flora (CITES), under *Euphorbia* spp. Therefore, its international trade requires a CITES certificate issued by the CITES management authority of Mexico (DGVS-SEMARNAT), and must be endorsed by the CONABIO. Since 2010 the Candelilla has the annotation number 4 to exempt "all the finished products of *Euphorbia antisyphilitica* packaged and ready for retail trade," in such a way that those who commercialized with finished products may internationally trade them with no need of a CITES permit (CONABIO, 2016; I.C., 2013; SEMARNAT, 2016).

## 21.3  SOCIOECONOMIC IMPACT

In Mexico, the most typical form of land property in rural communities is the community, which is an extension of land provided to a group of owners known as ejidatarios who are entitled to exploit the resources from this land, and toper form agricultural activities in their small properties within the limits of a given territory. This is based on the Political Constitution of the Mexican United States (Art.27) and the internal regulations of the communitarian assembly (Arato et al., 2014). Most of the rural inhabitants live in such settlements and most of the amenities present within the community such as elementary schools (in some cases secondary), rural shops and religious facilities are located there. The most stable activity is the recollection of candelilla plant and the extraction process of candelilla wax realized by male adult's member of the family. This process consists in placing the plant of candelilla in iron cauldrons called "pailas," adding a 0.3% (v/v) sulfuric acid solution, is heated up to the boiling point, the candelilla melted wax floats on the water surface as a foam, and then, immediately, it is removed from the "paila," this ceric material is knowing by the name of "cerote." Then the ceric material is refined with the same acid solution. This raw candelilla wax is sold at prices from 36 to 38 Mexican pesos per kg (DeLeón-Zapata, 2008; Rojas-Molina et al., 2011).

At the beginning of the 20th century, its commercialization started. During World War II, its demand increased as waterproof material and was used as protector for camping fabric tents against mosquitoes; also, to cover and prevent the deterioration of some parts of aircrafts and manufacturing of explosives (Dominguez-Lara, 2004; López et al., 2007; Rojas-Molina et al., 2011). In those years, the "candelilleros" were organized to achieve a better commercialization, but when that war came to an end, the development of the petrochemical industry affected significantly the market of the candelilla as many products made-up with its wax were replaced by others made from oil, such as: the components of some electronic devices, adhesive bases, coatings and cosmetics. Years later, this tendency was reverted as natural raw materials were preferred as the derivatives of oil became more expensive (Canales-Gutiérrez et al., 2006; De León-Zapata, 2008; López et al., 2007; Rojas-Molina et al., 2011; Romahn, 1992)

Another important use is the approach of Candelilla bagasse (without wax, after of extraction process) to make compost that could be used as an additive in the process to improve and strengthen the properties of plastics. This due to their composition of cellulose, hemicellulose and lignin. This opens another range of applications candelilla plant that could be used to produce fiber for making furniture from an extrusion process (I.C., 2016). However, it is an activity that takes place without security measures, but it involves a secure source of family income. Hence the importance of technological development to improve working conditions, safety, health effects and environmental damage.

By this way, in the last years, the Food Research Department (DIA) of the School of Chemistry (FCQ) of the Autonomous University of Coahuila (UAdeC) obtained the registration of two patents (MX/u/2011/000553 and MX/a/2010/002764). These patents are Machinery for the separation and model of candelilla wax refining (Rojas et al., 2011) and Candelilla wax extraction process (*Euphorbia antisyphilitica* Zucc.) (DeLeón-Zapata et al., 2010). Through of CONAFOR, the UAdeC has achieved three projects by technology transfer:

1. In 2012, the study of the feasibility of the implementation of the new extraction process candelilla wax-based citric acid and factors that

influence their appropriation in the community Estanque de Norias y La Reforma, Cuatrocienegas, Coahuila, Mexico.

2.  In 2014, the technology validation package extraction method Candelilla wax with organic acids in commercial scale equipment in the community Lucio Blanco, Cuatrocienegas, Coahuila, México.

3.  In 2015, the improved extraction process candelilla wax high quality organic acids in modern equipment in the community San Gerónimo, Melchor Ocampo, Zacatecas, México and Estanque de Palomas, Cuatrocienegas, Coahuila, México.

These three projects have improved the conditions of Candelilla wax extraction to replace the sulfuric acid organic acids, in addition to the modernization of the equipment and reducing the health risks and environmental contamination. In addition, the wax that is extracted is completely organically and opens a new market for the product that will impact its price. Also, it has trained staff for the preparation of products consisting of candelilla wax and achieve include mothers to generate extra income and improve the socioeconomically conditions.

## 21.4  CONCLUSIONS

The extraction process candelilla wax is an archaic methodology with more than 100 years it was invented. Although there is national and international regulation for exploitation is still used because it represents a reliable source of income for the family. These conditions under which it is performed are totally unsafe and pose a risk to health and the environment. However, after almost 100 years currently he is working on the design and implementation of an eco-friendly method to extract organic candelilla wax which improves working conditions and reduces health risks. This will improve the socioeconomic conditions of the producers and open the market for a new product and byproducts.

## KEYWORDS

- candelilla wax
- community
- regulation
- socioeconomic impact

## REFERENCES

Alvarez-Mitre, F. M., Morales-Rueda, J. A., Dibildox-Alvarado, E., Charó-Alonso, M. A., & Toro-Vazquez, J. F. (2012). Shearing as a variable to engineer the rheology of candelilla wax organogels. *Food Research International, 49*(1), 580–587. doi: http://dx.doi.org/10.1016/j.foodres.2012.08.025

Alvarez-Mitre, F. M., Toro-Vázquez, J. F., & Moscosa-Santillán, M. (2013). Shear rate and cooling modeling for the study of candelilla wax organogels' rheological properties. *Journal of Food Engineering, 119*(3), 611–618. doi: http://dx.doi.org/10.1016/j.jfoodeng.2013.06.009.

Arato, M., Speelman, S., & Van Huylenbroeck, G. (2014). The contribution of non-timber forest products towards sustainable rural development: The case of Candelilla wax from the Chihuahuan Desert in Mexico. *Natural Resources Forum, 38*(2), 141–153. doi: 10.1111/1477–8947.12043.

Canales-Gutiérrez, E., Canales-Martínez, V., & Zamarrón-Rodríguez, E. M. (2006). Candelilla, from the Mexican desert to the world. *Biodiversitas, 69*, 2–5.

Cites. (2016). What is CITES?. Convention on International Trade in Endangered Species of Wild Fauna and Flora. Disponible en: https://cites.org/eng/disc/what.php, *1*(1), 1–2.

Conabio. (2016). National Commission for the Knowledge and Use of Biodiversity. *CONABIO. Disponible en:* http://www.conabio.gob.mx/.

Conafor. (2016). National Forestry Commission. *CONAFOR. Disponible en:* http://www.conafor.gob.mx/web/.

Cruz, V., Rojas, R., Saucedo-Pompa, S., Martínez, D. G., Aguilera-Carbó, A. F., Alvarez, O. B., et al. (2015). Improvement of Shelf Life and Sensory Quality of Pears Using a Specialized Edible Coating. *Journal of Chemistry.*

De León-Zapata, M. A., Sáenz-Galindo, A., Rojas-Molina, R., Rodríguez-Herrera, R., Jasso-Cantú, D., & Aguilar, C. N. (2015). Edible candelilla wax coating with fermented extract of tarbush improves the shelf life and quality of apples. *Food Packaging and Shelf Life, 3*(0), 70–75. doi: http://dx.doi.org/10.1016/j.fpsl.2015.01.001.

DeLeón-Zapata, M. A. (2008). Technological improvements to the process of extraction of candelilla wax. Chemist Pharmaceutical Biologist, *Autonomous University of Coahuila, Saltillo, Coahuila, México.*

DeLeón-Zapata, M. A., Rojas, R., Saucedo-Pompa, S., Ochoa-Reyes, E., Garza, H. D. L., Rodriguez-Herrera, R. (2010). México Patent No. MX/a/2010/002764. Mexican Institute of Industrial Property: Autonomous University of Coahuila.

Dominguez-Lara, H. (2004). La Candelilla su cera y su cultura. "From Paila to the luster of the bolero and the glamor of the footbridges." *Chronicles of the Camino Real, 3*(28), 19–22.

Hagenmaier, R. D. (2000). Evaluation of a polyethylene–candelilla coating for 'Valencia' oranges. *Postharvest Biology and Technology, 19*(2), 147–154. doi: http://dx.doi.org/10.1016/S0925–5214(00)00087–9.

I. C. (2013). Applicable legislation. Institute of the candelilla. Available in: http://www.candelilla.org/?page_id=1321 & lang=es, *1*(1), 1–4.

I. C. (2016). Composed of Candelilla would have social and commercial benefits. Institute of the candelilla. Available in: http://candelilla.org/pdf/es/investigacion_candelilla_multiceras-itcm.pdf, *1*(1), 1–5.

Jan g, A., Bae, W., Hwang, H. S., Lee, H. G., & Lee, S. (2015). Evaluation of canola oil oleogels with candelilla wax as an alternative to shortening in baked goods. *Food Chemistry, 187*, 525–529. doi: http://dx.doi.org/10.1016/j.foodchem.2015.04.110.

López-Martínez, A., Morales-Rueda, J. A., Dibildox-Alvarado, E., Charó-Alonso, M. A., Marangoni, A. G., & Toro-Vazquez, J. F. (2014). Comparing the crystallization and rheological behavior of organogels developed by pure and commercial monoglycerides in vegetable oil. *Food Research International, 64,* 946–957. doi: http://dx.doi.org/10.1016/j.foodres.2014.08.029.

López, M., Carrillo., F., Gutiérrez, M. C., & Crespi, M. (2007). Alternative methods for the wool wax extraction from wool scouring wastes. *Fats and oils, 58*(4), 402–407.

Marshall, E., Schreckenberg, K., & Newton, A. C. (2006). Commercialization of non-timber forest products. Factors influencing success. Lessons learned from Mexico and Bolivia and policy implications for decision-makers. UNEP World Conservation Monitoring Centre, Cambridge, UK.

NOM-018-RECNAT. (1999). Official Mexican Standard. That establishes the procedures, criteria and technical and administrative specifications to realize the sustainable use of the grass of candelilla, transport and storage of the cerote. Official Journal of the Federation, *México., 1*(1), 1–11.

Ochoa, E., Saucedo-Pompa, S., Rojas-Molina, R., Heliodoro de la Garza., Charles-Rodríguez, A. V., & Aguilar, C. N. (2011). Evaluation of a Candelilla Wax-Based Edible Coating to Prolong the Shelf-Life Quality and Safety of Apples. *American Journal of Agricultural and Biological Sciences, 6*(1), 92–98.

PROFEPA. (2016). Federal Attorney for Environmental Protection. Profepa. Available in: http://www.profepa.gob.mx/.

Rocha, J. C. B., Lopes, J. D., Mascarenhas, M. C. N., Arellano, D. B., Guerreiro, L. M. R., & da Cunha, R. L. (2013). Thermal and rheological properties of organogels formed by sugarcane or candelilla wax in soybean oil. *Food Research International, 50*(1), 318–323. doi: http://dx.doi.org/10.1016/j.foodres.2012.10.043.

Rojas-Molina, R., León-Zapata, M. Á. d., Saucedo-Pompa, S., Aguilar-González, M. Á., & Aguilar-González, C. N. (2013). Chemical and structural characterization of Candelilla (*Euphorbia antisyphilitica* Zucc.). *Journal of Medicinal Plants Research, 7*(12), 702–705.

Rojas-Molina, R., Saucedo-Pompa, S., León-Zapata, M. Á. d., Jasso-Cantú, D., & Aguilar-González, C. N. (2011). Essay: Past, present and future of the candelilla. *Mexican Journal of Forestry Sciences, 2*(6), 7–18.

Rojas, R., DeLeón-Zapata, M. A., Ochoa-Reyes, E., Cruz-Antonio, F. V., Jasso-Cantú, D., Garza, H. d. l., et al. (2011). Saltillo, Coahuila Patent No. MX/u/2011/000553. México: I. M. d. l. P. Industrial.

Romahn, V. (1992). Los Non-Timber Forest Resources of Mexico: A Review. Bachelor's degree, National Autonomous University of Mexico, Mexico, D. F.

Saucedo-Pompa, S., Rojas-Molina, R., Aguilera-Carbó, A. F., Saenz-Galindo, A., Garza, H. d. L., Jasso-Cantú, D., et al. (2009). Edible film based on candelilla wax to improve the shelf life and quality of avocado. *Food Research International, 42*(4), 511–515. doi: http://dx.doi.org/10.1016/j.foodres.2009.02.017.

Scora, G. A., Ahmed, M., & Scora, R. W. (1995). Epicuticular hydrocarbons of candelilla (Euphorbia antisiphylitica) from three different geographical areas. *Industrial Crops and Products, 4*(3), 179–184. doi: http://dx.doi.org/10.1016/0926-6690(95)00030-G.

Semarnat. (2016). Directorate General of Wildlife of the Secretariat of Environment and Natural Resources. Ministry of the Environment and Natural Resources. Available in: http://www.gob.mx/semarnat.

Toro-Vazquez, J. F., Charó-Alonso, M. A., Pérez-Martínez, J. D., & Morales-Rueda, J. A. (2011). 6-Candelilla Wax as an Organogelator for Vegetable Oils An Alternative to

Develop Trans-free Products for the Food Industry A2-Marangoni, Alejandro G. In: N. Garti (Ed.), *Edible Oleogels* (pp. 119–148): AOCS Press.

Toro-Vazquez, J. F., Mauricio-Pérez, R., González-Chávez, M. M., Sánchez-Becerril, M., Ornelas-Paz, J. d. J., & Pérez-Martínez, J. D. (2013). Physical properties of organogels and water in oil emulsions structured by mixtures of candelilla wax and monoglycerides. *Food Research International, 54*(2), 1360–1368. doi: http://dx.doi.org/10.1016/j.foodres.2013.09.046.

Zhang, Y., Adams, M. J., Zhang, Z., Vidoni, O., Leuenberger, B. H., & Achkar, J. (2016). Plasticization of carnauba wax with generally recognized as safe (GRAS) additives. *Polymer, 86*, 208–219. doi: http://dx.doi.org/10.1016/j.polymer.2016.01.033.

# CHAPTER 22

# NATIONAL AND INTERNATIONAL CANDELILLA WAX MARKET

SAUL SAUCEDO-POMPA[1] and
GUILLERMO CRISTIAN G. MARTÍNEZ-ÁVILA[2*]

[1]Food Research Department. School of Chemistry, Autonomous University of Coahuila, 25280 Saltillo, Coahuila, Mexico

[2]Laboratory of Chemistry and Biochemistry, School of Agronomy, Autonomous University of Nuevo Leon, 66050 General Escobedo, Nuevo Leon, Mexico, E-mail: guillermo.martinezavl@uanl.edu.mx

## CONTENTS

## ABSTRACT

*Euphorbia antisyphilitica* is a plant used to obtain a material of commercial value: candelilla wax, which is exuded by this plant for protection from extreme environments Chihuahua Desert. This wax has unique properties that make it be a high demand material, these properties make it a material of high demand for applications in both food and drugs, its properties are highly appreciated by the cosmetic industry, is a natural material that has positioned itself between the next-generation materials. Their properties have placed this wax in a very wide range of products that are marketed around the world. Although its extraction presented great obstacles technological, social and environmental, their properties and vegetable origin make that maintains its status among the three natural waxes of greater commercial interests. It is expected that in the Mexican countryside problems that adversely affect the *Euphorbia Antisyphilitica* wax market. The objective of this chapter is to collect information about the factors that affect the market for this wax better known by its trade name like Candelilla wax, the factors that represent an advantage for the market of this wax as their weaknesses are presented, addition to the background that have shaped this market.

## 22.1 INTRODUCTION

The *Euphorbia Antisyphilitica*, better known by its common name as candelilla, is a perennial plant that grows in the Chihuahua desert, is produced by outbreaks of overhead and underground stems, well as by seed. The "candelillera region" (region where candelilla is found) is distributed mainly in the Mexican states: Durango, Zacatecas, Chihuahua, Nuevo Leon, Tamaulipas, San Luis Potosi and Coahuila. Being Coahuila the state where more abundant is candelilla and is the state where there is a greater commercial exploitation as it participates with 80% of the national production. There are 630,500.00 ha with the possibility to be exploited, however currently only take advantage 152,000 ha representing 24% of the total area (CONAFOR, 2015). Currently on the market there are substitutable candelilla wax products but yet, the demand for this material has been maintained with a fairly stable mostly for their physical and chemical properties. Mexico is the largest candelilla wax producer worldwide. The production of this wax is subject to international market demand and to internal factors that limit their use.

## 22.2 COMMERCIAL EXPLOITATION OF CANDELILLA PLANT

The candelilla wax market chain begins in rural communities through the process of collecting the plant, which grows in the wild uplands in extreme weather and low humidity.

To date collection method in the wild territory continues being the most used, even when there are some efforts associated with artificial plantations for commercial purposes (CONAFOR, 2015). The criteria, procedures and technical and administrative specifications that should be taken into consideration for the use of candelilla are established by the NOM-018-SEMARNAT-1999.

At the time the candelilla wax marketing activity has a volume of 200 tons per year, this represents less than 50% of sales of half a century ago, at the same time the number of candelilla wax producers has decreased due to social and economic problems associated with the abandonment of the field and the low financial and technological support to rural activities in Mexico since the late seventies to date. (Castro et al., 2010; Kehoe et al., 2011; Arato et al., 2013)

## 22.3   CANDELILLA WAX PRODUCTION

The University of Texas has conducted a study in relation to price and production of refined candelilla wax since the 30s, and shows that after World War II, higher candelilla wax production volumes were recorder, 11 million pounds were processed per year, 4–5 million kilograms, now produced 2 million kilograms per year. From the 70's the candelilla wax market was stabilized and wax prices have increased is currently more expensive to produce candelilla wax and is considered a high-value product. On the other hand, is said that if there were another similar product, would be questionable the candelilla wax commercialization by the price increase that has taken, however this product has unique properties, although customers are affected by the increase in price, candelilla wax continues to be a competitive product for their qualities (Eighteenth Meeting of the Plants Committee, 2009; Instituto de la Candelilla 2014).

## 22.4   CANDELILLA WAX DESCRIPTION

It is a 100% natural substance, hard, brittle and easy to pulverize, its color can be from light brown to yellow, depending on the refining and bleaching degree. Its surface can reach high brightness levels, which is one of the most valued properties for various specialty applications. Candelilla wax is recognized by the Food and Drug Administration of the United States (FDA), as a safe natural substance – GRAS (Generally Recognized as Safe) for industry application (Arato et al., 2013). México is the main candelilla wax producer in the world, it is almost the only producer of this material and production is subject to levels of international market demand.

## 22.5   NATURAL WAXES THAT COMPETE WITH CANDELILLA WAX ON THE MARKET

The natural waxes market is dominated by candelilla wax and other two materials: beeswax and carnauba wax. Beeswax is the produced in greater proportion by the facilities which has the production process of this wax.

## 22.5.1 BEESWAX

Wax is the material that bees use to build their hives. It is produced by young honeybees that secrete it as a liquid through their wax glands. The wax is produced by all species of honeybees, although waxes produced by different species of bees have slightly different chemical and physical properties.

Beeswax has many traditional uses. In some countries of Asia and Africa, is used to create Batik fabrics shall and manufacturing small metal ornaments by the method of molten wax. Beeswax is widely used as a waterproofing agent for wood and leather and for reinforcing yarns. It is used in village industries such as candle and as an ingredient in ointments, medicines, soaps and polishes. It has a great demand in the world market. There are over 300 industries that use it. The cosmetics and pharmaceutical industries are the main consumers, representing 70% of the global market and use wax class that cannot be overheated. Its price ranges between 4 and 8 USD per kg (FAO, 2015).

### 22.5.1.1 Beeswax World Production

Beeswax world production of is around 11,500 to 19,000 tons, the main importers are the United States (consume about 30% of the existing wax on the international market) followed by Germany, UK, Japan and France. The biggest exporters are: Chile, Tanzania, Brazil, Holland and Australia. (FAO, 2015).

## 22.5.2 CARNAUBA WAX

Carnauba wax comes from the carnauba palm, a Brazilian tree formally named *Copernicia prunifera*. In hot, dry weather, the plant secretes wax to protect the leaves from damage.

Carnauba wax is used for a wide array of products mostly falling into the categories of: Cosmetics, food products and polishes. Some of the products are: Candies/sweets, chewing gums, chocolates, confectionary sugar, fruit coating, polishing wax (for car, leather, floor, furniture), food packing, can coating, plastic film, matches, medicine/capsules, graphite pencils, paints, cosmetics, bullets coating, bar codes, dry batteries, computer chips, printing ink, carbon paper, toner, dehydrated vegetables, modeling flowers and fruits, dental wax, textile coatings, lubricants, skin care, hair care, shave creams (Portal da carnauba, 2015).

### 22.5.2.1    Carnauba Wax World Production

Carnauba wax world production is estimated at 22,409 tons per year of carnauba wax, Price is from $ 2.5 to $ 4. The largest importers of carnauba wax are USA (around 25%, Germany (10–15%), Japan (15–25%), Holland (around 5%) and Italy (around 5%). Brazil is the largest producer worldwide carnauba wax (Portal da carnauba, 2015).

## 22.6    MEXICAN COUNTRYSIDE BACKGROUND THAT THREATEN CANDELILLA WAX MARKET STABILITY

In the past century emerged in the Mexican countryside a movement for its activation, which began with a land distribution, subsequently was subjected to a modernization period, Currently there are tendencies to eradicate poverty and backwardness that currently affects people in rural areas after economic problems presented the past 40 years, the biggest challenge that have confronted the Mexican authorities is to balance development opportunities and growth compared to urban areas where other sectors of the economy provide better conditions and income sources. This has caused an increasing abandonment of rural areas. By 2014 approximately 48% of Mexico's population lives in poverty (CONEVAL, 2014), of this percentage the majority corresponds to rural areas where shortcomings in basic development occurs, this has caused that in recent years these economic sectors lose strength compared to other sectors of the population. This has generated the abandonment of the population in rural areas, the technological backwardness and lack of support have delayed it forestry activities in which certain species are exploited or example: lechuguilla for obtaining *ixtle*, maguey for the production of honey water and candelilla to produce wax, among many other species with potential for commercial exploitation. This kind of trends endangers this type of economic activities that hinders the use candelilla. (Arias et al., 2010; Cutler et al., 2002; Castro and Galindo, 2010; Kehoe et al., 2012; Arato et al., 2013).

Currently there are a great number of projects for rural development who have focused on boosting the market that develops in these areas based on candelilla wax and other forest products in communities Chihuahuan desert (CONAFOR, 2015).

## 22.7 CANDELILLA MARKET LIMITATIONS

There are limitations that surround the use of candelilla plant, wax of this species is a material for which the extraction process was implemented almost a century ago the extraction process has hardly changed since that time (Rojas-Molina, 2007.) so it is to be expected that the process present problems technological and social derivatives of the technology used which is very aggressive with the environment and its handlers, the candelilleros (people that extract candelilla wax) mostly non have the training or the right conditions to carry out a process of this type, so is to be expected damage to health and the environment by not taking into account the necessary steps for disposal of this type of wastes, as well as by ignorance in the treatment of this kind of waste and the lack of ecological monitoring by the authorities in these areas.

The exploitation of candelilla wax remained neglected by the state and federal authorities for decades, however currently a campaign emerged from the federal government in conjunction with some research centers and universities to improve the extraction process as social fields, have generated scientific and technological advances that allow use of candelilla plant in a more friendly way the environment and candelilleros. However the possibilities of using the candelilla are not limited to the use of the wax extracted, there are other possible applications of the plant as is the case of the use of molecules presents in candelilla with commercial interest, present in the candelilla stems of that are waste from the candelilla wax extraction process, these special molecules have antioxidant and antibacterial capacity (Burboa, 2014), which gives rise to the commercial use of a greater number of products from this plant new research in the environment of use of candelilla in extraction technologies in chemical composition and give rise to an improvement in future in marketing to give greater acceptance in socially responsible markets.

## 22.8 CANDELILLA MARKET ECOLOGICAL LIMITATIONS

One of the problems in the extraction process of candelilla wax is the use of sulfuric acid for removal, because they have generated pollution conditions caused by the deposition of sulfuric acid, affecting the flora in different ways such as affections in the flora as negatively affects plant development, besides representing a toxic threat to terrestrial and aquatic fauna.

Another limitation is the candelilla plant provision, because this is one of the critical points that mostly affect the candelilla wax market, because it is a limitation to accomplish the extraction of candelilla wax and therefore the commercializing, undoubtedly the biggest limitation in the market chain is found in the step; candelilla plant collection, this by all the regulations that surround the use of candelilla, because this plant is a protected species, is one of non-timber resources mostly used by forestry in the Chihuahuan Desert. Candelilla plant takes around seven years to reach its optimum point for the extraction of wax and the other side is the extraction yield is generally 10% so it needs to collect large volumes of candelilla plant for obtaining wax.

These factors have created the need for regulation of the plant, there is currently an effort by the federal government to achieve conserve species *Euphorbia Antisyphilitica* well as to preserve the ecological environment of the areas where it performs the extraction of wax, and to also ensure the sustainability of families who currently depend economically on this extraction process.

## 22.9　IMPACT OF TRADE CANDELILLA WAX IN THE CANDELILLERA REGION

In the candelillera region there is a great biodiversity of species of non-timber forest products, which are used by the residents of these areas of seasonally according to the availability of candelilla within the more stable activities throughout the year is found the processing of Candelilla plant (*Euphorbia Antisyphilitica*), which due to their market opportunities and availability in the field (which varies according to the rainy season) allows it to be utilized along the year, currently representing a source of employment for about 3,000 families who generate monthly income from the extraction of wax and allows them to have the flexibility to combine their activity with complementary sources of employment, which fosters a great opportunity to promote this activity as a means for economic development that generates benefits in communities today do not have stable employment sources and reduce poverty situations (Arato et al., 2013; CONABIO, 2015).

## 22.10 SOCIAL ISSUES SURROUNDING CANDELILLA MARKET

Currently the candelilla wax extraction is carried out in several rural communities in the Mexico northern states, most of them are difficult to access. There is currently an effort by Mexican authorities to avoid the use of sulfuric acid in the extraction of candelilla wax, currently the candelilleros as they are provided by refiners candelilla wax sulfuric acid, however the suppliers of this acid to candelilleros in the majority of cases not perform surveillance on the management of waste extraction process, wax extractors, due to their lack of resources and expertise to waste management the spread in areas near the places where they perform their extraction, which has caused environmental deterioration in wax extraction areas.

In addition, there is also a health problem because the wax removal is mainly effected by the traditional method that involves the use of sulfuric acid that threatens the health of producers because it generates toxic waste, thus causing damage inhalation or by direct contact. This situation has demanded the attention to this sector presented by the Mexican authorities to get better working conditions for producers and simultaneously obtaining a higher quality wax and market value that generates less damage to the health of producers and the environment.

## 22.11 NATIONAL AND INTERNATIONAL ENVIRONMENTAL REGULATIONS TO CANDELILLA

The Secretariat of Environment and Natural Resources (SEMARNAT) is the federal agency that through the General Law of Sustainable Forestry Development regulates and promotes sustainable use of natural resources. Within the same instrument tools are developed such as NOM-018-SEMARNAT-1999 which provides recommendations for the proper candelilla use in Mexico (PROFEPA, 2015). Within SEMARNAT there are various institutions like CONAFOR and PROFEPA collaborating each in its area of expertise and discipline to ensure the proper performance of the different actors of the process.

According to national regulations for the proper use of Candelilla, it is necessary to process a logging permit which is given after a general evaluation prepared by a licensed forester, who evaluates the status of the biomass and environmental impact at each site and defines the amount of resource that can be used as well as the period of use and recommendations for preservation (SEMARNT, 2015).

## 22.12   CITES MONITORING IN CANDELILLA WAX MARKET

CITES (Convention on International Trade in Endangered Species of Wild Fauna and Flora) is an international agreement between governments, it aims to ensure that international trade in specimens of wild animals and plants and ensure that their exploitation is not a threat to their survival (CITES, 2015). CITES regulates the export, re-export and import of plants and live animals or dead and their parts and derivatives of protected species, this regulation is based on a system of permits and certificates can only be issued if certain conditions are met, mainly the legal origin of the specimens and a Finding of No Damage Extraction (NDF) in which it is secured by studies in populations of the species that commercial exploitation is not a threat to the species (CONAFOR, 2015)

### 22.12.1   CITES DIVIDE THE SELECTED SPECIES INTO THREE APPENDICES (CITES, 2015)

#### 22.12.1.1   Appendix 1

Includes species in higher degree of endangered. Prohibits international trade in specimens of these species, except where the importation is made for non-commercial purposes.

#### 22.12.1.2   Appendix 2

Includes species that are not in endangered, but may become so, unless the trade to controlled. International trade in specimens of Appendix 2 may be authorized to grant an export license or re-export certificate.

#### 22.12.1.3   Appendix 3

Includes species for which a country requests that all Parties to cooperate in their protection. International trade is permitted but is controlled (certificate of origin).

*Euphorbia Antisyphilitica* (Candelilla) is in Appendix 2 because is located in monitoring to prevent that becomes endangered, the candelilla wax marketing is monitored and products that present in its content candelilla wax.

## 22.13   BARRIERS OF CITES IN CANDELILLA WAX MARKET

Formerly a barrier to candelilla wax trade was CITES monitoring in products that contain candelilla wax, because there was a regulation which complicated the exports of end-use products with candelilla wax in is composition. In 2008, a problem was presented in products containing candelilla wax in their formulation, because by the presence of this wax in the products caused his seizure by the regulatory authorities, this genre great damage to the cosmetics industry that used candelilla wax, which caused that many Asian and European companies replace candelilla wax by other materials in their products, this caused a drop in wax marketing of 20% (CONAFOR, 2015).

To this day are no longer necessary CITES permits to market internationally finished product and packaged that contain Candelilla wax, as cosmetics, paints, lubricants and food. However, candelilla wax crude commerce will remain regulated by the CITES. Export licenses are emitted by the department of wildlife of the Secretary of Environment and Natural Resources (SEMARNAT). The CITES monitoring is done to ensure sustainable use and that not endanger the species (CONABIO, 2015).

## 22.14   MAIN CANDELILLA WAX PROCESSING COMPANIES IN MEXICO

Companies actually have not problems on customer diversification, however these customers are mainly international wholesalers. To make these sales don't have contracts, as orders, answering and acceptance of the order acts as a contract. This product is not seasonal and sold throughout the year (CONAFOR, 2015). On average companies export 90% of its production and intended for the following countries: Germany, France, UK, Japan, USA, Belgium, Argentina, South America, Colombia, Venezuela, India, Egypt, South Africa, Australia, the Netherlands, Belgium, Ireland, Italy and

China, among other countries (CITES, 2015). The two main companies that have exported candelilla wax, are PRONAMEX and Multiceras and Ceras Coa as CENAMEX has had a low; Multiceras being the company that has specialized in some other resources (Hernandez-Garcia, 2013).

## 22.15   CENAMEX

Until 1992 the marketing of candelilla wax in Mexico was conducted by the National Bank of rural credit through a trust, however posterior this year the Mexican authorities disappeared Escrow and transferred its functions to a Mexican company called Ceras Naturales S.A. de C.V. of which shareholders were the 6000 candelilleros of Mexico, represented by 300 groups, for a period of two years Cenamex was the only company responsible for the marketing of wax, in 1994 after the signing of the FTA of North America (Free Trade Agreement), gave rise to the opening of new national and international companies, which forced Cenamex to be more competitive with quality product, however the company presented a technological backwardness that represent in a competitive disadvantage to new companies to date is Cenamex is closed due to poor management and organization (Rojas et al, 2011).

At the beginning of the 21st century, in 2006, Mexico produced 349 tons of natural waxes worth over $7,677,582. For 2007, the main consumers were Spain, Italy, Germany and China with 85.38% of the total of candelilla wax; the following year it was exported to Japan, United States and Germany. Nowadays the activity of more than 3,500 small producers moves this product around. (Rojas et al, 2011).

## 22.16   MULTICERAS

Multiceras S.A. de C.V. is a Mexican company dedicated to designing, manufacturing and marketing, synthetic and petroleum-related products for industrial use. Being a company created for the export of beeswax and Candelilla wax, currently manages a team of over 60 different products portfolio, which are marketed in more than 20 market sectors, both in Mexico as well as abroad. The factory and corporate offices Multiceras are located in the municipality of Garcia, Nuevo Leon at 20 km. of Monterrey city. The factory has a production capacity of 300 ton/

day and has modern equipment and facilities for processing products of high quality wax (Multiceras, 2015)

## 22.17 PRONAMEX

The company began operations under this name in 2002 however one of the partners had another company since 1995 with the same line. Only have eight clients, which are located 90% abroad, these industries to that wax is destined, are able to give other processes that have very specific applications.

## 22.18 PRODUCCION DE CERA S.A. DE C.V.

Produccion de Cera S.A. de C.V. began operations in 1995 dedicated to the production and direct marketing of candelilla wax refined. This company is dedicated to collecting candelilla plant, produce, refine and market candelilla wax; offices and production plant are located in Cuatro Cienegas, Coahuila. Other companies dedicated to refining wax are Ceras Coahuiltecas S.A. de C.V. Company located in Cuatro Cienegas Coahuila, company with just over 100 candelilla wax producers of the municipality and also in Cuatro Cienegas another company is located with the name of Ceras Naturales del Norte S.A. de C.V. dedicated to the candelilla wax extraction and refining.

## 22.19 MARKET PRICES OF CRUDE CEROTE

Until 2009 a price of 24 pesos was registered in the cerote kilogram, which is the price at which they are buying for the candelilleros such material. However in November 2014 in a press release the National Peasant Confederation (CNC) announced that authorize an increase from the federal government to 42 pesos per kilogram of cerote, with this action the federal government seeks to eliminate the "coyote" they are people dedicated to sale of cerote candelilla on informality resulting in abuse by those informal traders. Also reported in the same statement which currently has a monthly production of 120 tons of cerote and look create an organization similar to the background candelillero that formerly bought the cerote late last century (CNC, 2014).

However certain factors that cause fluctuations in the cost of present cerote because the price is moving depending on certain factors, be the main supply of cerote and demand from refiners wax, if the landowners are very focused on other crops such as oregano or ixtle, on the other hand also the demand that have the refiners causes changes in the price of cerote; if demand increases price of the raw material consequently increases, conversely when refining companies have inventories and in turn have no product demand cerote price tends to go down, the pricing has a dynamic of supply and demand, although it is important to mention that is not directly related to the price increase of the already refined wax, as it has tended indistinctly to increase (CITES, 2009).

## 22.20 NATIONAL CANDELILLA WAX PRICES

After refined candelilla wax increases its commercial value because with this process is given a higher value with new attributes, for example Multiceras SA de C.V. sells candelilla wax pills at a price of 6.78 USD and Produccion de Cera S.A. de C.V. at a price of 5.59 USD, these values are higher than those of cerote because it is a product which has been subjected to a refining process and given the characteristics necessary to compete in the market with other waxes in different areas of industry.

## 22.21 CANDELILLA WAX WORLD MARKET

At the global level the Strahl and Pitsch is the largest refiner of crude candelilla wax. They manage in their inventories very considerable amounts of this wax, in addition to mixtures which are the product of the demands of its customers, the sale of this wax is handled in two ways marketing International accounting for about 80% of marketing and domestic sales with 20%, the main countries where exports are the United States, Japan, Germany and France in order of importance.

## 22.22 CANDELILLA WAX COSTS ON THE WORLD MARKET

In the international market, candelilla wax is processed to highlight attributes sought for very specific markets, even is subjected to different certification processes to achieve considerably enhance its commercial value below a series of candelilla wax marketers that enhance their attributes and have higher market values are shown. Consultation of candelilla wax prices with dealers from different countries was conducted, mainly study the US market because it is the main candelilla wax consumer (Table 22.1).

This kind of candelilla wax sellers in different countries (Table 22.1), perform a series of marketing strategies which are reflected in the value of the candelilla wax production chain. One of the main strategies are the certifications, because certification is the process that ensures quality in some attribute of a product, verifying compliance with applicable regulations, according to the market destination of the product for large distributors, is a marketing tool, for those who purchase is a guarantee of confidence. Among the most outstanding candelilla wax certifications are Kosher certified and Vegecert, which we describe in the next section.

## 22.23 FEATURED CANDELILLA WAX CERTIFICATIONS APPLIED

Between the most interesting certifications that are found in the candelilla wax market are found Kosher and Vegecert, as they are directed to very specific markets. They are cited as an example of the large number of certifications that exist globally.

### 22.23.1 KOSHER CERTIFICATION

The kosher terminology means apt-appropriate. Are all those foods corresponding to the biblical and Talmudic rules of Jewish law. When a product or establishment is certified as Kosher, the buyers know it complies with a strict policy of the kosher foods laws, including cleanliness, purity and quality, Kosher means more than just food preparation responsible, refers to a set of intricate biblical laws that detail the types of food that a Jewish person can eat and the ways they can be prepared. For the Kosher certification, all the ingredients of each product and its manufacturing process must also be certified for compliance with kosher.

**TABLE 22.1** Candelilla Wax Seller in the International Market, Description of Their Product and Cost in US Dollars

| Candelilla wax seller | Location | Description | Kg USD |
|---|---|---|---|
| Making Cosmetics | Snoqualmie Washington USA | Natural vegetable wax extracted from the candelilla plant (*Euphorbia Antisyphilitica*) from Northeastern Mexico. Yellow prills, oderless. Soluble in alcohol, insoluble in water. Saponification value 43–65. Melting point 69–73°C (156–163°F). | **$17.38** |
| The Plant Guru | Plainfield, New Jersey, USA | Vegan alternative to beeswax, 100% natural vegetable | **$28.86** |
| Wax Bram Bleberry | Bellingham, Washington, USA | This unique vegetable wax helps provide gloss and lubricity ("slip") to lip balms and lotion bars. It's also excellent in furniture polish. Candelilla is obtained from the coating of the "wax slipper plant" (*Euphorbia Cerifera*) found mainly in Mexico. The wax is extracted by boiling the plant (to separate the wax and the plant material). The wax floats to the top of the water and is skimmed off and processed. This wax comes in granules in FDA food grade approved containers | **$13.8** |
| Aussie Soap Supplies | Kardinya, Australia | This unique vegetable wax helps provide gloss and lubricity (*slip*) to lip balms and lotion bars. It is also excellent in furniture polish. If substituting Candelilla Wax for Beeswax in an existing recipe, reduce the amount of wax by half since as Candelilla Wax has twice the stiffening power of Beeswax. | **$23.91** |
| Aroma Zone | Paris, France | Its properties are valued for achieving lipsticks which facilitates the release, and for the holding and drying mascaras. The Candelilla wax is also known for its excellent film-forming power. Very hard wax, interesting to harden sticks and lipsticks, facilitate their release and heat resistance. Adhesion agent facilitates the holding of mascaras and eyeliners. Excellent film forming. Thickener and stabilizer in emulsions | **$62.36** |
| Akoma Skincare | Derby, UK | Candellia wax is a great alternative to beeswax for your balm and body butter products. Try it yourself and see the great results that can achieved. Common Uses: lip balms, Soap Making. Extraction: Hand Collected/Hot Water Separated. | **$31.37** |

The audit takes place in the production plant by the supervisor of kosher certification to ensure that certain production lots meet the quality standards and specifications required (Kosher, 2015).

## 22.23.2   VEGECERT

VegeCert is a non-profit organization that certifies vegan and vegetarian food products. Vegans and Vegetarians represent approximately 5% of American and Canadian consumers, and VegeCert helps this key demographic make informed decisions about the food that they eat. VegeCert is unique in that it makes physical inspections of the certified food manufacturing facilities to ensure that they conform to VegeCert's high standards (VegeCert, 2015). With this type of certification marketers of natural waxes potentialize the candelilla wax value achieved in the international market.

## 22.24   CANDELILLA WAX VALUE CHAIN

### 22.24.1   SUBDIVISIONS OF VALUE CHAIN (FIGURE 22.1):

**Level 1:** Candelilla collectors and extraction of cerote. Participate directly residents of rural communities on the candelilla plant collection to later perform the cerote extraction, these actors called candelilleros sell the cerote to different refining companies. It is the first step of the candelilla wax market.

**Level 2:** Nationals wax refining companies and wax distributors, participate refiners who buy directly or indirectly cerote to candelilleros, at this level cerote is subjected to purification processes, including mixed with other waxes or different materials to give specific properties.

**Level 3:** Marketing companies international candelilla wax, participating natural waxes distributors, being between 85% and 90% produced candelilla wax earmarked for this level.

In the next picture we see a candelilla wax value chain diagram in the market, in which can be seen we seen graphically the three different levels and trends in national and international candelilla wax markets (Figure 22.1).

At different levels of the market value chain for candelilla wax we can distinguish the different actors in this market, there is an increase in the cost of candelilla wax, this is due to different processes apply to cerote who is the starting material such as refining, whitening, processing presentation, certifications and special features required by buyers, even a set of techniques, strategies and marketing applications are applied to enhance the attributes of candelilla wax, when adding new features to the candelilla wax an increase in the market value of candelilla wax is observed, in the following graph we can see the price behavior of the candelilla in the market, we can clearly see that when adding special attributes its commercial value is increased, being the international market where the highest prices of this wax is reached, in this market is largely involve marketing techniques, for example candelilla wax vegetable origin is emphasized making it suitable for a growing market for vegans, vegan certifications are applied in candelilla wax, thus ensuring

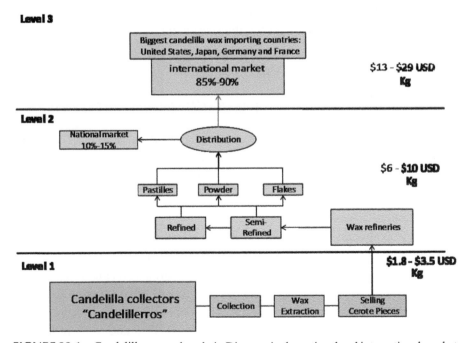

**FIGURE 22.1**    Candelilla wax value chain Diagram in the national and international market.

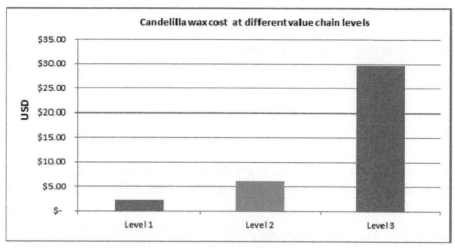

**FIGURE 22.2** Candelilla wax cost at different value chain levels.

its attributes and quality opening the possibility of its application in vegan products, because a producer of this kind of products requires this certification to certify their finished products and place them on a vegan market.

## 22.25 RISK FACTORS IN THE MARKET CANDELILLA WAX

It is very important to establish that there are some risks in the market candelilla wax such as:

### 22.25.1 THREATENED SPECIES

If not done an appropriate management of *Euphorbia Antisyphilitica* on their holdings,

It could get to be as a species in threat of extinction, which would generate prohibition of exploitation by regulatory authorities; therefore this would cause the disappearance of the candelilla market.

### 22.25.2 TECHNOLOGICAL DELAYS

Inadequate equipment, development of other similar products with lower cost, improper operation, can endanger the human health.

### 22.25.3   POLITICAL

There are several disadvantages and in obtaining governmental permissions and licenses for proper operation processes, which permits get adequate conditions for candelilla exploitation and commercialization.

### 22.5.4   REGULATIONS

Lack of rules in natural waxes, at present not having certifications products on the market is a competitive advantage over other materials.

### 22.25.5   CLIMATIC FACTORS

Weather conditions, droughts and floods can affect your ability to obtain candelilla and the subsequent extraction candelilla wax, affecting the availability on the market.

## 22.26   CONCLUSIONS

The exploitation of *Euphorbia Antisyphilitica* represents a sustainable and affordable activity, candelilla wax prices are directly linked to the wax quality and marketing techniques, were found selling prices 10 times more than the cost of the initial product (cerote) these marketing costs are possible due to the added candelilla wax attributes and to the properties presented in Candelilla wax, which are unique and unmatched, so much so that even with all the obstacles presented by the candelilla wax market this persists today. Current regulations on the use of candelilla, far from presenting a disadvantage, ensure the integrity of the species and therefore also ensure candelilla wax market. Currently on the market found principally candelilla wax as derived from this plant, however there is also the possibility of integrally exploiting candelilla plant taking advantage not only but also other derivatives such as the antioxidant compounds of commercial interest, which are in the stems of the plant and even the fibers which are extraction residue that are currently used as fuel in traditional candelilla wax extraction.

## KEYWORDS

- **candelilla wax market**
- **Chihuahuan Desert**
- *Euphorbia antisyphilitica*

## REFERENCES

Akoma Skin Care. (2015). Consulted January 2015 to Akoma Skin Care. http://www.akom-askincare.co.uk/

Arato, A., Speelman, S., & VanHuylenbroeck, G. (2013). Integration of private investment for development of the value chain of non-timber Mexican forest products. Conference of Agricultural Trade and Latin America, September 19-20, Buenos Aires, Argentina.

Arias, J., Azuara, O., Bernal, P., Heckman, J. J., & Villarreal, C. (2010). Policies to promote growth and economic efficiency in Mexico. National Bureau of Economic Research. (No. w16554).

Aroma Zone. (2015). Consulted January 2015 to Aroma Zone. http://www.aroma-zone.com/

Bazdresch C., & Levy S. (1991) Populism and Economic Policy in Mexico, 1970–1982. In The Macroeconomics of Populism in Latin America (Pp. 223–262). University of Chicago Press.

Bazdresch C., & Levy S. (1991). Populism and Economic Policy inMexico, 1970–1982.In The Macroeconomics of Populism in LatinAmerica (Pp. 223–262).

Brambleberry. (2015). Consulted January 2015 to Brambleberry http://www.brambleberry.com/

Burboa, Edgardo A., Ascacio-Valdés, Juan A., Zugasti-Cruz, Alejandro., Rodríguez-Herrera, Raúl., & Aguilar, Cristóbal N. (2014). Antioxidant and antibacterial capacities of candelilla waste extracts. *Revista Mexicana de Ciencias Farmacéuticas, 45*, 51–56.

Castro, G. A., & Galindo, J. O. (2010) Two recent economic crises in Mexico: causes and perspectives. *Asian Journal of Latin American Studies-IPN Vol, 23* No. 3.

Cites, (2015). Consulted Enero 2015 in Convention on International Trade in Endangered Species of Wild Flora and Fauna. www.cites.org.

CNC. (2014). Consulted November 2014 in National Confederation of Peasants. www.cnc.org.mx

Conabio. (2015). Consulted January 2015 to National Commission for the Knowledge and Use of Biodiversity. http://www.conabio.gob.mx/

Conafor. (2015). Consulted January 2015 to National Forestry Commission, http://www.conafor.gob.mx.

Coneval. (2015). Consulted January 2015 to National Council for the Evaluation of Social Development Policy. http://www.coneval.gob.mx/

Coria-Rivas, C. (2011) CONAFOR promotes candelilla wax production. Electronic Journalism. Consulted January (2013). http://www.periodismoelectronico.com/informacion-historicanotas-chihuahua/2301-impulsa-conafor-produccion-de-cera-de-candelilla.html.

Cutler, D. M., Knaul, F., Lozano, R., Méndez, O., & Zurita, B. (2002) Financial Crisis health outcomes and aging: Mexico in the 1980s and 1990s. *Journal of Public Economics,* *84*(2) 279–303.

FAO. (2015). Consulted January 2015 to in Food and Agriculture Organization of the United Nations. www.fao.org.

FIRA. (2011). FIRA en el Sector Forestal, Foro Temático de Candelilla, Chihuahua, Chih.

Institute of the Candelilla. (2014). Consulted January 2015 in Institute of the Candelilla. www.candelilla.org.

Kehoe, T. J., & Meza, F. (2012). Catch-up growth followed by stagnation: Mexico, 1950–2010. National Bureau of Economic Research Working Paper No. 17.

Kosher. (2015). Consulted Enero (2015). Kosher certification and compliance. www.ok.org

Madrigal Peña Carlos E. 2007.Thesis, Extraction of cane wax with ethanol from sludge of the sugar industry.

Making Cosmetics. (2015). Consulted on Juanary 2015 to Making Cosmetics http://www.makingcosmetics.com/

Miguel Arato, M., Speelman, S., & Van Huylenbroeck, G. (2013). Integración de la inversión privada en el desarrollo de la cadena de valor de los productos forestales no-maderables Mexicanos. Opportunity for the sustainable rural development of the communities of the Chihuahua Desert. Conference of Agricultural Trade and Latin America, September 19-20, Buenos Aires, Argentina.

Multiceras. (2014). Consulted November 2014 to Multiceras. http://www.multiceras.com.

Portal da Carnauba. (2015). Consulted January 2015 in Portal da carnauba. http://www.portaldacarnauba.org.br/

Prancacio Araújo, G., & Maria Alcobaça, M. (2008). Eco-efficiency in the Carnauva wax production in Campo Maior city, Piauí. Eco-eficiência na produção de cera de Carnaúba no município de Campo Maior, Piauí. *Journal of Economics and Rural Sociology*, 46,2.

Profepa. (2015). Consulted January 2015 in Federal Attorney for Environmental Protection. www.profepa.gob.mx

Sanchez Zavala, A. (1998). Half a century of agrarian policy seen through ten presidents, Bulletin of the General Agrarian Archive, no. 03,CIESAN-RAN, México, May-June, pp. 32–42.

Schneider, E. (2009) "Trade survey study on succulent *Euphorbia* species protected by CITES and used in cosmetic, food and medicine, with special focus on Candelilla wax." 18th meeting of the Plants Committee Commissioned by Bundesamtfür Naturschutz, CITES, Germany.

The Herbarie. (2015). Consulted January 2015 to The Herbarie. http://www.theherbarie.com/

The Plant Guru. (2015). Consulted January 2015 to The Plant Guru. http://www.theplantguru.com/

Vegecert. (2015). Consulted Enero (2015). Consulted January 2015 in Vegerecrt (Vegan & Vegetarian). http://www.vegecert.com/.

# CHAPTER 23

# DETERMINATION OF PHYSICAL AND CHEMICAL PROPERTIES OF NATURAL WAXES

M. L. RAMIREZ,[1] R. L. PERALTA,[1] G. A. SAENZ,*[1] L. L. LÓPEZ,[1]
B. L. BARAJAS,[1] S. L. CANTU,[1] M. P. GONZÁLEZ,[2] and G. J. A. VALDEZ[1]

[1]Department of Organic Chemistry, School of Chemistry, Autonomous University of Coahuila, 25280, Saltillo Coahuila, Mexico

[2]Department of Advanced Materials, Research Center of Applied Chemistry. C.P. 25294, Saltillo Coahuila, Mexico, E-mail: aidesaenz@ uadec.edu.mx

## CONTENTS

## ABSTRACT

This chapter describes the different physical and chemical properties of natural waxes, which are mostly formed of carbon, hydrogen and oxygen, have a highly hydrophobic nature. Also exposed some of its chemical and physical determinations by this natural materials. Specifically Candelilla wax, regional natural product of the state of Coahuila. Also are presents some uses and applications.

## 23.1   INTRODUCTION

The natural waxes are organic materials interesting, chemically formed by carbon, hydrogen and oxygen, forming functional groups like: grass acid, carboxilc acid, esters, ethers, alcohols. The natural waxes, show high molecular weight. Actually the chemical structure of the wax has not been found.However, thanks to the presence of carbon in its structure present highly hydrophobic natural, repulsion to water, this is interesting can form barriers against moisture.This is an important characteristic for their use and applications as food, cosmetics, electronics, home, school, mechanical, etc. There are different waxes like: candelilla, carnauba, bees waxes, etc., presenting similar chemical characteristic, usually they are not toxic to humans.

Specifically candelilla wax is our interest, the Candelilla plant is a small shrub and branched, that grows where it rains very little and temperatures are extreme, it has a pale green with a few leaves, the whole plant is covered with a waxy coating which prevents desiccation of their tissues. Candelilla wax grows in arid areas of Mexico, mainly in the desert of Chihuahua, located in the states of Durango, Nuevo Leon, Zacatecas, San Luis Potosi and Coahuila; the latter being the largest producer of Candelilla waxes. Today Candelilla wax has various applications such as cosmetics, polishes, lubricants, adhesives and skin protectants; chewing gum base, electrical insulators, etc. The finality of chapter is present different determinations of physical and chemical properties of natural waxes and some important applications.

## 23.2   CHARACTERISTICS OF NATURAL WAXES

Usually natural waxes are yellow color, waxy appearance, with low melting point and low solubility to organic solvents, in the Figure 23.1, is show an image of natural wax.

A characteristic of the Natural Waxes is its molecular weights. Physically are solid highly insoluble in aqueous, are solid at room temperature and flexible (Bosquez Molina, et al., 2008). Present important applications. This

**FIGURE 23.1**    Imagen of wax natural candelilla wax.

type of wax is so innocuous that one of your applications is as coatings fruit, (Saucedo et al., 2009).

Waxes are effective in blocking the migration of moisture, being candelilla wax the most resistant one in comparison with carnauba wax and beeswax, due to, highly chemical composition that prevents hydrophobic interactions with moisture, (Saucedo-Pompa et al., 2009), (De Léon Zapata et al., 2015).

There are many other applications and uses where Candelilla wax is currently used, like: cardboard coatings, manufacturing of crayons, paints, inks, candles, lubricants, adhesives, paper, waterproofing, corrosión and fireworks, (Cabello et al., 2013). Candelilla *E. antisyphilitica* Zucc. is an endemic specie of the semiarid regions of the borderline between Mexico and USA; from this plant is obtained a wax, which is a natural compound by carbon, hydrogen and oxygen principally generally recognized as safe (GRAS) by the Food and Drug Administration (FDA). In the Table 23.1, show some of the properties of the Candelilla wax, (Elzibieta and Hans Brand).

The typical composition is show in the Table 2, highlighting that mostly is constituted by hydrocarbons, (Instituto de la Candelilla), (Rojas et al. (2011)

**TABLE 23.1** Properties of the Candelilla Wax

| Properties | Value |
|---|---|
| Melting point, °C | 66–71 |
| Specific gravity 25°C | 0.9782–0.983 |
| Acid value, mg koh/g | 11–19 |
| Ester value, mg koh/g | 40–47 |
| Unsaponifiable, % | 47–62 |
| Iodine number, gi/100 g | 19–44 |

## 23.3 CHEMICAL DETERMINATIONS OF NATURAL WAXES

The chemical parameters, they are based on their chemical composition, exist different method for determiner their chemical. However so far has not been established exactly your chemical composition this due to their chemical complexity and high weigh molecular. However, some method used for determinate the chemical of wax as: spectroscopic of Infrared FT-IR, nuclear magnetic resonance (NMR) and solubility in organic solvent. The FT-IR This method is very important for identification of functional groups, it is used in organic compounds, usually it is considered a qualitative method.

The NMR, is a very specific method for identifying organic structure, bit is used for natural compounds their complexity structural. The solubility in organics solvents is a important factor for their uses, generally the waxes are poorly soluble in organic solvents, but frequently use temperature.

### 23.3.1 PHYSIC DETERMINATIONS OF NATURAL WAXES

Initially, the physical parameters as aspect, color, and melting point are determined in natural waxes and proved information about the general aspect of the waxes. Others physical characteristics of natural waxes have been determined by diverse methods and depend the information that is required in the analysis. For example, the hydrophobicity and water-barrier properties in natural waxes and specifically films waxes could be proportionated by using contact angle and water vapor permeability test. For the morphological characteristics of natural waxes the scanning electron microscopy (SEM) can be used. For instance, the physical determination in

**TABLE 23.2**   Typical Composition of the Candelilla Wax

| (Wt %) | Crude | Refined |
|---|---|---|
| Hydrocarbons | 46 | 57 |
| Free Alcohol | 13 | 14 |
| Free Acid | 7 | 7 |
| Esters | 2 | 21 |
| Hydroxylated Esters | 8 | 8 |
| Acid Esters | 10 | 0 |
| Diesters | 9 | 0 |

natural waxes will depend of the constituent of the mixture. In this part, we describe of general manner the physic determination of natural waxes and the information that provide it.

### 23.3.2   CONTACT ANGLE

The contact angle is defined as the angle formed by the intersection of the liquid-solid interface, considering a liquid drop resting on a flat, horizontal solid surface. Wettability studies usually involve the measurement of contact angles as the primary data, which indicates the degree of wetting when a solid and liquid interact, for example the grade of hydrophobicity of a waxes. A small contact angle is observed when the liquid spreads on the surface, while a large contact angle is observed when the liquid beads on the surface. More specifically, a contact angle less than 90° indicates that wetting of the surface is favorable, and the fluid will spread over a large area on the surface; while contact angles greater than 90° generally means that wetting of the surface is unfavorable so the fluid will minimize its contact with the surface and form a compact liquid droplet (Yuan and Lee, 2013).

As first described by Thomas Young in 1805 (Fowkes, 1962), the contact angle of a liquid drop on an ideal solid surface is defined by the mechanical equilibrium of the drop under the action of three interfacial tensions: liquid-vapor, solid-vapor, and solid-liquid interfacial tensions, and the contact angle ($\theta Y$), is usually referred to as Young's equation, and $\theta Y$ is Young's contact angle. The contact angle is measured in both optical and force tensiometry. Both static and dynamic contact angles can be measured by using Theta optical tensiometer. In practice, a droplet is placed on the solid surface and the image of the drop is recorded. Static contact angle is then defined by fitting Young-Laplace equation around the droplet, although other fitting methods such as circle and polynomial can also be used (Van Oss et al., 1988).

### 23.3.3   WATER VAPOR PERMEABILITY

Water vapor permeability (WVP) is a measure of the passage of water vapor through the material. It is also known as water vapor transmission rate (WVTR) or moisture vapor transmission rate (MVTR). It is the mass of water vapor transmitted through a unit area in a unit time under specified conditions of temperature and humidity (Turkan and Şahbazb, 2004). This measure is particularly important in films of waxes and natural waxes are not the exception. For instance, in the formulations of biofilm as coating in food (Bosquez-Molina et al., 2003). Water permeability of films is affected by many factors, depending on both the nature of barrier components, the film structure (homogeneous, emulsion, multilayer, etc.), crystal type, shape, size and distribution of lipids, and thermodynamics such as temperature, vapor pressure, or the physical state of water in contact to the films (Morillonad et al., 2002). Differences in permeabilities among the wax films are attributed mainly to differences in chemical composition, to make sure about the structure of composition the crystal type as determined by X-ray diffraction (Donhowe and Fennem, 1993). In this context, the film forming behavior and hydrophobicity of high amylose (HA) starch in the presence of three different natural waxes (beeswax, candelilla wax and carnauba wax) were studied in the presence and absence of Tween-80. The presence of these waxes in the presence and absence of Tween-80 resulted into different sorption isotherms and the water absorptivity and moisture diffusion coefficient values were also affected (Muscaf et al., 2013). A recent study showed that the whey protein-candelilla wax film/sheet the WVP of films from solution casting insignificantly decreased, at 5% level, as concentration of candelilla wax increased. The range of incorporating CAN was too low to get significant reduction effect on WVP of solution-cast film. Aside from amount of incorporated CAN, other factors were reported to affect WVPs of emulsion films, such as drying condition, pH and lipid particle size (Janjarasskul et al., 2014).

## 23.3.4  THERMAL PROPERTIES

The thermogravimetric analysis (TGA) is an analytical technique for study the thermal properties of diverse materials and waxes are not the exception. The TGA determinate the thermal stability of a material and its fraction of volatile components by monitoring the weight change that occurs as a sample is heated. The determinations are usually performed in air or in inert atmosphere, and the weight is recorded as a function of increasing temperature. On other occasions, the measurement is carried out in oxygen atmosphere to provide oxidation. There are two variants of thermal analysis, the differential thermal analysis (DTA) and differential scanning calorimetry (DSC) (Budai et al., 2012). For example, the crystallization and melting thermograms were determined by DSC of organogels formed by sugarcane or candelilla wax in soybean oil. In this study, the thermal behavior for sugarcane and candelilla wax was similar, especially during crystallization, differing only on enthalpy (Barbosa et al., 2013). Moreover, is reported in literature the thermal properties of organogels and water in oil emulsions structured by mixtures of candelilla wax and monoglycerides. DSC assays was performed for the samples and showed that the crystallization thermograms. These authors associated the exotherm at the higher temperature with the development of an α-lamellar organization, and the lower temperature exotherm with the crystallization of the aliphatic chains of the mono-glycerol (Toro-Vazquez et al., 2013).

## 23.3.5  SCANNING ELECTRON MICROSCOPE (SEM)

SEM Is a technique for examined surface characteristics in coating with waxes. For instance, SEM provide information about the morphology, size and distribution of the components of a mesquite gum-candelilla wax based edible emulsion coatings with potentials use in foods (Bosquez-Molina et al., 2003). In other example, the SEM images revealed differences in the surface morphology of biopolymer emulsion films in a study of effect of candelilla wax on functional properties (Kowalczyk and Baraniak, 2014). In a study of whey protein-candelilla wax film/sheet, the SEM images of solution-cast films revealed a continuous lipid layer over a compact protein matrix. This morphology is hypothesized to be the reason that solution-cast films were a better water vapor barrier than extruded sheet with elongated wax particles dispersed in the protein matrix. Furthermore, the surface view SEM images of solution-cast films show smaller pinholes on the surface than extruded sheet (Janjarasskul et al., 2014).

Other spectroscopy techniques may be used in the characterization of films of waxes. For example, the distribution of beeswax, candelilla wax and carnauba wax in high amylose starch-glycerol (HAG) films was determinate in the presence and absence of Tween-80 with Synchrotron Infrared Spectroscopy (S-FTIR) which provided 2D mapping along the thickness of the film is used. The results showns that the addition of these waxes to HAG increased the surface roughness and hydrophobicity of these films. The S-FTIR 2D images indicated that the distribution of wax in starch-wax films correlated with the roughness and hydrophobicity of the starch-wax films (Muscat et al., 2014).

## 23.4  CONCLUSIONS

In conclusion natural waxes, they are interesting and important compounds uses and applications, this due to their chemical and physical structure they present. Wax Candelilla is a one the most important natural waxes at national and international level because for their compositions. The most important determinations for such natural compounds are: spectroscopic of Infrared FT-IR, nuclear magnetic resonance (NMR), solubility in organic solvent, Contact angle, SEM, Thermal Properties, and Water Vapor Permeability, important for their applications.

## ACKNOWLEDGEMENTS

Authors thank CONACYT for the financial support through the research project SEP-CONACyT-CB-2011–01–0167903, SEP-CONACyT-2011–01–167668, and University Autonomy of Coahuila.

## KEYWORDS

- **candelilla waxes**
- **chemical properties**
- **physic properties**
- **waxes**

# REFERENCES

Barbosa, R. J. C., Dutra, L. J., Nucci, M. M., Barrera, A. D., Guerreiro, L. M., & Lopes da Cunha, R. (2013). Thermal and rheological properties of organogels formed by sugarcane or candelilla wax in soybean oil. *Food Research International, 50*, 318–323.

Bosquez, M. E. (2008). Development of edible coatings formulated with mesquite gum and candelilla wax for fruit preservation. Food World, 1, 28–31.

Bosquez, M. E., Guerrero, L. I., & Vernon, C. E. J. (2003). Moisture barrier properties and morphology of mesquite gum-candelilla wax based edible emulsion coatings. *Food Research International, 36*, 885–893.

Budai, L., Antal, I., Klebovich, I., & Budai, M. (2012). Natural oils and waxes: studies on stick bases. *Journal of Cosmetic Science, 63*, 93–101.

Cabello, A. Ch., Sáenz, G. A., Bermudez, B. L., Pérez, B. C., Ávila, O. C., & Váldes, G. J. (2013). Candelilla Wax and Applications. *Progress in Chemistry 8*(2), 105–110.

De León, Z. M., Sáenz, G. A., Rojas, M. R., Rodriguez, H. R., Cantú, J. D., & Aguilar, G. C. (2015). Edible candelilla wax coating with fermented extract of tarbush improves the shelf life and quality of apples. *Food Packaging and Shelf Life, 3*, 70–75.

Donhowe, G., & Fennem, A. O. (1993). Water vapor and oxygen permeability of wax films. *Journal of the American Oil Chemists, 70*, 867–873.

Elzbieta & Hans Brand. Elzbieta Cosmetics BV, The Netherlands http://www.elzbietacosmetics.com/lib/CANDELILLA_WAX.pdf. July 10th 2015.

Fowkes, F. M. (1962). Determination of interfacial tensions, contact angles, and dispersion forces in surfaces by assuming additivity of intermolecular interactions in surface. *Journal of Physical Chemistry, 66*, 382–382.

Instituto de la Candelilla. http://www.candelilla.org/?page_id=534 & lang=es. July 10th 2015.

Janjarasskul, T., Rauch, D. J., McCarthy, K. L., & Krochta, J. M. (2014). Barrier and tensile properties of whey protein-candelilla wax film/sheet. LWT. *Food Science and Technology, 56*, 377–382.

Kowalczyk, D., & Baraniak, B. (2014). Effect of candelilla wax on functional properties of biopolymer emulsion films-A comparative study. *Food Hydrocolloids, 41*, 195–209.

Morillonad, V., Debeaufortae, F., Blond, G., Capellef, M., & Voilley, A. (2002). Factors affecting the moisture permeability of lipid-based edible films: a review. *In critical reviews in Food Science and Nutrition, 42*, 67–89.

Muscat, D., Adhikari, R., McKnight, S., Guo, Q. P., & Adhikari, B. (2013). The physicochemical characteristics and hydrophobicity of high amylose starch-glycerol films in the presence of three natural waxes. *Journal of Food Engineering, 119*, 205–219.

Muscat, D., Tobin, M. J., Guo, Q. P., & Adhikari, B. (2014). Understanding the distribution of natural wax in starch-wax films using synchrotron-based FTIR (S-FTIR). Carbohydrate Polymers, *102*, 125–135.

Rojas, M. R., Saucedo P. S., De León, Z. M., Jasso, C. D., & Aguilar, C. (2011). Pasado, Presente y Futuro de la Cera de Candelilla, *2*, 7–18.

Saucedo, P. S., Rojas, M. R., Aguilera, C., A. F., Sáenz, G. A., Garza, H. L., & Jasso, C. D. (2009). Edible film based on candelilla wax to improve the shelf life and quality of avocado. *Food Research International, 42*(4), 511–515.

Toro-Vázquez, J. F., Mauricio-Pérez, R., González-Chávez, M., Sánchez-Becerril, M., Ornelas-Paz, J. J., & Pérez-Martínez, J. D. (2013). Physical properties of organogels and water in oil emulsions structured by mixtures of candelilla wax and monoglycerides. *Food Research International, 54,* 1360–1368.

Turhana, K. N., & Şahbazb, F. (2004). Water vapor permeability, tensile properties and solubility of methylcellulose-based edible films. *Journal of Food Engineering, 61,* 459–466.

Van Oss, C. J., Good, R. J., & Chaudhury, M. K. (1988). Additive and nonadditive surface tension components and the interpretation of contact angles, *Langmuir, 4,* 884–891.

Yuan, Y., & Lee T. R. (2013). Contact angle and wetting properties in Surface Science Techniques. Bracco, G., Holst, B. (eds.), Springer Series in Surface Sciences 51: Springer-Verlag Berlin Heidelberg.

**CHAPTER 24**

# SCANNING ELECTRON MICROSCOPY OF VARIABLE PRESSURE FOR *EUPHORBIA ANTISYPHILITICA* (CANDELILLA)

MIGUEL A. AGUILAR-GONZÁLEZ,[1] CRISTOBAL N. AGUILAR,[2] and J. A. AGUIRRE-JOYA[2]

[1]*Center for Research and Advanced Studies of Polythechnique National Institute. CINVESTAV Unidad Saltillo, Avenida Industria Metalьrgica # 1062, 25900, Ramos Arizpe, Coahuila, Mexico, E-mail: mgzlz@hotmail.com, miguel.aguilar@cinvestav.edu.mx*

[2]*Department of Food Science and Technology, School of Chemistry, Autonomous University of Coahuila, Unidad Saltillo, 25000, Saltillo, Coahuila, Mexico*

## CONTENTS

## ABSTRACT

This chapter describes a study through environmental scanning electron microscopy (ESEM) of the *Euphorbia antisyphilitica* (Candelilla). A historical overview of the evolution of this microscopic technique and its scientific impact is also presented. The study was conducted on the surface of the candelilla stems through the ESEM mode with conventional microscopic analysis and of compositional chemical contrast and for the first time that is presented the elemental chemical composition of candelilla wax before and after applying a purification treatment wax. The results indicated the presence of important chemical elements from the point of view of biotechnological and materials science areas in order to understand its behavior and reaction mechanisms in some of its many applications. Furthermore some important practices are detailed in the methodology of the preparation and operation of materials and methods.

## 24.1 INTRODUCTION

In the middle of the last century the most revolutionary invention of his time was generated by an optical engineer and a mechanical engineer. This device was called scanning electron microscope and by definition the word microscopy has its origins in the Greek roots *micro* and *scope* that mean small and look respectively. The divisions that are part of the area of microscopy are divided into the following modes: optical, scanning electron, transmission electron, confocal, tunneling, atomic force fluorescence, etc. Each of the above areas is in turn subdivided into other different modes and each have a specific application. The technique of scanning electron microscopy has been without doubt one of the techniques that have contributed in significantly way to the growth of knowledge of the properties of materials. Whereas this technique contributes to comprise 90% of knowledge (behavior-properties) in the materials science area that it has been obtained in the last 50 years.

Scanning electron microscopy technique offers the possibility of developing morphological characterization and of chemical composition of almost all materials. Solely exists restriction for the analysis of samples containing volatile organic and/or volatile hydrocarbon based material. In the spectroscopic technique of energy dispersive X-ray (EDS) and it is also possible to analyze quantitative and semi-quantitative amounts for more than 70 chemical elements of the periodic table. This analysis technique is inherent for the scanning electron microscope. It is possible to carry out quantifications using a high purity standard (99.7% of purity) of each chemical element that is necessary to quantify the standard and simultaneously to quantify the sample to be examined.

In this chapter it will be trying to address the study of samples of plant origin (Candelilla) by means of environmental scanning electron microscopy (ESEM) mode also some advantages and disadvantages of the technique as well as some secrets to make a practical and effective characterization are presented.

## 24.2 HISTORY

The environmental scanning electron microscope was introduced commercially from two decades ago. The possibilities to observe samples without preparation and non-conductive have made a momentous time for biological, natural and health areas, among many others. The possibility of watch images with greater magnification to 1000x in non-metallic and nor coated samples has strengthened the knowledge in the above areas. This objective has been achieved thanks to the ESEM was designed and built for observing characteristics in non-conductive samples, which undergo presence of polar charges that are the result of the interaction between the vacuum system and dehydration to which they are subjected, auto-balancing themselves in contact with the electron beam. These accessories are not included in the conventional scanning microscopes (Donald et al., 2003) therefore the samples assayed in high vacuum mode, need to be coated or plated with a conductive metal (Titze and Denk, 2013). The most interesting aspect is that environmental mode (ESEM) is also functional for samples conventionally analyzed by SEM that is under high vacuum.

One of the ESEM advantages is that detectors can also be used such as GSD (Gaseous Secondary Electrons) at the microscope. Figure 1 shows the different options of analysis that can be performed in an environmental scanning electron microscopy conditioning with the energy dispersive x-ray spectrometry technique. The EDS-WDS-EBSD acronyms correspond to: Energy Dispersive Spectrometry, Wavelength Dispersive Spectrometry and Electron Backscatter Diffraction, respectively.

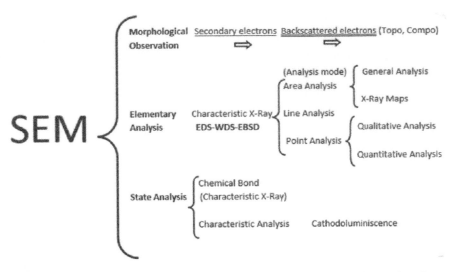

**FIGURE 24.1**   Classification of modes of analysis in an environmental scanning electron microscope.

An environmental scanning electron microscope is an instrument that works throughout an electron beam which is generated from a tungsten filament or Hexa-lanthanum boride, in the case of microscopes than are designed as field emission. In this kind of equipment it is possible imaging samples at different vacuum conditions (high-low-intermediate). The morphological features (SE Secondary Electrons) and also those formed from chemical compositional contrast (BSE- Backscatter Electrons) can be observed in polished or unpolished samples (Figure 24.1).

The first operation of an electron microscope was carried out by Knoll in 1935, who explored and imagined the surface of a solid material (Knoll, 1935). In 1938, von Ardenne reported the formation of the electron probe and its physical behavior as well as contributed to the basic principles of the SEM. The initial conditions of position and current intensity of the detectors were established by him in 1938. Zworykin and his team built an SEM which had several important original features. The first commercial appearance of a scanning electron microscope was conducted in 1965 and was launched by the brand Cambridge Instruments. It was not until the first decade of the 2000s when it appeared an electron microscopy commercially of environmental type marketed by the brand Philips, model XL30-ESEM.

The emergence of scanning electron microscopy since 1930 has represented a breakthrough for materials science and that since then was it possible to study the microstructural characteristics and of composition. The use of electromagnetic coils and capacitors lenses in combination with an electron beam interaction with the sample have resulted in a great depth of field in three dimensions and quantification of composition has been very close to the reality of material. Additionally one of the most important contributions was the observation by backscattered electrons which allows the observation of characteristics distinguishing the chemical compositional contrasts and thus decreasing doubts that by optical microscopy are unsolvable. The scanning electron microscope was originally designed for samples that contain outer surfaces which are electrically conductive or those samples than by application of an external metallic coating their surfaces acquire an electronic conductor character. The above is part of a series of requirements which are necessary for taking an image with a good approach and on the other hand for reliable semi-quantitative chemical analysis.

Figure 24.2 shows a summary of the most important historical facts presented in the field of scanning electron microscopy since 1930 and to the present date, it also presents the progress in resolution refers.

### 24.2.1 APPLICATION OF ESEM IN MATERIALS SCIENCE AND RELATED AREAS

Applications of ESEM technique are numerous and varied, this technique has allowed the SEM become more versatile in the area of materials

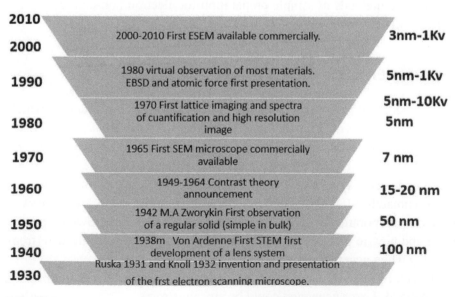

FIGURE 24.2 Summary of the history of the development of electron scanning microscopy.

characterization. The use of environmental microscopes has been applied in different areas of science and/or technology, among which we can mention the following: aerospace, medicine, orthodontics, biology, chemistry, architecture, archeology, genetics, astronomy atmospheric sciences, chemical, computing, mathematics, engineering, geochemistry- geophysics, materials science, oceanography, paleontology, physics, physics applied to planetary science, anatomy biomechanic, anthropology, biochemistry, botany, cell biology, development, ecology, epidemiology, evolution, genetics, immunology, medicine, microbiology, molecular biology, neuroscience, pharmacology/toxicology, physiology, psychology, and virology.

Except for materials with a chemical composition based on hydrocarbons and/or organic solvents ESEM can be analyzed in environmental scanning electron microscope almost all types of materials and their combinations together known. Types of materials can be grouped into six groups: (i) composites, (ii) polymers, (iii) ceramics, (iv) metals, (v) minerals, and (vi) biological.

Another very important aspect is the presentation of the materials can be grouped as follows: thin films, regular solids, prints, mounted on polymer samples polished fractured samples, biological tissues, etc.

### 24.3   PREPARATION METHODS

Among the methods of sample preparation for electron non-conductive are the following:

#### 24.3.1   METAL THERMAL EVAPORATION

The procedure is performed in an evaporator working in high vacuum which sublimes electrons of conductive metals on the surface of the samples. Being in high vacuum sublimation point of each metal is modified and with the help of a basket of tungsten and an electric arc generated by the electric current conductive layer is deposited. Sublimation of a metal conductor wire (approximately = 0.010 g). The layer thickness is a function of the amount of material, evaporation time and current applied. Figure 24.3 shows a tungsten cage and how to place small pieces of metal wire to prevent the wire from moving when generating the high vacuum.

C, Au, Pt, Ag, Al, Cu, Ni amongst other can used for to coat samples by this method. Layer generated could be any angstroms.

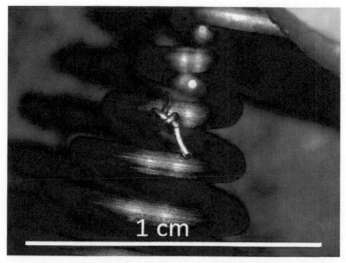

**FIGURE 24.3**   Digital Photograph of tungsten cage used for gold coating process.

## 24.3.2   ION SPUTTERING

In this equipment it is possible to obtain coatings with thicknesses from 50 to 300 A. A flow of argon ionizes a metal deposit called "Target" in order to generate a plasma which can produce thin film layers for samples to be studied by SEM. Gold, gold-palladium, platinum, nickel, silver, palladium and carbon are the items available for this technique. Operational tests can be carried out in a vacuum level of $10^{-1}$ mbar.

Both to minimize specimen damage (Royall et al., 2001) and to maximize contrast the candelilla specimens were coated with an Au layer (Figure 24.4).

Those materials are substances than can be evaporated in the absence of their corresponding vapor. Hence it, before imaging can take place, such samples require preparation in order to remove potentially volatile substances, and many procedures have been developed. Of course samples such as porous polymers, emulsions, vegetables and food systems and so on those than could contain water.

## 24.3.3   CANDELILLA SAMPLES PREPARATION

Candelilla stems were collected in Coahuila State, Mexico (geographical latitudes: 26° 01' 07. 45" N and 101° 18' 51. 69" O). Samples were cut

**FIGURE 24.4**    Gold Target used in ion sputtering processing.

following the Official Mexican Norm-018-SEMARNAT-1999 (NOM-018-SEMARNAT-1999), which prohibits removing the root of the soil. Stems were deposited in polyethylene bags and transported to our laboratory.

Candelilla stems were provided by CONAFOR (National Forestry Commission, Coahuila México) and were received no preparation. Samples were analyzed in an environmental scanning electron microscope brand Philips model XL-30 ESEM. Also a stereographic microscope brand Olympus and model SZ40 fitted with a digital camera Qimaging was used for stereographic observation. After the stereographic analysis, SEM study starts with the sample preparation for introduction into the ESEM. Pieces of 2 cm of and 0.5 cm length and diameter respectively of *Euphorbia antisyphilitica* were sectioned and placed on a sample holder bronze and the surface of each bush was scavenged with compressed air. Afterward semi-quantitative chemical analysis with the GSE detector in the ESEM mode were performed. Chemical semi-quantitative analyzes were carried out with the technique of energy dispersive X-ray spectroscopy, using a solid state detector brand Apollo and model Apex 2 with a window analysis of 30 mm of size and with a analysis software brand EDAX model genesis.

The photomicrographs were taken at different operating modes of the ESEM. Elemental compositional contrast maps were done in order to observe the elemental distribution throughout of the surface. Moreover punctual analysis were carried out on the surface of the candelilla. The conditions for the analysis of samples were: voltage = 20 keV, working distance of 7.5 to 10.0 mm for photomicrographs and microanalysis, respectively. Backscattered electron (BSE) mode was used for taking photomicrographs and thereby observe its chemical compositional contrast. The gold coating was applied in a vacuum evaporator brand JEOL and model JEE 400, this procedure is reported in the literature as metal thermal evaporation.

Samples must be perfectly dry (free of volatile matter organic type) and free from materials used for preparation such as diamond pastes, including solvents or colloidal silica or alumina. Interactions among the epicuticular wax layer, the intracuticular tissue and thickness of each internal tissue were measured by ESEM.

### 24.3.4 WORKING PRINCIPLE OF THE SCANNING ELECTRON MICROSCOPE

Figure 24.5 shows the main parts of an environmental scanning electron microscope. By analogy the total interaction between the electron beam and the sample in this type of equipment is similar to that of an electrical circuit which may be flowing various signals (Figure 24.6) In addition it is too important to note that the size of the interaction volume depends of kiloelectron-Volts ($kV$) and atomic weight ($Z$) and is also involved including the effect of physical ground (physical earth) on the equipment. Where:

$$Ip = IBSE + ISE + I\ earth$$

where $Ip = IBSE = ISE = I\ earth$.

### 24.3.5 FACTORS INVOLVED IN IMAGING SAMPLES IN THE ESEM MODE

Commonly always it thinks it essential for to have a very well-focused image and nice it depends to have a last generation microscope with a high resolution. Especially when these images are of variable pressure microscopy applications in non-conductive samples. Many factors environment for taking a picture such as the proper application of the technique, good calibration of the instrument, a team with a good preventive maintenance (and

Electron Beam
Tungsten Filament
GSE detector
BSE detector
Sample
Stage
SE detector

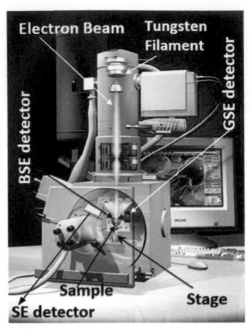

**FIGURE 24.5** Inside and outside of an environmental scanning electron microscope Brand Philips XL-30 Series. Philips® FEI Company. Photography Artwork by Martin Melis Studio.

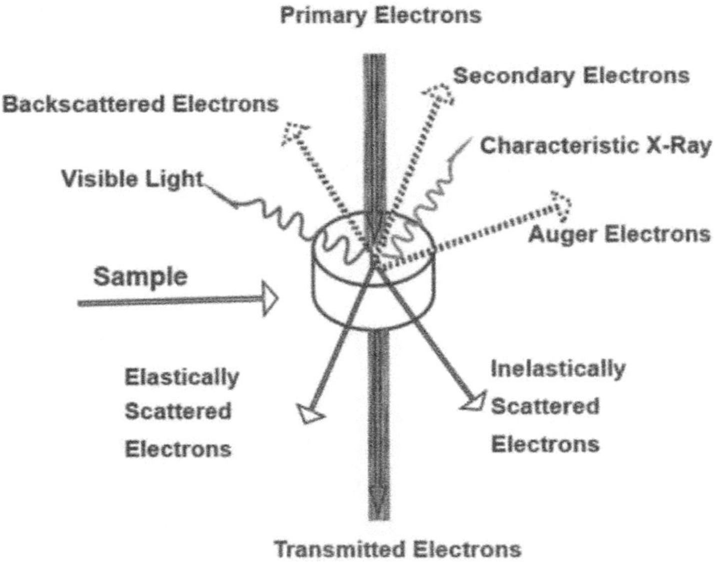

Primary Electrons
Secondary Electrons
Backscattered Electrons
Characteristic X-Ray
Visible Light
Auger Electrons
Sample
Elastically Scattered Electrons
Inelastically Scattered Electrons
Transmitted Electrons

**FIGURE 24.6** Schematic representation of the interactions that take place between the electron beam (tungsten filament) and the sample in an environmental scanning electron microscope.

regular), expertise and experience of the operator, noise, the preparation and/ or relevant sample mounting and the status of the filament.

## 24.3.6   VACUUM REQUIREMENTS FOR THE OPERATION OF VARIABLE PRESSURE SEM

ESEM was designed to analyze non-conductive samples that generate charge inside of the camera, these charges may be generated to be insulating or contain water or both of them as is the case of vegetable samples or ceramic materials (no sintered). Obtaining of an excellent approach of photomicrograph is the result of the balance of charges on the surface of the materials and a stability regulated equipment (physical earths, vacuum, power, etc.). The effect of electric charges in the samples has been well studied and described by some researchers (Titze and Denk, 2013). In the case of insulating materials such as are: glass, wood, polymers, ceramics and combinations thereof, it is necessary to apply a conductive coating on the surface. The above mentioned is the purpose of canceling the charges generated inside the chamber of the microscope.

On the other side low vacuum mode is preferred for insulating materials or inorganic surfaces with non-conductive surfaces and variable pressure mode either for samples with a relative moisture content and/ or biological samples.

Another important aspect is that every one of all the modes can be combined with a set of high contrast that is located independently from the vacuum chamber (is more functional in mixed modes).

In the case of using the variable pressure mode, the microscope is capable of regulating the water pressure vapor through the Peltier stage. This stage can regulate the temperature range from 0 to 65°C. With this device it is possible to perform hydration-dehydration dynamic experiments and vice-versa. Some environmental microscopes of new generation have an architecture and additional peripheral devices which can perform mechanical testing, fracture characterization, Electron micro-diffraction of X-ray patterns (EBSD-XRD) and those where it is possible to analyze thermomechanical tests. In all the above modes it is possible to watch videos if it is prepared a special device for this. In the case of candelilla which is a biological material (non-conductive), the ideal study condition is the observation in low vacuum mode with the electrolitically sample coated with gold or copper (Figure 24.4).

### 24.3.7   ANALYSIS MODE SELECTION IN ESEM

The environmental scanning electron microscope has a similar column to the conventional electron microscopes, but with the difference that it is possible to synchronize the conventional camera and column in three different modes: (a) low vacuum, (b) variable pressure with a Peltier stage, and (c) high vacuum. It is important to note that low vacuum mode is preferred when is used in insulating samples or inorganic samples without conductive surfaces and the variable pressure mode either for samples with a moisture content (for example: biological). In the environmental electron microscopes for to switch from high vacuum to low vacuum or variable pressure seems easy because only is needed pressing a switch and a bullet (vacuum compensator) will perform the work for which the mode will changed mechanically, but operationally (electronic and optical mechanical) this involves to change a whole independent vacuum system. In Figure 24.7 can be observed the two types of accessories for both high and low vacuum.

For the three modes results are obtained by a self-contained software to mechanical operation of the ESEM and its consist of a spectrogram (total peaks detected) and with a table with numerical values, the latter are contained in separated columns as: w% (weight percent), a% (atomic percent), and X-ray fluorescence and absorption parameters (ZAF-acronym) as well as for calculating the systematic software error. With all the aforementioned it is possible to predict something about of the composition of the present phases in the sample composition (Reed, 2005).

**FIGURE 24.7**   Digital Photograph of the bullets: (up) high vacuum mode bullet, and (bottom) low vacuum and variable pressure mode bullet.

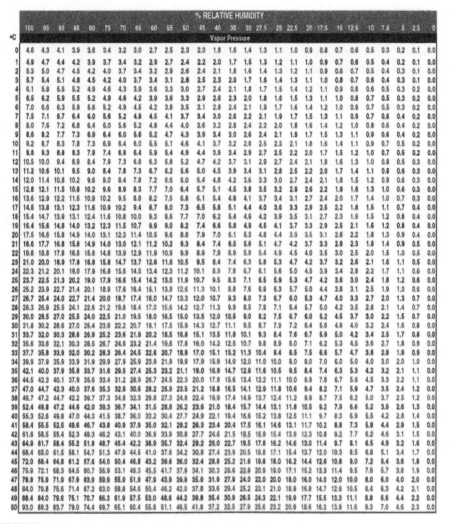

| °C | % RELATIVE HUMIDITY | | | | | | | | | | | | | | | | | | | | | | | | | | |
|----|-----|----|----|----|----|----|----|----|----|----|----|----|----|----|----|------|----|------|----|------|----|------|----|-----|----|-----|----|
|    | 100 | 95 | 90 | 85 | 80 | 75 | 70 | 65 | 60 | 55 | 50 | 45 | 40 | 35 | 30 | 27.5 | 25 | 22.5 | 20 | 17.5 | 15 | 12.5 | 10 | 7.5 | 5 | 2.5 | 0 |
|    | Vapor Pressure | | | | | | | | | | | | | | | | | | | | | | | | | | |
| 0 | 4.6 | 4.3 | 4.1 | 3.9 | 3.6 | 3.4 | 3.2 | 3.0 | 2.7 | 2.5 | 2.3 | 2.0 | 1.8 | 1.6 | 1.4 | 1.3 | 1.1 | 1.0 | 0.9 | 0.8 | 0.7 | 0.6 | 0.5 | 0.3 | 0.2 | 0.1 | 0.0 |
| 1 | 4.9 | 4.7 | 4.4 | 4.2 | 3.9 | 3.7 | 3.4 | 3.2 | 2.9 | 2.7 | 2.4 | 2.2 | 2.0 | 1.7 | 1.5 | 1.3 | 1.2 | 1.1 | 1.0 | 0.9 | 0.7 | 0.6 | 0.5 | 0.4 | 0.2 | 0.1 | 0.0 |
| 2 | 5.3 | 5.0 | 4.7 | 4.5 | 4.2 | 4.0 | 3.7 | 3.4 | 3.2 | 2.9 | 2.6 | 2.4 | 2.1 | 1.8 | 1.6 | 1.4 | 1.3 | 1.2 | 1.1 | 0.9 | 0.8 | 0.7 | 0.5 | 0.4 | 0.3 | 0.1 | 0.0 |
| 3 | 5.7 | 5.4 | 5.1 | 4.8 | 4.5 | 4.2 | 4.0 | 3.7 | 3.4 | 3.1 | 2.8 | 2.5 | 2.3 | 2.0 | 1.7 | 1.6 | 1.4 | 1.3 | 1.1 | 1.0 | 0.8 | 0.7 | 0.6 | 0.4 | 0.3 | 0.1 | 0.0 |
| 4 | 6.1 | 5.8 | 5.5 | 5.2 | 4.9 | 4.6 | 4.3 | 3.9 | 3.6 | 3.3 | 3.0 | 2.7 | 2.4 | 2.1 | 1.8 | 1.7 | 1.5 | 1.4 | 1.2 | 1.1 | 0.9 | 0.8 | 0.6 | 0.5 | 0.3 | 0.2 | 0.0 |
| 5 | 6.5 | 6.2 | 5.9 | 5.5 | 5.2 | 4.9 | 4.6 | 4.2 | 3.9 | 3.6 | 3.3 | 2.9 | 2.6 | 2.3 | 2.0 | 1.8 | 1.6 | 1.5 | 1.3 | 1.1 | 1.0 | 0.8 | 0.7 | 0.5 | 0.3 | 0.2 | 0.0 |
| 6 | 7.0 | 6.6 | 6.3 | 5.9 | 5.6 | 5.2 | 4.9 | 4.5 | 4.2 | 3.8 | 3.5 | 3.1 | 2.8 | 2.4 | 2.1 | 1.9 | 1.7 | 1.6 | 1.4 | 1.2 | 1.0 | 0.9 | 0.7 | 0.5 | 0.3 | 0.2 | 0.0 |
| 7 | 7.5 | 7.1 | 6.7 | 6.4 | 6.0 | 5.6 | 5.2 | 4.9 | 4.5 | 4.1 | 3.7 | 3.4 | 3.0 | 2.6 | 2.2 | 2.1 | 1.9 | 1.7 | 1.5 | 1.3 | 1.1 | 0.9 | 0.7 | 0.6 | 0.4 | 0.2 | 0.0 |
| 8 | 8.0 | 7.6 | 7.2 | 6.8 | 6.4 | 6.0 | 5.6 | 5.2 | 4.8 | 4.4 | 4.0 | 3.6 | 3.2 | 2.8 | 2.4 | 2.2 | 2.0 | 1.8 | 1.6 | 1.4 | 1.2 | 1.0 | 0.8 | 0.6 | 0.4 | 0.2 | 0.0 |
| 9 | 8.6 | 8.2 | 7.7 | 7.3 | 6.9 | 6.4 | 6.0 | 5.6 | 5.2 | 4.7 | 4.3 | 3.9 | 3.4 | 3.0 | 2.6 | 2.4 | 2.1 | 1.9 | 1.7 | 1.5 | 1.3 | 1.1 | 0.9 | 0.6 | 0.4 | 0.2 | 0.0 |
| 10 | 9.2 | 8.7 | 8.3 | 7.8 | 7.3 | 6.9 | 6.4 | 6.0 | 5.5 | 5.1 | 4.6 | 4.1 | 3.7 | 3.2 | 2.8 | 2.5 | 2.3 | 2.1 | 1.8 | 1.6 | 1.4 | 1.1 | 0.9 | 0.7 | 0.5 | 0.2 | 0.0 |
| 11 | 9.8 | 9.3 | 8.8 | 8.3 | 7.9 | 7.4 | 6.9 | 6.4 | 5.9 | 5.4 | 4.9 | 4.4 | 3.9 | 3.4 | 2.9 | 2.7 | 2.5 | 2.2 | 2.0 | 1.7 | 1.5 | 1.2 | 1.0 | 0.7 | 0.5 | 0.2 | 0.0 |
| 12 | 10.5 | 10.0 | 9.4 | 8.9 | 8.4 | 7.9 | 7.3 | 6.8 | 6.3 | 5.8 | 5.2 | 4.7 | 4.2 | 3.7 | 3.1 | 2.9 | 2.7 | 2.4 | 2.1 | 1.8 | 1.6 | 1.3 | 1.0 | 0.8 | 0.5 | 0.3 | 0.0 |
| 13 | 11.2 | 10.6 | 10.1 | 9.5 | 9.0 | 8.4 | 7.8 | 7.3 | 6.7 | 6.2 | 5.6 | 5.0 | 4.5 | 3.9 | 3.4 | 3.1 | 2.8 | 2.5 | 2.2 | 2.0 | 1.7 | 1.4 | 1.1 | 0.8 | 0.6 | 0.3 | 0.0 |
| 14 | 12.0 | 11.4 | 10.8 | 10.2 | 9.6 | 9.0 | 8.4 | 7.8 | 7.2 | 6.6 | 6.0 | 5.4 | 4.8 | 4.2 | 3.6 | 3.3 | 3.0 | 2.7 | 2.4 | 2.1 | 1.8 | 1.5 | 1.2 | 0.9 | 0.6 | 0.3 | 0.0 |
| 15 | 12.8 | 12.1 | 11.5 | 10.8 | 10.2 | 9.6 | 8.9 | 8.3 | 7.7 | 7.0 | 6.4 | 5.7 | 5.1 | 4.5 | 3.8 | 3.5 | 3.2 | 2.9 | 2.6 | 2.2 | 1.9 | 1.6 | 1.3 | 1.0 | 0.6 | 0.3 | 0.0 |
| 16 | 13.6 | 12.9 | 12.2 | 11.6 | 10.9 | 10.2 | 9.5 | 8.8 | 8.2 | 7.5 | 6.8 | 6.1 | 5.4 | 4.8 | 4.1 | 3.7 | 3.4 | 3.1 | 2.7 | 2.4 | 2.0 | 1.7 | 1.4 | 1.0 | 0.7 | 0.3 | 0.0 |
| 17 | 14.5 | 13.8 | 13.1 | 12.3 | 11.6 | 10.9 | 10.2 | 9.4 | 8.7 | 8.0 | 7.3 | 6.5 | 5.8 | 5.1 | 4.4 | 4.0 | 3.6 | 3.3 | 2.9 | 2.5 | 2.2 | 1.8 | 1.5 | 1.1 | 0.7 | 0.4 | 0.0 |
| 18 | 15.4 | 14.7 | 13.9 | 13.1 | 12.4 | 11.6 | 10.8 | 10.0 | 9.3 | 8.5 | 7.7 | 7.0 | 6.2 | 5.4 | 4.6 | 4.2 | 3.9 | 3.5 | 3.1 | 2.7 | 2.3 | 1.9 | 1.5 | 1.2 | 0.8 | 0.4 | 0.0 |
| 19 | 16.4 | 15.6 | 14.8 | 14.0 | 13.2 | 12.3 | 11.5 | 10.7 | 9.9 | 9.0 | 8.2 | 7.4 | 6.6 | 5.8 | 4.9 | 4.5 | 4.1 | 3.7 | 3.3 | 2.9 | 2.5 | 2.1 | 1.6 | 1.2 | 0.8 | 0.4 | 0.0 |
| 20 | 17.5 | 16.6 | 15.8 | 14.9 | 14.0 | 13.1 | 12.3 | 11.4 | 10.5 | 9.6 | 8.8 | 7.9 | 7.0 | 6.1 | 5.3 | 4.8 | 4.4 | 3.9 | 3.5 | 3.1 | 2.6 | 2.2 | 1.8 | 1.3 | 0.9 | 0.4 | 0.0 |
| 21 | 18.6 | 17.7 | 16.8 | 15.8 | 14.9 | 14.0 | 13.0 | 12.1 | 11.2 | 10.2 | 9.3 | 8.4 | 7.4 | 6.5 | 5.6 | 5.1 | 4.7 | 4.2 | 3.7 | 3.3 | 2.8 | 2.3 | 1.9 | 1.4 | 0.9 | 0.5 | 0.0 |
| 22 | 19.8 | 18.8 | 17.8 | 16.8 | 15.8 | 14.8 | 13.9 | 12.9 | 11.9 | 10.9 | 9.9 | 8.9 | 7.9 | 6.9 | 5.9 | 5.4 | 4.9 | 4.5 | 4.0 | 3.5 | 3.0 | 2.5 | 2.0 | 1.5 | 1.0 | 0.5 | 0.0 |
| 23 | 21.0 | 20.0 | 18.9 | 17.9 | 16.8 | 15.8 | 14.7 | 13.7 | 12.6 | 11.6 | 10.5 | 9.5 | 8.4 | 7.4 | 6.3 | 5.8 | 5.3 | 4.7 | 4.2 | 3.7 | 3.2 | 2.6 | 2.1 | 1.6 | 1.1 | 0.5 | 0.0 |
| 24 | 22.3 | 21.2 | 20.1 | 19.0 | 17.9 | 16.8 | 15.6 | 14.5 | 13.4 | 12.3 | 11.2 | 10.1 | 8.9 | 7.8 | 6.7 | 6.1 | 5.6 | 5.0 | 4.5 | 3.9 | 3.4 | 2.8 | 2.2 | 1.7 | 1.1 | 0.6 | 0.0 |
| 25 | 23.7 | 22.5 | 21.3 | 20.2 | 19.0 | 17.8 | 16.6 | 15.4 | 14.2 | 13.0 | 11.9 | 10.7 | 9.5 | 8.3 | 7.1 | 6.5 | 5.9 | 5.3 | 4.7 | 4.2 | 3.6 | 3.0 | 2.4 | 1.8 | 1.2 | 0.6 | 0.0 |
| 26 | 25.2 | 23.9 | 22.7 | 21.4 | 20.1 | 18.9 | 17.6 | 16.4 | 15.1 | 13.8 | 12.6 | 11.3 | 10.1 | 8.8 | 7.6 | 6.9 | 6.3 | 5.7 | 5.0 | 4.4 | 3.8 | 3.1 | 2.5 | 1.9 | 1.3 | 0.6 | 0.0 |
| 27 | 26.7 | 25.4 | 24.0 | 22.7 | 21.4 | 20.0 | 18.7 | 17.4 | 16.0 | 14.7 | 13.3 | 12.0 | 10.7 | 9.3 | 8.0 | 7.3 | 6.7 | 6.0 | 5.3 | 4.7 | 4.0 | 3.3 | 2.7 | 2.0 | 1.3 | 0.7 | 0.0 |
| 28 | 28.3 | 26.9 | 25.5 | 24.1 | 22.6 | 21.2 | 19.8 | 18.4 | 17.0 | 15.6 | 14.2 | 12.7 | 11.3 | 9.9 | 8.5 | 7.8 | 7.1 | 6.4 | 5.7 | 5.0 | 4.2 | 3.5 | 2.8 | 2.1 | 1.4 | 0.7 | 0.0 |
| 29 | 30.0 | 28.5 | 27.0 | 25.5 | 24.0 | 22.5 | 21.0 | 19.5 | 18.0 | 16.5 | 15.0 | 13.5 | 12.0 | 10.5 | 9.0 | 8.2 | 7.5 | 6.7 | 6.0 | 5.2 | 4.5 | 3.7 | 3.0 | 2.2 | 1.5 | 0.7 | 0.0 |
| 30 | 31.8 | 30.2 | 28.6 | 27.0 | 25.4 | 23.8 | 22.2 | 20.7 | 19.1 | 17.5 | 15.9 | 14.3 | 12.7 | 11.1 | 9.5 | 8.7 | 7.9 | 7.2 | 6.4 | 5.6 | 4.8 | 4.0 | 3.2 | 2.4 | 1.6 | 0.8 | 0.0 |
| 31 | 33.7 | 32.0 | 30.3 | 28.6 | 26.9 | 25.2 | 23.6 | 21.9 | 20.2 | 18.5 | 16.8 | 15.1 | 13.5 | 11.8 | 10.1 | 9.3 | 8.4 | 7.6 | 6.7 | 5.9 | 5.0 | 4.2 | 3.4 | 2.5 | 1.7 | 0.8 | 0.0 |
| 32 | 35.6 | 33.8 | 32.1 | 30.3 | 28.5 | 26.7 | 24.9 | 23.2 | 21.4 | 19.6 | 17.8 | 16.0 | 14.2 | 12.5 | 10.7 | 9.8 | 8.9 | 8.0 | 7.1 | 6.2 | 5.3 | 4.5 | 3.6 | 2.7 | 1.8 | 0.9 | 0.0 |
| 33 | 37.7 | 35.8 | 33.9 | 32.0 | 30.2 | 28.3 | 26.4 | 24.5 | 22.6 | 20.7 | 18.8 | 17.0 | 15.1 | 13.2 | 11.3 | 10.4 | 9.4 | 8.5 | 7.5 | 6.6 | 5.7 | 4.7 | 3.8 | 2.8 | 1.9 | 0.9 | 0.0 |
| 34 | 39.9 | 37.9 | 35.0 | 33.9 | 31.9 | 29.9 | 27.9 | 25.9 | 23.9 | 21.9 | 19.9 | 17.9 | 15.9 | 14.0 | 12.0 | 11.0 | 10.0 | 9.0 | 8.0 | 7.0 | 6.0 | 5.0 | 4.0 | 3.0 | 2.0 | 1.0 | 0.0 |
| 35 | 42.1 | 40.0 | 37.9 | 35.8 | 33.7 | 31.6 | 29.5 | 27.4 | 25.3 | 23.2 | 21.1 | 19.0 | 16.9 | 14.7 | 12.6 | 11.6 | 10.5 | 9.5 | 8.4 | 7.4 | 6.3 | 5.3 | 4.2 | 3.2 | 2.1 | 1.1 | 0.0 |
| 36 | 44.5 | 42.3 | 40.1 | 37.9 | 35.6 | 33.4 | 31.2 | 28.9 | 26.7 | 24.5 | 22.3 | 20.0 | 17.8 | 15.6 | 13.4 | 12.2 | 11.1 | 10.0 | 8.9 | 7.8 | 6.7 | 5.6 | 4.5 | 3.3 | 2.2 | 1.1 | 0.0 |
| 37 | 47.0 | 44.7 | 42.3 | 40.0 | 37.6 | 35.3 | 32.9 | 30.6 | 28.2 | 25.9 | 23.5 | 21.2 | 18.8 | 16.5 | 14.1 | 12.9 | 11.8 | 10.6 | 9.4 | 8.2 | 7.1 | 5.9 | 4.7 | 3.5 | 2.4 | 1.2 | 0.0 |
| 38 | 49.7 | 47.2 | 44.7 | 42.2 | 39.7 | 37.3 | 34.8 | 32.3 | 29.8 | 27.3 | 24.8 | 22.4 | 19.9 | 17.4 | 14.9 | 13.7 | 12.4 | 11.2 | 9.9 | 8.7 | 7.5 | 6.2 | 5.0 | 3.7 | 2.5 | 1.2 | 0.0 |
| 39 | 52.4 | 49.8 | 47.2 | 44.6 | 42.0 | 39.3 | 36.7 | 34.1 | 31.5 | 28.8 | 26.2 | 23.6 | 21.0 | 18.4 | 15.7 | 14.4 | 13.1 | 11.8 | 10.5 | 9.2 | 7.9 | 6.6 | 5.2 | 3.9 | 2.6 | 1.3 | 0.0 |
| 40 | 55.3 | 52.6 | 49.8 | 47.0 | 44.3 | 41.5 | 38.7 | 36.0 | 33.2 | 30.4 | 27.7 | 24.9 | 22.1 | 19.4 | 16.6 | 15.2 | 13.8 | 12.5 | 11.1 | 9.7 | 8.3 | 6.9 | 5.5 | 4.2 | 2.8 | 1.4 | 0.0 |
| 41 | 58.4 | 55.5 | 52.5 | 49.6 | 46.7 | 43.8 | 40.9 | 37.9 | 35.0 | 32.1 | 29.2 | 26.3 | 23.4 | 20.4 | 17.5 | 16.1 | 14.6 | 13.1 | 11.7 | 10.2 | 8.8 | 7.3 | 5.8 | 4.4 | 2.9 | 1.5 | 0.0 |
| 42 | 61.6 | 58.5 | 55.4 | 52.3 | 49.3 | 46.2 | 43.1 | 40.0 | 36.9 | 33.9 | 30.8 | 27.7 | 24.6 | 21.5 | 18.5 | 16.9 | 15.4 | 13.9 | 12.3 | 10.8 | 9.2 | 7.7 | 6.2 | 4.6 | 3.1 | 1.5 | 0.0 |
| 43 | 64.9 | 61.7 | 58.4 | 55.2 | 51.9 | 48.7 | 45.4 | 42.2 | 38.9 | 35.7 | 32.4 | 29.2 | 26.0 | 22.7 | 19.5 | 17.8 | 16.2 | 14.6 | 13.0 | 11.4 | 9.7 | 8.1 | 6.5 | 4.9 | 3.2 | 1.6 | 0.0 |
| 44 | 68.4 | 65.0 | 61.5 | 58.1 | 54.7 | 51.3 | 47.9 | 44.5 | 41.0 | 37.6 | 34.2 | 30.8 | 27.4 | 23.9 | 20.5 | 18.8 | 17.1 | 15.4 | 13.7 | 12.0 | 10.3 | 8.5 | 6.8 | 5.1 | 3.4 | 1.7 | 0.0 |
| 45 | 72.0 | 68.4 | 64.8 | 61.2 | 57.6 | 54.0 | 50.4 | 46.8 | 43.2 | 39.6 | 36.0 | 32.4 | 28.8 | 25.2 | 21.6 | 19.8 | 18.0 | 16.2 | 14.4 | 12.6 | 10.8 | 9.0 | 7.2 | 5.4 | 3.6 | 1.8 | 0.0 |
| 46 | 75.9 | 72.1 | 68.3 | 64.5 | 60.7 | 56.9 | 53.1 | 49.3 | 45.5 | 41.7 | 37.9 | 34.1 | 30.3 | 26.6 | 22.8 | 20.9 | 19.0 | 17.1 | 15.2 | 13.3 | 11.4 | 9.5 | 7.6 | 5.7 | 3.8 | 1.9 | 0.0 |
| 47 | 79.9 | 75.9 | 71.9 | 67.9 | 63.9 | 59.9 | 55.9 | 51.9 | 47.9 | 43.9 | 39.9 | 35.9 | 31.9 | 27.9 | 24.0 | 22.0 | 20.0 | 18.0 | 16.0 | 14.0 | 12.0 | 10.0 | 8.0 | 6.0 | 4.0 | 2.0 | 0.0 |
| 48 | 84.0 | 79.8 | 75.6 | 71.4 | 67.2 | 63.0 | 58.8 | 54.6 | 50.4 | 46.2 | 42.0 | 37.8 | 33.6 | 29.4 | 25.2 | 23.1 | 21.0 | 18.9 | 16.8 | 14.7 | 12.6 | 10.5 | 8.4 | 6.3 | 4.2 | 2.1 | 0.0 |
| 49 | 88.4 | 84.0 | 79.6 | 75.1 | 70.7 | 66.3 | 61.9 | 57.5 | 53.0 | 48.6 | 44.2 | 39.8 | 35.4 | 30.9 | 26.5 | 24.3 | 22.1 | 19.9 | 17.7 | 15.5 | 13.3 | 11.1 | 8.8 | 6.6 | 4.4 | 2.2 | 0.0 |
| 50 | 93.0 | 88.3 | 83.7 | 79.0 | 74.4 | 69.7 | 65.1 | 60.4 | 55.8 | 51.1 | 46.5 | 41.8 | 37.2 | 32.5 | 27.9 | 25.6 | 23.2 | 20.9 | 18.6 | 16.3 | 13.9 | 11.6 | 9.3 | 7.0 | 4.6 | 2.3 | 0.0 |

**FIGURE 24.8** Chart of numerical values for analysis in environmental mode of the ESEM, This includes the relationship of temperature (° C) versus the steam water pressure (Torr) present in different content of isobaric relative humidity (percentage).

To optimize the analysis of the conditions in the variable pressure mode so it is best to use the table in Figure 24.8 and it is including the chart of numerical values for analysis in environmental mode of the ESEM this includes the relationship of temperature (°C) versus the steam water pressure (Torr) present in different content of isobaric relative humidity (percentage).

Those materials are substances than can be evaporated in the absence of their corresponding vapor. Therefore, before imaging can take place, such samples require preparation in order to remove potentially volatile substances, and many procedures have been developed[@]. Of course, samples such as porous polymers, emulsions, vegetables and food systems and so on those than could contain water.

In the newest environmental microscopes can be found devices such as: GSD (Figure 24.9) detector with capacity up to 30 Torrs of pressure, cooling stages to –30°C in the cooling mode also heating at 1500°C, in conventional equipment (tungsten gun) and microscope based on a field emission gun.

In high vacuum mode a charged sample involves the inability to take both photomicrographs and analysis. In low vacuum mode if not are establish very well the optimal analysis conditions can occur the same. Most of the criteria about charged samples and their inherent effects are set in the work reported by (Milani et al., 2007). In contrast with this a well-prepared sample or the conditions of correct analysis lot of work in the formation of the images are saved.

A condition of some scanning electron microscopes that is not reported in the literature is the use of a set contrast expander. When this set is combined with other conditions enhanced images are obtained. This is an external device that is independent of the automated controls software and is located next to the camera analysis. In Figure 24.9 are presented an environmental detector and environmental bullet.

### 24.3.7  SIZE REQUIREMENT FOR ESEM ANALYSIS

The suggested size of the sample it should be of one cubic inch or smaller for introduction to SEM.

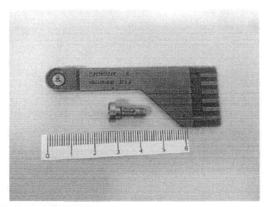

**FIGURE 24.9**   Digital photography: GSE Detector Gaseous Secondary Electrons and bullet for operation in both modes variable pressure and low vacuum.

But it is extremely important to mention than the smallest possible area for to analyze is 1 μm (diametric and of depth) for semi-cuantitative chemical composition. Where, as it was mentioned above the X-rays will result from the interaction between electron beam and sample thus the interaction volume or section analyzed will depend of the energy supplied and the corresponding atomic number of each chemical element (Figure 24.6).

### 24.3.9  CHARGING BEHAVIOR ON SAMPLES

From the chemical point of view, the water vapor is generated in the ESEM mode as a neutralizing agent and it acts mainly on the surface of non-conductive samples, this means that the sample will not present charges during its observation. Hence the sample preparation consists only of the cutting processing. In the Section 24.2.1 such applications areas where the samples do not require a conductive coating of metallic origin are written.

## 24.4  RESULTS

### 24.4.1  INSPECTION CANDELILLA BUSHES: STEREOMICROSCOPE

One of the most important factor in the candelilla study, is to conduct prestudy observations in a stereographic mode (stereoscope) in order to observe: micro-cracks, tears, porosity, endogenous and/or exogenous contaminations, roughness and topographies, it means that the surface defects which might be confused when the characterization by SEM is made. This is recommended because a three-dimensional image is blunt to determine the original natural condition and the most important factor is that it is possible to observe prepared and/or unprepared samples. The first and fundamental step in the observation of a biological sample such as in the case of bush candelilla as already mentioned, is the observation in a stereographic microscope in order to observe the condition of the local surface (three-dimensional) of the sample. This is important because it is believed that either mode can impair to the sample when it is under vacuum. Stereomicroscopes usually have a resolution of 0.1 mm and could observe up close 500x. Samples can be larger in size or volume, this makes it very useful in different areas such as: science industry or research, chiefly in applications where is necessary to manipulate the samples in examination or where a three-dimensional stereographic view is needed. These are the main reasons due to neither the optical microscopes nor the petrographic (ceramographic)/biologics can be used to obtain such images. One advantage concerning to the optical microscopes is that only one focal plane can be focused when the sample is analyzed and samples should be prepared by grinding- polishing processes rigorously.

In materials science the stereographic microscopes provide a clear idea of the nature of each sample and also of their resolution capacity. Although observations can be made from 5 and up to 110X conventionally. Figure 24.10 is shown a section of a candelilla stem observed in a stereomicroscope, this image was taken with normal yellow light to 18x, (a) and (b) to 50X respectively. On the surface it is possible to note a surface with reliefs this morphology is consistent with those reported by Howe and Williams in 1990. The surface shows an irregular porosity throughout of the entire section of the analyzed area. It is also possible to observe a transparent layer of uneven texture on the surface of the plant corresponding to waxy material. In the longitudinal section must not observe residual stresses which are generated by the cutting.

## 24.4.2  MICROSTRUCTURAL CHARACTERIZATION AND COMPOSITION

### 24.4.2.1   Analysis of the Candelilla in ESEM Mode

Below presents in more detail some of these conditions and selectivity that exists for the treatment of different types of materials placing emphasis on the preparation of the candelilla stems.

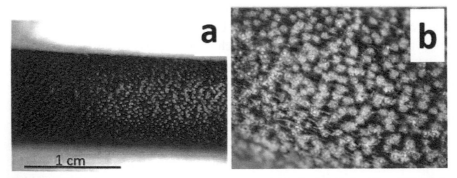

**FIGURE 24.10**   Stereographic photography of a candelilla bush, 12x in yellow light mode. Longitudinal section.

This chapter aims essentially indicate the methodology to observe samples in low vacuum condition, which is so different to how high vacuum and variable pressure (environmental). The required pressure is shown below. The first step in characterizing the candelilla is reviewing the different factors than they are involved intrinsically such as: an appropriate saturation of the electrons source (tungsten filament), appropriate maintenance and cleaning of peripheral parts and of (physical earth and regulated current) and appropriate in column (openings and PLCs) and in addition to this you need to have trained personnel to achieve this amendment. The required pressure is shown below:

High vacuum mode: 10–7 mBar.

Variable Pressure: 10–1 mBar.

Low vacuum or environmental: 10 mBar.

Work in the ESEM begins when power is applied to a tungsten filament in which the range of kV energy supply is located from 1 up to 30 kV and with a shelf life of 60 hours maximum. The next step is required strictly and is the saturation of the electron source. In Figure 24.10 is showed a diagram representative of the interaction of an electron beam, their interactions and their scope.

It is very important to note that the reason for the scientific name of the candelilla (*Euphorbia antisyphilitica*) is due to the fact that in ancient times this plant was used to treat venereal diseases, and the name of candelilla is that the morphological configuration of the wax above the surface looks like a candle (Howe and Williams, 1990).

Candelilla has been characterized previously by another analytical composition techniques for organic compounds mainly which provide results unique and exclusively from the polymeric groups information. The wax consists of hydrocarbon molecules with chains from 17 to 25 carbon atoms in long (Ballandrin et al, 1984). This section presents for the first time the chemical composition obtained by the technique of energy dispersive X-ray spectroscopy.

### 24.4.3   QUANTITATIVE ANALYSIS OF THE CANDELILLA

For microscopic analysis of the candelilla two procedures were employed: (1) surface analysis without preparation but with conductive coating (surface), and (2) analysis of the cross section to evaluate the thickness, prepared according to previously reported by Rojas Molina et al. (2013) and ASTM E-3 standard rule and with a conductive coating electrolytic gold based (cross section). In both cases of candelilla samples some micro-areas and representative spectrograms were selected, in both uncut samples and without preparation samples, as well as the cross-section coated with electrolytic gold were studied.

It is important to note that from the viewpoint of organic chemistry the composition of candelilla it is wax hydrocarbons based, resin esters, lactones, alcohols of resin wax-free, free waxes among other components (Cabello-Albarado et al, 2013). The proper form to perform quantitative analysis is through a certified standard for each of the elements that they will be analyzed and immediately after analyzing the region of interest so as well as verify microcurrent (microamperes). With similar conditions will be similar counts for each analysis.

In the correct order to calculate the concentration, sample counts and standard counts should be in a related concentration.

$$k_{ratio} = \text{Sample Intensity/Standard Intensity} = \text{Sample Concentration} \qquad (1)$$

In the practice software should correct this values. Software includes absorption and fluorescence corrections

$$k_{ratio} = \text{Sample Intensity/Standard Intensity}$$
$$= \text{Sample Concentration with ZAF corrections parameters} \qquad (2)$$

Another manner of presenting the obtained results in the ESEM consist in compositional contrast analysis and are commonly known as: X-ray maps, which consist show the distribution of each of the elements analyzed individually. For the case of candelilla X-ray mappings are performed on the surface as can be seen in the images of Figure 24.13.

Table 24.1 shows the chemical composition for candelilla samples before and after of a purification treatment. Figure 24.11 presents SEM Spectrogram of a candelilla bush after of the wax purification. Analysis was replicated five times and a mean value was obtained. Table 24.1 also show the results for microarea local analysis of candeilla surface, where C present a greater presence in the semiquantitative chemical composition. Candelilla stems were analyzed before and after of their purification processes. In these analyzes the presence of elements as: O, Al, Ca, Si, K, Na and Fe apart from C were detected.

**TABLE 24.1**    Semi Quantitative Chemical Analysis for Candelilla Before and After Purification Process

| Element | Before purification | | After purification | |
|---|---|---|---|---|
|  | Atomic (%) | Weight (%) | Atomic (%) | Weight (%) |
| C | 79.30 | 85.01 | 79.49 | 83.99 |
| O | 13.65 | 12.32 | 19.28 | 15.67 |
| Na | 0.21 | 0.14 | 0.05 | 0.03 |
| Mg | 0.20 | 0.06 | 0.02 | 0.01 |
| Al | 1.30 | 0.62 | 0.68 | 0.25 |
| Si | 1.84 | 0.81 | 0.10 | 0.05 |
| Ca | 2.67 | 0.83 | 0.11 | 0.00 |
| Fe | 0.83 | 0.21 | 0.00 | 0.00 |

*Result were obtained by energy dispersive X-ray spectroscopy.

Figure 24.12 shows four photomicrographs of a surface of a bush candelilla which they have been taken in three different modes of the ESEM. Figure 24.12a shows an image taken with a GSE detector where it is possible to observe any surface features of candelilla, nodules of about 70 microns are homogeneously distributed through the surface of the plant these nodules are the sites where the wax is released to the external part of the plant, also were observed with irregular texture in the form of a continuous matrix. The Figure 24.12c shows a photomicrograph of candelilla bush by the

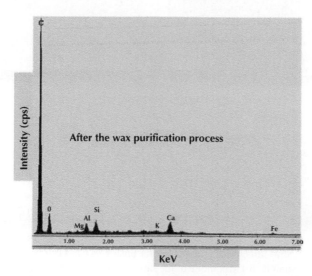

**FIGURE 24.11**    SEM Spectrogram of a candelilla bush after the wax purification.

BSE mode, in where areas of different tonality are observed, these tones are attributed to any possible segregation of some chemical of its composition. Figures 24.12b and 24.12d show micrographs taken with the mix mode (mixed), which allows us to generate an interposition of images using the detectors 50% GSE and 50% BSE with the main objective to observe surface relief and improved contrast and thus perceive all the details of the surface. Figure 24.12d shows the presence of micro-cracks mainly around the nodules, foveolar surfaces and jewel-kind morphologies were observed using a BSE Detector. In all micrographs is possible to observe closed and homogeneous porosity over the entire surface, this porosity probably allow traffic of substances into the plant.

In 2013 our working group published the chemical and structural characterization of candelilla in that investigation were characterized and measured different layers of plant tissue such as cuticular surface, polymeric matrix, lignin layer, phloem layer, xylem cambium and surface follicles. The chemical organic composition of the plant and its relation to their functional properties and application of edible films were studied. In addition, the following composition of all components as: esters, moisture, total solids, lipids, crude

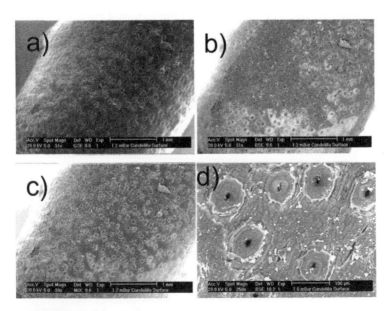

**FIGURE 24.12** SEM Images of: (a) Candelilla bush surface. GSE mode 31 X, (b) BSE mode of the same sample, (c) Mix mode of the same sample, and (d) Candelilla bush surface, GSE mode, 250 X.

fibers, proteins, ashes, total sugars, reducing sugars, hydrolysable tannins, ellagic acid, gallic acid, condensed, and catechin tannins were reviewed.

However, at that time were not analyzed element by element of the chemical composition, in this study is reported the advantage of reviewing that the plant does not have in its composition some harmful component to human health.

The interpretation of x-rays maps is that in a totally black screen (lower content of each chemical element) distribution of each element is recorded separately. The more intense and shiny color represents the higher content of each element. In the case of candelilla the elements C, O and Al were observed well distributed across the surface analyzed.

### 24.4.4 ANALYSIS METHODOLOGY USING THE TECHNIQUE OF ENERGY DISPERSIVE X-RAYS (EDX-S)

For the candelilla case the results consisted mainly in elements such as C (organic groups present on the surface), Mg, Al and Si, mainly. It is important to remark than in this research is the first time that the analysis of the elemental chemical composition containing candelilla plant is reported. This analysis inspired us to the consideration of the natural occurrence of $Al_2O_3$ nanoparticles present in candelilla wax due that the analyzed samples of in two different ways: (i) without treatment (in candelilla bush), and (ii) the presence of the same composition in wax after of a purification treatment. The Table 24.1 shows this comparison.

Those particles may contain an inert character of nature and thus reinforce the protective properties as reported by (Rojas-Molina et al, 2013). The possible cause of the presence of these elements can be attributed to circumstances of their geological and geographical nature and as well as to the mechanisms of how these elements leach into the soil where the plant grows. Another feature is that the candelilla is tolerant to salinity and drought species and is well adapted to calcareous soils and poor in terms of fertility (De la Garza et al., 1992; Flores et al., 1994). But these results will be presented in our next scientific issue.

### 24.5 CONCLUSIONS

This research explored the use of the Environmental Scanning Electron Microscopy (ESEM) for the characterization of candelilla stems under varying conditions. The ESEM technique affords a new and

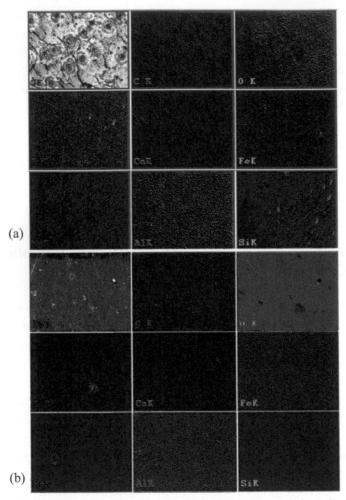

**FIGURE 24.13** Chemical compositional contrast for two samples of candelilla: (a) Candelilla sample before of a purification process, and (b) Candelilla sample after of a purification process.

powerful approach to characterize vegetable tissues for surface, interface and dynamic processes. The properties of candelilla wax are intrinsically linked to their nature and chemical composition. The morphological configuration plays an important role in its system of regulation of nutrient liquid contained in the plant. The potential for the use of the ESEM in candelilla research and development is now a reality and of high scientific and technological value.

## KEYWORDS

- candelilla
- environmental
- ESEM
- *Euphorbia antisyphilitica*
- microscopy

## REFERENCES

ASTM E 3–95. (1995). Standard Practice for Preparation of Metallographic Specimens.

Cabello-Alvarado, C. J., ASáenz-Galindo, A., Barajas-Bermúdez, L., C Pérez-Berumen, C., Ávila-Orta, C., Valdés-Garza, J. A. (2013). Cera de candelilla y sus aplicaciones (Spanisñ). *Avances en Química. 8*(2), 105–110.

De la Garza, De la P. F. E., Berlanga, R., C. A., & Tovar, V. F. J. (1992). Guía para el establecimiento y manejo de plantaciones de candelilla. Folleto Divulgativo Núm. 2. SARH. INIFAP. Campo Experimental La Sauceda. Saltillo, Coah. Méx. 14.

Donald, A. M. (2003). The use of environmental scanning electron microscopy for imaging wet and insulating materials. *Nature Materials, 2*, 511–516.

Flores, M. J. C., Ortega R., S. A., & Villavicencio R. E. (1994). El cultivo de la candelilla en la Comarca Lagunera. Folleto para Productores Núm. 2. INIFAP. Campo experimental de la Laguna. Matamoros, Coah. México. 12.

Howe, G. F., & Williams, E. L. (1990). *Euphorbia antisyphilitica* (The candelilla plant) Demonstrates Providence, Design and Typology in Creation. *CRSQ 27*(3), 86–91.

Joy D. C. (2008). Noise and Its Effects on the Low-Voltage SEM. In: H. Schatten & J. Pawley (Eds.), Biological Low-Voltage Electron Microscopy Scanning, (pp. 129–145). New York: Springer.

Knoll, M. (1935). Aufladepoteniel und sekundäremission elektronenbestrahlter Körper, Z. tech. *Phys., 16*, 467–475.

Milani, M., Drobne D., & Tatti, F. (2007). How to study biological samples by FIB/SEM?. In A. Méndez-Vilas & J. Díaz (Eds.), Modern Research and Educational Topics in Microscopy. (pp. 787–794). Milan: Formatex.

NOM-018-SEMARNAT-1999. Que establece los procedimientos, criterios y especificaciones técnicas y administrativas para realizar el aprovechamiento sostenible de la hierba de candelilla, transporte y almacenamiento del cerote (Spanish).

Reed, S. J. B. (2005) Electron Microprobe Analysis and Scanning Electron Microscopy in Geology (2nd ed.). New York: Cambridge University Press, (Chapter 3).

Rojas-Molina, R., De León-Zapata, M. A., Saucedo-Pompa, S., Aguilar-González, M. A. & Aguilar, C. N. (2013). Chemical and structural characterization of Candelilla (*Euphorbia antisyphilitica* Zucc.), *Journal of Medicinal Plants Research, 7*(12), 702–705.

Royall, C. P., Thiel. B. L., & Donald, A. M,. (2001). Radiation damage of water in environ-mental scanning electron microscopy. *Journal of Microscopy. 204*(3), 185–195.

Titze, B., & Denk, W. (2013). Automated in-chamber specimen coating for serial block-face electron microscopy. *Journal of Microscopy. 250*(2), 101–110.

Von Ardenne, M. (1938). Das Elektronen-Rastermikroskop. Theoretische Grundlagen, *Z. Tech. Phys, 109*, 553–572.

# CHAPTER 25

# RISKS AND CHALLENGES OF CANDELILLA INDUSTRY

JORGE A. AGUIRRE-JOYA,[1] ROMEO ROJAS,[2]
OMAR GARCIA-GALINDO,[1] RAUL RODRÍGUEZ-HERRERA[1] and
CRISTÓBAL N. AGUILAR[1]*

[1]*Department of Food Research, School of Chemistry, Autonomous University of Coahuila, 25280 Saltillo, Coahuila, Mexico*

[2]*Research Center and Development for Food Industries, School of Agronomy, Autonomous University of Nuevo Leon, Escobedo, Nuevo Leyn, 66050, Mexico, E-mail: cristobal.aguilar@uadec.edu.mx*

## CONTENTS

## ABSTRACT

Candelilla industry is well located in the Mexican states of Coahuila and Nuevo Leon, and it is of worldwide relevance because the candelilla wax is a natural product of excellent quality for multiple applications due its chemical, thermal, rheological, structural and biological properties. Candelilla wax has the required national and international certificates for its use free of risk in products used in several human activities. However, there are several conditions limiting the great potential of commercialization of this product. Candelilla industry can grows rapidly if certain risks are well identified, defined and documented. Quality and Quantity of candelilla wax are not the two main challenges of candelilla industry, because of the impact of this sector involves social, environmental, technological, scientific aspects, it is important to promote a renewed program of candelilla adapted under the conditions of the modern life and commercial requirements.

## 25.1 INTRODUCTION

The commercialization of the Candelilla plant for producing wax dates back to the beginning of the 20th century, and has since become one of the main economic activities of the Mexican region of the Chihuahuan Desert. There are more than 4,500 small producers of wax from this plant, who are known locally and nationally as "Candelilleros." Some of these producers complement their income by gathering other plant species native to the desert, or by agriculture, raising cattle and even mining.

In recent years and thanks to the efforts of several institutions like CONAFOR and UAdeC, the manufacturing of Candelilla wax can be carried out in modern conditions avoiding the use of the old and inefficient technique which remain virtually unchanged since this activity first began. The risk is to continue without changes due the lack of government and social program to promote the modernization of this activity, while the challenge is to extend the use of the new method for extractions the low yields and high costs in producing the wax under traditional process to extract the candelilla wax.

The exploitation and distribution of the Candelilla plant require real changes to substitute the actual system where the harvesters travel by burros (or by truck) to the areas where the Candelilla plant is to be found in abundance. After gathering all the plants found in the area, they load up and move to other nearby fields. There are various factors influencing the production volume of Candelilla wax, as well as indicators of the yield in the extraction and final quality of the wax. Among these are:

- Time of year.
- Climatic conditions.
- Level of rains.
- Soil composition.
- Frequency of exploitation.
- Altitude of the region
- Collection, extraction and refining processes.

## 25.2   CONSOLIDATION OF THE CANDELILLA WAX MARKET

During the second war (1939–1945) the demand of candelilla wax was estimated in six or seven thousands ton per year, the army of the US used the wax to cover and protect some parts of the air planes, also to develop explosives (Aguilar et al., 2008) the UCPCC decides that the 61% of the total production will be performed in Coahuila and the other 39% between Chihuahua, Durango and Nuevo Leon. Thereafter due to a politic movement in 1940 the UCPCC loses their autonomy and controls only the 80% of the exportations. In 1943 the taxes for candelilla were incremented to 5.85 Mexican pesos for ton of candelilla plant and 292.50 Mexican pesos per ton of crude wax (CIQA-CONAZA, 1981).

One of the first attempts to modify the contaminating and dangerous wax extraction method with sulfuric acid was made in 1953, when was proposed the use of solvents for the wax extraction. In the same year Bancomex takes the control of wax commercialization and the illegal sale of candelilla wax was incremented. Thereafter in 1954 were integrated, with help of the government, the union of ranches producers of candelilla wax (UEPCC in Spanish) and they decree that the national found for ranch development concentrates the production of wax and deliver it to Bancomext for its distribution. It creates diverse committees pro-development of arid zones of Mexico; the committee was in charge to invest in candelilla zone, with the gains of wax sale.

In 1955 the South West Research Institute (San Antonio, Texas) designs a portrait wax extractor; they reports good wax yields. In this year the national bank for rural credit will take the control of the wax production but Bancomex stills controlling the blanching and distribution of wax and destinies the 2.5% of the gain to elaborate a grant to develop the candeliller region but is in 1956 that the grant is created. In 1957 the federal taxes for ton of crude wax was of 428.00 Mexican pesos. Due to the price that they pay for candelilla wax the candelilleros start a strike calling for an increment in the price of the wax, finally the increment is approved (CENAMEX, 2007).

After the Second World War ending, and during the decades of 50's and 60's, the demand for candelilla wax decreases and the exploitation was wobbling. But a cause of the scares or null economic opportunities of the people of the arid zones of Mexico they were taken in the candelilla wax exploitation as the only economic activity to solvent their basic human needs like food, home and clothing.

In 1960 the Mexican government declines to keep buying candelilla wax so the candelilleros start a political movement to reincorporate the wax in the national market, the government try to stop the movement with false promises of solution. In 1961 the movement starts again and the government and candelilleros found a solution; increase the pay per kilogram of crude wax and establish a limit of production of 3,000 tons per year.

## 25.3   CANDELILLA WAX RESURGENCE

At the middle of 1970 resurge the interest for natural and renewable commodities, principally because the crisis of petroleum and their sub products market, and surge the need and interest for innovate their production processes. During these time candelilla wax was obtained by the same and archaic method of extraction intact since beginning of the century (CIQA-CONAZA, 1981)

To 1979 Bancomex loses the control of distribution of the wax; Banrural takes the control of the production and distribution of the wax, it is permitted the candelilla exploitation to smallholders. Table 1 summarizes the most relevant scientific and/or technical researches about candelilla, either plant or wax that have been reported during the candelilla exploitation period

During the next years the candelilla exploitation experienced a series of changes until the actual system of exploitation, where the candelilleros approach the plant located in the immediate hills or neighbor properties, were they travel in donkeys or trucks to gather the wild plant. They can stay 15 days or more collecting the plant, the collecting process consist in take manually the braches and pull it from the ground. After they had collected enough plant they back to the places were the pailas are, to start the extraction process, in this point all family members (if is necessary) including wife, sons or child participates. When the wax is cool it is ready to be sold. Actually the market of candelilla wax is controlled by a small number of companies which recollect the wax in collection centers in the cities near to the rural places that produces the wax or collect the wax directly in the production areas, for later purify it by diverse processes and offer different and specific products for a wide market of waxes either in Mexico or other countries like Japan, USA, France, Germany, etc.

## 25.4 CONCLUSIONS

Candelilla wax extraction is a hard work that represents for 3,500 candelilleros of 33 localities of the Chihuahuan desert, the main economic source. Most of them are located in medium or high marginalization zones, with scarcest of water and other sources and some of them do not have access to education and the only way to solve economic needs is by candelilla exploitation, however, national market in Mexico pays candelilla wax very low compared with the price of exportation, resulting in poverty conditions for candelilleros, to say nothing of the actual process is very dangerous for the health of the producers due to the use of sulfuric acid and it is dangerous to environment too. At the present time, there is an effort by the Mexican forestry commission, Comisión Nacional Forestal (CONAFOR) and the Food Research Department of the Autonomous University of Coahuila (DIA-UAdeC) to change the dangerous and contaminating process with sulfuric acid by a extraction process with organic acids that permits count with a secure process for the candelillero, innoxiously for the environment, and a completely organic and traceable wax in an ecofriendly and sustainable process.

## KEYWORDS

- candelilla
- challenges
- extraction
- industry
- modern technology
- risks

## REFERENCES

Aguilar, C. N., Rodríguez-Herrera, R., Saucedo-Pompa, S., & Jasso-Cantú, D. (2008). Outstanding phytochemicals from Mexican desert: from plants to the natural chemicals and biotechnology. *Autonomous University of Coahuila*, Mexico.

Aguilera-Carbo, A. F., Augur C., Prado-Barragan L. A., Aguilar C. N., & Favela-Torres E. (2008) Extraction and analysis of ellagic acid from novel complex sources. *Chemical Papers*, *62* (4), 440–444. DOI: 10.2478/s11696-008-0042-y.

Arato, M., Speelman, S., & Van Huylenbroeck, G. (2014). The contribution of non-timber forest products towards sustainable rural development: The case of candelilla wax from the chihuahuan desert in Mexico. *Natural Resources Forum*.

Ascacio-Valdés, J., Burboa, E., Aguilera-Carbo, A. F., Aparicio, M., Pérez-Schmidt, R., Rodríguez, R., et al. (2013). Antifungal ellagitannin isolated from *Euphorbia antisyphilitica* Zucc. *Asian Pacific Journal of Tropical Biomedicine, 3* (1), 41–46.

Ascacio-Valdés, J. A., Aguilera-Carbó, A., Martínez-Hernández, J. L., Rodríguez-Herrera, R., & Aguilar, C. N. (2010) *Euphorbia antisyphilitica* residues as a new source of ellagic acid. *Chemical Papers*. 64 (4): 528–532. DOI: 10.2478/s11696-010-0034-0036.

Baldinus, J. G. (1949). Candelilla wax: an x-ray and a chromatographic study. *Thesis (Ph. D.), University of Wisconsin, Madison*, 109.

Bosquez-Molina, E., Guerrero-Legarreta, I., & Vernon-Carter, E. J. (2003). Moisture barrier properties and morphology of mesquite gum-candelilla wax based edible emulsion coatings. *Food Research International. 36*(9), 885–893.

Cenamex. (2007). Mexican natural waxes Corp., Informative bulletin, printed in Mexico.

Ciqa. (2007). Research Center of Applied Chemistry. http://www.ciqa.mx/

Ciqa-Conaza. (1981). Candelilla. Desert Series. Vol 5. Mex.

Conafor. (2013). National Forestry Commission of Mexico. www.conafor.gob.mx

Dosseti-Romero, V., Méndez-Bermúdez, J. A., & López-Cruz, E. (2002). Thermal diffusivity, thermal conductivity and resistivity of candelilla wax. *Journal of Physics Condensed Matter. 14*(41), 9725–9732.

Ochoa, E., Saucedo-Pompa, S., Rojas-Molina, R., De la Garza, H., Charles-Rodríguez, A. V., & Aguilar, C. N. (2011). Evaluation of a Candelilla Wax-Based Edible Coating to Prolong the Shelf-Life Quality and Safety of Apples. *American Journal of Agricultural and Biological Sciences. 6* (1), 92–98.

Ochoa-Reyes, E., Saucedo-Pmpa, S., De la Garza, H., Martínez, D. G., Rodríguez, R., & Aguilar-Gonzalez, C. N. (2010). Traditional extraction of *Euphorbia antysiphilitica* wax. *Mexican Chemical Acta, 2* (3).

Paredes-López, O., Camargo-Rubio, E., & Gallardo-Navarro, Y. (1974). Use of coatings of candelilla wax for preservation of limes. *Journal of the Science of Food and Agriculture*. 25(10): 1207–1210.

Ritter, B., Schulte, E., & Their, H. P. (1999). Thin layer chromatography with FID-Analysis or the detection of paraffin and wax coating in apples (In german). *German Food Journal (Deutsche Lebensmittel-Rundschau)*, *95*(3), 90–94.

Rojas-Molina, R., De León-Zapata, M. A., Saucedo-Pompa, S., Aguilar-Gonzalez, M. A., & Aguilar C. N. (2013). Chemical and Structural characterization of Candelilla (*Euphorbia antisyphilitica* Zucc.). *Journal of Medicinal Plants Research, 7* (12): 702–702. DOI: 10.5897/JMPR11.321.

Romahn, C. F. (1992). Top non-timber forestry products from Mexico. Thesis. *Autonomous University of Chapingo, México*, 376.

Saucedo-Pompa, S., Jasso-Cantu, D., Ventura-Sobrevilla, J., Sáenz-Galindo, A., Rodríguez-Herrera R., & Aguilar A. C. (2007). Effect of Candelilla Wax with Natural Antioxidants on the Shelf Life Quality of Fresh-cut Fruits. *Journal of Food Quality, 30,* 823–836.

Saucedo-Pompa S., Rojas-Molina R., Aguilera-Carbó A. F., Saenz-Galindo A., De la Garza H., Jasso-Cantú D., et al. (2009) Edible film based on candelilla wax to improve the shelf life and quality of avocado. *Food Research International, 42*(4), 511–515. DOI: 10.1016/j.foodres.2009.02.017.

Schuette, H. A., & Baldinus, J. G. (1948). Studies on candelilla wax. Its n-acids and n-alcohols. *Journal of the American Oil Chemists Society, 26*(10), 530–532.

Scora, G. A., Ahmed, M., & Scora, R. W. (1995). Epicuticular hydrocarbons of candelilla (Euphorbia antisiphylitica) from three different geographical areas. *Industrial Crops and Products, 4*(3), 179–184.

Taes. (1911). Texas Agricultural Experimental Station, Annual report of the Office of Experiment Stations. U.S. Agricultural Department, 242.

Toro-Vazquez, J. F., Mauricio-Perez, R., González-Chávez, M. M., Sánchez-Becerril, M., Ornelas-Paz, J. J., & Pérez-Martínez, J. D. (2013). Physical properties of organogels and wather in oil emulsions structured by mixtures of candelilla wax monoglycerides. *Food Research International, 54,* 1360–1368.

Ventura-Sobrevilla, J. M., Aguilar-Gonzalez, C. N., Belmares-Cerda, R. E., Aguilera-Carbó, A. F., & Saucedo-Pompa, S. (2005). Effect of Mexican Desert Plants Extracts Creams in the Cicatrisation of Skin Wounds in Rabbits. Marzo, La Habana Cuba.

Ventura-Sobrevilla, J. M., Aguilar-Gonzales, C. N., Belmares-Cerda, R. E., Aguilera-Carbo, A. F., & Saucedo-Pompa S. (2005). Effect of Mexican Desert Plants Extracts Creams in the Cicatrization of Skin Wounds in Rabbits. March, La Habana, Cuba.

Villa-Castorena, A., Catalán-Valencia, E. A., Inzunza-Ibarra, M. A., González-López, M. de L., & Arreola-Ávila, J. G. (2010). Production of Candelilla Seedlings (*Euphorbia antisiphyllitica* Zucc.) by Cuttings. *Chapingo Journal. Series Forestry and Environmental Sciences, 16*(1), 37–47.

Wagner,, F. T., & Brown, J. B. (1953). The compositions of some unhydrolyzed naturally occurring waxes, calculated using functional group analysis and fractionation by molecular distilation, with a note on the saponification of carnauba wax and the composition of the resulting fractions. *Journal of the American Oil Chemists Society, 30*(7), 291–298.

# INDEX

# T

Printed and bound by CPI Group (UK) Ltd, Croydon, CR0 4YY
23/10/2024
01777701-0020